Electroactive Polymeric Materials

I0061003

Electroactive Polymeric Materials

Edited by
Inamuddin, Mohd Imran Ahamed, Rajender Boddula,
and Adil A. Gobouri

CRC Press
Taylor & Francis Group
Boca Raton London New York

CRC Press is an imprint of the
Taylor & Francis Group, an informa business

First edition published 2022
by CRC Press
6000 Broken Sound Parkway NW, Suite 300, Boca Raton, FL 33487-2742

and by CRC Press
2 Park Square, Milton Park, Abingdon, Oxon, OX14 4RN

CRC Press is an imprint of Taylor & Francis Group, LLC

© 2022 selection and editorial matter, Inamuddin, Mohd Imran Ahamed, Rajender Boddula, Adil A. Gobouri; individual chapters, the contributors

Reasonable efforts have been made to publish reliable data and information, but the author and publisher cannot assume responsibility for the validity of all materials or the consequences of their use. The authors and publishers have attempted to trace the copyright holders of all material reproduced in this publication and apologize to copyright holders if permission to publish in this form has not been obtained. If any copyright material has not been acknowledged please write and let us know so we may rectify in any future reprint.

Except as permitted under U.S. Copyright Law, no part of this book may be reprinted, reproduced, transmitted, or utilized in any form by any electronic, mechanical, or other means, now known or hereafter invented, including photocopying, microfilming, and recording, or in any information storage or retrieval system, without written permission from the publishers.

For permission to photocopy or use material electronically from this work, access www.copyright.com or contact the Copyright Clearance Center, Inc. (CCC), 222 Rosewood Drive, Danvers, MA 01923, 978-750-8400. For works that are not available on CCC please contact mpkbookspermissions@tandf.co.uk

Trademark notice: Product or corporate names may be trademarks or registered trademarks and are used only for identification and explanation without intent to infringe.

Library of Congress Cataloging-in-Publication Data
Names: Inamuddin, 1980– editor. | Ahamed, Mohd Imran, editor. |
Boddula, Rajender, editor. | Gobouri, Adil A., 1975– editor.
Title: Electroactive polymeric materials / edited by Inamuddin,
Mohd Imran Ahamed, Rajender Boddula, Adil A. Gobouri.
Description: First edition. | Boca Raton, FL : CRC Press, [2022] |
Includes bibliographical references and index.
Identifiers: LCCN 2021050132 | ISBN 9781032002804 (hbk) |
ISBN 9781032002842 (pbk) | ISBN 9781003173502 (ebk)
Subjects: LCSH: Electronic polymers–Industrial applications. |
Conducting polymers–Industrial applications.
Classification: LCC TK7871.15.P6 E56 2022 |
DDC 621.381–dc23/eng/20211228
LC record available at https://lccn.loc.gov/2021050132

ISBN: 9781032002804 (hbk)
ISBN: 9781032002842 (pbk)
ISBN: 9781003173502 (ebk)

DOI: 10.1201/9781003173502

Typeset in Times
by Newgen Publishing UK

Contents

Preface

Electroactive polymers (EAPs) are smart materials that can undergo size or shape structural deformations in the presence of an electrical field. These polymeric materials possess properties, such as flexibility, light weight, cost-effectiveness, rapid response time, easy controllability (especially physical to electrical), and low power consumption. Therefore, EAPs are widely used in a diverse range of technologies, such as sensors, actuators, microelectromechanical systems (MEMS), biomedical catheters, smart windows, energy storage, packaging, conductive textile, corrosion protection, and industrial sectors.

This book will explore an overview of the history and progress on EAPs and provides in-depth literature on their synthesis, properties, and diverse applications. It brings together panels of world-renowned experts that work in the field of EAP materials. It encompasses basic studies and addresses topics of novel issues and presents all aspects in one place that concern the multi-functional applications. This book is an invaluable guide to newcomers, academics, scientists, and R&D industrial experts that work in polymer technologies in EAPS. The reported literature in each chapter is summarized as follows.

Chapter 1 provides an overview of EAPs, which considers what they are and the role they play in society. These polymers are divided into electronic and ionic EAPs, polyaniline (PANI), polythiophene (PTh), polypyrrole (PPy), and polyacetylene (PAc) are further highlighted in this chapter that considers their great contribution to several fields of science and technology, detailing their relevant applications.

Chapter 2 presents the different classes of EAPs. Their key properties are highlighted. These polymers present the processability of conventional polymers with the advantage of responding to electrical stimuli; therefore, they can be used in different fields, which are discussed in detail in this chapter.

Chapter 3 examines the properties of EAPs in detail, including their advantages and disadvantages. Depending on the properties of the EAPs, information is given about what kind of studies they will be suitable for, and sample studies are included.

Chapter 4 focuses on intelligent EAPs, which are innovative and smart materials. They are composed of the various types of conductive EAPs, such as polyaniline (PANI), polypyrrole (PPy), poly(3,4-ethylenedioxythiophene) (PEDOT), and some other functionalized conducting polymers for their applications in several biomedical fields.

Chapter 5 describes various types of EAPs. The major focus of this chapter is the properties of different types of EAP. Moreover, a brief outline of trends in EAPs is included.

Chapter 6 describes the relevant characteristics of the most studied conjugated polymers with electrochromic properties. Strategies to modify and adjust the performance of these materials based on their smart window responses are discussed. Finally, the future perspectives section gives a summary of the emerging trends.

Chapter 7 focuses on EAPs. EAPs have exhibited a promising performance when utilized in microelectromechanical systems (MEMS). A few studies are available that give a comprehensive picture of the utilization of EAPs in MEMS. Therefore, this chapter discusses the utilization of EAPs in MEMS, specifically targeting EAPs in energy systems.

Chapter 8 conducts a brief introduction to EAP-based sensors that uses several examples, which are divided into two categories according to different sensing principles. In addition, the sensing mechanism of different EAP-based sensors is discussed.

Chapter 9 describes the recent progress made in the preparation methods for conductive polymers (CPs) and reports their electrodeposition techniques as solid supports. In addition, this chapter discusses the strategies to incorporate CPs for the construction of electrochemical sensors.

In addition, the analytical characteristics of sensing organic or inorganic compounds and cancer biomarkers that use a CP-based sensor are described.

Chapter 10 presents the use of EAPs as functional materials for artificial muscles. The classification, working principle, and research status of EAPs are summarized. Specifically, the application of EAPs as artificial muscles is summarized and discussed. In addition, a perspective of EAPs for further investigation and practical application is provided.

Chapter 11 focuses on several electrochromic (EC) conjugated polymers that include PANIs, PPy, and PTh; moreover, polyamides, polycarbazole, and viologens in addition to their polymer composites as their metal coordination complexes, carbon-based materials, and metal oxide films that improve their thermal stability, mechanical strength, and electrochromic performance are discussed.

Chapter 12 provides a background on the use of EAPs in batteries, which provides a discussion on the innovative technologies that will bring even more efficiency to this type of device. Among these technologies, new methods of electrode nano assembling are mentioned, and new techniques for the characterization and prediction of composites.

Chapter 13 discusses the EAP, which is a distinctive class of materials that have low moduli high strain abilities that can conform to various shapes. It discusses the major types of electroactive material used in sensing phenomena, drug delivery, tissue regeneration, and antimicrobial and antifouling applicability.

Chapter 14 discusses the roles of ionic and electronic EAPs in remediating the environment. The discussions focus on two aspects, such as fabrication and application. Their applications in energy harvesting, nanogenerators, and corrosion protection systems are discussed.

Chapter 15 discusses EAPs that are used in space applications. It provides a background on the space environment, specifically the low earth orbit and then the type of EAPs that are under development and potential applications in the space environment. In addition, it highlights recent advances in EAPs in aerospace.

Chapter 16 contains a wide discussion on the uses of EAPs in industry with their types and further classifications. The role of sensors and actuators in different industries are discussed in detail. In addition, imprinting lithography, innovative power generators for energy harvesting, and their use as electroacoustic transducers are presented.

Chapter 17 describes the need for EAPs in biomedicine. It details the different EAP products using PANI, PEDOT, and PPy. A brief overview of applications of polymers in medicine is mentioned.

Chapter 18 discusses the classification and basic principles of various EAPs and their significance over other materials that enable the development of advanced and cost-effective polymeric materials along with modern digital technologies. In addition, it discusses the unique properties of EAP materials and application strategies for packaging technologies.

Chapter 19 describes the application of electroactive polymers as intelligent carriers for drug delivery. Their conducting mechanisms and various synthetic approaches for the production of the polymers are discussed in this chapter. In addition, the performance of the electroactive conducting polymers was also discussed in the development of smart drug delivery.

Editors

Mohd Imran Ahamed received his PhD degree on the "Synthesis and characterization of inorganic-organic composite heavy metals selective cation-exchangers and their analytical applications", from Aligarh Muslim University, Aligarh, India in 2019. He has published several research and review articles in the journals of international recognition. Springer (UK), Elsevier, CRC Press Taylor & Francis Asia Pacific, and Materials Research Forum LLC (USA). He has completed his BSc (Hons) Chemistry from Aligarh Muslim University, Aligarh, India, and MSc (Organic Chemistry) from Dr Bhimrao Ambedkar University, Agra, India. He has coedited more than 20 books with Springer (UK), Elsevier, CRC Press Taylor & Francis Asia Pacific, Materials Research Forum LLC (USA), and Wiley-Scrivener (USA). His research work includes ion exchange chromatography, wastewater treatment, and analysis, bending actuators, and electrospinning.

Rajender Boddula works in the Chinese Academy of Sciences-President's International Fellowship Initiative (CAS-PIFI) at the National Center for Nanoscience and Technology (NCNST, Beijing). He obtained a Master of Science in Organic Chemistry from Kakatiya University, Warangal, India, in 2008. He received his Doctor of Philosophy in Chemistry with the highest honors in 2014 for the "Synthesis and Characterization of Polyanilines for Supercapacitor and Catalytic Applications" at the Council of Scientific and Industrial Research (CSIR)-Indian Institute of Chemical Technology (CSIR-IICT) and Kakatiya University (India). Before joining the NCNST as CAS-PIFI research fellow, China, he worked as a senior research associate and Postdoc at the National Tsing-Hua University (Taiwan) in biofuel and CO_2 reduction applications, respectively. His academic honors include a UCR National Fellowship and many merit scholarships, study-abroad fellowships from the Australian Endeavour Research Fellowship, and CAS-PIFI. He has published many scientific articles in international peer reviewed journals and has authored approximately 20 book chapters, and he serves as an editorial board member and a referee for reputed international peer reviewed journals. He has published edited books with Springer (UK), Elsevier, Materials Science Forum LLC (USA), Wiley-Scrivener, (USA), and CRC Press Taylor & Francis group. His specialized areas of research are energy conversion and storage, which include sustainable nanomaterials, graphene, polymer composites, heterogeneous catalysis for organic transformations, environmental remediation technologies, photoelectrochemical water-splitting devices, biofuel cells, batteries, and supercapacitors.

Adil A. Gobouri is Professor of Organic Chemistry, at Taif University, Kingdom of Saudi Arabia. He is currently Dean of the Faculty of Science. He completed his BSc at Umm Al-Qura University in 1997. He gained his Masters in Organic Chemistry from King Abdulaziz University in 2005. In 2011 he obtained his PhD in Organic Chemistry from the University of Manchester, UK, working with Dr John M Gardiner on the synthesis of new stereoisomeric dendrimers. His research interests include synthetic chemistry that encompasses methodology, target synthesis, heterocyclic chemistry, nanomaterials synthesis, and their applications.

Inamuddin is Assistant Professor at the Department of Applied Chemistry, Aligarh Muslim University, Aligarh, India. He obtained a Master of Science degree in Organic Chemistry from Chaudhary Charan Singh University, Meerut, India, in 2002. He received his Master of Philosophy (2004) and Doctor of Philosophy (2007) degrees in Applied Chemistry from Aligarh Muslim University. He has extensive research experience in the multidisciplinary fields of analytical chemistry, materials chemistry, and electrochemistry, and, more specifically, renewable energy and the environment. He has worked on different research projects as project fellow and senior research fellow funded by the University Grants Commission and the Council of Scientific and Industrial Research (CSIR), Government of India. He received the Fast Track Young Scientist Award from the Department of Science and Technology, India, to work in the area of bending actuators and artificial muscles. He has completed four major research projects sanctioned by the University Grant Commission, Department of Science and Technology, Council of Scientific and Industrial Research, and Council of Science and Technology, India. He has published 196 research articles in international journals of repute and 19 book chapters. He has also published 145 edited books and is a member of various journals' editorial boards. He is Associate Editor for *Environmental Chemistry Letter, Applied Water Science and Euro-Mediterranean Journal for Environmental Integration*, Frontiers Section Editor for Current Analytical Chemistry, Editorial Board Member for Scientific Reports at Nature, Editor of the *Eurasian Journal of Analytical Chemistry*, and Review Editor of Frontiers in Chemistry. He has worked as a Postdoctoral Fellow, leading a research team at the Creative Research Initiative Center for Bio-Artificial Muscle, Hanyang University, South Korea, in the field of renewable energy, especially biofuel cells. He has also worked as a Postdoctoral Fellow at the Center of Research Excellence in Renewable Energy, King Fahd University of Petroleum and Minerals, Saudi Arabia, in the field of polymer electrolyte membrane fuel cells and computational fluid dynamics of polymer electrolyte membrane fuel cells. He is a life member of the *Journal of the Indian Chemical Society*. His research interests include ion exchange materials, a sensor for heavy metal ions, biofuel cells, supercapacitors and bending actuators.

Contributors

Heba M. Abdallah
Department of Polymers and Pigments, Chemical Industries Research Institute, National Research Center, Dokki, Giza,12622, Egypt

Henni Abdellah
Lab. Dynamic Interactions and Reactivity of Systems, Kasdi-Merbah University, 30000, Ouargla, Algeria

Mahmoud H. Abu Elella
Chemistry Department, Faculty of Science, Cairo University, Giza, 12613, Egypt

Simran Aggarwal
Amity Institute of Click Chemistry Research and Studies (AICCRS), Amity University Uttar Pradesh, India

Khuram Shahzad Ahmad
Department of Environmental Sciences, Fatima Jinnah Women University, The Mall, 46000, Rawalpindi, Pakistan

Larissa Bach-Toledo
Grupo de Pesquisa em Macromoléculas e Interfaces (GPMIn), Departamento de Química, Universidade Federal do Paraná, PO Box 19061, 81531-980, Curitiba (PR), Brazil

Tarun Kanti Bandyopadhyay
Department of Chemical Engineering, National Institute of Technology Agartala, Jirania, 799046, India

Biswanath Bhunia
Department of Bio Engineering, National Institute of Technology Agartala, Jirania, 799046, India

Kai Cai
School of Medical Instrument and Food Engineering, University of Shanghai for Science and Technology, Shanghai, China

Ernesto Chaves Pereira
Center for the Development of Functional Materials, Department of Chemistry, Federal University of São Carlos, Mailbox 676, CEP 13565-905, São Carlos – SP, Brazil

Gabriela de Alvarenga
GPMIn, Departamento de Química, Universidade Federal do Paraná, PO Box 19061, 81531-980, Curitiba (PR), Brazil

Jean Gustavo de A. Ruthes
GPMIn, Departamento de Química, Universidade Federal do Paraná, PO Box 19061, 81531-980, Curitiba (PR), Brazil

Jessica I.S. de Paula
GPMIn, Departamento de Química, Universidade Federal do Paraná, PO Box 19061, 81531-980, Curitiba (PR), Brazil

Andrei Deller
GPMIn, Departamento de Química, Universidade Federal do Paraná, PO Box 19061, 81531-980, Curitiba (PR), Brazil

Shaimaa Elyamny
Electronic Materials Research Department, Advanced Technology and New Materials Research Institute, City of Scientific Research and Technological Applications (SRTA-City), New Borg El-Arab City, 21934, Alexandria, Egypt

Benmoussa Fateh
Laboratory of Valorisation and Promotion of Saharian Resources (VPSR), Kasdi-Merbah University, Ouargla, 30000, Algeria

Santiago P. Fernandez Bordín
Laboratorio de Materiales Poliméricos (LAMAP), Departamento de Química Orgánica, Universidad Nacional de Córdoba, Instituto Investigación en Ingeniería de Procesos y Química Aplicada (IPQA-CONICET), Haya de la Torre y Medina Allende, Córdoba, Argentina

Achi Fethi
Laboratory of VPSR, Kasdi-Merbah University

Heba Gamal
Home Economy Department, Faculty of
Specific Education, Alexandria University,
Alexandria, Egypt, Ouargla, 30000, Algeria

Qiulong Gao
State Key Laboratory of Solid Lubrication,
Lanzhou Institute of Chemical Physics,
Chinese Academy of Sciences, Lanzhou
730000, China
Center of Materials Science and
Optoelectronics Engineering, University
of Chinese Academy of Sciences, Beijing
100049, China

Kübra Gençdağ Şensoy
Department of Food Processing, Köşk
Vocational High School, Aydın Adnan
Menderes University, Aydın 09570, Turkey

Emad S. Goda
Organic Nanomaterials Lab, Department of
Chemistry, Hannam University, Daejeon
34054, Republic of Korea

Roger Gonçalves
Federal University of São Carlos, Graduate
Program in Materials Science and
Engineering, Rodovia Washington Luiz, km
235 SP-310, CEP 13565-905, São Carlos,
SP, Brazil

Dong Guo
School of Medical Instrument and Food
Engineering, University of Shanghai for
Science and Technology, Shanghai,
China

Kashish Gupta
Noida International University, Yamuna Expy,
Greater Noida-203201

Bruna M. Hryniewicz
GPMIn, Departamento de Química,
Universidade Federal do Paraná,
PO Box 19061, 81531-980, Curitiba (PR),
Brazil

Zembouai Idris
Laboratoire des Matériaux Polymères
Avancés (LMPA), Université de Bejaia,
06000, Algeria

Sapana Jadoun
Department of Analytical and Inorganic
Chemistry, Faculty of Chemical Sciences,
University of Concepción, Concepción,
Chile

Shaan Bibi Jaffri
Department of Environmental Sciences, Fatima
Jinnah Women University, The Mall, 46000,
Rawalpindi, Pakistan

Aniruddha Jaiswal
School of Materials Science and Technology,
Indian Institute of Technology (Banaras
Hindu University), Varanasi-221005,
India

Rafi Ullah Khan
Institute of Polymer and Textile Engineering,
University of the Punjab, Lahore, 54590,
Pakistan

Roya Khosrokhavar
Food and Drug Laboratory Research Center,
Food and Drug Administration, MOH&ME,
Tehran, Iran

Vanessa Klobukoski
GPMIn, Departamento de Química,
Universidade Federal do Paraná, PO Box
19061, 81531-980, Curitiba (PR), Brazil

Chaoyue Li
Key Laboratory of Materials Physics, Ministry
of Education, School of Physics and
Microelectronics, Zhengzhou University,
Zhengzhou 450001, China

Zhangpeng Li
State Key Laboratory of Solid Lubrication,
Lanzhou Institute of Chemical Physics,
Chinese Academy of Sciences, Lanzhou
730000, China
Qingdao Center of Resource Chemistry & New
Materials, Qingdao 266000, China

Yantai Zhongke Research Institute of Advanced Materials and Green Chemical Engineering, Yantai 264006, China
Center of Materials Science and Optoelectronics Engineering, University of Chinese Academy of Sciences, Beijing 100049, China

Dan Ling
Key Laboratory of Materials Physics, Ministry of Education, School of Physics and Microelectronics, Zhengzhou University, Zhengzhou 450001, China
Fire Protection Laboratory, National Institute of Standards, 136, Giza 12211, Egypt

Ria Majumdar
Department of Civil Engineering, National Institute of Technology Agartala, Jirania, 799046, India

Yanchao Mao
Key Laboratory of Materials Physics, Ministry of Education, School of Physics and Microelectronics, Zhengzhou University, Zhengzhou 450001, China

Rita Martins
Chemical Engineering Department of the Faculty of Sciences and Technology, University of Coimbra, Rua Silvio Lima, Polo II, 3030-790, Coimbra

Vivek Mishra
AICCRS, Amity University Uttar Pradesh, India

Mehdi Mogharabi-Manzari
Pharmaceutical Sciences Research Center, Mazandaran University of Medical Sciences, Sari, Iran

Ali Motaharian
Food and Drugs Control Laboratory, Food and Drugs Administration, Birjand University of Medical Sciences, Birjand, Iran

Kaci Mustapha
LMPA, Université de Bejaia, 06000, Algeria

Pinku Chandra Nath
Department of Bio Engineering, National Institute of Technology Agartala, Jirania, 799046, India

Zahra Pakdin-Parizi
Department of Nuclear Medicine and Molecular Imaging, Razavi Hospital, Mashhad, Iran

Shubham Pandey
AICCRS, Amity University Uttar Pradesh, India

Shahla Rezaei
Pharmaceutical Sciences Research Center, Mazandaran University of Medical Sciences, Sari, Iran

Ufana Riaz
Material Research Laboratory, Department of Chemistry, Jamia Millia Islamia, New Delhi, India

Marcero R. Romero
LAMAP, Departamento de Química Orgánica, Universidad Nacional de Córdoba
IPQA-CONICET, Haya de la Torre y Medina Allende, Córdoba, Argentina

Biplab Roy
Department of Chemical Engineering, National Institute of Technology Agartala, Jirania, 799046, India

Samson Rwahwire
Department of Polymer, Textile and Industrial Engineering, Faculty of Engineering and Technology, Busitema University, P. O Box 236, Tororo, Uganda
Institute of Materials and Energy Engineering, Busitema University, P. O Box 236, Tororo, Uganda

Aneela Sabir
Institute of Polymer and Textile Engineering, University of the Punjab, Lahore, 54590, Pakistan

Parastou Sadeghi
Chemical Engineering Department of the
  Faculty of Sciences and Technology,
  University of Coimbra, Rua Silvio Lima,
  Polo II, 3030-790, Coimbra

Masoud Salehipour
Department of Biology, Parand Branch, Islamic
  Azad University, Parand, Iran

Goreti Sales
Chemical Engineering Department of the
  Faculty of Sciences and Technology,
  University of Coimbra, Rua Sílvio Lima,
  Pólo II, 3030-790, Coimbra

Esraa Samy Abu Serea
BCMaterials, Basque Center for
  Materials, Applications and
  Nanostructures, Martina Casiano,
  UPV/EHU Science Park, Barrio
  Sarriena s/n, Leioa 48940, Spain

Muhammad Shafiq
Institute of Polymer and Textile Engineering,
  University of the Punjab, Lahore, 54590,
  Pakistan

Ahmed Esmail Shalan
Central Metallurgical Research and
  Development Institute (CMRDI), P.O. Box
  87, Helwan, Cairo 11421, Egypt
BCMaterials, Basque Center for Materials,
  Applications and Nanostructures, Martina
  Casiano, UPV/EHU Science Park, Barrio
  Sarriena s/n, Leioa 48940, Spain

Rafael J. Silva
GPMIn, Departamento de Química,
  Universidade Federal do Paraná,
  PO Box 19061, 81531-980, Curitiba (PR),
  Brazil

Xiupeng Sun
Key Laboratory of Materials Physics, Ministry
  of Education, School of Physics and
  Microelectronics, Zhengzhou University,
  Zhengzhou 450001, China

Ana P.M. Tavares
Chemical Engineering Department of the
  Faculty of Sciences and Technology,
  University of Coimbra, Rua Silvio Lima,
  Polo II, 3030-790, Coimbra

Kaique Afonso Tozzi
Federal University of São Carlos, Graduate
  Program in Materials Science and
  Engineering, Rodovia Washington Luiz, km
  235 SP-310, CEP 13565-905, São Carlos,
  SP, Brazil

Marcio Vidotti
GPMIn, Departamento de Química,
  Universidade Federal do Paraná, PO Box
  19061, 81531-980, Curitiba (PR), Brazil

Jinqing Wang
State Key Laboratory of Solid Lubrication,
  Lanzhou Institute of Chemical Physics,
  Chinese Academy of Sciences, Lanzhou
  730000, China
Center of Materials Science and
  Optoelectronics Engineering, University
  of Chinese Academy of Sciences, Beijing
  100049, China

Maria Wasim
Institute of Polymer and Textile Engineering,
  University of the Punjab, Lahore, 54590,
  Pakistan

Samm Okinyi Youma
Department of Polymer, Textile and Industrial
  Engineering, Faculty of Engineering and
  Technology, Busitema University, P. O Box
  236, Tororo, Uganda

Wentao Zheng
Key Laboratory of Materials Physics, Ministry
  of Education, School of Physics and
  Microelectronics, Zhengzhou University,
  Zhengzhou 450001, China

1 State-of-the-Art and Perspectives for Electroactive Polymers

Rita Martins, Parastou Sadeghi, Ana P.M. Tavares, and Goreti Sales

CONTENTS

1.1 INTRODUCTION

Certain polymers can be stimulated to change their physical and chemical properties by an environmental stimulus. This may be chemical, optical, magnetic, or electric (Bar-Cohen and Anderson, 2019). In particular, electrical fields are one of the most attractive stimulators and have received great attention from the scientific community (Kumar *et al.*, 2007; Gonzalez, Garcia, and Newell, 2019; Wei *et al.*, 2017; Bajpai *et al.*, 2011; Aguilar and Román, 2019).

Polymers that show some property change that face an electric field are known as electroactive polymers (EAPs). These polymers are very important, due to their functional and structural properties, such as large strain, relatively low manufacturing costs, fast response, and reasonable elastic energy density (Carpi *et al.*, 2011; Bar-Cohen, 2004; Bar-Cohen *et al., 2017*). For example, these polymers can transfer electrons or ions when an electric field is applied; therefore, they can convert electrical energy into a mechanical response that might reversibly change the shape or size. This type of reaction is characteristic of an EAP that is used as an actuator. However, they can be used as sensors or even generators if the EAP exhibits an inverse effect, which depends on the electromechanical properties of each EAP (Bar-Cohen and Zhang, 2008).

DOI: 10.1201/9781003173502-1

In brief, EAPs undergo a structural change in material under the application of a given electrical condition; therefore, generating a mechanical actuation force (Bar-Cohen, 2004; Gonzalez, Garcia, and Newell, 2019; Cardoso, Ribeiro, and Lanceros-Mendez, 2017; Nicolau-Kuklinska et al., 2018; Nasri-Nasrabadi *et al.,* 2018). These polymers response is characterized according to their electrical properties (e.g., dielectric relaxation, permittivity, and electrical breakdown), and mechanical properties (Young's modulus). Therefore, voltage of operation (V), stress (MPa) and strain (%) induced by an electric field, efficiency (%), response time, and power density (W/cm^3) are parameters to take into account when comparing EAPs (Bar-Cohen and Zhang, 2008; Bar-Cohen and Anderson, 2019; Bar-Cohen *et al.*, 2002). However, these intrinsic properties can be improved using different methods to synthesize new structural designs, such as polymer/composite and blended polymer, which makes these materials suitable for generating innovative sensor-actuator technology (Figure 1.1) (Ahmed, Ounaies, and Lanagan, 2017; Ning *et al.,* 2018; Bashir and Rajendran, 2018; Palza, Zapata, and Angulo-Pineda, 2019; Thetpraphi *et al.,* 2020).

The concept of EAPs was documented for the first time in 1880 by Roentgen, which showed that a natural rubber film changed its shape in response to an electrical stimulus (Roentgen, 1880). Almost 20 years later, Sacerdote formulated a theory that related the strain response of the polymer to the activation of the electric field of electrically stretchable materials (Sacerdote, 1899). In 1925, an important milestone was achieved by Eguchi, who discovered electrets, a polymer with piezoelectricity that was obtained by solidifying carnauba wax, rosin, and beeswax, under the application of an electric field (Eguchi, 1925). Following these studies, in 1969, a substantial piezoelectric activity in polyvinylidene fluoride (PVDF) was observed in research, which involved the EAPs by Kawai (1969). In 1977, conductive polymers emerged as a novel type of organic polymers that displayed electrical properties, Shirakawa reported a conductivity increase in trans- polyacetylene (PAc) films after its exposure to vapors from halogen, which included chlorine, bromine, or iodine (Shirakawa *et al.,* 1977). In the early 90s, a great advance was observed in the area of EAPs, in which dielectric elastomers could generate strains of >100% in >0.1 s (Bar-Cohen and Zhang, 2008). Since then, significant progress has been made in improving the performance and processing features of the materials, applications in several areas, including biomedicine and energy harvesting/storage, and robotics (Bar-Cohen, 2004; Bar-Cohen *et al.,* 2017; Ahmed, Ounaies, and Lanagan, 2017; Ning *et al.,* 2018; Rasmussen *et al.,* 2019).

This chapter aims to detail the activation mechanisms of EAPs. Another topic that deserves being mentioned is the wide range of applications of EAPs in diverse fields. Particular attention is given

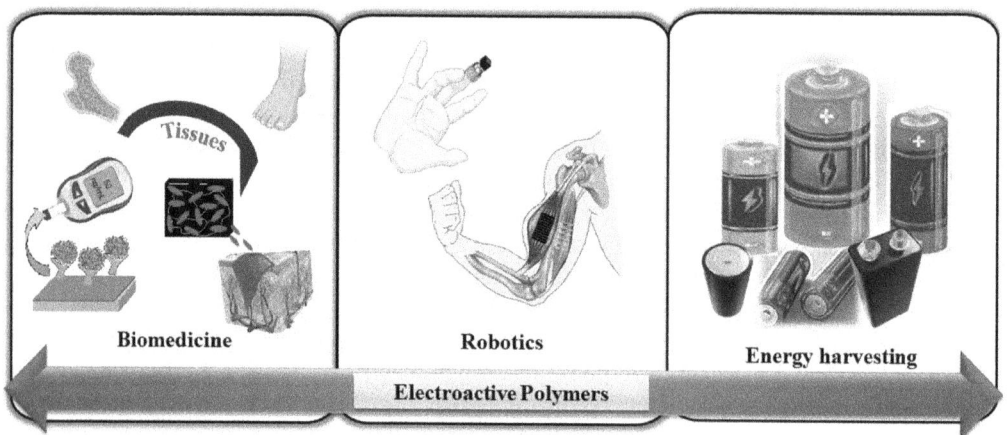

FIGURE 1.1   Main applications areas of electroactive polymers.

to conductive polymers, due to their easy and low cost methods of production, desirable electrical properties, and low power requirement. In addition, this class of polymer has compatibility with the biological environment, which makes these polymers very attractive for applications in the biomedical field (Ning *et al.,* 2018; Tavares *et al.* 2019a; Tavares, Truta, Moreira, Carneiro, *et al.* 2019b).

1.2 TYPES OF ELECTROACTIVE POLYMER

Historically, EAPs have been distinguished by their activation mechanisms, being categorized in two different groups, which are electronic and ionic (Bar-Cohen *et al.,* 2017; Biggs *et al.,* 2013; Carpi *et al.,* 2011; Bar-Cohen, 2007a).

1.2.1 Electronic Electroactive Polymers

Electronic EAPs are insulating materials that are activated by Coulomb forces, when two electrodes that have a polymeric film in the middle are under an electric field (Gonzalez, Garcia, and Newell, 2019; Bar-Cohen *et al.,* 2017; Bar-Cohen, 2007a). In general, an electronic EAP can hold the induced displacement by the activation of a given DC voltage (Kim and Tadokoro, 2007; Bar-Cohen *et al.,* 2017). Although these polymers require high electrical conduction fields for actuation (≤100 V/μm), electrical power consumption is low, with current values from 100 mA to 2,000 mA. In addition, electronic EAPs can work for a long time at room temperature with quick responses (Cardoso, Ribeiro, and Lanceros-Mendez, 2017; Gonzalez, Garcia, and Newell, 2019; Carpi *et al.,* 2011; Bar-Cohen, 2007a). Examples of electronic EAPs include dielectric elastomers, ferroelectric polymers, electrostrictive graft elastomers, and liquid crystal elastomers (LCEs).

1.2.1.1 Dielectric Elastomers

Dielectric polymers are characterized by low levels of elastic stiffness, high levels of dielectric breakdown strength, and high energy densities. Therefore, they are suitable materials for several applications, which have been reported in the literature (Dong *et al.,* 2018; Gonzalez, Garcia, and Newell, 2019; Li *et al.,* 2015; Li *et al.,* 2019; Youn *et al.,* 2020; Ellingford *et al.,* 2020; Zhang *et al.,* 2020; Bar-Cohen and Anderson, 2019).

In general, dielectric polymers operate as shown in Figure 1.2. When the electrodes undergo the application of a high voltage (1–10 kV), the dielectric polymers are electrically charged and cause a mechanical response in the elastomers. In response, the elastomer film stretches in length and width (i.e., in the film plane direction) as they contract in the thickness direction by electrostatic forces that are derived from Maxwell pressure (Hau, Rizzello, and Seelecke, 2018; Hill, Rizzello, and Seelecke, 2017; Hau, York, and Seelecke, 2016). The Maxwell stress (σ_{Max}) generated may be calculated by the following equation:

$$\sigma_{Max} = \varepsilon_0 \varepsilon_r E^2 \tag{1.1}$$

where compressive stress is proportional to vacuum permittivity ($\varepsilon_0 = 8.85 \times 10^{-12} F/m$), relative permittivity or dielectric constant of the polymer (ε_r), the square of the electric field ($E = V/t$), and imposed voltage (V) across the thickness (t) of the elastomer film (Hau, Rizzello, and Seelecke, 2018; Hill, Rizzello, and Seelecke, 2017; Shahinpoor, 2020).

Recently, research has focused on solving critical topics about dielectric polymers, including the reduction of the operating voltage. Several studies reported the development of new dielectric polymers that show an enhancement of their properties according to the need of each application. The most explored features of these polymers to achieve the high-performance actuation of dielectric materials in various fields (Sheima, Caspari, and Opris, 2019; Brochu and Pei, 2010; Tan,

Thangavel, and Lee, 2019; Zhang *et al.*, 2020; Poulin, Rosset, and Shea, 2015; Qiu *et al.*, 2019; Quinsaat *et al.*, 2015; Zhao *et al.*, 2018) include enhancing the ε_r and lowering the elastic modulus and *t* of the elastomer films. Depending on the application field, the selection of the type of dielectric material is one of the crucial factors in the performance of the dielectric polymer.

Dielectric polymers are currently widely used in silicone rubber, acrylate, and thermal plastic polyurethane (Jiang *et al.*, 2020; Zhao *et al.*, 2018; Youn *et al.*, 2020; Jiang *et al.*, 2018; Gupta *et al.*, 2019). Compared with acrylic elastomers, silicones have a lower ε_r between 2.8 and 3.7, and the others are present 4.8 at ambient conditions (Shankar, Ghosh, and Spontak, 2007). This means that acrylic elastomers require a lower voltage to achieve a mechanical response (Youn *et al.*, 2020). However, thermoplastic polyurethane has recently been considered a very relevant material, because of its good mechanical properties and a high ε_r (Ning *et al.*, 2019; Tian *et al.*, 2014).

In general, these materials might be employed for different applications, such as actuators, generators, or sensors (Benslimane, Kiil, and Tryson, 2010; Mao and Qu, 2018). In general, actuators convert electrical energy into mechanical energy. Generators store mechanical energy in the dielectric film and later convert it into electrical energy. Sensors change in capacitance, after dimensional mechanical changes of the dielectric film (i.e., sensor mode) (Benslimane, Kiil, and Tryson, 2010; Carpi *et al.*, 2008; Newell *et al.*, 2016; Zhao *et al.*, 2020).

1.2.1.2 Ferroelectric Polymers

Ferroelectricity is a property associated with spontaneous electric polarization that occurs in a non-conducting crystal or dielectric material. Polymers with ferroelectric behavior are called ferroelectric polymers, typified by a change in the spontaneous/reversible polarization of the material, under exposure to an external electric field (Bar-Cohen and Anderson, 2019; Kim and Tadokoro, 2007; Qian *et al.*, 2015; Bar-Cohen *et al.*, 2017; Katsouras *et al.*, 2016). Besides ferroelectricity, ferroelectric materials can exhibit a piezoelectric and pyroelectric behavior, in which it can convert mechanical vibrations or thermal fluctuations into electrical energy, respectively (Bar-Cohen *et al.*, 2017). Although ferroelectrics are termed with the prefix *ferro* which means iron (Fe), these materials do not contain iron atoms in their chemical structure. However, they exhibit analogous characteristics to ferromagnetics, and therefore, the origin of the name ferroelectric (Li and Wang, 2016; Huang and Scott, 2018).

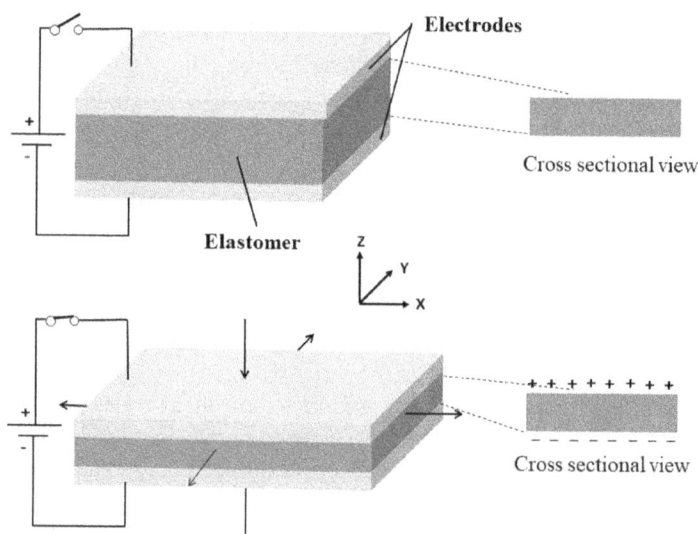

FIGURE 1.2 Operating principle of a dielectric polymer under an electric field.

The most widely ferroelectric polymer explored by research is poly(vinylidene fluoride) (PVDF), along with several copolymers obtained from it (Bar-Cohen and Anderson, 2019; Mai *et al.*, 2015; Furukawa, Takahashi, and Nakajima, 2010; Bae and Chang, 2019; Qian *et al.*, 2015). Its ferroelectricity originates from the dipolar moment of the monomer of difluoroethylene (CH_2CF_2) units due to the proper orientation of the hydrogen and fluorine atoms, which have opposite electronegativity features (Furukawa, Takahashi, and Nakajima, 2010; Das, Bhowmik, and Meikap, 2017). This polymer exhibits a piezoelectric behavior under an electrical stimulus, with a strain response that displays linear behavior to the electric field that generates it (Bar-Cohen *et al.*, 2017).

As mentioned previously, the piezoelectricity of PVDF was discovered in 1969 by Kawai, and 2 years later, Bergman discovered pyroelectricity in this polymer (Kawai, 1969). PVDF polymers are characterized by being semicrystalline with an inactive amorphous phase and with a Young's modulus in the range of 1–10 GPa, which offers superior mechanical energy density (Bar-Cohen *et al.*, 2017; Bae and Chang, 2019). However, when the electric field applied is too large (approximatley 200 MV/m), electrostrictive strains are induced ≤2%, which is close to dielectric breakdown. In 1998, Zhang observed a high electrostrictive response with an increase in dielectric constant from PVDF's copolymer with trifluoroethylene, when defects were introduced by the irradiation of high energy electrons (Zhang, Bharti, and Zhao, 1998). Since then, PVDF and its derived materials have been explored and widely applied in fields, such as piezoelectric or pyroelectric sensors (Han *et al.*, 2016; Bae and Chang, 2016; Pullano *et al.*, 2017), non-volatile memory devices (Kim *et al.*, 2016; Zhu *et al.*, 2018; Li *et al.*, 2020) and electric energy harvesting and self-powered electronics (Kim *et al.*, 2019; Anand and Bhatnagar, 2019; Karan, Mandal, and, Khatua 2015).

1.2.1.3 Electrostrictive Graft Elastomer

A polymer is considered electrostrictive when its behavior demonstrates a quadratically based relationship between the strain and the electric fields. Since Zhenyl first reported electrostrictive properties in polyurethane in 1994 (Zhenyl *et al.*, 1994), several advances have been reported to understand the electrostrictive effect in polymers. Then, 5 years later, Su developed a graft elastomer that offered approximately 4% induced strain with a high mechanical modulus of approximately 550 MPa. The authors compared the new material with polyurethane and observed that the induced strain was smaller than polyurethane (11%) (Su *et al.*, 1999). The mechanical module was also larger than polyurethane, 550 MPa versus 17 MPa, which made it a potential material that could be used as an actuator that offered high output power (Wang *et al.*, 2003). In addition, the dielectric constant presented by the elastomer graft was greater than that of polyurethane, which gave the advantage to use this material for specific functions that require actuation with lower voltages.

In general, graft elastomers demonstrate electrostriction behavior, which shows a large mechanical displacement by increasing the electric field. It entails a flexible backbone polymer structure with a non-crystalline chain, cross-linked with grafted crystalline moieties (Tohluebaji, Putson, and Muensit, 2019; Bar-Cohen *et al.*, 2017). According to the computational models reported by Wang, the mechanical response (deformation) of electrostrictive graft elastomers is probably attributed to a combined effect of the rotation of the crystal unit and the reorientation of the backbone chain (Wang *et al.*, 2003).

1.2.1.4 Liquid Crystal Elastomers

LCEs are active materials that combine the self-organization of the liquid crystalline phase with polymer elasticity, resulting in a cross-linked liquid crystalline polymer network. Usually, they have a large and reversible shape change through a transition between isotropic and liquid crystal (nematic) states, in response to an external stimulus (Wang *et al.*, 2016; Bar-Cohen, 2002; Ula *et al.*, 2018; Barnes and Verduzco, 2019; Yuan *et al.*, 2017; Jiang *et al.*, 2012). This feature was predicted for the first time by Degennes (1975), and 6 years later Finkelmann developed the first liquid crystalline polymer (Finkelmann, Kock, and Rehage, 1981). Since this pioneering work, the scientific community has focused on producing new materials to improve actuation mechanisms, using the selection of suitable liquid crystalline phase, crosslink density, flexibility of structure of the polymer, and coupling the liquid crystal group to the structure of the LCEs (Ambulo *et al.*, 2020; Bar-Cohen, 2004).

These materials have suitable features for other potential applications. Therefore, the implementation of LCEs as actuators in artificial muscles has been highlighted. They have been widely explored by researchers (Ohm, Brehmer, and Zentel, 2010; He *et al.*, 2019; Thomsen *et al.*, 2001; Shahinpoor, 2000; Roach *et al.* 2019; Chambers *et al.*, 2009; Yuan *et al.*, 2017) since De Gennes proposed theoretical studies on the possibility of using LCEs, such as in synthetic muscles, in 1997 (Li and Keller, 2006; Jiang, Li, and Huang, 2013; Yu and Ikeda, 2006).

The basic principle of operation depends on the alignment directions and phase transitions. These materials are electrically activated when a conducting material is added to the LCE. The application of an electrical field to these materials causes heat, due to the current flow. Consequently, the nematic phase of the material switches to isotropic. Then, the liquid crystal goes from an ordered to disordered phase transition; therefore, changing the shape of the LCE (Figure 1.3). The transition from the nematic to the isotropic phase occurs in <1 s; however, the reverse process becomes longer, because it requires a cooling process, which takes approximately 8 s for the elastomers to expand back to their original length (He *et al.*, 2019; Shahinpoor, 2000).

1.2.2 IONIC ELECTROACTIVE POLYMERS

The electromechanical response of ionic polymers accounts for the mobility of ions inside the polymeric layer or their diffusion through this layer. It requires an electrolyte for the actuation mechanism that is sandwiched between two electrodes. The activation of ionic EAPs requires a very low electric field, usually a few volts (1–10 V), to induce a bending displacement (Bar-Cohen and Anderson, 2019; Bar-Cohen and Zhang, 2008; Bar-Cohen *et al.*, 2017). Examples of ionic EAPs include ionomeric polymer–metal composites (IPMC), ionic gels, carbon nanotubes (CNT), and conductive polymers (CP).

1.2.2.1 Carbon Nanotubes

CNTs are ordered structures in the form of tubes made from carbon (C) in nanometer scale, which can be formed as single-walled carbon nanotubes (SWNT) or as multi walled carbon nanotubes (MWNT) (Qu *et al.*, 2008; Carpi *et al.*, 2011). Typically, SWNT are from 0.4 to 3 nm in diameter. MWNT show much higher diameter values, between 1.4 and 100 nm, which depends on the number of layers in the tube (Carpi *et al.*, 2011; Eatemadi *et al.*, 2014).

The discovery of nanotubes by Iijima and Ichihashi (1993) contributed a great advance in science and in nanotechnology. Then, 6 years later, the nanotubes were used as an actuator for the first time by Baughman *et al.* (1999), which used sheets of SWNT as electrodes that contained an electrolyte

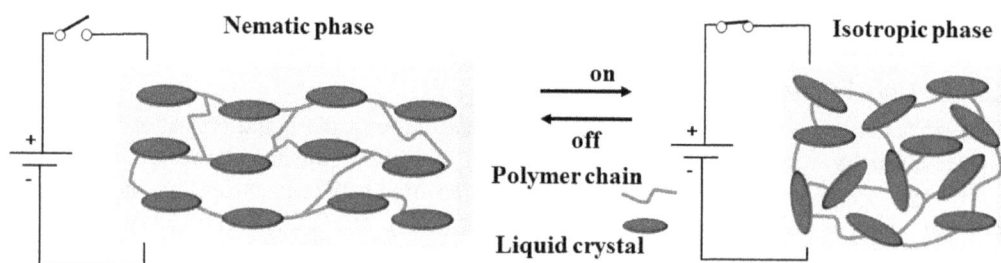

FIGURE 1.3 Operating principle of a liquid crystal elastomer, under an electric field.

in the middle. In this study, it was estimated that nanotubes could generate improved work densities in each cycle and generate higher currents at lower voltages than other actuators, such as ferro-electric. Since then, several works that announced new developments in CNT materials have been reported, including a wide range of synthetic and chemical modification processes. In 2008 Liangti in "Carbon Nanotube Electroactive Polymer Materials: Opportunities and Challenges" summarized the electromechanical actuation generated by CNT as EAPs and corroborated the good electromech-anical properties of CNTs as EAPs (Qu *et al.*, 2008). These advantages of CNTs make them useful for several potential applications, especially in biosensors and actuators, on a micro and nanoscale (Bar-Cohen, 2007a; Qu *et al.*, 2008).

In general, CNTs require only 1 V to cause strains between 0.1% and 1% along the length of a nanotube (Carpi *et al.*, 2011; Bar-Cohen, 2007a). The electrical stimulus on these nanomaterials produces a mechanical effect associated with the length change of the C–C bond in the CNTs. This actuation principle results from contact between an electrolyte and two CNTs that act as electrodes. Under a voltage effect, the ionic charges accumulate on each CNT electrode surface and are balanced by the injection of large electronic charges through the double layer formed at the CNT and the electrolyte interface (Carpi *et al.*, 2011; Bar-Cohen *et al.*, 2017; Bar-Cohen, 2007a; Torop and Baughman ,2016; Qu *et al.*, 2008).

1.2.2.2 Ionic Polymer Gels

Ionic polymer gels contain a polymeric network with ionized groups that are bound at the end of its internal chain and a fluid that fills the interstitial areas of the polymeric web (Wallmersperger *et al.*, 2001; Kwon, Osada, and Gong, 2006). These polymers allow ion exchange inside the polymer under the influence of counter ion and its shape and size can be affected by an environmental stimulus. According to a review "Ion exchange" (Boyd, 1951), the relationship between ion exchange and dimensional changes in polymers (ion exchange polymers) has been discussed since 1939. However, the mechanical mechanisms in a polymer gel in response to a stimulus (chemical) were reported for the first time in Katchalsky and Zwick (1955). This study evaluated the performance of a polycarboxylic acid compound, in this case polymethacrylic acid. However, Tanaka *et al.* (1982) studied first studied electrical actuation on a polymer gel using a polyacrylic acid gel.

Therefore, ionic polymer gels placed between two electrodes (cathode and anode) and subjected to an electrical stimulus respond by mechanical changes. There is a mobility of ions at the anode and cathode that results in a concentration gradient. The mobility of ions within the gel, under an electrostatic force and osmotic pressure, causes a reversible volume change of the gel (Jo, Naguib, and Kwon, 2011; Wallmersperger *et al.*, 2001; Doi, Matsumoto, and Hirose, 1992). This behavior demonstrates that these polymers are very promising actuators for applications in bionic robots (Chang *et al.*, 2018) and biological muscles (Bar-Cohen, 2004).

1.2.2.3 Ionic Polymer–Metal Composite

Ionic polymer–metal composite (IPMC) materials consist of an ionomeric membrane ≤100 μm thick that have a thin layer of a noble metal electrode (i.e., platinum) coating it (Pugal *et al.*, 2010; Baglio and Bulsara, 2006; Kim and Tadokoro, 2007; MohdIsa, Hunt, and HosseinNia, 2019). Similar to ionic gels, IPMCs contain an ionic group covalently bonded at the polymer chain (Unal *et al.*, 2019; Park *et al.*, 2008). The membrane usually contains as the ionic group a sulfonic acid ($-SO_3^-$) or a carboxylic acid ($-COO^-$), which is aimed at cation exchange. The most used polymer for IPMC is perfluorinated, with sulfonated (Nafion™) and carboxylated (Flemion™) moieties (Unal *et al.*, 2019; Tiwari and Garcia, 2011; Park *et al.*, 2008). Its electromechanical actuation accounts for the electrostatic interaction between the ionic groups and water molecules, and therefore, creates a pathway for the migration of ions. Under an applied voltage (usually 2–7 V), the ions and water molecules move within the polymer and cause a shape change in the IPMC (Figure 1.4) (Kim and Tadokoro, 2007; MohdIsa, Hunt, and HosseinNia, 2019; Bar-Cohen *et al.*, 2002; Yang *et al.*, 2019).

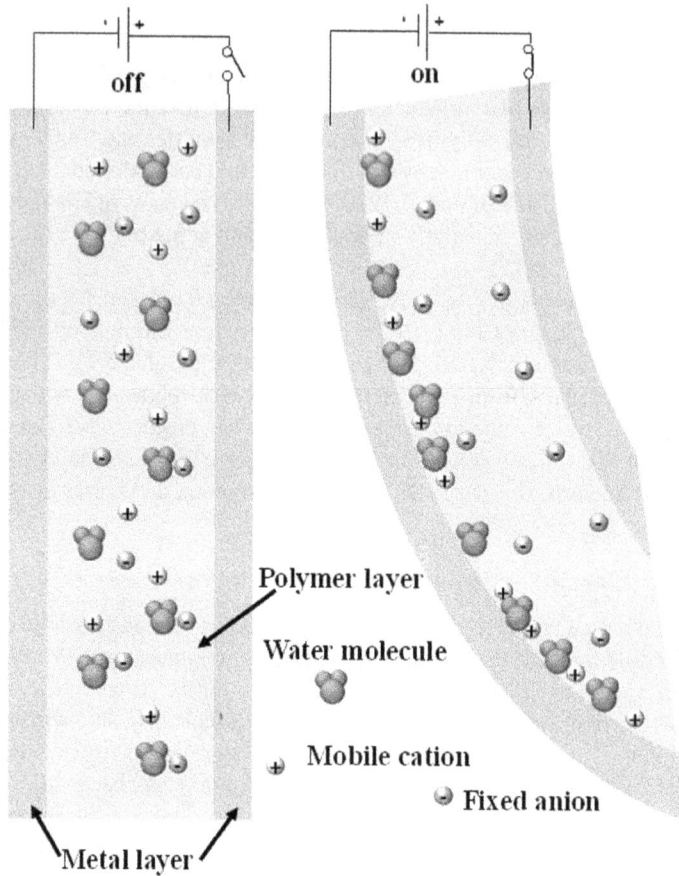

FIGURE 1.4 Operating principle of an ionic polymer–metal composition under an electric field.

The operation as an actuator of an IPMC was first documented by Oguro, Kawami, and Takenaka (1992), who subsequently introduced the patent for the prototype of an ion exchange membrane as an actuator element. Since then, many studies have been carried out to understand and improve the electromechanical response of IPMCs. In 2000, the first theoretical study on the electromechanical effect of IPMCs was reported (de Gennes *et al..* 2000). Nemat-Nasser, 2 years later discussed the activity displayed by cations and the degree of membrane hydration, on the electromechanical behavior of IPMCs (Nemat-Nasser, 2002). Therefore, the scientific community has given particular attention to the improvement in speed and voltage in the electromechanical actuation of IPMCs in actuator or sensor applications, as reported in the most recent papers (Zhu *et al..* 2016; Wang *et al..* 2014; Ma *et al..* 2020; Zhang *et al..* 2020; Zhu *et al..* 2019).

1.2.2.4 Conducting Polymers

Polymeric materials are well-known as insulating materials, there are polymeric structures that hold conductivity features. Therefore, conducting polymers (CPs) are polymeric materials with intrinsic conductivity (Skotheim, 2006). These polymers combine the electronic features of inorganic compounds with specific beneficial features of organic polymeric structures, such as processability, thermal stability, low cost fabrication, and various optical and mechanical properties (Skotheim, 2006). Therefore, over the last decades, intensive research has been devoted into conductive polymers, which explores new theoretical and technological possibilities, which has led to innumerous applications. Electrochromic displays, organic photovoltaic devices, sensors, organic capacitors, tissue engineering scaffolds, and drug delivery systems are among the current applications in this new field of organic electronics (Ning *et al.*, 2018; Jiang *et al.*, 2013; Bagheri *et al.*, 2019; Wibowo *et al.*, 2020; Vijeth *et al.*, 2018; Hocevar *et al.*, 2016). Many conjugated polymers are known as smart biomaterials (Serra Moreno *et al.*, 2009)

The first CP was prepared using an ancient mode of electropolymerization (Letheby, 1862), which could conduct electrical current. In 1975, a novel form of PAc was discovered by Shirakawa and coworkers. They worked on acetylene's polymerization mechanism using Ziegler–Natta catalysts (Ito, Shirakawa, and Ikeda, 1974). Since then, interest in CPs has been increasing among the scientific community. A notable discovery in this type of polymer occurred in 1977, when two chemists, Shirakawa and MacDiarmid, and a physicist, Heeger, collaborated and successfully oxidized iodine-doped PAc with high conductivity (Shirakawa *et al.*, 1977). These researchers received a Nobel prize by discovering and developing conductive polymers in 2000 (MacDiarmid 2001; Heeger, 2003).

In general, CPs are flexible, inexpensive, lightweight, and offer a low elastic modulus. Most of them can be synthesized easily, via cost-effective methods (Shklovsky *et al.*, 2018; Nezakati *et al.*, 2018; Ning *et al.*, 2018). These polymers are categorized as ionic EAPs. Compared with the other ionic polymers, CPs can bear constant displacement under the application of a DC current (Bar-Cohen, 2007b; Ning *et al.*, 2018). Over the last decades, many articles and books have been published in this field. Most of them focused on various methods to synthesize CPs to improve their properties.

The mechanism by which an electric current passes through the conductive polymer remains unconsensual, but all CPs share a crucial property: they consist of extended π-conjugated systems, with a pattern of single or double alternating bonds along the backbone of the polymer (Inzelt, 2008). This allows the free movement of electrons due to electron delocalization through the polymer chain, which provides the pathway for charge mobility (Mishra, Sharma, and Tandon, 2011). In addition, conductive polymers need to be doped by external charge carriers.

Electromechanical actuation might be observed in CPs when two electrodes that contain a conducting polymer and an electrolyte inside are subject to a voltage. This effect induces an electrochemical redox reaction in the CP that then causes the migration of ions through the polymer chain and leads to its volume expansion or contraction (Carpi *et al.*, 2011; Bar-Cohen *et al.*, 2017). This effect occurs in a low voltage range, typically from 1 to 5 V. The mechanical response speed is conditioned by voltage and the thickness of the polymer layer (Bar-Cohen *et al.*, 2017).

Well-known CPs are shown in Figure 1.5, which include polypyrrole (PPy), polyaniline (PANI), polythiophene (PTh), and polyacetylene (PAc). The conductivity of each polymer will be different and is based on its properties. This variance in the conductivity intensity is based on the difference

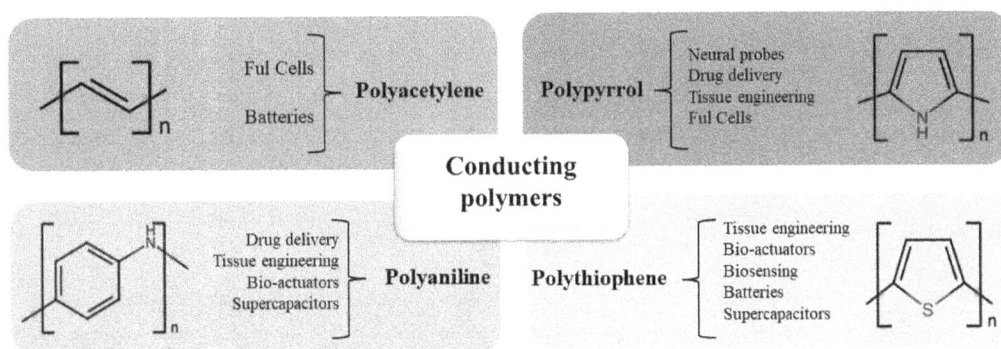

FIGURE 1.5 Molecular structures of some conductive polymers and their typical applications.

in the energy levels of the valence and conduction bands, which is known as energy gap; and lower bandgap energy leads to higher conductivity (Guimard, Gomez, and Schmidt, 2007).

Overall, due to their promising optical and electrical properties, CPs have gained great attention in different areas of research. Therefore, they have become a primary material in biological investigations, including biomedical and environmental applications. Relevant CP polymers are highlighted in the following sections, which detail the relevant applications they have been involved in.

1.3   POLYANILINE

One of the longest and well-known conducting organic materials described is PANI (Figure 1.6), which was initially known as black aniline. Although its exact date of discovery remains unknown, four scientists are known as pioneers in the field: F. Ferdinand Runge (1794–1867), Carl Fritzsche (1808–1871), John Lightfoot (1831–1872), and Henry Letheby (1816–1876) (Rasmussen, 2017). PANI can be obtained by polymerization, which is initiated by a chemical or electrical stimulus. Typically, this is carried out in an acidic aqueous solution, because the acidic media is a favorable condition for its production (MacDiarmid, 1997).

There are several well-defined oxidation states in PANI, which were first described by Green and Woodhead (1910). This includes a reduced form (leucoemeraldine), partially oxidize form (emeraldine form), and a fully oxidized form (pernigraniline). These are associated with color change due to the doping or undoping process, which leads to a change within the electronic structure of the polymer (which is translated by band gap alterations), and therefore, its conductivity (Csahok, Vieil, and Inzelt, 2000; Ning *et al.*, 2018).

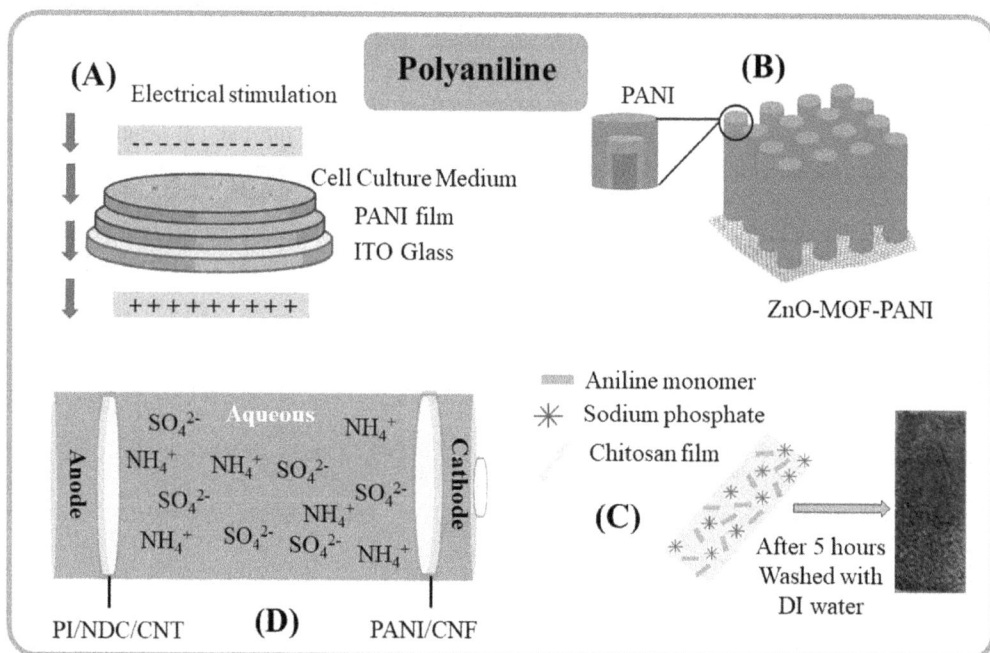

FIGURE 1.6   Examples of the application of polyaniline in: (a) tissue engineering; (b) energy harvesting (supercapacitor); (c) drug delivery; and (d) energy harvesting (battery).

In general, PANI presents several interesting properties that make it a very commonly used polymer in many different fields. It is cheap, environmentally stable, and easy to synthesize (Ning et al., 2018), and can produce EAPs with good conductivity features. The conductivity of PANI increases with doping, from $<10^{-10}$ S/cm in the undoped insulating base form, to >10 S/cm in the conducting acid form, when fully doped. Doping processes are usually carried out chemically or electrochemically, with common acids, such as hydrochloric acid and bases, such as ammonium hydroxide, or by applying an electrical potential (Huang, Humphrey, and Macdiarmid, 1986). PANI may be used alone or in combination with other nanomaterials, as nanowires, nanotubes, nanorods, or nanofibers, which yielding new functional materials with unique physicochemical properties (Figure 1.6). Therefore, PANI has found many applications to date.

In biomedical engineering, PANI may be used in tissue engineering, which could open new doors to the regeneration of damaged or dead tissues. In 2015, Wang's team synthesized a film of PANI on indium tin oxide glass to evaluate the impact of an electrical stimuli on the growth of a neuronal cell line and the length of neurites cells cultured on glass (Figure 1.6(a)). The synthesized nanostructure appeared to be very favorable to the outgrowth and proliferation of neurites. The coating of a nerve with this conducting polymer to yield electrical stimulation allowed favorable neurite differentiation (Wang, Huang, and Wang, 2015). A novel biomaterial was investigated by Sarvari et al. (2017) who synthesized a conducting scaffold with biocompatible and porous features to grow osteoblasts using polycaprolactone and PANI. The resulting polymeric structure displayed good electroactivity features upon electrical stimulation, by improving cell adhesion and growth, and cell proliferation. In addition to biocompatibility, the star-like copolymer of polycaprolactone–PANI (S–PCL–PANI) displayed great solubility and mechanical features (Sarvari et al., 2017). Bagheri et al. (2019) developed an electroactive hydrogel that was composed of a composite material with agarose and an oligomer of chitosan/aniline, which displayed self-gelling features that corresponded to a platform that could release a drug on-demand (Figure 1.6(b)). It is the electrical stimulation that promotes this very important on-demand feature. The application of a given specific voltage condition resulted in the release of large quantities of drugs. The biocompatibility of this platform was confirmed by cell studies that confirmed improved cell growth. This was related to the conductivity features produced by the aniline-based oligomers. In another work on PANI and dedicated to bone tissue engineering, a scaffold of polycaprolactone was loaded with this conductive material, using three-dimensional (3D) printing (Wibowo et al., 2020). The conductivity of these scaffolds increased when PANI was present, which accounted for its conductivity. Of note, this improved its biocompatibility by displaying low cytotoxic values. Moreover, this PANI-based scaffold helped to guide cell growth, which was based on electrical stimulation (Wibowo et al., 2020). An electroconductive polymeric patch was developed by Rahman, Bilal, and Shah (2020), which investigated tissues that responded to electrical stimulation for regenerative medicine. They created a network of PANI with good electrical activity features that operated in a stable manner under physiological conditions (Figure 1.6(c)).

PANI has been used for sensor development, detecting, or measuring, or both for a physical quantity of a compound and providing a readable output for it. Therefore, it was possible to carry out analysis on numerous analytes under mild conditions and using simple methods (Bai and Shi, 2007; Sen, Mishra, and Shimpi, 2016). In general, by interacting with the analytes, PANI's redox properties change, which leads to variations in its intrinsic conductivity and operating voltage.

PANI has a predominant position in energy storage and is one of the common materials used to produce supercapacitors, which can improve the capacitance, or maximize cell voltage, or both, which leads to a high energy storage density. Recently, Zhu, He, et al. (2019) developed a supercapacitor with core-shell nanorod arrays, which used an organic and metal base framework and PANI (Figure 1.6(b)). The composite obtained in this work provided enhanced electroactivity features and increased energy storage capacity. Recently, an aqueous ammonium dual ion battery was developed in China based on an anode of a flexible polyimide/nitrogen-doped C/CNTs and a cathode of PANI/C nanofiber (PANI/CNF) (Figure 1.6(d)). This overall configuration enhanced the discharge or charge processes, and therefore, favored the analytical features of the battery (Zhou *et al.*, 2020).

1.4 POLYTHIOPHENE

PTh was first synthesized in the 1980s. It was obtained by polymerizing 2,5-dibromothiophene, using a metal-catalyzed polycondensation procedure that involving magnesium (Yamamoto, Sanechika, and Yamamoto, 1980). However, the obtained material showed very little solubility, which hampered the determination of its chemical structure. Since then, PTh, shown in Figure 1.7, has become a widely used polymer in charge transport studies among CPs. This polymer has excellent thermal and environmental stability, and high conductivity when doped, which makes it a very promising material among CPs (Roncali, 1992).

PTh can be obtained using different methods, such as metal-catalyzed polycondensation polymerization of 2,5-diiodothiophene and Wurtz coupling of 2,5-dilithiothiophene through other compounds, and electrochemical polymerization by the oxidation of thiophene monomers (Berlin, Pagani, and Sannicolo, 1986; Kobayashi *et al.*, 1984). The charge of these polymers can be modulated according to the pH of the solution, as demonstrated by Murugavel and Malathi (2016). For pH<8.5, PTh shows positive potential, versus the negative potential demonstrated for pH>8.5, according to dynamic light scattering with Zeta potential data. In addition, PTh has excellent thermal stability and good conductivity features (from 10^{-4} to 10^2 S/cm), which makes it a highly desirable material (Kobayashi *et al.*, 1984).

PTh derivatives are very interesting, with particular emphasis on poly(3,4-ethylenedioxythiophene) (PEDOT). PEDOT is obtained by polymerizing 3,4-ethylenedioxythiophene (EDOT) and has more appealing properties than other PTh derivatives for a wide range of applications. It has a lower bandgap energy and a better electrochemical performance (Temmer *et al.*, 2013).

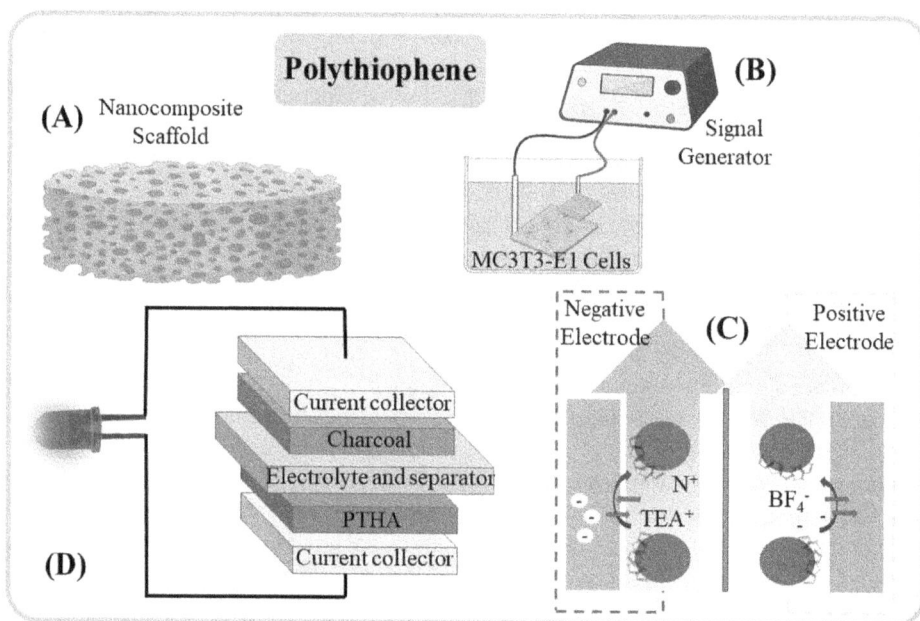

FIGURE 1.7 Examples of the application of polythiophene in: (a) and (b) tissue engineering; (c) energy harvesting battery and (d) energy harvesting (supercapacitor).

Research on PTh and other conducting polymers for biomedical applications has dramatically increased in the last decades, due to their compatibility with biological molecules. From the wide range of applications involving these materials, their use in tissue engineering, biosensors, and energy are highlighted in the following sections.

A conductive hydrogel that combines PTh, chitosan and carboxymethyl chitosan was used to produce an artificial muscle. This composite polymer structure was obtained by a redox polymerization performed in chloroform and demonstrated bending features under the application of an electric field. The bending angle observed was dependent of the intensity of the electric field applied (Pattavarakorn et al., 2013). Based on another work, bone healing was significantly expedited by applying an electrical stimulus to an injured region. Shahini *et al.* (2014) developed a 3D conductive scaffold with ceramic materials, which could locally deliver the electrical stimuli for the correction of large-sized damaged tissues. The conductivity features of these 3D scaffolds were produced by a composite material containing PEDOT and poly(4-styrene sulfonate), among other components (PEDOT:PSS) (Figure 1.7(a)). In an optimized composition, the final composite enhanced cell metabolism. Overall, the electric field established in a cell membrane that is polarized impacts upon the adjacent cells via the ion flux that is generated, which in turn allows smooth electrical communication between the different cells present. This was observed in conductive scaffolds with mesenchymal stem cells, in which the conductivity properties of the scaffold improved cell performances (Shahini *et al.*, 2014). Liu *et al.* (2021) synthesized an electroactive composite (MnO_2–PEDOT) to enhance the electrical stimulation of osteoblasts (MC3T3-E1 osteoblastic cells had their adhesion, proliferation, and differentiation properties enhanced when subjected to an electrical stimulus (Figure 1.7(b)). Moreover, coupling the electroactive propertied of the MnO_2-PEDOT composite nano coating the electrical stimulus was considered a very promising approach for the regeneration of bone tissue (Liu *et al.*, 2021).

In 2016, a potentiometric glucose sensor was developed employing a very special PTh derivative, which was manufactured with a hydroxyl substituent (Hocevar *et al.,* 2016). Due to the presence of the resulting CP, the sensor did not involve the typical catalytic materials required in glucose sensing. The great electroactive properties of poly(hydroxymethyl-3,4-ethylendioxythiophene) played a key role on the activation of hydroxyl-based catalytic effects, which promoted the catalytic-free oxidation of glucose molecules (Hocevar *et al.,* 2016). Recently, the PTh derivative 4-[2,5-di(thiophen-2-yl)-1H-pyrrol-1-yl]aniline ($SNS–NH_2$) was used in an electrochromic cell to provide self-signaling potential in a biosensing device operating under an electrochemical principle (Tavares *et al.,* 2019). This polymer received an electrical signal from a dye sensitized solar cell that integrated a sensing unit. The current produced by this solar cell depended on the concentration of the target analyte bound to the sensing unit. Overall, the energy generated by the dye sensitized solar cell decreased with the increase in the analyte [carcinoembryonic antigen (CEA)], and it was possible to visualize a gradient color change, enhanced by $SNS–NH_2$ (Tavares *et al.,* 2019).

PTh and its derivates have an important role in energy harvesting. In 2012, it contributed to the production of a novel organic redox flow battery, which offered high potentials without being restricted to the typical metals required for this purpose. PTh microparticles were employed as redox couples (Figure 1.7(c)), which were dispersed in an electrolyte solution. Overall, PTh reduced the electrical resistance of the cell, which had promising results without requiring to use of metals in the battery, which posed sustainability and environmental safety issues (Oh *et al.*, 2014).

In addition to batteries, CPs of PTh are good materials for capacitors. CPs can be used to make electrochemical supercapacitors (Vijeth *et al.*, 2018). They designed an asymmetric supercapacitor (ASC). This ASC used an anode electrode with a PTh/aluminum oxide nanocomposite and a cathode electrode of charcoal (Figure 1.7(d)). The ASC acted as a portable and efficient energy storage tool, which offered the capability of illuminating for some time in a light-emitting diode (Vijeth *et al.,* 2018).

1.5 POLYPYRROLE

PPy is synthesized from pyrrole monomers by chemical or electrochemical polymerizations (Ateh, Navsaria, and Vadgama, 2006; Ning *et al.*, 2018). Pyrrole was discovered by Dall'olio *et al.* (1968) when it was known as pyrrole black. However, it only become a famous primary material for many investigations, when Diaz, Kanazawa, and Gardini (1979) published the report about forming a PPy film under electrochemical conditions that offered good conductivity and stability features. Since then, PPy and its derivatives have emerged as very promising materials. The most important features of PPy include ease of synthesis and simple preparation (Saidman and Bessone, 2002), ion exchange capacity, proper electrical conductivity, even in a wide pH range (Mert *et al.*, 2011), stability under different environmental and physiological conditions (Bendrea, Cianga, and Cianga, 2011) and bio-compatibility (Ramanaviciene *et al.*, 2007).

The structure of PPy is charged positively during polymerization, which leads to the absorption of anionic species to reach electroneutrality. Therefore, researchers have attempted to find the appropriate doping anions to modify PPy's chemical and physical properties (Serra Moreno *et al.*, 2009). Nezakati *et al.* (2018)reported advances in its conductive features, and their potential use in biomedical applications, among others.

Globally, PPy has been investigated in different biomedical areas, including drug delivery systems (Jiang *et al.*, 2013), tissue regeneration and repairing damaged tissues including bone, neural, myocardial, and cartilage (Jiang *et al.*, 2013; Ning *et al.*, 2018) due to its excellent mechanical, electrical, and stimulus characteristics (Figure 1.8(a)). Therefore, it has been considered one of the best EAP for biomaterials (Khan *et al.*, 2019). PPy as a conductive polymer can provide an electrical stimulus for tissue genesis, cell growth, and precise control over time.

Wong, Langer, and Ingber (1994) used PPy in cell culture to control cell function. In this study, the electrochemical changes of PPy under an electrical potential provided a method to the control growth, shape, and differentiation of cells. Then, Schmidt *et al.* (1997) used oxidized PPy for nerve regeneration. They determined that PPy's electrical stimulation altered the adsorption of fibronectin, and therefore, affected the performance of the cells. Their research has become a topic of increased interest in tissue regeneration via CPs.

PPy has been focused on in Schwann cell cultures. Fonner *et al.* (2008). discovered that PPy doped with poly(styrene sulfonate) resulted in ideal properties to culture Schwann cells. Huang *et al.* (2010) used PPy/chitosan (2.5:97.5) films to stimulate Schwann cells under an electrical stimulus. This electrical stimulus intensified the production of nerve growth factor and brain-derived neuro-trophic factor. Forciniti *et al.* (2014) studied the behavior of Schwann cell on PPy, using electrical stimulation to regenerate and recover neural tissue. Their results revealed that electrical stimulation could increase cell average displacement and started the extensive effect on migration speed. This considered that the protein adsorption after the film's oxidation and the increase in cell migration to the injury's distal end could lead to neural regeneration.

In addition, Ning *et al.* (2018) reviewed the biocompatibility features of PPy and its use in the regeneration of cardiac and bone tissues. However, few studies have been carried out on bone regeneration with CPs and using PPy, compared with the use of CPs in nerve and cardiac tissue regeneration. Liao *et al.* (2014) suggested that doping biomolecules with PPy improved osteogenic differentiation, which confirmed that PPy may be an encouraging material to regenerate bone tissue.

PPy combined with nanostructures, especially nanowire networks, has attracted great attention in a wide range of applications. This includes biosensors, energy (Figure 1.8(b)), and electronics (Li, Bai, and Shi, 2009), and drug delivery. Jiang *et al.* (2013) reported that micro and nanogaps between the individual nanowires could be employed as reservoirs to store drugs. This study concluded that the quantity of drug released depended on the scan rate of the cyclic voltametric conditions applied, and on the overall thickness of the PPy layer (Figure 1.8(c)).

PPy has been used in batteries for energy storage and fuel cells for energy conversion (Kausar, 2017). Zhou *et al.* (2013) reported that electroactive organic anion-doped PPy was an appropriate

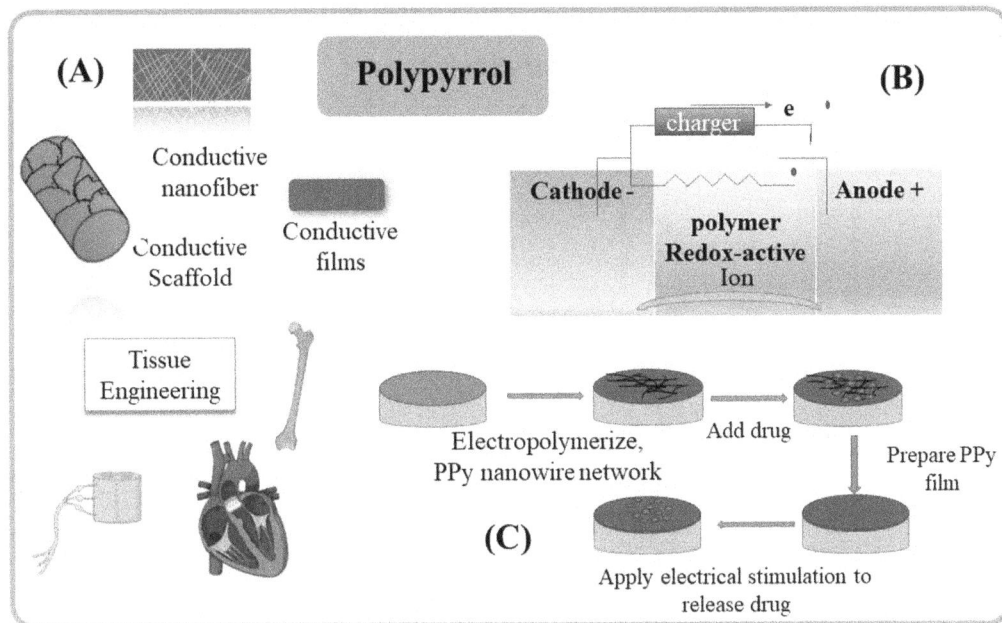

FIGURE 1.8 Examples of the application of polypyrrole in: (a) tissue engineering; (b) energy harvesting (supercapacitor); and (c) drug delivery.

cathode for sodium ion batteries due to its inexpensive and renewable features. The redox-active polymers were the suitable hosts for sodium ion batteries; however, they have some disadvantages due to the large anionic's low doping level. This problem was solved by activation of the PPy chains, with the addition of redox anionic species (Zhou *et al.*, 2013).

1.6 POLYACETYLENE

PAc is known as acetylene black. (Heeger, 2003; Liu *et al.*, 2019; Nezakati *et al.*, 2018; Margolis, 1989). Ito, Shirakawa, and, Ikeda (1974) used PAc and its high conductivity features on doping, after being oxidized by iodine vapor. PAc can be doped with metallic species in a reversible manner, by undergoing a partial oxidation or reduction that is initiated by chemical or electrochemical stimulus (MacDiarmid 2001); therefore, being considered an EAP. PAc has a higher electrical conductivity comparison with PPy, PANI, PTh, or PEDOT, which is the most well-known PTh derivative (Saxman, Liepins, and Aldissi, 1985; MacDiarmid, 2001; Ning *et al.*, 2018). PAc has been used in number of applications (Yang, Wang, and Hu, 2007; Wang *et al.*, 2010; Bahceci and Esat, 2013; Aguilo-Aguayo *et al.*, 2014), among which its use in fuel cells and supercapacitors are shown in Figure 1.9.

Shacklette *et al.* (1987) reported that PAc was a promising EAP for non-aqueous secondary batteries. PAc can behave as doped *p*-type material (oxidized one) or doped *n*-type material (reduced one), and therefore, involved in the production of anode or cathode electrodes in various types of batteries (MacInnes *et al.*, 1981). For instance, Bahceci and Esat (2013) utilized a PAc derivative in a rechargeable battery against a lithium (Li) anode. This derivative contained a pendant TEMPO radical and acted as cathode electrode. The results of this study revealed that this new composition could improve the recharging features of batteries that are Li ion.

PAc has been used to produce bioelectrodes, which aimes to improve their electrochemical features. Wang *et al.* (2010) developed a new electrochemical sensor by modifying glassy carbon

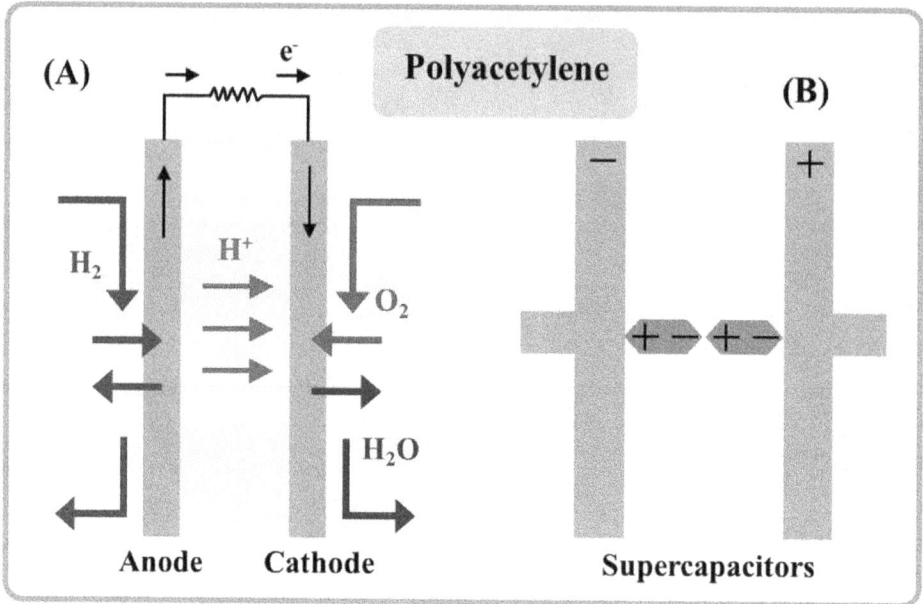

FIGURE 1.9 Examples of the application of polyacetylene's in energy harvesting including: (a) fuel cells; and (b) supercapacitors.

electrodes with an acetylene–chitosan composite film to detect methyl parathion. The currents generated by the redox alteration of methyl parathion were improved when the electrodes contained this acetylene–chitosan film. Overall, this film provided more active sites for the redox reaction, which resulted in a higher sensitivity to monitor methyl parathion.

Although PAc is an electroactive CP that is a conjugated polymer, it is not an appropriate choice in many applications that directly depend on its electroactivity. Therefore, research has mainly focused on the other CPs as EAPs, and less research has been carried out with PAc.

1.7   OUTLOOK AND FUTURE PERSPECTIVES

This chapter aimed to support and confirm the promising research that involves EAPs. This includes a wide range of applications, in which EAPs might act as actuators, sensors, or energy generators, which was supported by some examples, selected from the extensive studies available in the literature. The advances presented were important evidence, which shows that the evaluation of EAPs is required in the search for new polymers that require a lower operating voltage and higher dielectric constant, to improve electromechanical responses. Over time, the performance of EAPs has been enhanced especially by exploring different processing methods, although advances in computational chemistry models and electromechanical analytical tools are very important in this field.

The selection of EAPs depends on the intended application, some of which might require high electrical conduction fields, a fast response, and others might need a low operating voltage. For an application that requires relatively large actuation forces with quick responses, electronic EAPs are an ideal choice. However, this type of EAP requires high voltages, which can compromise the performance as an actuation material in certain applications, such as biomimetic applications. In contrast to electronic EAPs, ionic materials are selected for applications that require a low operation voltage. A typical application example involves tissue engineering, which is supported by its biocompatibility.

Future research is expected to expand and intensify the use of CPs in biomedical applications. This should consider investigations into more biocompatible and degradable materials, and on the adjustment of the electrical stimuli applied. Each tissue has a particular response when faced with an electrical stimulus; therefore, applications in this field should be personalized. Further investigations should consider the modulation of the type of electrical signal and the range of the electrical field, in vitro and in vivo, and evaluate possible collateral damages as a result of the electrical stimulation.

Overall, the scientific community is continually making improvements for the synthesis and fabrication of different types of EAPs to maximize the actuation capability. In addition, other interesting properties that have been the subject of investigations include electrochemical properties. Electrochromic properties relate to the alterations in the optical properties of a given material that is subjected to an electrical stimulus, and these are among the most studied areas in the field of CP. These optical changes could be considered to be a response mechanism, and their use as colorimetric transducers in biosensing have potential in EAPs.

Overall, EAPs have a wonderful and intense story, recently and in the past, and the future appears to be more promising than the past. The application of EAPs is currently vast, which has been enhanced recent and solid developments, mainly in nanotechnology. In combination, the technology, engineering, and EAP materials might generate new hyphenated approaches, or more efficient processes, or both than those known today, with strong benefits for society.

REFERENCES

Aguilar, M.R., and Román, J.S. (2014) *Smart Polymers and Their Applications*, 2nd edn. Cambridge: Woodhead Publishing.

Aguilo-Aguayo, I.M. *et al.* (2014) 'Pulsed electric fields pre-treatment of carrot purees to enhance their polyacetylene and sugar contents', *Innovative Food Science & Emerging Technologies*, 23, pp. 79–86. doi:10.1016/j.ifset.2014.02.010.

Ahmed, S., Ounaies, Z., and Lanagan, M.T. (2017) 'On the impact of self-clearing on electroactive polymer (EAP) actuators', *Smart Materials and Structures*, 26(10). doi:10.1088/1361-665X/aa87c7.

Ambulo, C.P. *et al.* (2020) 'Processing advances in liquid crystal elastomers provide a path to biomedical applications', *Journal of Applied Physics*, 128(14). doi:10.1063/5.0021143.

Anand, A., and Bhatnagar, M.C. (2019) 'Role of vertically aligned and randomly placed zinc oxide (ZnO) nanorods in PVDF matrix: Used for energy harvesting', *Materials Today Energy*, 13, pp. 293–301. doi:10.1016/j.mtener.2019.06.005.

Ateh, DD., Navsaria, H.A., and Vadgama P. (2006) 'Polypyrrole-based conducting polymers and interactions with biological tissues', *Journal of the Royal Society Interface*, 3(11), pp. 741–752. doi:10.1098/rsif.2006.0141.

Bae, J.H., and Chang, S.H. (2016) 'A new approach to fabricate poly(vinylidene fluoride-trifluoroethylene) fibers using a torsion-stretching method and characterization of their piezoelectric properties', *Composites Part B-Engineering*, 99, pp. 112–120. doi:10.1016/j.compositesb.2016.06.037.

Bae, J.H., and Chang, S.H. (2019) 'PVDF-based ferroelectric polymers and dielectric elastomers for sensor and actuator applications: A review', *Functional Composites and Structures*, 1, 1. doi:10.1088/2631-6331/ab0f48.

Bagheri, B. *et al.* (2019) 'Self-gelling electroactive hydrogels based on chitosan-aniline oligomers/agarose for neural tissue engineering with on-demand drug release', *Colloids and Surfaces B-Biointerfaces*, 184, doi:10.1016/j.colsurfb.2019.110549.

Baglio, S., and Bulsara, A. (2006) *Device Applications of Nonlinear Dynamics*. New York: Springer.

Bahceci, S., and Esat, B. (2013) 'A polyacetylene derivative with pendant TEMPO group as cathode material for rechargeable batteries', *Journal of Power Sources*, 242, pp. 33–40. doi:10.1016/j.jpowsour.2013.05.051.

Bai, H., and Shi, G.Q. (2007) "Gas sensors based on conducting polymers." *Sensors*, 7 (3):267–307. doi:10.3390/s7030267.

Bajpai, A.K., Bajpai, J., Saini, R., and Gupta, R. (2011) 'Responsive Polymers in Biology and Technology,' *Polymer Reviews*, 51(1), pp. 53–97. doi:10.1080/15583724.2010.537798.

Bar-Cohen, Y. (2002) "Electroactive polymers as artificial muscles: A review." *Journal of Spacecraft and Rockets*, 39 (6):822–827. doi:10.2514/2.3902.

Bar-Cohen, Y. (2004) *Electroactive Polymer (EAP) Actuators as Artificial Muscles: Reality, Potential and Challenges.* SPIE Digital Library.

Bar-Cohen, Y. (2007a) 'Electroactive polymers as an enabling materials technology', *Proceedings of the Institution of Mechanical Engineers Part G-Journal of Aerospace Engineering*, 221(G4), pp. 553–564. doi:10.1243/09544100jaero141.

Bar-Cohen, Y. (2007b) 'Focus issue on Biomimetics Using Electroactive Polymers as Artificial Muscles', *Bioinspiration & Biomimetics*, 2(2). doi:10.1088/1748-3190/2/2/e01.

Bar-Cohen, Y., and Anderson, I.A. (2019) 'Electroactive polymer (EAP) actuators: Background review', *Mechanics of Soft Materials*, 1, 5.

Bar-Cohen, Y., and. Zhang, Q.M (2008) 'Electroactive polymer actuators and sensors', *MRS Bulletin*, 33(3), pp. 173–181. doi:10.1557/mrs2008.42.

Bar-Cohen, Y., Bao, X.Q., Sherrit, S., and Lih, S.S. (2002) 'Characterization of the electromechanical properties of ionomeric polymer-metal composite (IPMC)', *Proc. SPIE 4695, Smart Structures and Materials 2002: Electroactive Polymer Actuators and Devices (EAPAD)*, 11 July. doi.org/10.1117/12.475173.

Bar-Cohen, Y., Cardoso, V.F., Ribeiro, C., and Lanceros-Mendez, S. (2017) 'Electroactive Polymers as Actuators', in Uchino, K. (ed.) *Advanced Piezoelectric Materials: Science and Technology*, 2nd edn. Cambridge: Woodhead Publishing, pp. 319–352. doi:10.1016/b978-0-08-102135-4.00008-4.

Barnes, M., and Verduzco, R. (2019) 'Direct shape programming of liquid crystal elastomers', *Soft Matter*, 15 (5), pp. 870–879. doi:10.1039/c8sm02174k.

Bashir, M., and Rajendran, P. (2018) 'A review on electroactive polymers development for aerospace applications', *Journal of Intelligent Material Systems and Structures*, 29(19), pp. 3681–3695. doi:10.1177/1045389x18798951.

Baughman, R.H. *et al.* (1999) 'Carbon nanotube actuators', *Science*, 284(5418), pp. 1340–1344. doi:10.1126/science.284.5418.1340.

Bendrea, A.D., Cianga, L., and Cianga, I. (2011) 'Progress in the field of conducting polymers for tissue engineering applications', *Journal of Biomaterials Applications*, 26(1), pp. 3–84. doi:10.1177/0885328211402704.

Benslimane, M.Y., Kiil, H.E., and Tryson M.J. (2010) 'Dielectric electro-active polymer push actuators: performance and challenges', *Polymer International*, 59(3), pp. 415–421. doi:10.1002/pi.2768.

Berlin, A., Pagani, G.A., and Sannicolo, F. (1986) 'New synthetic routes to electroconductive polymers containing thiophene units', *Journal of the Chemical Society-Chemical Communications*, 1986, pp. 1663–1664. doi:10.1039/c39860001663.

Biggs, J. *et al.* (2013) 'Electroactive Polymers: Developments of and Perspectives for Dielectric Elastomers', *Angewandte Chemie-International Edition*, 52 (36), pp. 9409–9421. doi:10.1002/anie.201301918.

Boyd, G.E. (1951) 'Ion exchange", *Annual Review of Physical Chemistry*, 2, pp. 309–342. doi:10.1146/annurev.pc.02.100151.001521.

Brochu, P., and Pei, Q.B. (2010) 'Advances in dielectric elastomers for actuators and artificial muscles', *Macromolecular Rapid Communications*, 31(1), pp. 10–36. doi:10.1002/marc.200900425.

Cardoso, V.F., Ribeiro, C., and Lanceros-Mendez, S. (2017) 'Metamorphic biomaterials', *Bioinspired Materials for Medical Applications*, 121, pp. 69–99. doi:10.1016/b978-0-08-100741-9.00003-6.

Carpi, F., De Rossi, D., Kornbluh, R., Pelrine, R., and Sommer-Larsen, P. (2008) *Dielectric elastomers as electromechanical transducers: Fundamentals, materials, devices, models and applications of an emerging electroactive polymer Technology.* Amsterdam: Elsevier. doi:10.1016/b978-0-08-047488-5.00033-2.

Carpi, F., Kornbluh, R., Sommer-Larsen, P., and Alici, G. (2011) 'Electroactive polymer actuators as artificial muscles: are they ready for bioinspired applications?' *Bioinspiration & Biomimetics*, 6(4). doi:10.1088/1748-3182/6/4/045006.

Chambers, M., Finkelmann, H., Remskar, M., Sanchez-Ferrer, A., Zalar, B., and Zumer, S. (2009) 'Liquid crystal elastomer-nanoparticle systems for actuation', *Journal of Materials Chemistry*, 19(11), pp. 1524–1531. doi:10.1039/b812423j.

Chang, L.F. *et al.* (2018) 'Ionic electroactive polymers used in bionic robots: A review', *Journal of Bionic Engineering*, 15(5), pp. 765–782. doi:10.1007/s42235-018-0065-1.

Csahok, E., Vieil, E., and Inzelt, G. (2000) 'In situ DC conductivity study of the redox transformations and relaxation of polyaniline films', *Journal of Electroanalytical Chemistry*, 482(2), pp. 168–177. doi:10.1016/s0022-0728(00)00044-9.

Dall'olio, A., Dascola, G., Vacara, V., and Bocchi V. (1968) 'Resonance paramagnetique electronique et conductivité d'un noir d'oxypyrrol electrolytique', *Comptes Rendus de l'Académie des Sciences Paris*, 267, pp. 433–435.

Das, A.K., Bhowmik, R., and Meikap, A.K. (2017) 'Surface functionalized carbon nanotube with polyvinylidene fluoride: Preparation, characterization, current-voltage and ferroelectric hysteresis behaviour of polymer nanocomposite films', *AIP Advances*, 7(4). doi:10.1063/1.4980051.

de Gennes, P.G., Okumura K., Shahinpoor, M., and Kim, K.J. (2000) 'Mechanoelectric effects in ionic gels', *Europhysics Letters*, 50(4), pp. 513–518. doi:10.1209/epl/i2000-00299-3.

Degennes, P.G. (1975) 'One type of nematic polymers', *Comptes Rendus Hebdomadaires Des Seances De L Academie Des Sciences Serie B*, 281(5–8), pp. 101–103.

Diaz, A.F., Kanazawa, K.K., and Gardini, G.P. (1979) 'Electrochemical polymerization of pyrrole', *Journal of the Chemical Society-Chemical Communications*, pp. 635–636. doi:10.1039/c39790000635.

Doi, M., Matsumoto, M., and Hirose, Y. (1992) 'Deformation of ionic polymer gels by electric fields', *Macromolecules*, 25(20), pp. 5504–5511. doi:10.1021/ma00046a058.

Dong, L. *et al.* (2018) 'Resonant frequency tuning of electroactive polymer membranes via an applied bias voltage', *Smart Materials and Structures*, 27(11). doi:10.1088/1361-665X/aacdc0.

Eatemadi, A. *et al.* (2014) 'Carbon nanotubes: properties, synthesis, purification, and medical applications', *Nanoscale Research Letters*, 9, 393. doi:10.1186/1556-276x-9-393.

Eguchi, M. (1925) 'On the permanent electret', *Philosophical Magazine*, 49(289), pp. 178–192. doi:10.1080/14786442508634594.

Ellingford, C. *et al.* (2020) 'Self-healing dielectric elastomers for damage-tolerant actuation and energy harvesting', *ACS Applied Materials & Interfaces*, 12(6), pp. 7595–7604. doi:10.1021/acsami.9b21957.

Finkelmann, H., Kock, H.J., and Rehage, G. (1981) 'Investigations on liquid-crystalline polysiloxanes 3. Liquid-crystalline elastomers: A new type of liquid-crystal material', *Makromolekulare Chemie-Rapid Communications*, 2(4), pp. 317–322.

Fonner, J.M. *et al.* (2008) 'Biocompatibility implications of polypyrrole synthesis techniques', *Biomedical Materials*, 3(3). doi:10.1088/1748-6041/3/3/034124.

Forciniti, L., Ybarra, J., Zaman, M.H., and Schmidt C.E. (2014) 'Schwann cell response on polypyrrole substrates upon electrical stimulation', *Acta Biomaterialia*, 10(6), pp. 2423–2433. doi:10.1016/j.actbio.2014.01.030.

Furukawa, T., Takahashi, Y., and Nakajima, T. (2010) 'Recent advances in ferroelectric polymer thin films for memory applications', *Current Applied Physics*, 10(1), pp. E62-E67. doi:10.1016/j.cap.2009.12.015.

Gonzalez, D., Garcia, J., and Newell, B. (2019) 'Electromechanical characterization of a 3D printed dielectric material for dielectric electroactive polymer actuators', *Sensors and Actuators a-Physical*, 297, p. 111565. doi:10.1016/j.sna.2019.111565.

Green, A.G., and Woodhead, A.E. (1910) 'Aniline-black and allied compounds. Part I', *Journal of the Chemical Society*, 98, pp. 2388–2403.

Guimard, N.K., Gomez, N., and Schmidt, C.E. (2007) 'Conducting polymers in biomedical engineering', *Progress in Polymer Science*, 32(8–9), pp. 876–921. doi:10.1016/j.progpolymsci.2007.05.012.

Gupta, U. *et al.* (2019) 'Soft robots based on dielectric elastomer actuators: a review', *Smart Materials and Structures*, 28 (10), doi:10.1088/1361-665X/ab3a77.

Hall, N. (2003) 'Twenty-five years of conducting polymers', *Chemical Communications*, 2(1), pp. 1–4.

Han, X. *et al.* (2016) 'Flexible polymer transducers for dynamic recognizing physiological signals', *Advanced Functional Materials*, 26(21), pp. 3640–3648. doi:10.1002/adfm.201600008.

Hau, S., York, A., and. Seelecke, S. (2016) 'High-force dielectric electroactive polymer (DEAP) membrane actuator', *SPIE Conference on Electroactive Polymer Actuators and Devices (EAPAD)*, Las Vegas, NV, 21–24 March.

Hau, S., Rizzello, G., and Seelecke, S. (2018) 'A novel dielectric elastomer membrane actuator concept for high-force applications', *Extreme Mechanics Letters*, 23, pp. 24–28. doi:10.1016/j.eml.2018.07.002.

He, Q.G. *et al.* (2019) 'Electrically controlled liquid crystal elastomer-based soft tubular actuator with multimodal actuation', *Science Advances*, 5(10). doi:10.1126/sciadv.aax5746.

Hill, M., Rizzello, G., and Seelecke, S. (2017) 'Development and experimental characterization of a pneumatic valve actuated by a dielectric elastomer membrane', *Smart Materials and Structures*, 26(8). doi:10.1088/1361-665X/aa746d.

Hocevar, M.A. *et al.* (2016) 'Nanometric polythiophene films with electrocatalytic activity for non-enzymatic detection of glucose', *European Polymer Journal*, 79, pp. 32–139. doi:10.1016/j.eurpolymj.2016.04.032.

Huang, H., and Scott, J.F. (2018) *Ferroelectric Materials for Energy Applications*. Berlin: Wiley-VCH.

Huang, J.H. *et al.* (2010) 'Electrical regulation of Schwann cells using conductive polypyrrole/chitosan polymers', *Journal of Biomedical Materials Research Part A*, 93,(1), pp. 164–174. doi:10.1002/jbm.a.32511.

Huang, W.S., Humphrey, B.D., and Macdiarmid, A.G. (1986) 'Polyaniline, a novel conducting polymer, morphology and chemistry of its oxidation and reduction in aqueous-electrolytes', *Journal of the Chemical Society-Faraday Transactions I*, 82, p. 2385. doi:10.1039/f19868202385.

Iijima, S., and Ichihashi, T. (1993) 'Single-shell carbon nanotubes of 1 nm diameter', *Nature*, 363(6430), pp. 603–605. doi:10.1038/363603a0.

Inzelt, G. (2008) 'Conducting polymers: A new era in electrochemistry', in *Conducting Polymers: A New Era in Electrochemistry*. Berlin: Springer.

Ito, T., Shirakawa, H., and Ikeda, S. (1974) 'Simultaneous polymerization and formation of polyacetylene film on the surface of concentrated soluble Ziegler-type catalyst solution', *Journal of Polymer Science Part A: Polymer Chemistry*, 12, pp. 11–20.

Jiang, H.R., Li, C.S., and Huang, X.Z. (2013) 'Actuators based on liquid crystalline elastomer materials', *Nanoscale*, 5, pp. 5225–5240. doi:10.1039/c3nr00037k.

Jiang, L. *et al.* (2018) 'Fabrication of dielectric elastomer composites by locking a pre-stretched fibrous TPU network in EVA', *Materials*, 11(9). doi:10.3390/ma11091687.

Jiang, L. *et al.* (2020) 'Fabrication of dielectric elastomers with improved electromechanical properties using silicone rubber and walnut polyphenols modified dielectric particles', *Materials & Design*, 192.

Jiang, Y. *et al.* (2012) 'Electrothermally driven structural colour based on liquid crystal elastomers', *Journal of Materials Chemistry*, 22(24), pp. 11943–11949. doi:10.1039/c2jm30176h.

Jo, C., Naguib, H.E., and Kwon, R.H. (2011) 'Fabrication, modeling and optimization of an ionic polymer gel actuator', *Smart Materials and Structures*, 20(4). doi:10.1088/0964-1726/20/4/045006.

Karan, S.K., Mandal, D., and Khatua, B.B. (2015) 'Self-powered flexible Fe-doped RGO/PVDF nanocomposite: an excellent material for a piezoelectric energy harvester', *Nanoscale*, 7(24), pp. 10655–10666. doi:10.1039/c5nr02067k.

Katchalsky, A., and Zwick, M. (1955) 'Mechanochemistry and ion exchange', *Journal of Polymer Science*, 16(82), pp. 221–234. doi:10.1002/pol.1955.120168212.

Katsouras, I. *et al.* (2016) 'The negative piezoelectric effect of the ferroelectric polymer poly(vinylidene fluoride)', *Nature Materials*, 15(1), p.78. doi:10.1038/nmat4423.

Kausar, A. (2017) 'Overview on conducting polymer in energy storage and energy conversion system', *Journal of Macromolecular Science Part a-Pure and Applied Chemistry*, 54(9), pp. 640–653. doi:10.1080/10601325.2017.1317210.

Kawai, H. (1969) 'Piezoelectricity of poly(vinylidene fluoride)', *Japanese Journal of Applied Physics*, 8(7), p. 975. doi:10.1143/jjap.8.975.

Khan, M. *et al.* (2019) 'A review on biomaterials for 3D conductive scaffolds for stimulating and monitoring cellular activities', *Applied Sciences-Basel*, 9(5). doi:10.3390/app9050961.

Kim, H.S. *et al.* (2019) 'Ferroelectric-polymer-enabled contactless electric power generation in triboelectric nanogenerators', *Advanced Functional Materials*, 29(45). doi:10.1002/adfm.201905816.

Kim, K.L. *et al.* (2016) 'Epitaxial growth of thin ferroelectric polymer films on graphene layer for fully transparent and flexible nonvolatile memory', *Nano Letters*, 16(1), pp. 334–340. doi:10.1021/acs.nanolett.5b03882.

Kim., K.J., and Tadokoro, S. (2007) *Electroactive Polymers for Robotic Applications: Artificial Muscles and Sensors*. London: Springer.

Kobayashi, M. *et al.* (1984) 'Synthesis and properties of chemically coupled poly(thiophene)', *Synthetic Metals*, 9(1), pp. 77–86. doi:10.1016/0379-6779(84)90044-4.

Kumar, A., Srivastava, A., Galaev, I.Y., and Mattiasson, B. (2007) 'Smart polymers: Physical forms and bioengineering applications', *Progress in Polymer Science*, 32(10), pp. 1205–1237. doi:10.1016/j.progpolymsci.2007.05.003.

Kwon, H.J., Osada, Y., and Gong, J.P. (2006) 'Polyelectrolyte gels-fundamentals and applications', *Polymer Journal*, 38(12), pp. 1211–1219. doi:10.1295/polymj.PJ2006125.

Letheby, H. (1862) 'On the production of a blue substance by the electrolysis of sulyhate of aniline' *Journal of the Chemical Society*, 15, pp. 161–163. doi:10.1039/JS8621500161.

Li, C., Bai, H., and Shi, G.Q. (2009) 'Conducting polymer nanomaterials: electrosynthesis and applications', *Chemical Society Reviews*, 38(8), pp. 2397–2409. doi:10.1039/b816681c.

Li, H.L., Wang, R.P., Han, S.T., and Zhou, Y. (2020) 'Ferroelectric polymers for non-volatile memory devices: A review', *Polymer International*, 69(6), pp. 533–544. doi:10.1002/pi.5980.

Li, J.R., Liu, L.W., Liu, Y.J., and Leng, J.S. (2019) 'Dielectric elastomer spring-roll bending actuators: Applications in soft robotics and design', *Soft Robotics*, 6(1), pp. 69–81. doi:10.1089/soro.2018.0037.

Li, M.H., and Keller, P. (2006) 'Artificial muscles based on liquid crystal elastomers', *Philosophical Transactions of the Royal Society A - Mathematical Physical and Engineering Sciences*, 364(1847), pp. 2763–2777. doi:10.1098/rsta.2006.1853.

Li, Q., and Wang, Q. (2016) 'Ferroelectric polymers and their energy-related applications', *Macromolecular Chemistry and Physics*, 217(11), pp. 1228–1244.

Li, Q. *et al.* (2015) 'Flexible high-temperature dielectric materials from polymer nanocomposites', *Nature*, 523(7562), pp. 576. doi:10.1038/nature14647.

Liao, J.W. *et al.* (2014) 'Surface-dependent self-assembly of conducting polypyrrole nanotube arrays in template-free electrochemical polymerization', *ACS Applied Materials & Interfaces*, 6(14), pp. 10946–10951. doi:10.1021/am5017478.

Liu, H.R., Ge, J., Ma, E., and Yang, L. (2019) 'Advanced biomaterials for biosensor and theranostics', *Biomaterials in Translational Medicine*. Cambridge: Woodhead Publishing.

Liu., S. *et al.* (2021) 'Electroactive MnO2-poly(3,4-ethylenedioxythiophene) composite nanocoatings enhance osteoblastic electrical stimulation', *Applied Surface Science*, 545, 148827.

Ma, S.Q. *et al.* (2020) 'High-performance ionic-polymer-metal composite: Toward large-deformation fast-response artificial muscles', *Advanced Functional Materials*, 30(7). doi:10.1002/adfm.201908508.

MacDiarmid, A.G. (1997) 'Polyaniline and polypyrrole: Where are we headed?' *Synthetic Metals*, 84(1–3), pp. 27–34. doi:10.1016/s0379-6779(97)80658-3.

MacDiarmid, A.G. (2001) '"Synthetic metals": A novel role for organic polymers (Nobel lecture)', *Angewandte Chemie-International Edition*, 40 (14), pp. 2581–2590. doi:10.1002/1521-3773(20010716)40:14<2581:aid-anie2581>3.0.co;2-2.

MacInnes, D. *et al.* (1981) 'Organic batteries, reversible n-type and p-type electrochemical doping of poly-acetylene (CH)X', *Journal of the Chemical Society-Chemical Communications*, 7(1), pp. 317–319. doi:10.1039/c39810000317.

Mai, M.F., Ke, S.M., Lin, P., and Zeng X.R. (2015) 'Ferroelectric polymer thin films for organic electronics', *Journal of Nanomaterials*. doi:10.1155/2015/812538.

Mao, G.Y., and Qu, S.X. (2018) 'Review on instability of dielectric elastomer structures and devices subjected to electromechanical loading', *Scientia Sinica-Physica Mechanica & Astronomica*, 48(9). doi:10.1360/sspma2018-00187.

Margolis, J. (1989) *Conductive Polymers and Plastics*. Boston, MA: Springer.

Mert, B.D., Solmaz, R., Kardas, G., and Yazici, B. (2011) 'Copper/polypyrrole multilayer coating for 7075 aluminum alloy protection', *Progress in Organic Coatings*, 72(4), pp. 748–754. doi:10.1016/j.porgcoat.2011.08.006.

Mishra, R., Sharma, A., and Tandon, P. (2011) 'Conducting polymers—modern semiconductors: A theoretical overview', *Advancements and Futuristic Trends in Material Science*, Bareilly, India, 26–27 March.

MohdIsa, W., Hunt, A., and Hossein-Nia, S.H. (2019) 'Sensing and self-sensing actuation methods for ionic polymer-metal composite (IPMC): A review', *Sensors*, 19(18). doi:10.3390/s19183967.

Murugavel, S., and Malathi, M. (2016) 'Structural, photoconductivity, and dielectric studies of polythiophene-tin oxide nanocomposites', *Materials Research Bulletin*, 81, pp. 93–100. doi:10.1016/j.materresbull.2016.05.004.

Nasri-Nasrabadi, B. *et al.* (2018) 'An electroactive polymer composite with reinforced bending strength, based on tubular micro carbonized-cellulose', *Chemical Engineering Journal*, 334, pp. 1775–1780. doi:10.1016/j.cej.2017.11.140.

Nemat-Nasser, S. (2002) 'Micromechanics of actuation of ionic polymer-metal composites', *Journal of Applied Physics*, 92(5), pp. 2899–2915. doi:10.1063/1.1495888.

Newell, B., Krutz, G., Garcia-Bravo, J., and Harmeyer, K. (2016) 'Industrial capacitance sensors and actuators', *Proceedings of 2016 Future Technologies Conference (Ftc)*, pp. 450–457.

Nezakati, T., Seifalian, A., Tan, A., and Seifalian, A.M. (2018) 'Conductive polymers: Opportunities and challenges in biomedical applications', *Chemical Reviews*, 118(14), pp. 6766–6843. doi:10.1021/acs. chemrev.6b00275.

Nicolau-Kuklinska, A. *et al.* (2018) 'A new electroactive polymer based on carbon nanotubes and carbon grease as compliant electrodes for electroactive actuators', *Journal of Intelligent Material Systems and Structures*, 29(7). doi:10.1177/1045389x17740979.

Ning, C.Y. *et al.* (2018) 'Electroactive polymers for tissue regeneration: Developments and perspectives', *Progress in Polymer Science*, 81, pp. 144–162. doi:10.1016/j.progpolymsci.2018.01.001.

Ning, N.Y. *et al.* (2019) 'Improved dielectric and actuated performance of thermoplastic polyurethane by blending with XNBR as macromolecular dielectrics', *Polymer*, 179. doi:10.1016/j.polymer.2019.121646.

Oguro, K., Kawami, Y., and Takenaka, H. (1992) 'An actuator element of polyelectrolyte gel membrane-electrode composite', *Bulletin of the Government Industrial Research Institute, Osaka*, 43, pp. 21–24.

Oh, S.H. *et al.* (2014) 'A metal-free and all-organic redox flow battery with polythiophene as the electroactive species', *Journal of Materials Chemistry A*, 2(47), pp. 19994–19998. doi:10.1039/c4ta04730c.

Ohm, C., Brehmer, M., and Zentel, R. (2010) 'Liquid crystalline elastomers as actuators and sensors', *Advanced Materials*, 22(31), pp. 3366–3387. doi:10.1002/adma.200904059.

Palza, H., Zapata, P.A., and Angulo-Pineda, C. (2019) 'Electroactive smart polymers for biomedical applications', *Materials*, 12(2). doi:10.3390/ma12020277.

Park, I.S. *et al.* (2008) 'Physical principles of ionic polymer-metal composites as electroactive actuators and sensors', *MRS Bulletin*, 33(3), pp. 190–195. doi:10.1557/mrs2008.44.

Pattavarakorn, D. *et al.* (2013) 'Electroactive performances of conductive polythiophene/hydrogel hybrid artificial muscle', *10th Eco-Energy and Materials Science and Engineering Symposium*, Thailand, 34, pp. 673–681. doi:10.1016/j.egypro.2013.06.799.

Poulin, A., Rosset, S., and Shea, H.R. (2015) 'Printing low-voltage dielectric elastomer actuators', *Applied Physics Letters*, 107(24). doi:10.1063/1.4937735.

Pugal, D., Jung, K., Aabloo, A., and. Kim, K.J. (2010) 'Ionic polymer-metal composite mechanoelectrical transduction: review and perspectives', *Polymer International*, 59(3), pp. 279–289. doi:10.1002/pi.2759.

Pullano, S A., Mahbub, I., Islam, S.K., and Fiorillo, A.S. (2017) 'PVDF sensor stimulated by infrared radiation for temperature monitoring in microfluidic devices', *Sensors*, 17(4). doi:10.3390/s17040850.

Qian, X.S. *et al.* (2015) 'Ferroelectric polymers as multifunctional electroactive materials: Recent advances, potential, and challenges', *MRS Communications*, 5(2), pp. 115–129. doi:10.1557/mrc.2015.20.

Qiu, Y., Zhang, E., Plamthottam, R., and Pei, Q.B. (2019) 'Dielectric elastomer artificial muscle: Materials innovations and device explorations', *Accounts of Chemical Research*, 52(2), pp. 316–325. doi:10.1021/acs.accounts.8b00516.

Qu, L.T. *et al.* (2008) 'Carbon nanotube electroactive polymer materials: Opportunities and challenges', *MRS Bulletin*, 33(3), pp. 215–224. doi:10.1557/mrs2008.47.

Quinsaat, J.E. *et al.* (2015) 'Highly stretchable dielectric elastomer composites containing high volume fractions of silver nanoparticles', *Journal of Materials Chemistry A*, 3(28), pp. 14675–14685. doi:10.1039/c5ta03122b.

Rahman, S.U., Bilal, S., and Shah, A.U.A. (2020) 'Synthesis and characterization of polyaniline-chitosan patches with enhanced stability in physiological conditions', *Polymers*, 12(12). doi:10.3390/polym12122870.

Ramanaviciene, A., Kausaite, A., Tautkus, S., and Ramanavicius, A. (2007) 'Biocompatibility of polypyrrole particles: An in-vivo study in mice', *Journal of Pharmacy and Pharmacology*, 59(2), pp. 311–315. doi:10.1211/jpp.59.2.0017.

Rasmussen, L. *et al.* (2018) 'Synthetic muscle (TM) electroactive polymer (EAP) based actuation and pressure sensing for prosthetic and robotic gripper applications', *21st Conference on Electroactive Polymer Actuators and Devices (EAPAD) XXI*, Denver, CO, 4–7 March.

Rasmussen, S.C. (2017) 'The early history of polyaniline: Discovery and origins', *Substantia*, 1(12), pp. 99–109.

Roach, D.J. *et al.* (2019) 'Long liquid crystal elastomer fibers with large reversible actuation strains for smart textiles and artificial muscles', *ACS Applied Materials & Interfaces*, 11(21), pp. 19514–19521. doi:10.1021/acsami.9b04401.

Roentgen, W.C. (1880) 'About the changes in shape and volume of dielectrics caused by electricity', *Annual Review of Physical Chemistry*, 11, pp. 771–786.

Roncali, J. (1992) 'Conjugated poly(thiophenes): Synthesis, functionalization, and applications', *Chemical Reviews*, 92(4), pp. 711–738. doi:10.1021/cr00012a009.

Sacerdote, M.P. (1899) 'On the electrical deformation of isotropic dielectric solids', *Journal of Physics*, 3 (1), pp. 282–285.

Saidman, S.B., and Bessone, J.B. (2002) 'Electrochemical preparation and characterisation of polypyrrole on aluminium in aqueous solution', *Journal of Electroanalytical Chemistry*, 521(1–2), pp. 87–94. doi:10.1016/s0022-0728(02)00685-x.

Sarvari, R. *et al.* (2017) 'Conductive and biodegradable scaffolds based on a five-arm and functionalized star-like polyaniline-polycaprolactone copolymer with a D-glucose core', *New Journal of Chemistry*, 41(14), pp. 6371–6384. doi:10.1039/c7nj01063j.

Saxman., A.M., Liepins, R., and Aldissi, M. (1985) 'Polyacetylene: Its synthesis, doping and structure', *Progress in Polymer Science*, 11(1–2), pp. 57–89.

Schmidt, C.E., Shastri, V.R., Vacanti, J.P., and Langer, R. (1997) 'Stimulation of neurite outgrowth using an electrically conducting polymer', *Proceedings of the National Academy of Sciences of the United States of America*, 94(17), pp. 8948–8953. doi:10.1073/pnas.94.17.8948.

Sen, T., Mishra, S., and Shimpi, N.G. (2016) 'Synthesis and sensing applications of polyaniline nanocomposites: a review', *RSC Advances*, 6(48), pp. 42196–42222. doi:10.1039/c6ra03049a.

Serra Moreno, J. *et al.* (2009) 'Polypyrrole-polysaccharide thin films characteristics: Electrosynthesis and biological properties', *Journal of Biomedical Materials Research Part A*, 88(3), pp. 832–840. doi:10.1002/jbm.a.32230.

Shacklette, L.W. *et al.* (1987) 'Secondary batteries with electroactive polymer electrodes', *Synthetic Metals*, 18(1–3, pp. 611–618. doi:10.1016/0379-6779(87)90949-0.

Shahini, A. *et al.* (2014) '3D conductive nanocomposite scaffold for bone tissue engineering', *International Journal of Nanomedicine*, 9, pp. 167–181. doi:10.2147/ijn.s54668.

Shahinpoor, M. (2000) 'Electrically-activated artificial muscles made with liquid crystal elastomers', Smart Structures and Materials 2000 Conference, Newport Beach, CA, March 5–9.

Shahinpoor, M. (2020) *Fundamentals of Smart Materials*. Cambridge: Royal Society of Chemistry.

Shankar, R., Ghosh, T.K., and Spontak, R.J. (2007) 'Dielectric elastomers as next-generation polymeric actuators', *Soft Matter*, 3(9), pp. 1116–1129. doi:10.1039/b705737g.

Sheima, Y., Caspari, P., and Opris, D.M. (2019) 'Artificial muscles: Dielectric elastomers responsive to low voltages', *Macromolecular Rapid Communications*, 40(16). doi:10.1002/marc.201900205.

Shirakawa, H. *et al.* (1977) 'Synthesis of electrically conducting organic polymers: Halogen derivatives of polyacetylene, (CH)X', *Journal of the Chemical Society-Chemical Communications*, 1(16), pp. 578–580. doi:10.1039/c39770000578.

Shklovsky, J. *et al.* (2018) 'Towards fully polymeric electroactive micro actuators with conductive polymer electrodes', *Microelectronic Engineering*, 199, pp. 58–62. doi:10.1016/j.mee.2018.07.012.

Skotheim, T.A., and Reynolds, J.R. (2006) *Handbook of Conducting Polymers*, 3rd edn, Vol. 1. Boca Raton: CRC Press.

Su, J. *et al.* (1999) 'Electrostrictive graft elastomers and applications', *Symposium on Electroactive Polymers at the 1999 MRS Fall Meeting*, Boston, MA, 29 November–1 December.

Tan, M.W.M., Thangavel, G., and Lee, P.S. (2019) 'Enhancing dynamic actuation performance of dielectric elastomer actuators by tuning viscoelastic effects with polar crosslinking', *NPG Asia Materials*, 11. doi:10.1038/s41427-019-0147-5.

Tanaka, T. *et al.* (1982) 'Collapse of gels in an electric-field', *Science*, 218(4571), pp. 467–469. doi:10.1126/science.218.4571.467.

Tavares, A.P.M. *et al.* (2019a) 'Self-powered and self-signalled autonomous electrochemical biosensor applied to cancinoembryonic antigen determination', *Biosensors and Bioelectronics*, 140, 111320. doi:10.1016/j.bios.2019.111320.

Tavares, A.P.M. *et al.* (2019b) 'Photovoltaics, plasmonics, plastic antibodies and electrochromism combined for a novel generation of self-powered and self-signalled electrochemical biomimetic sensors', *Biosensors and Bioelectronics*, 137, pp. 72–81. doi:10.1016/j.bios.2019.04.055.

Temmer, R. *et al.* (2013) 'In search of better electroactive polymer actuator materials: PPy versus PEDOT versus PEDOT-PPy composites', *Smart Materials and Structures*, 22(10). doi:10.1088/0964-1726/22/10/104006.

Thetpraphi, K. *et al.* (2020) 'Advanced plasticized electroactive polymers actuators for active optical applications: Live mirror', *Advanced Engineering Materials*, 22(5). doi:10.1002/adem.201901540.

Thomsen, D.L. *et al.* (2001) 'Liquid crystal elastomers with mechanical properties of a muscle', *Macromolecules*, 34(17), pp. 5868–5875. doi:10.1021/ma001639q.

Tian, M. *et al.* (2014) 'Largely improved actuation strain at low electric field of dielectric elastomer by combining disrupting hydrogen bonds with ionic conductivity', *Journal of Materials Chemistry C*, 2(39), pp. 8388–8397. doi:10.1039/c4tc01140f.

Tiwari, R., and Garcia E. (2011) 'The state of understanding of ionic polymer metal composite architecture: a review', *Smart Materials and Structures*, 20(8). doi:10.1088/0964-1726/20/8/083001.

Tohluebaji, N., Putson, C., and Muensit, N. (2019) 'High electromechanical deformation based on structural beta-phase content and electrostrictive properties of electrospun poly(vinylidene fluoride-hexafluoropropylene) nanofibers', *Polymers*, 11(11). doi:10.3390/polym11111817.

Torop., J. *et al.* (216) 'Electrochemically driven carbon-based materials as EAPs: Fundamentals and device configurations', in F. Carpi (ed.), *Electromechanically Active Polymers*. Cham: Springer.

Ula, S.W. *et al.* (2018) 'Liquid crystal elastomers: an introduction and review of emerging technologies', *Liquid Crystals Reviews*, 6(1), pp. 78–107. doi:10.1080/21680396.2018.1530155.

Unal, F.A. *et al.* (2019) 'Ionic polymer-metal composite actuators operable in dry conditions', in Inamuddin and A. Asiri (eds), *Ionic Polymer Metal Composites for Sensors and Actuators*. Cham: Springer.

Vijeth, H. *et al.* (2018) 'Flexible and high energy density solid-state asymmetric supercapacitor based on polythiophene nanocomposites and charcoal', *RSC Advances*, 8(55), pp. 31414–31426. doi:10.1039/c8ra06102e.

Wallmersperger, T., Kroplin, B., Holdenried, J., and Gulch, R.W. (2001) 'A coupled multi-field-formulation for ionic polymer gels in electric fields', *Smart Structures and Materials 2001 Conference*, Newport Beach, CA, 5–8 March.

Wang, L.P., Huang, Q.W., and Wang, J.Y. (2015) 'Nanostructured polyaniline coating on ITO glass promotes the neurite outgrowth of PC 12 cells by electrical stimulation', *Langmuir*, 31(44), pp. 12315–12322. doi:10.1021/acs.langmuir.5b00992.

Wang, T.S. *et al.* (2016) 'Electroactive polymers for sensing', *Interface Focus*, 6(4). doi:10.1098/rsfs.2016.0026.

Wang, Y.J. *et al.* (2014) 'Effects of preparation steps on the physical parameters and electromechanical properties of IPMC actuators', *Smart Materials and Structures*, 23(12). doi:10.1088/0964-1726/23/12/125015.

Wang, Y.Q., Sun, X.K., Sun, C.J., and Su, J. (2003) 'Two-dimensional computational model for electrostrictive graft elastomer', *Smart Structures and Materials 2003 Conference*, San Diego, CA, 2–6 March.

Wang, Y.Z., Qiu, H.X., Hu, S.Q., and Xu, J.H. (2010) 'A novel methyl parathion electrochemical sensor based on acetylene black-chitosan composite film modified electrode', *Sensors and Actuators B-Chemical*, 147(2), pp. 587–592. doi:10.1016/j.snb.2010.03.034.

Wei, M.L., Gao, Y.F., Li, X., and Serpe, M.J. (2017) 'Stimuli-responsive polymers and their applications', *Polymer Chemistry*, 8(1), pp. 127–143. doi:10.1039/c6py01585a.

Wibowo, A. *et al.* (2020) '3D Printing of polycaprolactone-polyaniline electroactive scaffolds for bone tissue engineering', *Materials*, 13(3). doi:10.3390/ma13030512.

Wong, J.Y., Langer, R., and Ingber, D.E. (1994) 'Electrically conducting polymers an noninvasively control the shape and growth of mammalian-cells', *Proceedings of the National Academy of Sciences of the United States of America*, 91(8), pp. 3201–3204. doi:10.1073/pnas.91.8.3201.

Yamamoto, T., Sanechika, K., and Yamamoto, A. (1980) 'Preparation of thermostable and electric-conducting poly(2,5-thienylene)', *Journal of Polymer Science Part C-Polymer Letters*, 18(1), pp. 9–12. doi:10.1002/pol.1980.130180103.

Yang, D. *et al.* (2019) 'Ionic polymer-metal composites actuator driven by the pulse current signal of triboelectric nanogenerator', *Nano Energy*, 66. doi:10.1016/j.nanoen.2019.104139.

Yang, X.F., Wang, F., and Hu, S.S. (2007) 'High sensitivity voltammetric determination of sodium nitroprusside at acetylene black electrode in the presence of CTAB', *Colloids and Surfaces B-Biointerfaces*, 54(1), pp. :60–66. doi:10.1016/j.colsurfb.2006.09.003.

Youn, J.H. *et al.* (2020) 'Dielectric elastomer actuator for soft robotics applications and challenges', *Applied Sciences-Basel*, 10(2). doi 10.3390/app10020640.

Yu, Y.L., and Ikeda, T. (2006) 'Soft actuators based on liquid-crystalline elastomers', *Angewandte Chemie-International Edition*, 45(33), pp. 5416–5418. doi:10.1002/anie.200601760.

Yuan, C. *et al.* (2017) '3D printed reversible shape changing soft actuators assisted by liquid crystal elastomers', *Soft Matter*, 13(33), pp. 5558–5568. doi:10.1039/c7sm00759k.

Zhang, C.L. *et al.* (2020) 'Energy harvesting from a novel contact-type dielectric elastomer generator', *Energy Conversion and Management*, 205. doi:10.1016/j.enconman.2019.112351.

Zhang, J. *et al.* (2020) 'Improving actuation strain and breakdown strength of dielectric elastomers using core-shell structured CNT-Al2O3', *Composites Science and Technology*, 200. doi:10.1016/j.compscitech.2020.108393.

Zhang, Q.M., Bharti, V., and Zhao, X. (1998) 'Giant electrostriction and relaxor ferroelectric behavior in electron-irradiated poly(vinylidene fluoride-trifluoroethylene) copolymer', *Science*, 280(5372), pp. 2101–2104. doi:10.1126/science.280.5372.2101.

Zhang, X.J. *et al.* (2020) 'Fabrication of macroporous nafion membrane from silica crystal for ionic polymer-metal composite actuator', *Processes*, 8(11). doi:10.3390/pr8111389.

Zhao, Y. *et al.* (2018) 'Constructing advanced dielectric elastomer based on copolymer of acrylate and poly-urethane with large actuation strain at low electric field', *Polymer*, 149, pp. 39–44. doi:10.1016/j.polymer.2018.06.065.

Zhao, Y. *et al.* (2020) 'Review of dielectric elastomers for actuators, generators and sensors', *IET Nanodielectrics*, 3(4), pp. 99–106.

Zhenyl, M., Scheinbeim, J.I., Lee, J.W., and Newman, B.A. (1994) 'High-field electrostrictive response of polymers', *Journal of Polymer Science Part B-Polymer Physics*, 32(16), pp. 2721–2731. doi:10.1002/polb.1994.090321618.

Zhou, G.Y. *et al.* (2020) 'Highly porous electroactive polyimide-based nanofibrous composite anode for all-organic aqueous ammonium dual-ion batteries', *Composites Communications*, 22. doi:10.1016/j.coco.2020.100519.

Zhou, M., Xiong, Y., Cao, Y.L., Ai, X.P., and Yang, H.X. (2013) "Electroactive organic anion-doped polypyrrole as a low cost and renewable cathode for sodium-ion batteries." *Journal of Polymer Science Part B-Polymer Physics*, 51 (2):114–118. doi:10.1002/polb.23184.

Zhu, C.M. *et al.* (2019) 'ZnO@MOF@PANI core-shell nanoarrays on carbon cloth for high-perform-ance supercapacitor electrodes', *Journal of Energy Chemistry*, 35, pp. 124–131. doi:10.1016/j.jechem.2018.11.006.

Zhu, H. *et al.* (2018) 'Resistive non-volatile memories fabricated with poly(vinylidene fluoride)/poly(thiophene) blend nanosheets', *RSC Advances*, 8(15), pp. 7963–7968. doi:10.1039/c8ra01143e.

Zhu, Z.C. *et al.* (2016) 'Influence of ambient humidity on the voltage response of ionic polymer-metal com-posite sensor', *Journal of Physical Chemistry B*, 120(12), pp. 3215–3225. doi:10.1021/acs.jpcb.5b12634.

Zhu, Z.C. *et al.* (2019) 'Rapid deformation of IPMC under a high electrical pulse stimulus inspired by action potential', *Smart Materials and Structures*, 28(1). doi:10.1088/1361-665X/aadc38.

2 Overview of Electroactive Polymers

Types and Their Applications

Gabriela de Alvarenga, Larissa Bach-Toledo, Vanessa
Klobukoski, Andrei Deller, Jean Gustavo de A. Ruthes,
Rafael J. Silva, Jessica I.S. de Paula, Marcio Vidotti, and
Bruna M. Hryniewicz

CONTENTS

2.1 INTRODUCTION

Electroactive polymers (EAPs) are materials that respond to an external electrical stimulus, resulting in mechanical changes. EAPs are often divided in two categories: ionic EAPs and electronic EAPs. In the former, the mechanical changes are driven by ionic diffusion, such as in polyelectrolyte gels, conducting polymers, and carbon nanotubes. In the latter, the changes are activated by an electric field, such as in piezoelectric materials, liquid crystals, ferroelectric polymers, and electrostrictive polymers.

Depending on their specific properties, these materials can be applied in a variety of fields. The high conductivity of conducting polymers makes them suitable for sensors, electrocatalysis, and energy conversion or storage, and their ability to incorporate and expel ions in response to an external stimulus allows their application in drug delivery and artificial muscles. The stimuli-responsive changes of polyelectrolyte gels, which are triggered by pH, temperature, or electric field changes, can be used in numerous applications, and have been explored as solid state electrolytes for energy storage devices. Liquid crystal polymers have attracted considerable attention due to their unique properties, and can be applied as sensors, in energy related applications, and muscle-like materials. Finally, piezoelectric polymers are very versatile, and have been the subject of studies in the food industry, sensors, biomaterials, and energy harvesting. In this chapter, some types of EAPs will be introduced and their main applications in different fields will be described and explored.

DOI: 10.1201/9781003173502-2

2.2 CONDUCTING POLYMERS

Conducting polymers (CPs) were discovered by chance in the mid-1970s, during the synthesis of polyacetylene, which had its conductivity increased by 10 orders of magnitude when oxidized by iodine vapor. This discovery was awarded the Nobel Prize in Chemistry for the American physicist Alan J. Heeger, and the chemists Alan G. MacDiarmid and Hideki Shirakawa in 2000 (Shirakawa *et al.,* 1977). Also referred to as synthetic metals, these polymers have optical, magnetic, and electrical properties similar to those of metals (Shirakawa *et al.,* 1977; Medeiros *et al.,* 2012).

The main studied CPs are polypyrrole (PPy), polythiophene (PT), poly(3,4-ethylenedioxythiophene) (PEDOT), and polyaniline (PANI). These polymers have conjugated double bonds arranged in their network, which are the reason for their intrinsic conductivity. However, these π-electrons are not the sole cause of CPs conductivity, which occurs due to a process called doping (Lima *et al.,* 2018; Medeiros *et al.,* 2012). During doping, charges arise in the polymer backbone due to reversible oxidation, or reduction reactions of the conjugated π system, which are balanced due to the addition of dopants (counterions) (Medeiros *et al.,* 2012). The amount of dopants added to the polymer will determine the final conductivity of the material, by either p-doping (oxidative) or, more rarely, n-doping (reductive) processes (Lange, Roznyatovskaya, and Mirsky, 2008).

CPs have enough free electrons to be shared during conduction; therefore, their enhanced conductivity is related to the increasing number of free electrons as a function of the doping agents addition. The band model theory describes this conductivity, which is similar to inorganic semiconductor materials. Of note, there are significant differences in how this theory explains the conductivity of CPs and metals, because the mechanisms are not the same (Lima *et al.,* 2018).

Therefore, half a century after their discovery, CPs and their derivatives continue to attract a lot of attention, especially due to the unique combination of electrical and optical characteristics that are similar to semiconductors and metals, with the classic processability and ease of synthesis of polymers (Balint, Cassidy, and Cartmell, 2014). These unique characteristics make CPs important in several areas, such as energy storage and electrochromic devices, biosensors, and in biomedical areas, such as drug delivery systems and tissue engineering (Hong and Marynick, 1992; Lima *et al.,* 2018). These applications will be discussed in more detail in the following sections.

2.2.1 STIMULI-RESPONSIVE APPLICATIONS

The conductivity of CPs is an interesting property itself; however, CPS can alternate between conductive and non-conducting states, by simple oxidation or reduction processes. This means that applying an external stimulus (e.g., anodic or cathodic current) results in a change in the polymer oxidation state through the loss of π-electrons, which promotes a change in their conductivity, which is associated with changes in other properties, such as volume, wettability, color, and stored charge (Otero, Martinez, and Arias-Pardilla, 2012). This behavior is considered stimuli-responsive, and the exploration of this property expanded the application of CPs in many areas, especially in the biomedical field.

The redox reactions in CPs are usually followed by ion and solvent exchange, and one of the most exploited properties change is the expansion or contraction that result from these processes, which is commonly referred to as electro–chemo–mechanical change. This can happen in two main ways, as shown in Figure 2.1 (Alvarenga *et al.,* 2020).

When a CP is oxidized, π-electrons are removed from the polymer backbone, which creates positive charges. To counterbalance these positive charges, anions from the media will move into the polymeric matrix, resulting in a volume expansion (Figure 2.1 A→B) upon oxidation. If the polymer is reduced again, the positive charges disappear by the electron gain, and the anions are expelled, maintaining the polymer electroneutrality (Figure 2.1 B→A). When a CP undergoes this type of transition is usually called a anion exchanger, and this process prevails when small anions are present in the media (Otero and Sansiñena, 1995).

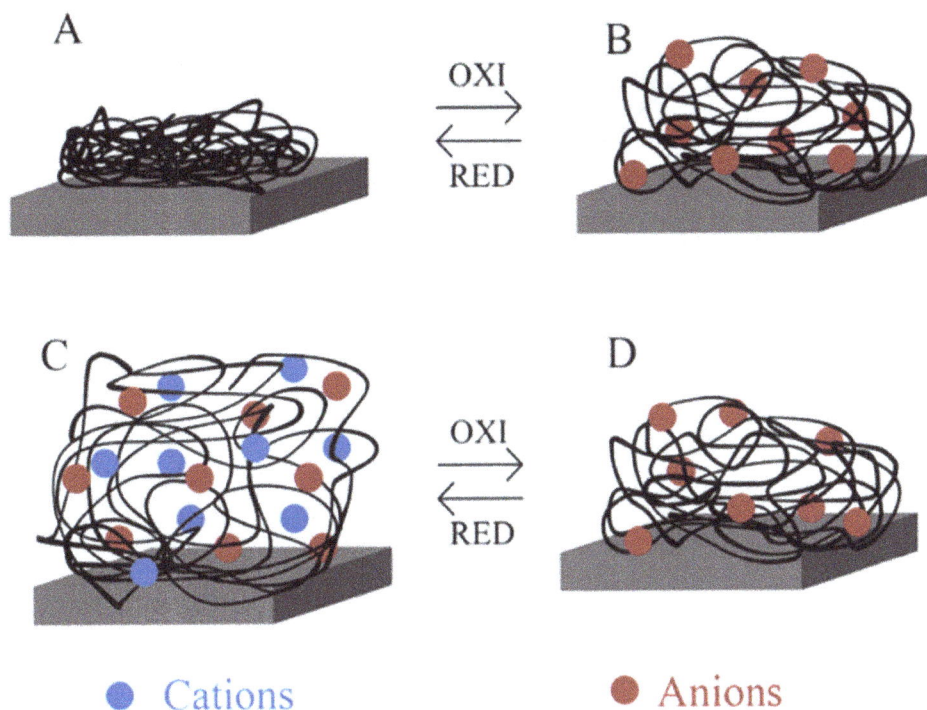

FIGURE 2.1 Electro–chemo–mechanical processes associated with redox reactions in CPs.

(From Alvarenga *et al.*, 2020. With permission.)

However, if a large anion is incorporated into the polymeric matrix during synthesis, these ions remain trapped in the polymer independently of the oxidation state. Therefore, when a polymer that contains large anions is reduced and the positive charges on the polymer disappear, there is an excess of negative charges in the polymer backbone due to the presence of the entrapped anions. Therefore, instead of anion expulsion cation incorporation is observed (Figure 2.1 D→C), to maintain charge neutrality, which results in a volume expansion upon reduction. When the polymer is oxidized the cations are expelled (Figure 2.1 C→D), and the polymer contracts. In this case, the polymer is often called a cation exchanger (Fujisue *et al.*, 2007).

The first application based on this stimuli responsiveness of CPs was the development of artificial muscles, which has been reported since the 1990's (Otero *et al.*, 1993). These systems try to replicate the brain, neuron, and muscle multi-sensing and actuating properties. In the human body, an electric signal sent by the brain, triggers the transformation of the chemical energy stored in ATP ions into mechanical energy, sensing signals, and heat, which starts the muscle contraction. These brain and muscle actuating and sensing properties have been partially transported and mimicked in the artificial muscles systems using CPs. Films of CPs can respond to electric pulses in liquid electrolytes, undergo ionic exchange to maintain the charge balance, which finally leads to volume change (e.g., expansion or contraction), and mechanical displacement or movement of the actuator (Otero, 2021). In addition, the displacement of the CP actuators is linked to the incoming current, and devices with different dimensions, with a different mass of CPs gave the same movement rate under the flow of the same specific current (Valero *et al.*, 2011). Therefore, these materials have potential for the development of multi-sensing and self-aware artificial tools.

In addition, the stimuli-responsive characteristics of CPs have been exploited to fabricate controlled drug delivery devices. For this application, anionic or cationic drugs can be incorporated

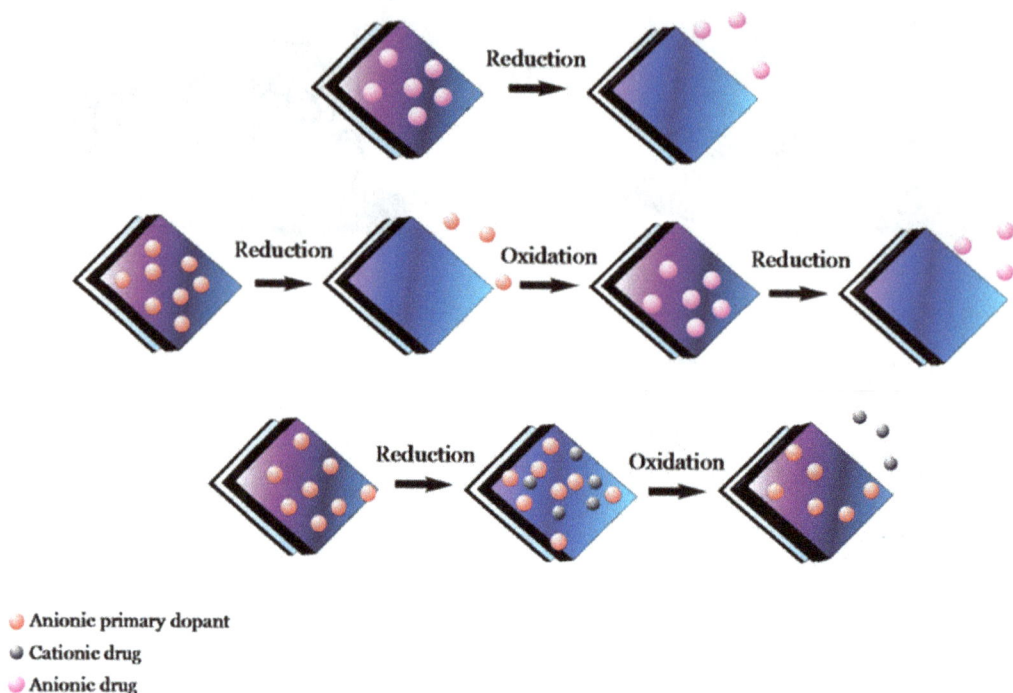

• Anionic primary dopant
• Cationic drug
• Anionic drug

FIGURE 2.2 Drug incorporation and release from CP films.

(From Puiggalí-Jou, del Valle, and Alemán, 2019. With permission.)

in CP films via three different mechanisms, as shown in Figure 2.2 (Puiggalí-Jou, del Valle, and Alemán, 2019). An anionic drug can be directly incorporated into the polymer film during polymerization and then released upon the reduction of the polymer. Otherwise, the incorporation of an anionic drug can be performed using a three-step methodology, which uses a primary anionic dopant to form the polymeric film and then expel this primary dopant by reduction. This is then replaced for the anionic drug via an oxidation process. Then, the anionic drug will be released upon reduction. It is possible to incorporate a cationic drug into CPs, using a two-step methodology where a bulky primary anionic dopant is incorporated during synthesis, and remains trapped upon reduction, which allows the incorporation of the cationic drug, which is further released upon oxidation.

Finally, due to the development of composites and the advent of nanotechnology, these polymers have been exploited in tissue engineering, especially in conducting polymeric hydrogels (CPHs). Hydrogels are crosslinked polymeric systems with hydrophilic characteristics. They have found applications in several areas, in particular in the biomedical field, ranging from wound healing to tissue scaffolds. However, their use for neural, cardiac, and skeletal muscle tissues, for example, is limited due to their non-conductive nature, and therefore, the fabrication of conducting hydrogels is very appealing. Here, the CPHs act due to the combination of the electric conductivity of CPs and the ionic exchange nature of the hydrogels, which results in a material that is suitable for the electroresponsive modulation cells, such as neurons, cardiac tissue cells, endothelial cells, and mesenchymal stem cells (Bhat, Rather, and Shalla, 2021).

2.2.2 ENERGY APPLICATIONS

CPs have been extensively used in energy applications, either in energy conversion or energy storage devices. Some of the important features of CPs applied in these fields will be highlighted in this section.

For energy conversion, CPs have found a remarkable performance as component of solar cells, which are photovoltaic devices that convert light into electricity. The main steps to produce electricity from light are the absorption of photons, which produces excitons, which diffuse and dissociate to create charge carriers, and are then transported and collected by the electrodes, which produces a photocurrent.

CPs have been used in different types of solar cells, such as in dye-sensitize solar cells (DSSCs), perovskites solar cells, and organic solar cells (Ibanez *et al.*, 2018). In DSSCs, the electron transport material is formed by a semiconductor that is deposited into a conductive glass with a molecule monolayer attached to its surface, which creates the photoactive layer to absorb the light. The charges are transported to the electron and to the hole conducting material; the latter is a redox-coupled electrolyte that is in contact with a counter electrode where the reduction of the donor occurs. The optical and electrical properties of CPs make them suitable to be used as counter electrodes, electron transport material, hole transport materials, or sensitizers.

For organic solar cells, the light is absorbed by organic molecules or CPs, if they meet the requirements, such as high conductivity, low band gap, stability and appropriate highest-occupied molecular orbital/lowest occupied molecular orbital (HOMO/LUMO) levels (Ibanez *et al.*, 2018). The mechanism involves four main steps: (1) the absorption of photons and promotion of electrons in the donor from the HOMO to the LUMO level, leaving holes in the HOMO, which creates the excitons; (2) the diffusion of the excitons to the donor–acceptor interface; (3) the transport of electrons from the LUMO of the donor to the LUMO of the acceptor, where the electrons and holes are strongly joined by coulombic attraction; and (4) the dissociation of the pair in free holes and electrodes and their migration to the anode and cathode, respectively, to be collected by the electrodes (Murad *et al.*, 2020). In perovskites solar cells, the perovskite absorbs the light, which induces the formation of the electron–hole pairs. CPs have been used as electron and hole transport materials and for moisture protection (Ibanez *et al.*, 2018).

CPs have been extensively applied in energy storage devices, mainly in the development of supercapacitors (Bryan *et al.*, 2016). Supercapacitors have intermediate behavior between batteries and conventional capacitors for power and energy density and they are often divided in two groups: electrical double layer capacitors (EDLC) and pseudo capacitors. The mechanism of EDLCs relies on the charge accumulation in the electrochemical double layer of high surface area materials, as carbon materials, and for pseudo capacitors, the electroactive materials undergo faradaic reaction, which achieve higher energy densities than EDLCs.

CPs are examples of pseudo capacitive materials, which show high capacitance values due to their reversible redox reactions at the electrode–electrolyte interface and the possibility to have high surface area, which contributes to the double layer capacitance. In addition, the flow of the counterions through the polymeric matrix to balance the charges formed during the redox processes, cause swelling and shrinking of the polymeric lattice, which affects the morphology causing an increase in the interfacial charge–transfer resistance (Bryan *et al.*, 2016). This lack of stability leads to a limitation in the use CPs for long-cycle devices. To overcome this issue, some strategies have been employed, such as the synthesis of CPs nanostructures that can reduce the path length for ion transport and cause the better accommodation of the deformations that are induced by the electrochemical reaction (Shi *et al.*, 2015), and the synthesis of hybrid materials. For example, the association of a CP with a carbon (C) material can lead to a high specific capacitance that is related to the redox reactions of CPs and a high stability, which is associated with the rigid structure of the C material (Dubal *et al.*, 2015).

CPs have found interesting applications in flexible solid state supercapacitors (FSSCs) due to their high mechanical flexibility. FSSCs are a new class of energy storage devices, which are formed by two flexible electrodes (e.g., positive and negative) that can be modified with flexible electroactive materials; (1) a solid state gel electrolyte that is used as separator of the electrodes; and (2) a packaging material, that needs to be flexible (Dubal *et al.*, 2018; Lu *et al.*, 2014). These devices show interesting features for the construction of wearable and portable devices. However, they must present high stability under mechanical stress, such as bending, twisting, and stretching. FSSCs can have symmetric or asymmetric configurations; in the first, both electrodes have the same capacitance and the second is the opposite, and usually combines electrodes with different charge storage mechanisms (i.e., EDLC and pseudo capacitive material). These electrodes can be assembled in a sandwich-like design, or as wires, fibers, plates, and cables. (Dubal *et al.*, 2018)

CPs, such as PANI, PPy, and PEDOT are extensively applied in FSSCs, with working potentials in positive and negative voltages. Kurra *et al.* (2015) developed an asymmetric all conducting polymer, with PEDOT deposited on gold-coated (Au) polyethylene naphthalate (PEN) plastic substrate as a negative electrode, and PANI, deposited on an Au/PEN substrate, as a positive electrode (Kurra, Wang, and Alshareef, 2015). The cyclic voltammetry in 1 mol/L H_2SO_4 in a three-cell configuration (Figure 2.3(a)) shows the positive working potential of PANI and the negative window of PEDOT, and Figure 2.3(b) shows the cyclic voltammetry of the asymmetric device, fabricated with PANI in the positive electrode, PEDOT in the negative and separated with a gel electrolyte.

Their high electrochemical performance, processability, and flexibility make CPs good candidates to be used in many types of batteries. CPs are often used as cathode materials since they are usually p-doped. Lithium (Li) and Li-ion batteries (LIBs) are currently the dominant energy supply systems, and many works have focused on using CPs as cathodes for LIBs, and a C material can be used as the anode (Jia *et al.*, 2019). In this case, the discharge process is related to the reduction of CPs and the release of anions, and the anode is oxidized, which releases Li ions. During charging, the CP is oxidized, which incorporates the anions to counterbalance the charges and the Li cations are reduced and stored in the anode (Figure 2.4) (Jia *et al.*, 2019). CPs alone can be used in LIBs, although practical applications are limited by their poor capacity. However, they can be combined with inorganic compounds to improve their performance (Sengodu and Deshmukh, 2015).

2.2.3 ELECTROCATALYSIS AND SENSOR APPLICATIONS

The International Union of Pure and Applied Chemistry (IUPAC), defines a chemical sensor as being "a device that transforms chemical information, ranging from the concentration of a specific sample component to total composition analysis, into an analytically useful signal" (Hulanicki, Glab, and Ingman, 1991). Some definitions were added to the original definition that sensors must operate dynamically and reversibly, which were related to the fact that sensors operate in real time as the sample undergoes changes (e.g., in concentration or in some physicochemical property) (Gründler, 2007).

EAPs are widely exploited for use in sensors because their electromechanical properties make them respond to external stimuli, which can be measured and analyzed as useful signals for sensors. In addition, the use of EAPs for the construction of sensors allows greater control of the size, flexibility, and weight of the sensor, which allows the miniaturization of the device, which in addition leads to cost reductions (Bar-Cohen and Zhang, 2008; Cardoso, Ribeiro, and Lanceros-Mendez, 2017). Other CPs properties are of interest for sensors development, such as, corrosion resistance, and good electrical, mechanical, and optical properties (Wang *et al.*, 2020). The most exploited CPs for sensors development are PANI, PPy, PT, and PEDOT (Janata and Josowicz, 2003; Balint, Cassidy, and Cartmell, 2014).

For sensors development, CPs can act as a selective layer or as the transducer itself. To act as a selective layer, the CPs are directly involved in the recognition reaction, which might be a chemical or electrochemical reaction. The mechanism of sensing and detection of CPs might involve redox reactions, adsorption and desorption of analytes, changes in mass, conformational changes in chains, and charge transfer. When the CP acts as the transducer, its role is to provide the electrical conductivity necessary to transduce the recognition reaction between the receptor and the analyte. In this case, the conducting polymer should not interact with the analyte, and the signal transduced is proportional to the intensity of the interaction between the receptor and the analyte. The range of

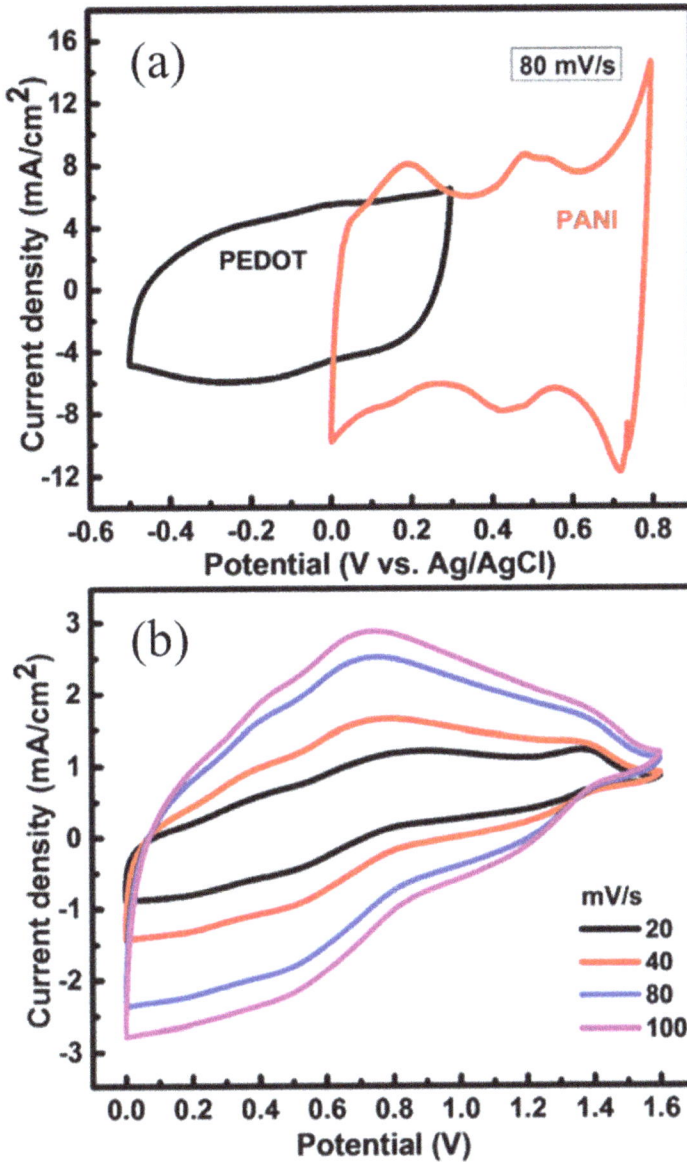

FIGURE 2.3 Showing: (a) cyclic voltammetry of PEDOT and PANI in different potential windows, measured in a three-electrode cell in 1 mol/L H_2SO_4; and (b) cyclic voltammetry of the solid state device.

(From Kurra *et al.*, 2015.)

application is quite wide, and materials based on CPs for sensing gases, humidity, and environmental pollutants, and for food purposes and for the detection and control of diseases have been identified (Shukla, Kushwaha, and Singh, 2017; Tsakova and Seeber, 2016; Janata and Josowicz, 2003).

In addition to their interesting physicochemical properties, it is possible to add other compounds to the structure of CPs to form hybrid materials, which further enhances their properties. Compounds, such as metallic particles and nanostructured doping agents have been extensively explored to form hybrids with CPs, which shows their wide potential for use in sensors (Le, Kim, and Yoon, 2017). Another advance that led to further sensor development was the advent of CPs nanostructures. The

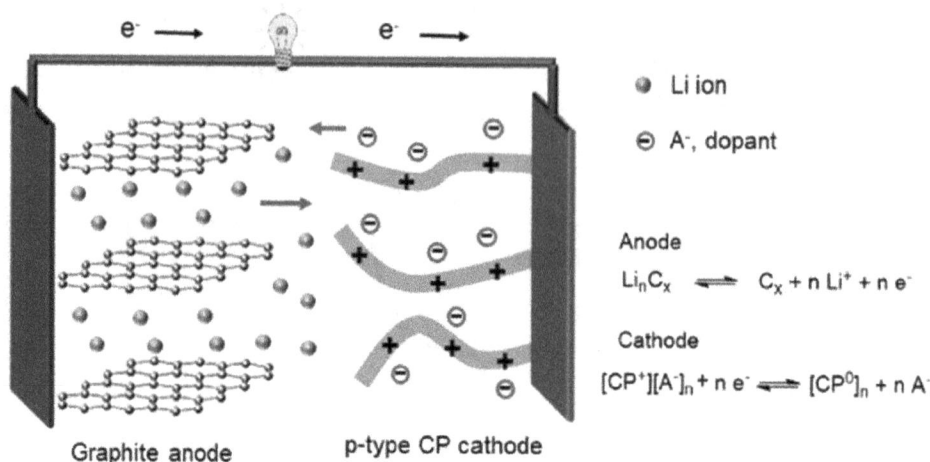

FIGURE 2.4 Working principal of a LIB using a p-type conducting polymer as cathode and a carbon material as anode. A represents the doping anions.

(From Jia *et al.* 2019.)

high surface area of nanostructures leads to a higher exposure of the electroactive sites and to a decrease in the diffusional layer, which improves the transport of ions and charges (Killard, 2010).

Among the properties of CPs, their ability to catalyze some electrochemical reactions has gained interest, due to their intrinsic electrocatalytic properties (Malinauskas, 1999; Mu and Zhang, 2010; Ibanez *et al.*, 2018). A small amount of a CP deposited onto an electrode surface favors the kinetics of the electrode processes of some analytes. In this process, the electrode plays a fundamental role, since it is the donor or acceptor of electrons and it provides sites for adsorption (i.e., of the species undergoing electrocatalysis) and electron transfer (Li, Bai, and Shi, 2009; Menzel *et al.*, *2012*; Zinola *et al.*, 2012).

Electrocatalysis based on CPs has been used in several areas, which range from the detection and degradation of micropollutants (Kumar *et al.*, 2007; Hryniewicz, Bach-Toledo, and Vidotti, 2020; Ferreira *et al.*, 2011; Zhou and Shi, 2016), the development of catalysts (Pandey and Lakshminarayanan, 2012; Ferreira *et al.*, 2011; Zhou and Shi, 2016), and dye sensitized solar cells (Nogueira, Longo, and De Paoli, 2004; Koh *et al.*, 2011; Saranya, Rameez, and Subramania, 2015), to the production of fuels (Esteban *et al.*, 1989; Gonzalez, 2000; Dreyse *et al.*, 2015; Mitraka *et al.*, 2019).

The surface area is a critical factor in catalytic systems and electrodes, which interfere the most with the efficiency of the process. In this context, the surface area of CPs can be modulated through the synthesis of nanostructures, such as nanosheets, nanowires, and nanotubes, usually through templated assisted synthesis, where the template guides the polymerization of CPs monomers (Menzel *et al.*, 2012; Mu and Zhang, 2010; Malinauskas, 1999; Zhou and Shi, 2016; Li, Bai, and Shi, 2009).

In addition, the incorporation of heterogeneous catalysts, such as metal nanoparticles and metal oxides, onto the conducting polymer chain to improve the electrocatalytic activity of the material is common (Yoneyama, Shoji, and Kawai ,1989; Qin Zhou *et al.*, 2008; Malinauskas, 1999; Kumar *et al.*, 2007; Menzel *et al.*, 2012). This association could improve the previously mentioned properties and minimize the limitations of the individual components. In some cases, the CP act only as solid polymeric supports for active species; however, in other hybrid systems, the activity of organic and inorganic species combine to reinforce and modify each other (Mu and Zhang, 2010).

2.3   POLYELECTROLYTE GELS

Another interesting class of electroactive polymers are the polyelectrolytes (PEs), which are known as ionic polymers and can be defined as polymers composed of macromolecules that contain ionic or ionizable groups (Kikuchi and Tsuchitani, 2019; McNaught and Wilkinson, 1997). These polymers received great attention in the last decades because of their promising applications in artificial muscle-like actuators (Zarras and Irvin, 2003; Nemat-Nasser and Thomas, 2004). PEs are used to form two different classes of materials, ionic polymer–metal composites (IPMCs) and ionic polymeric gels. In this section we will limit the discussion in ionic polymeric gels.

The definition of a PE is known; therefore, to understand what a PE gel is, the definition of a gel must be known. This term was first used in 1861 by Thomas Graham and over time several definitions have been presented. The difficulty in finding a complete definition for gels can be summed up by the words written by Dorothy Jordan Lloyd in 1926 "[…] the 'gel', is one which it is easier to recognize than to define" (Nishinari, 2009; Almdal *et al.*, 1993). Although there are still some discussions on the subject, the current definition of gel according to IUPAC is a "non-fluid colloidal network or polymer network that is expanded throughout its whole volume by a fluid" (Alemán *et al.*, 2007).

Therefore, a gel that has its polymeric network formed by PE is considered a PE gel. These materials can absorb several times their weight of water (i.e., other solvents, without dissolving) which makes the material swell and changes its volume. This can occur due to other environmental conditions, such as pH, light, temperature, and electric field. Of note, PE gels can respond to external stimuli by either swelling or shrinking. These mechanisms will be addressed in the following sections (Wallmersperger and Leichsenring, 2016; Kwon, Osada, and Gong, 2006).

The stimuli-responsive volume change of PE gels is very attractive, because small variations in external conditions can generate a large response in their volume; therefore, these materials have been employed in applications, such as tissue engineering and drug release (Chu *et al.*, 1995; Kwon, Osada, and Gong, 2006). Examples of the most used ionic polymers in PE gels synthesis are: poly(acrylate) (PA); poly(acrylic acid) (PAA); poly(acrylamide) (PAM); poly(2-acrylamido-2-methyl-1-propanesulfonic acid) (PAMPS); poly(styrene sulfonate) (PSS); poly(vinyl alcohol) (PVA); and poly(vinyl sulfonate) (PVS) (Hansson, 2006).

The synthesis of PE gels generally consist in free radical polymerization with or without a cross-linker agent and uses an initiator (Glazer *et al.*, 2012; Roy, Haldar, and De, 2014). A cross-linker can improve the network density, and the molar ratio between the cross-linker and the monomer has a great influence on this parameter.

Before discussing the stimuli-responsive mechanism it is necessary to understand the composition of a swollen PE gel (Figure 2.5). To explain this, the swollen PE gels will be treated

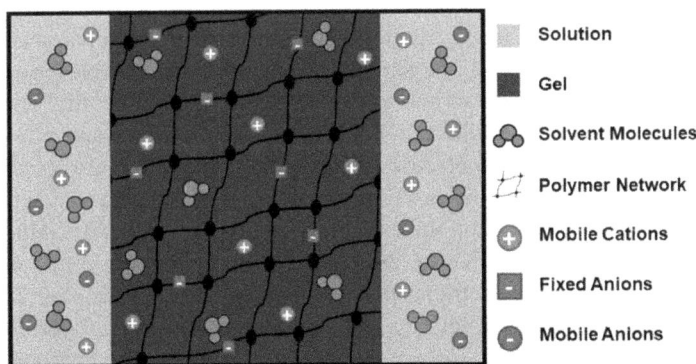

FIGURE 2.5   Composition of a swollen PE gel.

as a combination of the gel and the solution. The solution is composed of a solvent and mobile ion (e.g., cations and anions). Meanwhile, the gel composition is the polymer network with fixed charges, pores, and mobile ions (e.g., cations and anions). Of note, both systems are in electrical equilibrium.

In a PE gel, the fixed charges come from the polymer and the mobile ions stabilize these charges. When the gel is combined with the solution, the last one fills the pores of the gel and a new equilibrium is achieved (Jo, Naguib, and Kwon, 2011).

The gel shown in Figure 2.5 is composed of an anionic polymer, in which the fixed charges are negative. In addition, there are cationic polymers in which the fixed charges are positive and these polymers will respond differently to some stimuli, such as the electric field (Calvert, 2004).

As mentioned previously, a PE gel can react to various external stimuli, and although there are differences between how these responses occur, in general, the mechanisms are governed by osmotic pressure, electrostatic, and elastic forces (Keller *et al.*, 2011; Landsgesell and Holm, 2019; Chu *et al.*, 1995). The elastic forces are related to the structure of the material and the osmotic pressure comes from the difference in the concentration between the inner and outer sides of the gel.

Several efforts were made in the last few decades to model the PE gel responses (Wallmersperger *et al.*, 2013; Rumyantsev *et al.*, 2016; Landsgesell and Holm, 2019; Jo, Naguib, and Kwon, 2011; Keller *et al.*, 2011; Chu *et al.*, 1995; Wallmersperger and Leichsenring, 2016). Therefore, a lot of experiments were carried to better understand how these responses occur. In this section, the response mechanisms for variations in pH, temperature, and electric field will be discussed.

The PE gel swelling behavior in which the fixed charge comes from a weak basic electrolytic group (i.e., an amin group) can be explained by the PAM example (Chu *et al.*, 1995). Assume the PAM gel is in equilibrium with a solution with pH_1 and this pH changes to a lower pH that is known as pH_2 ($pH_1 > pH_2$).

When this change occurs, H^+ ions can diffuse into the gel, which increases the H^+ concentration on the inside; therefore, the amine group is protonated and forms an ammonium group, which increases the charge density, and therefore the electrostatic repulsion. The polymeric chain responds to this disturbance, to decrease the electrostatic repulsion forces, by swelling. This swelling is limited by the increase in the elastic retraction force, and when the forces reach a new equilibrium, the swelling ceases (Chu *et al.*, 1995). If the group in the PE gel is acid, the mechanism must be opposite to this, and the swelling occurs when increasing the pH (Tang *et al.*, 2020).

Temperature is especially important for the degree of swelling, because it influences many physicochemical parameters, such as ion mobility, viscosity, and osmotic pressure. Studies show that the increase in temperature increases the degree of swelling; in contrast, decreasing the temperature made the gel shrink (Guenther *et al.*, 2007; Keller *et al.*, 2011).

To simplify the understanding of how temperature can influence the degree of swelling, it is possible to treat the swelling as a phase equilibrium between the swollen phase and shrunk phase. As shown by Guenther *et al.* (2007), the shrunk gel phase is more stable than the swollen gel phase. Therefore, the transition from shrunk to swollen phase is endothermic. Therefore, with the increasing temperature, the phase equilibrium is shifted for the swollen phase.

Finally, the PE gel response to an electric field will be discussed, more specifically, the electrical bending. The mechanism by which a PE gel bends when exposed to an electric field depends on the solution, the type of gel, and on the experimental configuration (Blyakhman *et al.*, 2015; Glazer *et al.*, 2012; Jo, Naguib, and Kwon, 2011). There are three main configurations: (1) a gel immersed in a solution with both electrodes placed far away from the sample; (2) the gel in solution with both electrodes in contact with the gel surface; and (3) the gel placed in the air with both electrodes touching the sample surface (Glazer *et al.*, 2012). The configurations with one electrode touching the gel surface and the other far away, are less but possible to be employed (Glazer et al., 2012).

Although there are other mechanisms, and the understanding of electrical response of PE gel is not complete, many phenomena can be explained by the dynamic enrichment or depletion mechanism, which is based mainly on ionic mobility and osmotic pressure. The gel bending arises due to differences between the osmotic pressure in the gel side that faces the anode and the side that faces the cathode. Therefore, one side is swollen and the other shrinks (Liu *et al.,* 2018; Jo, Naguib, and Kwon, 2011). When the electric field is applied, it generates a flow of cations from the anodic side to the cathodic side, which makes the last one have a greater concentration of ions. In addition, the ions in the solution migrate to try to stabilize the charge differences. Therefore, the cathode side has a greater number of ions inside and outside the gel (Liu *et al.,* 2018).

Depending on the solution composition, or the type of PE gel, or both (e.g., anionic or cationic), the gel might bend toward the cathode or anode. Since the difference between concentrations inside and outside the gel induces the osmotic pressure, two scenarios are possible. When there is a greater difference in ion concentration between the internal and the external media on the anodic side, they swell more, which leads to the curvature toward the cathode (Glazer *et al.,* 2012), however, the opposite bending might occur (Jo, Naguib, and Kwon, 2011).

2.3.1 Applications for Polyelectrolyte Gels

As mentioned previously, PE gels are used in a lot of applications due to their stimuli-responsive mechanisms. For example, their smart sensitivity can be used for various biomedical applications (Kwon and Gong, 2006), which have special potential in the design of materials for tissue engineering and controlled drug delivery systems.

The great enthusiasm for the use of these materials in the biomedical field is due to some characteristics of these materials, such as biodegradability and biocompatibility (Ning *et al.,* 2018), which makes them suitable materials for the development of scaffolds for cellular fixation and proliferation (Kwon *et al.,* 2014; Blyakhman *et al.,* 2015). In addition, gels share common features with natural tissues, such as softness, moisture, and elasticity; however, PEs are usually found in biological tissues , such as polysaccharides and proteins (Kwon *et al.,* 2014; Ning *et al.,* 2018).

Polymers that can be used for biological purposes include polysaccharides, proteins, heparin, chitosan, and hyaluronic acid. For example, a chitosan hydrogel with alkaline lignin was prepared by Ravishankar *et al.* (2019), which provided a suitable surface for cell fixation and proliferation. The mouse fibroblast cells showed good cell migration, which suggested that the gel might be suitable for applications in wound healing. In addition, gel actuators that were prepared with the negatively charged polyacrylonitrile electrolyte (PAN) are another example. The pH-sensitive PAN fibers developed by Lee *et al.* (2007) showed a change in the length of the PAN gel fiber when immersed repeatedly in acid and basic solutions, with the resistance of these fibers being comparable with those of human muscle.

The ability of PE gels to organize their structures, or alter their physicochemical characteristics, or both in response to a specific external stimulus, such as those previously mentioned, have made them materials of great interest for drug delivery systems (Bithi *et al.,* 2020; Peers, Montembault, and Ladavière, 2020).

The pH-responsive volume phase transition is one of the important properties of these materials, and this feature has been widely explored for the control and release of molecules in response to specific physiological changes. In addition, PE gels have other interesting bioactive properties, such as antimicrobial, antioxidant, antitumor, and anti-inflammatory, which are attractive for their advanced therapeutic potential (Bithi *et al.,* 2020).

Zhang *et al.* (2018) used chitosan and hyaluronic acid to produce a gel, which demonstrated an example of a PE gel used for the pH-sensitive release of doxorubicin. In this study, the authors observed that doxorubicin release was triggered under acid conditions (pH = 4.00) and that the higher the hyaluronic acid concentration in the hydrogel formulation, the slower the release of doxorubicin.

Another way to control the release of medicine is through the application of an electric field, which leads to a sensitive electrical release (Peers, Montembault, and Ladavière, 2020). This mechanism was demonstrated by Sangsuriyonk, Paradee, and Sirivat (2020) in which a carboxymethyl cellulose (CMC) was used for the transdermal release of 5-fluorouracil, a non-ionic anticancer medicine, through the application of an electric field. The release mechanism of the drug was through as the matrix deswelling and electroosmosis, in addition to pure diffusion without an electric field. CMC hydrogel has shown potential to be used in transdermal controlled drug release; therefore, enhancing the drug release rate and concentration.

PE gels have been applied in the construction of flexible energy storage devices. Their advantages over aqueous electrolytes include being leak-free and enabling the development of flexible devices. In addition, they can be used as separators, which efficiently reduces the cell's internal resistivity (Xuan *et al.*, 2021).

Another advantage is the development of repairable PEs, which was demonstrated by Wang *et al.* (2017), who prepared a PE with vinyl imidazole copolymer and hydroxypropyl acrylate used as an electrolyte in a supercapacitor that was assembled with activated carbon electrodes. The supercapacitor showed good performance for self-healing and could repair itself nine times. Xuan *et al.* (2021) prepared a poly(2-acrylamide-2-methyl-1-propanesulfonic) (PAMPSA) polyvinylidezole (PVI)/LiCl PE gel, which had a good performance on self-healing and showed mechanical resistance. This performance was attributed to the combination of hydrogen bonds and the electrostatic interaction between PAMPSA and PVI.

PE gels have been used in batteries (Kalapala *et al.*, 2005) where a polymeric gel was synthesized from methyl methacrylate and the Li salt of 2-acrylamide-2-methylpropanesulfonic acid (LiAMPS), for use in rechargeable batteries, which obtained ionic conductivity in the range of $0.2 \leq \sigma \leq 0.8$ mS/cm at room temperature ($20°C \pm 1°C$).

2.4 LIQUID CRYSTAL POLYMERS

Liquid crystalline (or crystal) polymers (LCPs) have attracted considerable interest in interdisciplinary research, which includes polymer and materials science and the chemistry and physics of liquid crystals. LCPs have a significant scientific value and an extensive application area, due to their unique combination of properties. Similar to polymers they have good mechanical properties, and have similar electroactive properties to liquid crystals (Cox, 1987; Shibaev, 2016). LCPs are classified as electronic electroactive polymers, and are free from any electrolyte medium, which means that no ion migration is required for electromechanical coupling (Wang *et al.*, 2016).

To comprehend an LCP is important to understand the basic theory on liquid crystals or liquid crystalline (LCs), which has been widely studied since the 1940s. LCs are thermotropic, which means that their structure is partially maintained in the liquid phase, for instance, above the melting point. However, there are lyotropic LCs, which change behavior due to its concentration in a solvent, which is outside the temperature influence (Fink, 2014; Brostow, 1990). In summary, LC compounds are named for their mesomorphic or mesophase state (i.e., from Greek, *meso* means intermediate), and the main structure characteristic of LC compounds are their shape anisotropy and the rigidity of their molecules. These molecules can be classified according to their type and degree of ordering (Fink, 2014; Shibaev, 2016; Wang *et al.*, 2016; Brostow, 1990). Some of these mesophase types are shown in Figure 2.6.

In addition, there are two main categories of LCP polymerization Picken (1996). One corresponds to the side-chain LCP, where the mesogenic groups are attached to a polymer backbone. The other corresponds to the main-chain LCP, where the polymer chain contains the mesogenic groups. In addition, it is possible to have a continuous semiflexible polymer chain, or an alternating sequence of rigid aromatic mesogens and flexible aliphatic spacers, it is common to incorporate different monomers in LCPs structure (Shibayev and Byelyayev, 1990; Picken, 1996; Fink, 2014;

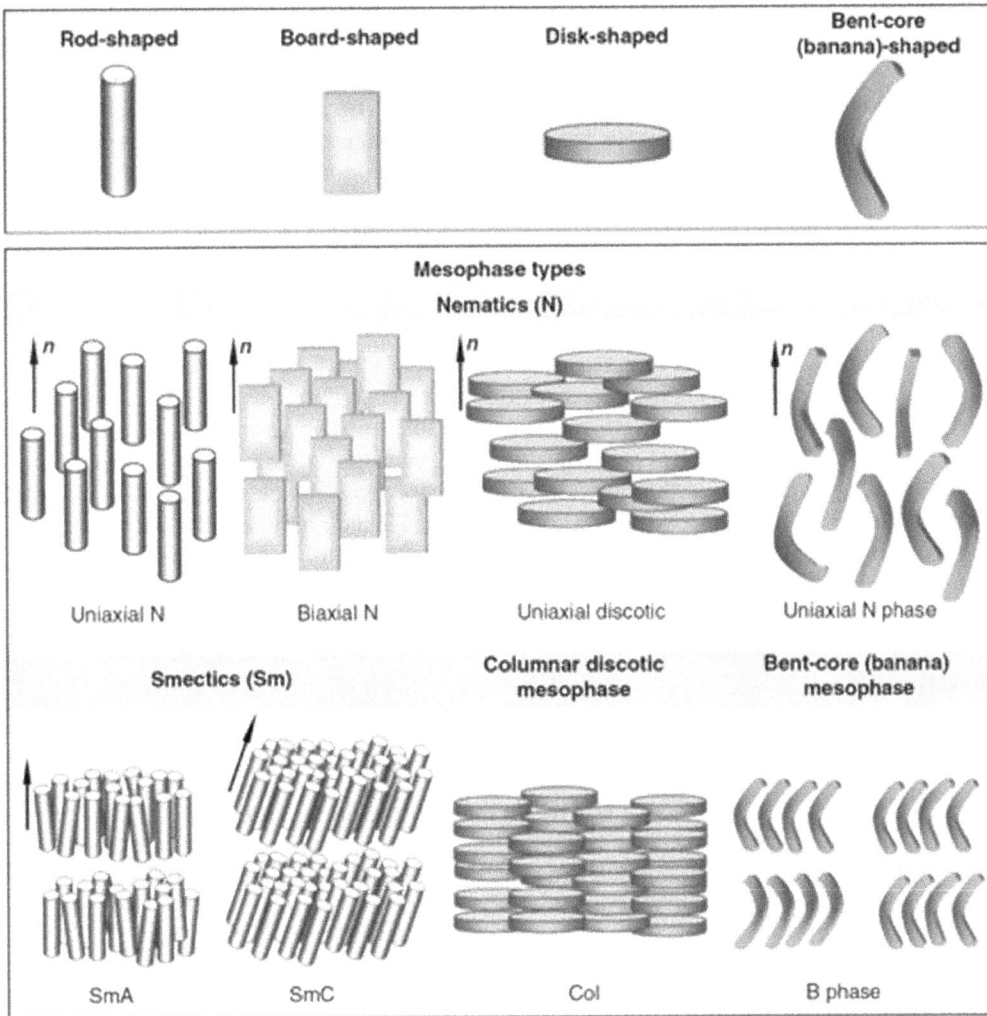

FIGURE 2.6 Different mesophase types that form general LCs.

(From Shibaev, 2016.)

Brostow, 1990). Therefore, the organization of the monomers directly influences the LCPs properties and their applications. In general, LCPs exhibit basic properties, such as low melt viscosity, excellent mechanical properties, low thermal expansion, and good solvent resistance. Shibaev (2016) described the properties and applications of side-chain and main-chain LCPs.

Among LCPs properties, their electro-optical properties have achieved notoriety. Brostow (1990), described that the majority of mesogens have strong and permanent dipole moments, which are easily orientable under an external electric field. Moreover, the extent of orientational freedom is a result of the dipole vector of polymer chains, which determines the degree of polarization, and these groups achieve reorientation freedom at different temperatures, which results in various transition or relaxation processes.

Some LCs have spontaneous electric polarization, which influences the structural orientation, even in the absence of an external electric field, which are known as ferroelectric liquid crystals (Fink 2014; T. Wang *et al.,* 2016). Ferroelectric liquid crystals must fulfill certain symmetry requirements, and due to their rapid switching times, they have been widely studied in the fabrication of displays (Doane *et al.*, 1988; De Filpo *et al.*, 2008; Dutta *et al.*, 1990; Bouteiller and Lebarny, 1996). Moreover, they can potentially show piezoelectric effects, which depend on their incorporation into the liquid crystal matrix (T. Wang *et al.*, 2016). In addition, it is possible to insert a different material (e.g., carbonaceous nanomaterials, magnetic particles, and metal oxides) into non-ferroelectric liquid crystal structures to improve the performance of the material in electrical actuation (Ji, Marshall, and Terentjev, 2012; Chambers *et al.*, 2009). However, in these examples, caution must be taken to ensure the mechanical properties and avoid its stiffness.

2.4.1 APPLICATIONS FOR LIQUID CRYSTAL POLYMERS

The stimuli-responsive character of materials based on LCPs, such as thermoresponsive (Li *et al.*, 2006; Thomsen *et al.*, 2001), photoresponsive (Takeshima *et al.*, 2015; Kubo *et al.*, 2019; M.H. Li *et al.*, 2003), and electroresponsive (Huang, Zhang, and Jákli, 2003; Yusuf *et al.*, 2005; Chambers *et al.*, 2009) are the most exploited properties in their applications. It is possible to create stimuli-responsive LC networks due to their high molecular anisotropy, and elasticity of the crosslinked networks, which can therefore, reversibly deform and change their birefringence under an external stimulus, such as temperature or light (De Bellis *et al.*, 2019). An interesting approach of the use of stimuli-responsive LCPs is in the development of micro or nano artificial muscles. Li *et al.* (2006) proposed a muscle-like material based on a network of side-on nematic LC homopolymers, instead of a main-chain LCP, and therefore, the material showed thermo and photoresponsive characteristics.

In addition, due to their stimuli-responsive characteristics, LCPs are a suitable class of materials that are used for sensing, and can be classified as electronic electroactive polymers (Wang *et al.*, 2016). Many sensors that use LCs are based on changes in their optical and electrochemical properties that are caused by the presence of analytes in the LC structure (Shibaev *et al.*, 2015; Lai, Kuo, and Yang, 2014). Furthermore, LCPs are used in electrochemical devices, battery current collectors, high efficiency radio frequency interference shielding, automotive communication, and fuel cell applications (Anilkumar *et al.*, 2020; Fink, 2014).

2.5 PIEZOELECTRIC POLYMERS

Few materials have the ability of electromechanical transduction – or piezoelectricity – and are mainly used to generate an electrical signal that responds to a mechanical stresses (Sappati and Bhadra, 2018). Figure 2.7 shows piezoelectricity, with different variations in electrical polarization.

Studies based on polymers' piezoelectricity highlighted some advantages of these materials, such as low cost of fabrication, lightweight, high flexibility, large area, and high deformability (Bauer and Bauer, 2008). There are three major categories of piezoelectric polymers: (1) bulk piezopolymers; (2) piezocomposites; and (3) cellular polymer films or void charged polymers (VCP) (Hamdi, Mighri, and Rodrigue, 2020) Figure 2.8 shows the classification of piezoelectric polymers.

Bulk polymers exhibit piezoelectricity due to the presence of diploes in their molecular structure. Through poling process, these dipoles can suffer reorientation and retain their most stable orientation state (i.e., remanent polarization). (Hamdi, Mighri, and Rodrigue, 2020). Poling is triggered by the application of an external field and an elevated temperature and depends on the strength and time of the applied electrical field.

Cellular polymer films (i.e., VCPs) can exhibit an elevated piezoelectric constant compared with traditional piezoceramics. The mechanism is based on gas molecules in the thin films that are ionized by the application of electric fields (Sappati and Bhadra, 2018). The formation of VCPs

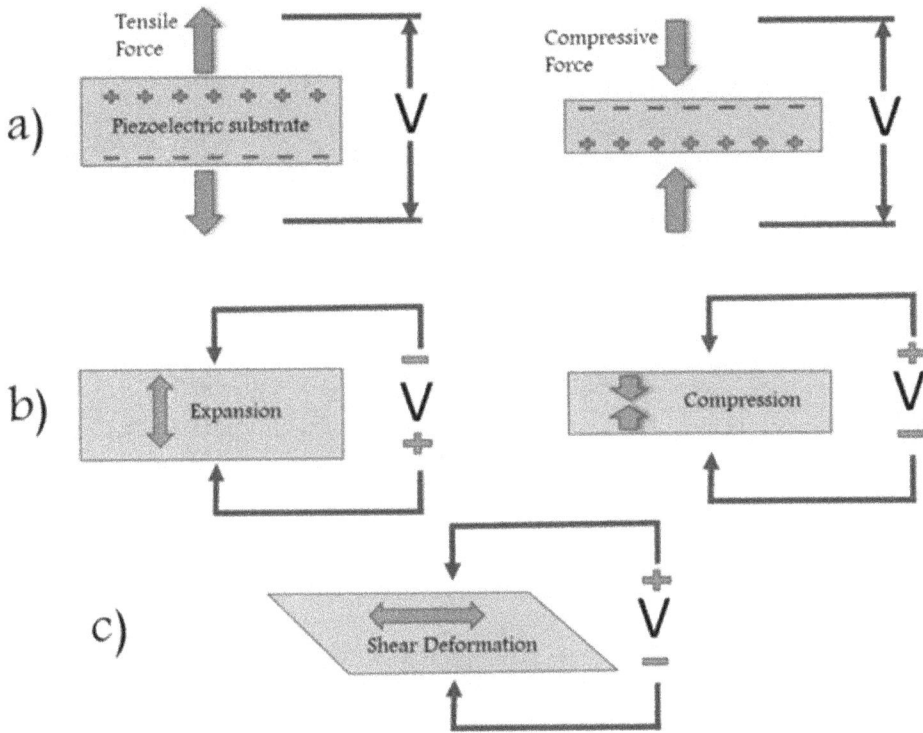

FIGURE 2.7 Different variations in electrical polarization in piezoelectric materials: (a) longitudinal direction; (b) converse; and (c) shear piezoelectricity.

(From Sappati and Bhadra, 2018.)

is based on gas molecules that are ionized by the application of an electric field. Due to the electric field, the opposite charges accelerate and implant on either side of the voids, which forms internal dipoles. Any deformation of the void can induce a piezoelectric effect in the material, such as density, shape, type, and pressure of the gas. These parameters affect the piezoelectric coefficient of the VCP (Sappati and Bhadra, 2018).

Piezoelectric polymer composites (i.e., piezocomposites) consist of a polymeric material embedded with inorganic piezoelectric material (i.e., ceramic). The piezoelectric properties are directly affected by the connectivity between the different materials that are used to form the piezocomposite, where the ceramics are the active piezoelectric material and the polymers are passive, because they show no piezoelectricity effect in this example (Ramadan, Sameoto, and Evoy, 2014; Bauer and Bauer, 2008).

2.5.1 Applications for Piezoelectric Polymers

Due to their versatility, piezoelectric polymer applications have been researched, mainly as an alternative to lead-based piezoceramics (Mishra *et al.*, 2019). The applications include areas of technology, such as the food industry, energy harvesting, and biomaterials.

Khot, Panigrahi, and Lin (2011) reported the use of a piezoelectric polymer thin film sensor to detect low concentrations of volatile organic compounds (VOCs) in food safety applications. In an attempt to control the contamination in food samples by *Salmonella typhimurium*, they used piezopolymers in a quartz crystal microbalance electrode, which provided responses to alcohol

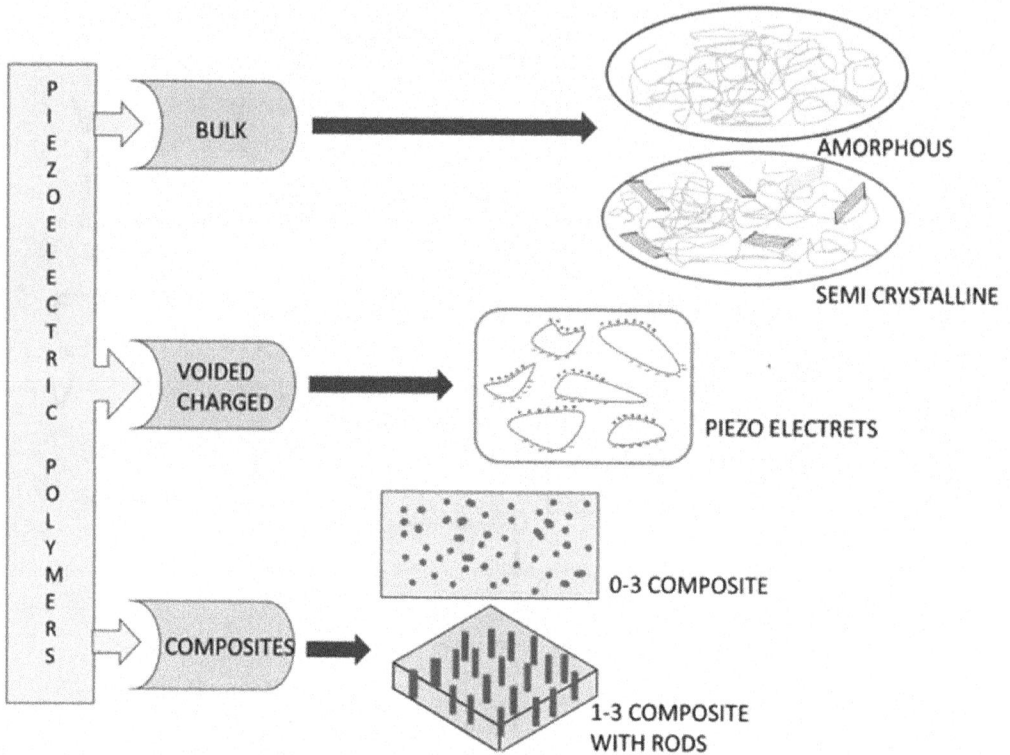

FIGURE 2.8 Classification of piezoelectric polymers.

(From Sappati and Bhadra, 2018.)

VOCs, and could distinguish different types of alcohols, which were from methyl-1-butanol to 1-hexanol in the samples.

In biomaterials, Chiu *et al.* (2013) evaluated the use of the piezoelectric polymer polyvinylidene fluoride (PVDF) as the base for a patch sensor for heartbeat and breath monitoring. The working principle of the device was based on the detection of pulsatile vibrations and periodical deformations on the human chest during heartbeats and breath. Their results showed the potential use for piezopolymers in biomedical applications. In addition, in biomedicine the piezopolymers electro-mechanical properties were employed on tissue engineering, such as scaffolds for supporting cell differentiation and growth (Ribeiro *et al.*, 2015).

Another parameter of piezopolymers that has been explored is their flexibility, which has led to their use in textile applications, such as self-powering electronic textiles. Ponnamma *et al.* (2019) reported the use of piezopolymers on smart and robust electrospun fabrics for design of piezo-electric nanogenerators (PENG) that used PVDF– hexafluoropropylene (PVDF–HFP). The results showed that, in all sorts of movements, the nanogenerator showed a good output in electrical performance, excellent mechanical properties, and was considered suitable for wearable electronic devices.

Another example of the versatility of piezo-polymers was reported by Azimi *et al.* (2021), where the parameters of a self-powered cardiac pacemaker that used a piezoelectric polymer were evaluated. They developed an encapsulated, flexible, and biocompatible nanogenerator that used a PVDF-composite fiber, and the final device was implanted on the heart of an adult dog. The device harvested energy from the animal pacing and delivered 0.487 µJ for the device operation.

2.6 FINAL REMARKS

Electroactive materials offer several possibilities and applications, due to their unique properties. CPs are some of the most traditional electroactive polymers and after 50 years of research, there is an increasing demand to understand more of the fundamental characteristics and properties of CPs and their derivatives, such as composites and hybrids. The applications of CPs are as numerous as their newly developed materials, and this versatility could expand into many new areas.

The stimuli-responsive property of PE gels toward variations in pH, temperature, and electric fields leads to volume changes (expansion and contraction), which makes these materials suitable to be employed in different applications, such as tissue engineering and controlled drug delivery systems. Furthermore, their leak-free ability and flexibility enable their use in different energy storage devices, such as solid state supercapacitors and batteries.

In addition, LCPs offer a unique combination of properties, which allow many modifications, and they can be used with various transduction techniques, which enables the construction of highly efficient materials and devices. In addition, the possibilities for system design (e.g., rigidity, mesogens, homopolymers, copolymers, and functionality) allow the formation of the most diverse materials that have properties specifically designed for the desired application, whether it is industrial, environmental, medicinal, or academic.

Finally, piezoelectric polymers are innovative materials, especially for energy harvesting. The possibilities highlighted, provide the potential for new devices in several areas of science. Piezopolymers have significant potential and require research for their improvement and development.

REFERENCES

Alemán, J.V. *et al.* (2007) 'Definitions of terms relating to the structure and processing of sols, gels, networks, and inorganic-organic hybrid materials (IUPAC Recommendations 2007)', *Pure and Applied Chemistry*, 79(10), pp. 1801–1829. doi.10.1351/pac200779101801.

Almdal, K., Dyre, J., Hvidt, S., and Kramer, O. (1993) 'Towards a phenomenological definition of the term 'Gel,'" *Polymer Gels and Networks*, 1(1), pp. 5–17. doi.10.1016/0966-7822(93)90020-I.

Alvarenga, G. *et al.* (2020) 'Recent trends of micro and nanostructured conducting polymers in health and environmental applications', *Journal of Electroanalytical Chemistry*, 879, pp. 114754. doi.10.1016/j.jelechem.2020.114754.

Anilkumar, T., Madhav, B.T.P., Venkateswara Rao, M., and Prudhvi Nadh, B. (2020) 'Bandwidth reconfigurable antenna on a liquid crystal polymer substrate for automotive communication applications', *AEU - International Journal of Electronics and Communications*, 117, pp. 153096.doi.10.1016/j.aeue.2020.153096.

Azimi, S. *et al.* (2021) 'Self-powered cardiac pacemaker by piezoelectric polymer nanogenerator implant', *Nano Energy*, 105781. doi.10.1016/j.nanoen.2021.105781.

Balint, R., Cassidy, N.J., and Cartmell, S.H. (2014) 'Conductive polymers: Towards a smart biomaterial for tissue engineering', *Acta Biomaterialia*, 10(6), pp. 2341–2353. doi.10.1016/j.actbio.2014.02.015.

Bar-Cohen, Y., and Zhang, Q. (2008) 'Electroactive polymer actuators and sensors', *MRS Bulletin*, 33(3), pp. 173–181. doi.10.1557/mrs2008.42.

Bauer, S., and Bauer, F. (2008) 'Piezoelectric polymers and their applications', in *Piezoelectricity*, 114. Berlin: Springer, pp. 157–177. doi.10.1007/978-3-540-68683-5_6.

Bellis, I. *et al.* (2019) 'Modulation of optical properties in liquid crystalline networks across different length scales', *Journal of Physical Chemistry C*, 123(43), pp. 26522–26527. doi.10.1021/acs.jpcc.9b06973.

Bhat, M.A., Rather, R.A., and Shalla, A.H. (2021) 'PEDOT and PEDOT:PSS conducting polymeric hydrogels: A report on their emerging applications', *Synthetic Metals*, 273, p. 116709. doi.10.1016/j.synthmet.2021.116709.

Bithi, K.A. *et al.* (2020) 'Cationic polyelectrolyte grafted mesoporous magnetic silica composite particles for targeted drug delivery and thrombolysis', *Materialia* ,11, p. 100676. doi.10.1016/j.mtla.2020.100676.

Blyakhman, F.A., Safronov, A.P., Shklyar, T.F., and Filipovich, M.A. (2015) 'To the mechanism of polyelectrolyte gel periodic acting in the constant DC electric field', *Sensors and Actuators, A: Physical*, 229, pp. 104–109. doi.10.1016/j.sna.2015.03.030.

Bouteiller, L., and Lebarny, P. (1996) 'Polymer-dispersed liquid crystals: Preparation, operation and application', *Liquid Crystals*, 21(2), pp. 157–174. doi.10.1080/02678299608032820.

Brostow, W. (1990) 'Properties of polymer liquid crystals: Choosing molecular structures and blending', *Polymer*, 31(6), pp. 979–995. doi.10.1016/0032-3861(90)90242-Q.

Bryan, A.M, Santino, L.M., Lu, Y., Acharya, S., and Arcy, J.M.D. (2016) 'Conducting polymers for pseudocapacitive energy storage', *Chemistry of Materials*, 28, pp. 5989–5998. doi.10.1021/acs.chemmater.6b01762.

Calvert, P. (2004) 'Electroactive polymer gels' in *Electroactive Polymer (EAP) Actuators as Artificial Muscles: Reality, Potential, and Challenges*, 2nd edn. Bellingham: SPIE, pp. 151–170. doi.10.1117/3.547465.ch5.

Cardoso, V.F., Ribeiro, C., and Lanceros-Mendez, S. (2017) 'Metamorphic biomaterials', in *Bioinspired Materials for Medical Applications*, pp. 69–99. Cambridge: Woodhead Publishing. doi.10.1016/B978-0-08-100741-9.00003-6.

Chambers, M. *et al.* (2009) 'Liquid crystal elastomer–nanoparticle systems for actuation', *Journal of Materials Chemistry*, 19(11), pp. 1524–1531. doi.10.1039/B812423J.

Chiu, Y. *et al.* (2013) 'Development of a piezoelectric polyvinylidene fluoride (PVDF) polymer-based sensor patch for simultaneous heartbeat and respiration monitoring', *Sensors and Actuators A: Physical*, 189. doi.10.1016/j.sna.2012.10.021.

Chu, Y., Varanasi, P.P., McGlade, M.J., and Varanasi, S. (1995) 'PH-induced swelling kinetics of polyelectrolyte hydrogels', *Journal of Applied Polymer Science*, 58(12), pp. 2161–2176. doi.10.1002/app.1995.070581203.

Cox, M.K. (1987) 'The application of liquid crystal polymer properties', *Molecular Crystals and Liquid Crystals Incorporating Nonlinear Optics*, 153(1), pp. 415–422. doi.10.1080/00268948708074556.

Doane, J.W. *et al.* (1988) 'Polymer dispersed liquid crystals for display application', *Molecular Crystals and Liquid Crystals Incorporating Nonlinear Optics*, 165(1), pp. 511–532. doi.10.1080/00268948808082211.

Dreyse, P., Honores, J., Quezada, D., and Isaacs, M. (2015) 'Electrocatalytic transformation of carbon dioxide into low carbon compounds on conducting polymers derived from multimetallic porphyrins', *ChemSusChem*, 8(22), pp. 3897–3904. doi.10.1002/cssc.201500816.

Dubal, D.P., Ayyad, O., Ruiz, V., and Gómez-Romero, P. (2015) 'Hybrid energy storage: The merging of battery and supercapacitor chemistries', *Chemical Society Reviews*, 44(7), pp. 1777–1790. doi.10.1039/C4CS00266K.

Dubal, D.P., Chodankar, N.R., Kim, D.-H., and Gomez-Romero, P. (2018) 'Towards flexible solid-state supercapacitors for smart and wearable electronics', *Chemical Society Reviews*, 47, pp. 2065–2129. doi.10.1039/C7CS00505A.

Dutta, D., Fruitwala, H., Kohli, A., and Weiss, R.A. (1990) 'Polymer blends containing liquid crystals: A review', *Polymer Engineering & Science*, 30(17), 1005–1008. doi.10.1002/pen.760301704.

Esteban, P.O., Leger, J.-M., Lamy, C., and Genies, E. (1989) 'Electrocatalytic oxidation of methanol on platinum dispersed in polyaniline conducting polymers', *Journal of Applied Electrochemistry*, 19(3), pp. 462–464. doi.10.1007/BF01015254.

Ferreira, V.C., Melato, A.I., Silva, A.F., and. Abrantes, L.M. (2011) 'Attachment of noble metal nanoparticles to conducting polymers containing sulphur-Preparation conditions for enhanced electrocatalytic activity', *Electrochimica Acta*, 56(10), pp. 3567–3574. doi.10.1016/j.electacta.2010.11.033.

Filpo, G. *et al.* (2008) 'UV tuning of the electro-optical and morphology properties in polymer-dispersed liquid crystals', *Liquid Crystals*, 35(1), pp. 45–48. doi.10.1080/02678290701769915.

Fink, J.K. (2014) 'Liquid crystal polymers', *High Performance Polymers*, pp. 381–400. doi.10.1016/B978-0-323-31222-6.00017-0.

Fujisue, H., Sendai, T. Yamato, K., and Takashima, W. (2007) 'Work behaviors of artificial muscle based on cation driven polypyrrole', *Bioinspiration and Biommetics*, 2, pp. S1–5. doi.10.1088/1748-3182/2/2/S01.

Glazer, P.J. *et al.* (2012) 'Role of PH gradients in the actuation of electro-responsive polyelectrolyte gels', *Soft Matter*, 8(16), pp. 4421–4426. doi.10.1039/c2sm07435d.

Gonzalez, E.R. (2000) 'Eletrocatálise e poluição ambiental', *Química Nova*, 23(2), pp. 262–266. doi.10.1590/S0100-40422000000200019.

Gründler, P. (2007) 'Chemical sensors: An introduction for scientists and engineers', *Chemical Sensors: An introduction for scientists and engineers*, Vol. 66. Berlin: Springer-Verlag.

Guenther, M. *et al.* (2007) 'Application of polyelectrolytic temperature-responsive hydrogels in chemical sensors', *Macromolecular Symposia*, 254, pp. 314–321. doi.10.1002/masy.200750846.

Hamdi, O., Mighri, F., and Rodrigue, D. (2020) 'Piezoelectric polymer films: Synthesis, Applications, and modeling', *Polymer Nanocomposite-Based Smart Materials*, 79–101. doi.10.1016/B978-0-08-103013-4.00005-4.

Hansson, P. (2006) 'Interaction between polyelectrolyte gels and surfactants of opposite charge', *Current Opinion in Colloid and Interface Science*, 11(6), pp. 351–362. doi.10.1016/j.cocis.2006.11.005.

Hong, S.Y., and Marynick, D.S. (1992) 'Understanding the conformational stability and electronic structures of modified polymers based on polythiophene', *Macromolecules*, 25(18), pp. 4652–4657. doi.10.1021/ma00044a029.

Hryniewicz, B.M., Bach-Toledo, L., and Vidotti, M. (2020) 'Harnessing energy from micropollutants electrocatalysis in a high-performance supercapacitor based on PEDOT nanotubes', *Applied Materials Today*, 18, pp. 100538. doi.10.1016/j.apmt.2019.100538.

Huang, C., Zhang, Q., and Jákli, A. (2003) 'Nematic anisotropic liquid-crystal gels-self-assembled nanocomposites with high electromechanical response', *Advanced Functional Materials*, 13(7), pp. 525–529. doi.10.1002/adfm.200304322.

Hulanicki, A., Glab, S., and Ingman, F. (1991) 'Chemical sensors: definitions and classification', *Pure and Applied Chemistry*, 63(9), pp. 1247–1250. doi.10.1351/pac199163091247.

Ibanez, J.G. *et al.* (2018) 'Conducting polymers in the fields of energy, environmental remediation, and chemical–chiral sensors', *Chemical Reviews*, 118(9), pp. 4731–4816. doi.10.1021/acs.chemrev.7b00482.

Janata, J., and Josowicz, M. (2003) 'Conducting polymers in electronic chemical sensors', *Nature Materials*, 2, pp. 9–24. doi.10.1038/nmat768.

Ji, Y., Marshall, J.E., and Terentjev, E.M. (2012) 'Nanoparticle-liquid crystalline elastomer composites', *Polymers*, 4(1), pp. 316–40. doi.10.3390/polym4010316.

Jia, X. *et al.* (2019) 'Tunable conducting polymers: Toward sustainable and versatile batteries', *ACS Sustainable Chemistry & Engineering*. 7, pp. 14321–14340. doi.10.1021/acssuschemeng.9b02315.

Jo, C.,. Naguib, H.E., and Kwon, R.H. (2011) 'Fabrication, modeling and optimization of an ionic polymer gel actuator', *Smart Materials and Structures*, 20(4). doi.10.1088/0964-1726/20/4/045006.

Kalapala, S., and Easteal, A.J (2005) 'Novel poly(methyl methacrylate)-based semi-interpenetrating polyelectrolyte gels for rechargeable lithium batteries', *Journal of Power Sources*, 147(1–2), pp. 256–259. doi.10.1016/j.jpowsour.2005.01.020.

Keller, K. *et al.* (2011) 'Modeling of temperature-sensitive polyelectrolyte gels by the use of the coupled chemo-electro-mechanical formulation', *Mechanics of Advanced Materials and Structures*, 18(7), pp. 511–523. doi.10.1080/15376494.2011.605006.

Khot, L.R., Panigrahi, S., and Lin, D. (2011) 'Development and evaluation of piezoelectric-polymer thin film sensors for low concentration detection of volatile organic compounds related to food safety applications', *Sensors and Actuators B: Chemical*, 153(1). doi.10.1016/j.snb.2010.05.043.

Kikuchi, K., and Tsuchitani, S. (2019) 'Ionic conductive polymers', in *Soft actuators*, Singapore: Springer, pp.151–169.

Killard, A.J. (2010) 'Nanostructured conducting polymers for (electro) chemical sensors', in *Nanostructured conductive polymers*. Hoboken, NJ: John Wiley & Sons, pp. 563–598.

Koh, J.K. (2011) 'Highly efficient, iodine-free dye-sensitized solar cells with solid-state synthesis of conducting polymers', *Advanced Materials*, 23(14), pp. 1641–1646. doi.10.1002/adma.201004715.

Kubo, S., Kumagai, M., Kawatsuki, N., and Nakagawa, M. (2019) 'Photoinduced reorientation in thin films of a nematic liquid crystalline polymer anchored to interfaces and enhancement using small liquid crystalline molecules', *Langmuir*, 35(44), pp. 14222–14229. doi.10.1021/acs.langmuir.9b02673.

Kumar, S.S., Kumar, C.S., Mathiyarasu, J., and Phani, K.L. (2007) 'Stabilized gold nanoparticles by reduction using 3,4-ethylenedioxythiophene-polystyrenesulfonate in aqueous solutions: Nanocomposite formation, stability, and application in catalysis', *Langmuir*, 23(6), pp. 3401–3408. doi.10.1021/la063150h.

Kurra, N., Wang, R., and Alshareef, H.N. (2015) 'All conducting polymer electrodes for asymmetric solid-state supercapacitors', *Journal of Materials Chemistry A*, 3(14), pp. 7368–7364. doi.10.1039/c5ta00829h.

Kwon, H.J., and Gong, J.P. (2006) 'Negatively charged polyelectrolyte gels as bio-tissue model system and for biomedical application', *Current Opinion in Colloid and Interface Science*, 11(6), pp. 345–350. doi.10.1016/j.cocis.2006.09.006.

Kwon, H.J., Osada, Y., and Gong, J.P. (2006) 'Polyelectrolyte gels-fundamentals and applications', *Polymer Journal*, 38(12), pp. 1211–1219. doi.10.1295/polymj.PJ2006125.

Kwon, H.J., Yasuda, K., Gong, J.P., and Ohmiya, Y. (2014) 'Polyelectrolyte hydrogels for replacement and regeneration of biological tissues', *Macromolecular Research*, 22(3), pp. 227–235doi.10.1007/s13233-014-2045-6.

Lai, Y.T, Kuo, J.C., and Yang, Y.J. (2014) 'A novel gas sensor using polymer-dispersed liquid crystal doped with carbon nanotubes', *Sensors and Actuators A: Physical*, 215, pp. 83–88. doi.10.1016/j.sna.2013.12.021.

Landsgesell, J., and Holm, C. (2019) 'Cell model approaches for predicting the swelling and mechanical properties of polyelectrolyte gels', *Macromolecules*, 52(23), pp. 9341–9353. doi.10.1021/acs.macromol.9b01216.

Lange, U., Roznyatovskaya, N.V., and Mirsky, V.M. (2008) 'Conducting Ppolymers in chemical sensors and arrays', *Analytica Chimica Acta*, 614(1), pp. 1–26. doi.10.1016/j.aca.2008.02.068.

Le, T.-H., Kim, Y., and Yoon, H. (2017) 'Electrical and electrochemical properties of conducting polymers', *Polymers*, 9(12), pp. 150. doi.10.3390/polym9040150.

Lee, S.J., Lee, D.Y., Song, Y.S., and Cho, N.I. (2007) 'Chemically driven polyacrylonitrile fibers as a linear actuator', *Solid State Phenomena*, 124–126, pp. 1197–1200. doi.10.4028/www.scientific.net/ssp.124-126.1197.

Li, C., Bai, H., and Shi, G. (2009) 'Conducting polymer nanomaterials: Electrosynthesis and applications', *Chemical Society Reviews*, 38(8), p. 2397. doi.10.1039/b816681c.

Li, M.H. *et al.* (2003) 'Light-driven side-on nematic elastomer actuators', *Advanced Materials*, 15 (7–8), pp. 569–572. doi.10.1002/adma.200304552.

Li, M.H. *et al.* (2006) 'Artificial muscles based on liquid crystal elastomers', *Philosophical Transactions of the Royal Society A: Mathematical, Physical and Engineering Sciences*, 364, pp. 2763–2777. doi.10.1098/rsta.2006.1853.

Lima, P.H.C., Fonseca, D.F., Braz, C.J.F., and Cunha, C.T.C. (2018) 'Polímeros condutores com propriedades eletrocrômicas: Uma revisão', *Revista Eletrônica de Materiais e Processos*, 13(1), pp. 1–17.

Liu, Q. *et al.* (2018) 'Electroresponsive homogeneous polyelectrolyte complex hydrogels from naturally derived polysaccharides', *ACS Sustainable Chemistry & Engineering*, 6(5), pp. 7052–7063. doi.10.1021/acssuschemeng.8b00921.

Lu, X. *et al.* (2014) 'Flexible solid-state supercapacitors: Design, fabrication and applications', *Energy and Environmental Science*, 7(7), pp. 2160–2181. doi.10.1039/c4ee00960f.

Malinauskas, A. (1999) 'Electrocatalysis at conducting polymers', *Synthetic Metals*, 107(2), pp. 75–83. doi.10.1016/S0379-6779(99)00170-8.

McNaught, A.D., and Wilkinson, A. (1997) 'IUPAC. Compendium of Chemical Terminology', in *Gold Book*, 2nd edn. Oxford: Blackwell Scientific Publications.

Medeiros, E.S., Oliveira, J.E., Paterno, L.G., and Mattoso, L.H.C. (2012) 'Uso de polímeros condutores em sensores. Parte 1: Introdução aos polímeros condutores', *Revista Eletrônica de Materiais e Processos*, 7(2), pp. 62–77.

Menzel, N., Ortel, E., Kraehnert, R., and Strasser, P. (2012) 'Electrocatalysis using porous nanostructured materials', *ChemPhysChem*, 13, pp. 1385–1394. doi.10.1002/cphc.201100984.

Mishra, S., Unnikrishnan, L., Nayak, S.K., and Mohanty, S. (2019) 'Advances in piezoelectric polymer composites for energy harvesting aApplications: A systematic review', *Macromolecular Materials and Engineering*, 304(1), p. 1800463. doi.10.1002/mame.201800463.

Mitraka, E. *et al.* (2019) 'Electrocatalytic production of hydrogen peroxide with poly(3,4-ethylenedioxythiophene) electrodes'. *Advanced Sustainable Systems*, 3(2), p. 1800110. doi.10.1002/adsu.201800110.

Mu, S., and Zhang, Y. (2010) 'Electrocatalysis by nanostructured conducting polymers', in *Nanostructured conductive polymers*. Hoboken, NJ: John Wiley & Sons.

Murad, A.R., Iraqi, A., Aziz, S.B., and Abdullah, S.N. (2020) 'Conducting polymers for optoelectronic devices and organic solar cells: A review', *Polymers*, 12, p. 2627. doi.10.3390/polym12112627.

Nemat-Nasser, S., and Thomas, C. (2004) 'Ionomeric polymer-metal composites', in *Electroactive polymer (EAP) actuators as artificial muscles: Reality, potential, and challenges*, 2nd edn. Bellingham: SPIE.

Ning, C. *et al.* (2018) 'Electroactive [olymers for tissue regeneration: Developments and perspectives', *Progress in Polymer Science*, 81, pp. 144–162. doi.10.1016/j.progpolymsci.2018.01.001.

Nishinari, K. (2009) 'Some thoughts on the definition of a gel', in *Gels: Structures, properties, and functions*, 136:87–94. Berlin, Heidelberg: Springer Berlin Heidelberg.

Nogueira, A.F., Longo, C., and De Paoli, M.-A. (2004) 'Polymers in dye sensitized solar cells: Overview and perspectives', *Coordination Chemistry Reviews*, 248(13–14), pp. 1455–1468. doi.10.1016/j.ccr.2004.05.018.

Otero, T.F. (2021) 'Towards artificial proprioception from artificial muscles constituted by self-sensing multi-step electrochemical macromolecular motors', *Electrochimica Acta*, 368, p. 137576. doi.10.1016/j.electacta.2020.137576.

Otero, T.F., and Sansiñena, J.M. (1995) 'Artificial muscles based on conducting polymers', *Bioelectrochemistry and Bioenegetics*, 38, pp. 411–414. doi.10.1016/0302-4598(95)01802-L.

Otero, T.F., Rodigues, J.M., Angulo, E.M., and Santamaria, C. (1993) 'Artificial muscles from bilayer structures', *Synthetic Metals*, 57, pp. 3713–3717.

Otero, T.F., Martinez, J.G., and Arias-Pardilla, J. (2012) 'Biomimetic electrochemistry from conducting Polymers. A Review: Artificial muscles, smart membranes, smart drug delivery and computer/neuron interfaces', *Electrochimica Acta*, 84, pp. 112–128. doi.10.1016/j.electacta.2012.03.097.

Pandey, R.K., and Lakshminarayanan, V. (2012) 'Ethanol electrocatalysis on gold and conducting polymer nanocomposites: A study of the kinetic parameters', *Applied Catalysis B: Environmental*, 125, pp. 271–281. doi.10.1016/j.apcatb.2012.06.002.

Peers, S., Montembault, A., and Ladavière, C. (2020) 'Chitosan hydrogels for sustained drug delivery', *Journal of Controlled Release*, 326, pp. 150–63. doi.10.1016/j.jconrel.2020.06.012.

Picken, S.J. (1996) 'Applications of liquid crystal polymers: Part 1: Fibre spinning', *Liquid Crystals Today*, 6 (1), pp. 12–15. doi.10.1080/13583149608047635.

Ponnamma, D., Parangusan, H., Tanvir, A., and AlMa'adeed, M.A.A. (2019) 'Smart and robust electrospun fabrics of piezoelectric polymer Nnanocomposite for self-powering electronic textiles', *Materials & Design*, 184, p. 108176. doi.10.1016/j.matdes.2019.108176.

Puiggalí-Jou, A., del Valle, L.J., and Alemán, C. (2019) 'Drug delivery systems based on intrinsically conducting polymers', *Journal of Controlled Release*, 309, pp. 244–264. doi.10.1016/j.jconrel.2019.07.035.

Ramadan, K.S., Sameoto, D., and Evoy, S. (2014) 'A review of piezoelectric polymers as functional materials for electromechanical transducers', *Smart Materials and Structures*, 23(3), p. 33001.doi.10.1088/0964-1726/23/3/033001.

Ravishankar, K. *et al.* (2019) 'Biocompatible hydrogels of chitosan-alkali lignin for potential wound healing applications', *Materials Science and Engineering C*, 102, p. 447–457. doi.10.1016/j.msec.2019.04.038.

Ribeiro, C., Sencadas, V., Correia, D.M., and Lanceros-Méndez, S. (2015) 'Piezoelectric polymers as biomaterials for tissue engineering applications', *Colloids and Surfaces B: Biointerfaces*, 136, pp. 46–55. doi.10.1016/j.colsurfb.2015.08.043.

Roy, S.G., Haldar, U., and De, P. (2014) 'Remarkable swelling capability of amino acid based cross-linked polymer networks in organic and aqueous medium', *ACS Applied Materials & Interfaces*, 6(6), pp. 4233–4241. doi.10.1021/am405932f.

Rumyantsev, A.M. *et al.* (2016) 'Polyelectrolyte gel swelling and conductivity vs counterion type, cross-linking density, and solvent polarity', *Macromolecules*, 49(17), pp. 6630–6643. doi.10.1021/acs.macromol.6b00911.

Sangsuriyonk, K., Paradee, N., and Sirivat, A. (2020) 'Electrically controlled release of anticancer drug 5-Fluorouracil from carboxymethyl cellulose hydrogels ', *International Journal of Biological Macromolecules*, 165, pp. 865–873. doi.10.1016/j.ijbiomac.2020.09.228.

Sappati, K.K., and Bhadra, S. (2018) 'Piezoelectric polymer and paper substrates: A review', *Sensors*. doi.10.3390/s18113605.

Saranya, K., Rameez, M., and Subramania, A. (2015) 'Developments in conducting polymer based counter electrodes for dye-sensitized solar cells – An overview', *European Polymer Journal*, 66, pp. 207–227. doi.10.1016/j.eurpolymj.2015.01.049.

Sengodu, P., and Deshmukh, A.D. (2015) 'Conducting polymers and their inorganic composites for advanced Li-ion batteries: A review', *RSC Advances*, 5, pp. 42109–42130. doi.10.1039/c4ra17254j.

Shi, Y. *et al.* (2015) 'Nanostructured conductive polymers for advanced energy storage', *Chemical Society Reviews*, 44(1), pp. 6684–6696. doi.10.1039/c5cs00362h.

Shibaev, V. (2016) 'Liquid crystalline polymers', in Hashmi, S. (ed) *Reference module in materials science and materials engineering*. Oxford: Elsevier. doi.10.1016/B978-0-12-803581-8.01301-1. pp. 1–46.

Shibaev, P.V. *et al.* (2015) 'Rebirth of liquid crystals for sensoric applications: Environmental and gas sensors', *Advances in Condensed Matter Physics*, 2015, pp. 1–8. doi.10.1155/2015/729186.

Shibayev, V.P., and Byelyayev, S.V. (1990) 'Prospects for the use of functional liquid crystal polymers and composites. Review', *Polymer Science U.S.S.R.*, 32(12), pp. 2384–2428. doi.10.1016/0032-3950(90)90414-2.

Shirakawa, H. *et al.* (1977) 'Synthesis of electrically conducting organic polymers: Halogen derivatives of polyacetylene, (CH)X', *Journal of the Chemical Society, Chemical Communications*, 16, pp. 578–580. doi.10.1039/C39770000578.

Shukla, S.K., Kushwaha, C.S., and Singh, N.B. (2017) 'Recent developments in conducting polymer based composites for sensing devices', *Materials Today: Proceedings*, 4(4), pp. 5672–5681. doi.10.1016/j.matpr.2017.06.029.

Takeshima, T. *et al.* (2015) 'Photoresponsive surface wrinkle morphologies in liquid crystalline polymer films', *Macromolecules*, 48(18), pp. 6378–644. doi.10.1021/acs.macromol.5b01577.

Tang, J. *et al.* (2020) 'Swelling behaviors of hydrogels with alternating neutral/highly charged sequences', *Macromolecules*, 53(19), pp. 8244–8254. doi.10.1021/acs.macromol.0c01221.

Thomsen, D.L. *et al.* (2001) 'Liquid crystal elastomers with mechanical properties of a muscle', *Macromolecules*, 34(17), pp. 5868–5875. doi.10.1021/ma001639q.

Tsakova, V., and Seeber, R. (2016) 'Conducting polymers in electrochemical sensing: Factors influencing the electroanalytical signal', *Analytical and Bioanalytical Chemistry*, 408(26), pp. 7231–721. doi.10.1007/s00216-016-9774-7.

Valero, L. *et al.* (2011) 'Characterization of the movement of polypyrrole-dodecylbenzenesulfonate-perchlorate/tape artificial muscles. Faradaic control of reactive artificial molecular motors and muscles', *Electrochimica Acta*, 56(10), pp. 3721–3726. doi.10.1016/j.electacta.2010.11.058.

Wallmersperger, T., and Leichsenring, P. (2016) 'Polymer gels as EAPs: Models', in *Electromechanically active polymers*. Cham: Springer International Publishing.

Wallmersperger, T. *et al.* (2013) 'Modeling and simulation of hydrogels for the application as bending actuators', in *Intelligent hydrogels*. Cham: Springer International Publishing. doi.10.1007/978-3-319-01683-2_15. pp. 189–204.

Wang, T. *et al.* (2016) 'Electroactive polymers for sensing', *Interface Focus*, 6(4), p. 1–19. doi.10.1098/rsfs.2016.0026.

Wang, J., Liu, F., Tao, F., and Pan Q. (2017) 'Rationally designed self-healing hydrogel electrolyte toward a smart and sustainable supercapacitor', *ACS Applied Materials and Interfaces*, 9 (33), pp. 27745–27753. doi.10.1021/acsami.7b07836.

Wang, Y., Liu, A., Han, Y., and Li, T. (2020) 'Sensors based on conductive polymers and their composites: A review', *Polymer International*, 69(1), pp. 7–17. doi.10.1002/pi.5907.

Xuan, R. *et al.* (2021) 'Stretchable and self-healable polyelectrolytes for flexible and sustainable supercapacitor', *Journal of Power Sources*, 487, pp. 229394. doi.10.1016/j.jpowsour.2020.229394.

Yoneyama, H., Shoji, Y., and Kawai, K. (1989) 'Electrochemical synthesis of polypyrrole films containing metal oxide particles', *Chemistry Letters*, 18(6), pp. 1067–1070. doi.10.1246/cl.1989.1067.

Yusuf, Y. *et al.* (2005) 'Low-voltage-driven electromechanical effects of swollen liquid-crystal elastomers', *Physical Review E - Statistical, Nonlinear, and Soft Matter Physics*, 71(6), pp.1–8. doi.10.1103/PhysRevE.71.061702.

Zarras, P., and Irvin, J. (2003) 'Electrically active polymers', on *Encyclopedia of polymer science and technology*. Hoboken: John Wiley & Sons. doi.10.1002/0471440264.pst107. pp. 121–190.

Zhang, W. *et al.* (2018) 'Injectable and body temperature sensitive hydrogels based on chitosan and hyaluronic acid for pH sensitive drug release', *Carbohydrate Polymers*, 186, pp. 82–90. doi.10.1016/j.carbpol.2018.01.008.

Zhou, Q., and Shi, G. (2016) 'Conducting polymer-based catalysts', *Journal of the American Chemical Society*, 138(9), pp. 2868–2876. doi.10.1021/jacs.5b12474.

Zhou, Q. *et al.* (2008) 'Electrocatalysis of template-electrosynthesized cobalt–porphyrin/polyaniline nanocomposite for oxygen reduction', *The Journal of Physical Chemistry C*, 112(47), pp. 18578–18583. doi.10.1021/jp8077375.

Zinola, C.F., Martins, M.E., Tejera, E.P., and Neves, N.P. (2012) 'Electrocatalysis: Fundamentals and applications', *International Journal of Electrochemistry*, 2012, pp. 1–2. doi.10.1155/2012/874687.

3 Properties of Electroactive Polymers

Kübra Gençdağ Şensoy

CONTENTS

3.1 INTRODUCTION

Polymers that can change shape and size when an electric field is applied are called electroactive polymers (EAPs). EAP materials can be produced easily and quickly by different methods, making them versatile.

EAPs are generally categorized according to their activation mode, either electronic or ionic (1). In electronic EAPs, attractive forces are applied to the electrodes by the electric field. Due to these forces, their shape and size change (2). In ionic EAPs, shape change occurs due to the mobility and diffusion of ions. EAP materials are particularly suitable for actuators that are used to control and move mechanisms. Ionic polymer–metal composite (IPMC), an ionic EAP, can achieve large bending strains at low voltages (3).

EAPs can be characterized in many different ways. Stress–strain curves, dynamic mechanical, and dielectric thermal analysis is commonly used.

3.2 ELECTROACTIVE POLYMERIC MATERIALS

3.2.1 IONIC POLYMER–METAL COMPOSITES

Synthetic composite nanomaterials that act as artificial muscles due to the electric field effect are called ionic polymer–metal composites (IPMCs). The surfaces of IPMCs consist of an ionic polymer. Due to the stress applied across the IPMC strip, ion migration and redistribution cause bending deformation. The applied voltage can cause various deformations, such as bending, rolling, torsion, rotation, rotational, and unsymmetrical bending deformation. When these deformations are applied to IPMC strips, they generate an output voltage signal. Therefore, IPMCs are known EAPs. They work well in air and liquid media. They can generate a peak force of approximately 40 times their weight and have a wide bandwidth (4-11).

DOI: 10.1201/9781003173502-3

3.2.2 Ion Gels

An ion gel is a material that consists of an inorganic or polymer matrix immobilized ionic liquid (12, 13). An ion gel can be obtained by mixing or synthesizing the solid matrix and ionic liquid in situ or by using a block copolymer polymerized in solution with the ionic liquid. The aim is to create a self-assembled nanostructure in which ions can be dissolved selectively. Ion gels can be synthesized using materials such as oxides, non-copolymer polymers, or boron nitride. Ion gels can be polymeric and inorganic. The main purpose of ion gel applications is to electrically insulate the matrix components to provide ionic conductivity (14).

Ion gels have been used as insulators (15), dielectrics (16), and electrolytes (12) in many electrical device systems. Solid-state and flexible ion gels are especially preferred in mobile devices (17). The high viscosity of ion gels makes electrolytes and separators between anodes and cathodes, especially in battery applications. In addition, the viscoelastic flow that occurs in gels under stress creates a quality electrode or electrolyte interface contact, which highlights ion gels.

Ion gels can withstand ≤300°C before they degrade (18). Due to their high-temperature capability, they have high thermal stability (19) and this stability is well above the capacity of existing commercial electrolytes. For example, it has been used in lithium-ion batteries (LIBS) to run cells at 175°C (20).

3.2.3 Carbon Nanotubes

Carbon nanotubes (CNTs) have attracted significant attention since their discovery due to their broad mechanical and electrical properties. They are single-walled carbon nanotubes (SWCNTs) approximately 1 nm in diameter. There are multi-walled carbon nanotubes (MWCNTs) composed of interlocking SWCNTs that are weakly bonded due to Van der Waals interactions (21).

CNTs have excellent tensile strength (22) and thermal conductivity (23, 24) dues to the strength of the bonds in carbon atoms and their nanostructure. Their electrical conductivity is high (25), and chemical modifications are possible (26). Because of these properties they are valuable in many fields, such as optics, electronics, and nanotechnology. A report was recently published on super elastic CNT air gel muscles (27). CNT air gel sheets have 220% anisotropic linear elongation and strain rates. Airgel can be permanently frozen at temperatures between 80 and 1900 K. Unlike conventional CNT electrolytes actuators are not required, and actuation is accomplished by applying a positive voltage to the counter electrode. When CNTs are used as reinforcements in polymers, they are first randomly dispersed in a solvent or polymer fluid/melt by shear mixing or sonication; then further processing is performed to form the composite (28, 29). In addition, poor dispersion is generally specified as a processing limitation (30) and the critical reduction factor (31–33). Improving the CNT dispersion might be possible in the presence of surfactants (34), oxidation, or chemical functionalization of the surface (35, 36)

3.2.4 Polymer Dots

Polymer dots (PDs) were synthesized using the grass hydrothermal route (37). PDs are new nanomaterials composed of conjugated organic polymers with conjugated PDs. Semiconductor PDs have emerged as a fluorescent carbon-based material class.

PDs are from linear polymers or monomers from the clustered or crosslinked polymer. In addition, a carbon core and bonded polymer can form PDs. The synthesis of PDs from polymers is shown in Figure 3.1.

PDs show essential potential applications in drug delivery and therapeutics, biological imaging, and detection due to the advantages of their structure, good biocompatibility, optical properties, and accessible surface modification. These nanomaterials have the potential for fluorescent imaging and optical detection (39, 40).

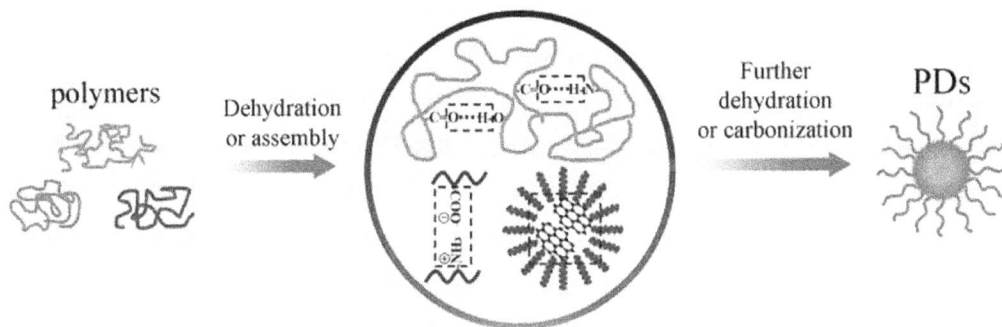

FIGURE 3.1 PDs synthesis from polymers (38).

Fluorescent polymer nanoparticles were prepared to determine nitro-explosive picric acid (41). PDs were prepared by polymerizing carbon tetrachloride and ethylenediamine (42). In another study, a path that was not linearly conjugated from polymers to fluorescent PDs was used (43).

3.2.5 MOLECULARLY IMPRINTED POLYMERS

Molecularly imprinted polymers (MIPs) are prepared by the copolymerization of monomers (44). First, polymerization of functional monomers occurs in the presence of a template molecule. Then, highly selective gaps are formed by removing the template molecule (45, 46). MIPs have several advantages, such as stability at different pHs and temperatures compared with natural recognition receptors, such as antibodies (47, 48).

MIPs are used in several nanosensor applications, such as antioxidants (49), antibiotics (50, 51), toxic compounds (52, 53), and drugs (54, 55). MIPs can be synthesized by different methods, such as electropolymerization, photopolymerization, and free radical polymerization (56). In electropolymerization, they can control the polymer film. This makes them more advantageous than other methods (57). A thin MIP layer and carbon-based materials formed on the electrode surface can increase the sensor's conductivity (58–61). In addition, the electrode combined with MIP can help prevent impurities, which are the biggest problem in electrochemical determination.

Based on the properties of the polymers, MIPs have different applications. The suspension (62), two-step swelling (63), bulk (64), precipitation, emulsion, and core-shell polymerization (65) are examples. Scanning electron microscopy (SEM) provides excellent resolution and can be used to study the morphology of MIPs.

In Figure 3.2, the printed polymer surface is shown at different magnifications. The polymer has a homogeneous microsphere structure.

3.2.6 CONDUCTIVE POLYMERS

Conductive polymers (CPs) were discovered approximately 20 years ago and remain of interest. CPs are an important invention that can replace metallic and semiconductors.

CPs of polyenes polyaromatics have been studied extensively. Polyaniline (PANI) attracts great attention due to its different transmission mechanisms and good environmental stability. It is one of the oldest known CPs. PANI has magnetic, electronic, and optical properties similar to metals. PANI can be used in various fields, such as supercapacitors, electromagnetic shielding devices, battery electrodes, anti-corrosion coatings, light-emitting diodes, non-linear optical devices, molecular sensors, electrochromic displays, and microelectronic devices (67, 68). PANI is an excellent active cathode material for LIBs (69). The polymerization scheme of aniline is shown in Figure 3.3 (70).

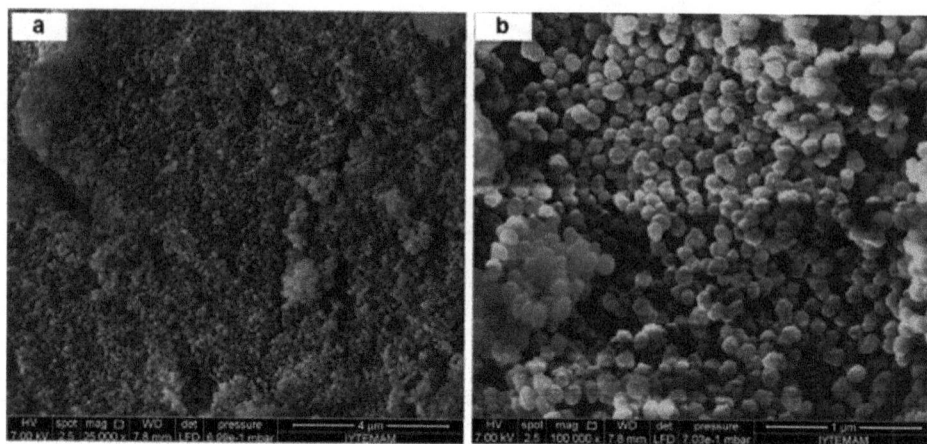

FIGURE 3.2 SEM images of molecularly imprinted polymers with magnitude of: (a) 25,000 ×; and (b) 100,000 × (66).

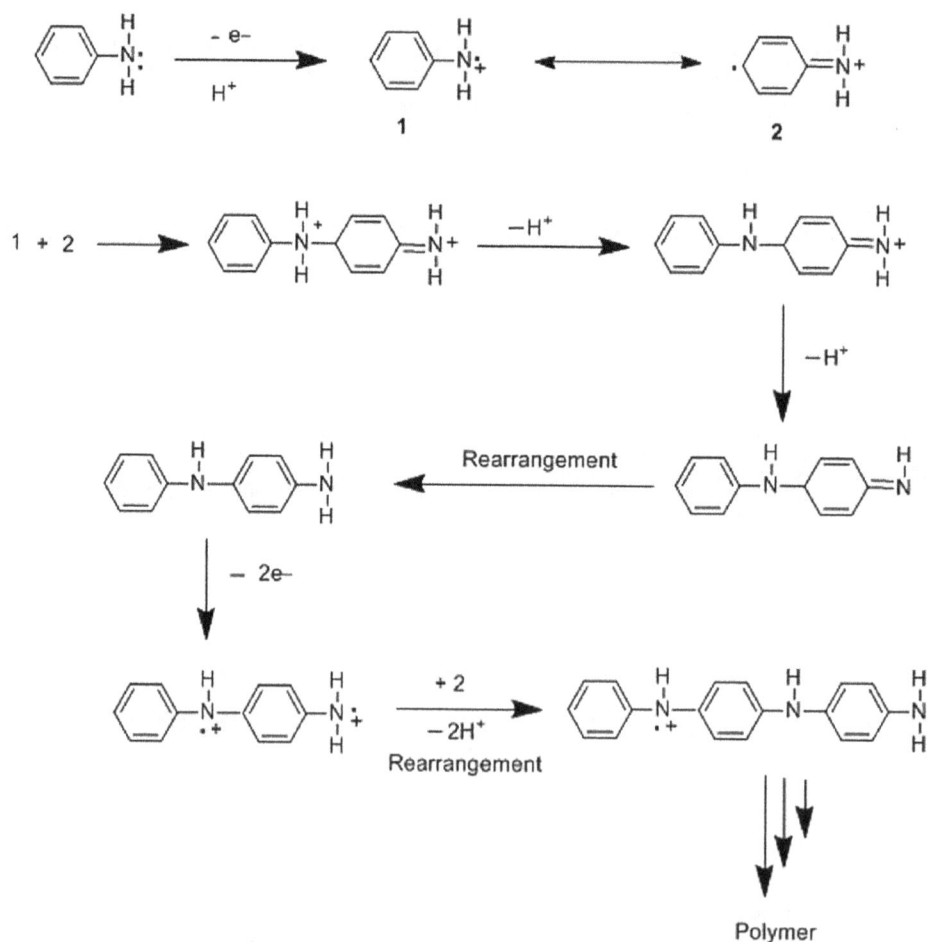

FIGURE 3.3 Aniline polymerization (70).

Pyrrole's well-known feature is its high primary step velocity after cation radical formation. After a sufficiently long chain is formed, a thin film is formed on the electrode surface with pyrrole oligomers (71–73). Electronic devices and chemical sensors are two main application areas of pyrrole-related CPs (74).

CPs have an operational efficiency of approximately 1% (75). Because of the resistance between the electrolyte and the polymer (76) and their mechanism based on the migration of ions, the actuation speeds are limited (77). Due to the electrolyte requirement, the CP actuator might need to be encapsulated (78).

3.2.7 BISTABLE ELECTROACTIVE POLYMERS

Bistable electroactive polymers (BSEPs) are newly developed EAPs (79). BSEPs exhibit soft elastomer behavior above the glass transition temperature. Depending on the shape of the material, there might be changes in their movements. If it is a thin film compressed between the electrodes, it can act as a rigid capacitor, or a variable capacitor if it is in an elastomeric state. Poly(tert-butyl-acrylate) (PTBA) is a BSEP material that has gained much attention and research (79, 80). It shows a significant glass transition at 50°C, and exhibits outstanding strain stability and stress recovery. Operational features are available with a fault field strength of >250 MV/m.

BSEP has two balanced actuation characteristics based on a high voltage and specific power density. Dielectric materials consume energy when activated due to current leakage through the film. BSEPs can maintain their shape without effort and time. Therefore, BSEPs are important EAPs. (81). Figure 3.4 shows the variable stiffness and activation of BSEPs (82).

BSEPs contain crystalline clusters of long alkyl side chains in a crosslinked polymer matrix. This crosslinked polymer matrix makes the polymer film translucent (83–85). The abrupt and reversible phase transition of the crystalline aggregates of polymers causes a rapid shift between the solid and

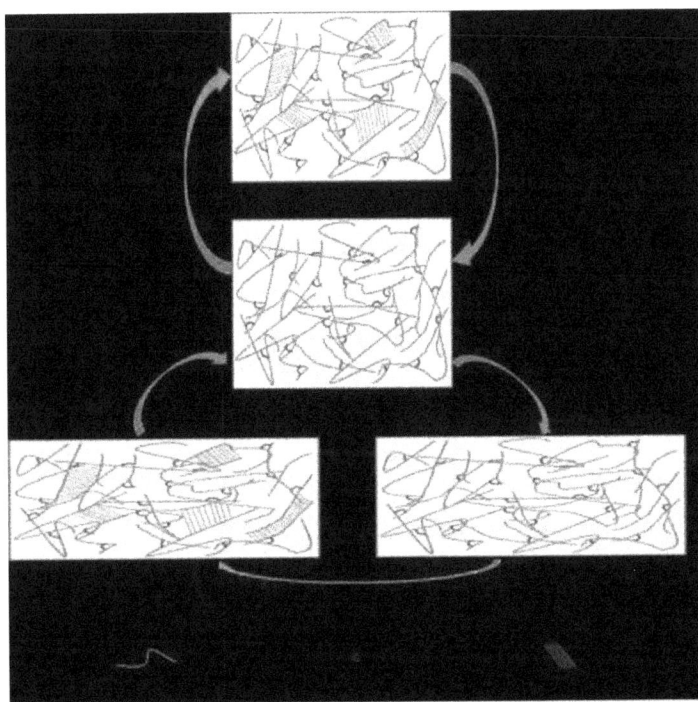

FIGURE 3.4 Rigid-to-rigid actuation mechanism of BSEP (82).

rubbery states of the polymers during temperature cycles. Cooling the material can maintain this deformation. The polymer regains its original shape on reheating.

3.2.8 FERROELECTRIC POLYMERS

Ferroelectric polymers are especially crystalline polar polymers that can provide permanent electrical polarization (86, 87). These properties are exhibited by poly(vinylidene difluoride) (PVDF), PVDF copolymers, polyamides (88), and mixtures of these (89, 90). It is not sufficient for polymers to have only polar side groups to exhibit ferroelectric behavior. In addition, they need to maintain molecular configurations when there is no polarity.

Polymers, such as poly (vinyl chloride), do not exhibit ferroelectric behavior since the bond must regulate itself in the helical conformation, which results from the relatively large steric influences of Van der Waals forces. Due to internal steric and electrostatic interactions, potential energy is generated in the chains.

PVDF consists of a repeating unit ($-CH2CF2-$). It is associated with negatively charged fluorine atoms and positively charged hydrogen atoms. (Figure 3.5). The repeating unit's ($-CH2CF2-$) dipole moment is $\mu = 7\times10^{-30}$ Cm (2.1 D). Since these dipoles are closely attached to the main chain, their direction varies depending on the conformation and packaging of the molecules. The resulting crystal has a large self-polarization (P_s) responsible for the ferroelectricity of PVDFs. The values of a, b, and c are lattice constants (91).

Ferroelectric polymers have been included in many applications and they are undergoing further research. For example, new ferroelectric polymer composites with a high dielectric constant are being developed. These are essential for many applications because they exhibit good pyroelectric and piezoelectric responses and have low acoustic impedance.

In addition, ferroelectric materials are used as sensors. High-pressure sensors are examples of these (92). They exhibit piezoluminescence against stress (93). They can be applied in the robotic and biomedical fields.

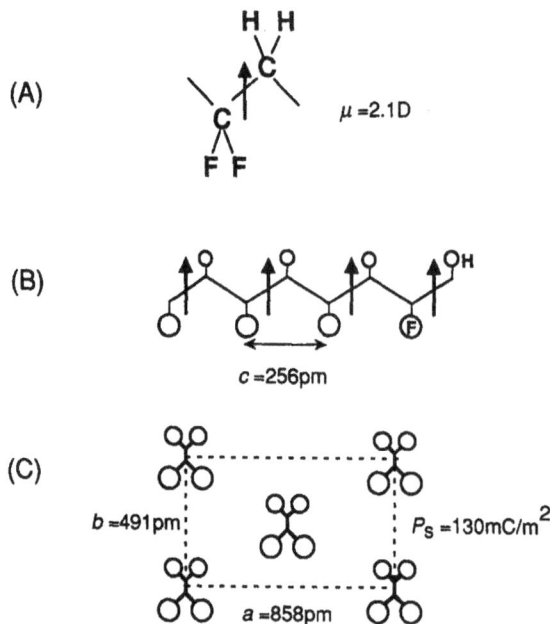

FIGURE 3.5 Showing: (a) molecular; (b) chain; and (c) crystal structures of PVDF (91).

3.2.9 DIELECTRIC ELASTOMERS

Dielectric elastomers (DEs) are innovative materials that cause significant stress. They are EAPs that can convert electrical energy into mechanical work. They have a fairly good elastic energy density, and have many prototype implementations.

DE actuators are compatible variable capacitors (Figure 3.6). They consist of a thin elastomeric film coated on both sides with electrodes. When an electric field is applied to the electrodes, the electrostatic attraction between opposite charges and the repulsion between similar charges places pressure on the film. Most of the elastomers used are incompressible. If there is any reduction in their thickness, this causes an increase at the same time (81).

Many elastomeric materials were investigated, including silicones, isoprene, polyurethanes, and fluoroelastomers between 1990 and 2000 (94–97). Polyurethanes, acrylics, and silicones have been identified as promising material groups (98).

Grease films loaded with carbon powder or carbon black as electrodes are potential applications for DEAs. However, the reliability of these materials is not very good, and if more advanced properties are desired, graphene sheets, liquid metal, embedded metallic nanoclusters, corrugated or patterned metal films, and carbon nanotube coatings could be used (99, 100). Acrylic elastomers and silicones are other alternatives.

Elastomer materials should have the following properties: (1) low material hardness, (2) high dielectric constant, and (3) high electrical breakage resistance. The mechanically prestressed elastomer film improves the electrical breaking strength. The film thickness is reduced and requires lower voltage to achieve the same pressure. The electrode must be compatible and conductive. This is important so that the elastomer is not restricted mechanically.

Significant research has been carried out on DEs that are based on silicon and natural rubber (101). When comparing them, acrylic elastomer based DEs are more advantageous due to their efficiency and fast response times (102).

3.2.10 POLYMER ELECTRETS

Electrets are insulating materials that show piezoelectric effects. Unequal area load distributions cause them to exhibit these effects (103, 104). Current polymer electrets consist of a highly porous polymer. The porous films' charging voltages (5–10 kV) vary. For electrical discharge, charges accumulate at the polymer–gas interface. There are charges on opposite sides of the pores that form macroscopic dipoles according to the direction of the applied area (105). Pores in the films can be distorted when a compression force is applied. The applied voltage will cause a change in the thickness of the materials.

Polymer electrets can be used as sensors or actuators. Compared with solid PVDF ferroelectric polymers, the converter coefficient (d33) is higher (106).

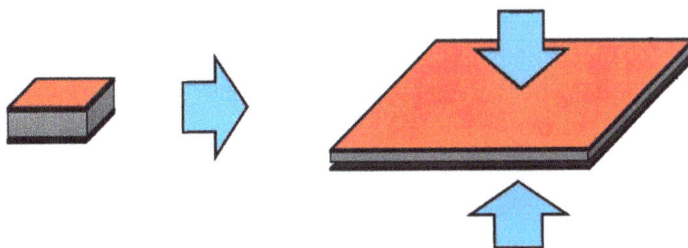

FIGURE 3.6   Sandwich structure of two compatible electrode layers and dielectric elastomer membrane.

Due to the increase in performance demands, research into polymer electret mixes is ongoing. Recently, studies based on mixtures with poly(2,6-dimethyl-1,4-phenylene ether) and polystyrene have been reported (107). Good performance with the new electrets has been achieved.

3.2.11 ELECTROSTRICTIVE POLYMERS

Recently, the development of electrostrictive polymers has created new research areas for high voltage actuators. Research on the use of electrostrictive polymers for the conversion of mechanical to electrical energy or energy harvesting has some potential applications.

Electrostrictive polymers have an inherent electrical polarization. The change in the dipole density of the material causes electrostriction. These polymers contain nanocrystalline or molecular polarizations from the applied electric field effect.

Other materials discussed in this section are electrostrictive graft copolymers, PVDF copolymers with nano-sized crystalline areas, and liquid crystal elastomers (LCEs). LCEs were first used in artificial muscles (108), where the polymer network allowed sufficient movement. LCEs are nematic and smectic and have different mechanisms.

Ferroelectric polymers are based on PVDF copolymers, and they have two main limitations. First, electrically induced paraelectric-ferroelectric transition only allows operation above the Curie temperature. Second, the presence of strong hysteresis makes control difficult (105).

REFERENCES

1. Prahlad H, Kornbluh R. Pelrine R. Stanford, Eckerle J, Oh S. Polymer power: dielectric elastomers and their applications in distributed actuation and power generation. Proc ISSS. 2005; 13:8.
2. Biddiss E, Chau T. Dielectric elastomers as actuators for upper limb prosthetics: challenges and opportunities. Med Eng Phys. 2008; 30(4):403–18.
3. Pelrine R, Kornbluh R, Joseph J, Heydt R, Pei Q, Chiba S. High-field deformation of elastomeric dielectrics for actuators. Mater Sci Eng C. 2000; 11(2):89–100.
4. Segalman DJ, Witkowski WR, Adolf DB, Shahinpoor M. Theory and application of electrically controlled polymeric gels. Int J Smart Mater Struct. 1992; 1:95–100.
5. Shahinpoor M. 1992. Conceptual Design, Kinematics and Dynamics of Swimming Robotic Structures Using Ionic Polymeric Gel Muscles. Int J Smart Mater Struct; 1:91–4.
6. Osada Y, Okuzaki H, Hori H. A polymer gel with electrically driven motility. Nature. 1992; 355: 242–4.
7. Oguro K, Kawami Y, Takenaka H. Bending of an ion-conducting polymer film electrode composite by an electric stimulus at low voltage. Trans Micro Machine Soc. 1992; 5:27–30.
8. Doi M, Marsumoto M, Hirose Y. Deformation of ionic gels by electric fields. Macromolecules. 1992; 25:5504–11.
9. Oguro K, Asaka K, Takenaka H. Polymer film actuator driven by low voltage. In Proceedings of the 4th International Symposium of Micro Machines and Human Science, Nagoya. 1993.
10. Adolf D, Shahinpoor M, Segalman D, Witkowski W. Electrically controlled polymeric gel actuators. US Patent Office US 5250167. October 13, 1993.
11. Oguro K, Kawami Y, Takenaka H. Actuator element. US Patent Office US 5268082. December 7, 1993.
12. Chen N, Zhang H, Li L, Chen R, Guo S. Ionogel electrolytes for high-performance lithium batteries: A review. Adv Energy Mater. 2018; 8 (12):1702675.
13. Osada I, de Vries H, Scrosati B, Passerini. Ionic-liquid-based polymer electrolytes for battery applications. Angew Chem Int Ed. 2016; 55 (2):500–13.
14. Guyomard-Lack A, Abusleme J, Soudan P, Lestriez B, Guyomard D, Bideau JL. Hybrid silica-polymer ionogel solid electrolyte with tunable properties. Adv Energy Mater. 2014; 4 (8):1301570.
15. Lodge TP. Materials science: A unique platform for materials design. Science . 2008; 321(5885):50–1.
16. Yong H, Park H, Jung C. Quasi-solid-state gel polymer electrolyte for a wide temperature range application of acetonitrile-based supercapacitors. J Power Source. 2020; 447:227390.

17. Palchoudhury S, Ramasamy K, Gupta RK, Gupta A. Flexible supercapacitors: A materials perspective. Front Mater. 2019; 83: 5.

18. Zhao K, Song H, Duan X, Wang Z, Liu J, Ba X. Novel chemical cross-linked ionogel based on acrylate terminated hyperbranched polymer with superior ionic conductivity for high performance lithium-ion batteries. Polymers. 2019;11(3):444.

19. Lewandowski A, Swiderska-Mocek A. Ionic liquids as electrolytes for Li-ion batteries—An overview of electrochemical studies. J Power Source 2009;194(2)601–9.

20. Hyun WJ, de Moraes ACM, Lim JM, Downing JR, Park KY, Tan MTZ *et al*. High-modulus hexagonal boron nitride nanoplatelet gel electrolytes for solid-state rechargeable lithium-ion batteries. ACS Nano. 2019;13(8):9664–72.

21. Iijima S. Helical microtubules of graphitic carbon. Nature. 1991; 354:56.

22. Yu M, Lourie O, Dyer MJ, Moloni K, Kelly TF, Ruoff RS. Strength and breaking mechanism of multiwalled carbon nanotubes under tensile load. Science. 2000; 287(5453):637–40.

23. Berber S, Kwon YK, Tomanek D. Unusually high thermal conductivity of carbon nanotubes. Phys Rev Lett. 2000; 84 (20):4613–16.

24. Kim P, Shi L, Majumdar A, McEuen PL. Thermal transport measurements of individual multiwalled nanotubes. Phy Rev Lett. 2001; 87 (21): 215502.

25. Torres-Dias AC. From mesoscale to nanoscale mechanics in single-wall carbon nanotubes. Carbon. 2017; 123:145–50.

26. Karousis N, Tagmatarchis N, Tasis D. Current progress on the chemical modification of carbon nanotubes. Chem Rev. 2010; 110 (9): 366–97.

27. Aliev AE, Oh J, Kozlov ME, Kuznetsov AA, Fang S, Fonseca AF. *et al*. Giant-stroke, superelastic carbon nanotube aerogel muscles. Science. 2009; 323:1575.

28. Biercuk MJ, Llaguno MC, Radosvljevic M, Hyun JK, Johnson AT. Carbon nanotube composites for thermal management. App Phys Lett. 2002; 80: 15.

29. Bai JB, Allaoui A. Effect of the length and the aggregate size of MWNTs on the improvement efficiency of the mechanical and electrical properties of nanocomposites–experimental investigation. Composites Part A: Appl Sci Manufact. 2003; 34(8)689–94.

30. Ko F, Gogotsi Y, Ali A, Naguib N, Ye H, Yang G. *et al*. Electrospinning of continuous carbon nanotube-filled nanofiber yarns. Adv Mater. 2003; 15(14):1161–65.

31. Odegard GM, Gates TX, Wise KE, Park C, Siochi EJ. Constitutive modeling of nanotube-reinforced polymer composites. Compos Sci Technol. 2003; 63:1671–87.

32. Bai JB, Allaoui A. Effect of the length and the aggregate size of MWNTs on the improvement efficiency of the mechanical and electrical properties of nanocomposites–experimental investigation. Composites Part A: Appl Sci Manufact. 2003. 34 (8):689–94.

33. Bai J. Evidence of the reinforcement role of chemical vapour deposition multi-walled carbon nanotubes in a polymer matrix. Carbon. 2003; 41:1309–28.

34. Dalton AB, Collins S, Munoz E, Razal JM, Ebron VH, Ferraris JP. *et al*. Super-tough carbon-nanotube fibres. Nature. 2003; 423:703.

35. Thostenson ET, Chou TW. Aligned multi-walled carbon nanotube-reinforced composites: processing and mechanical characterization. J Phys D: Appl Phys. 2002; 35:77–80.

36. Bae J, Jang J, Yoon SH. Cure behavior of the liquid-crystalline epoxy/carbon nanotube system and the effect of surface treatment of carbon fillers on cure reaction. Macromol Chem Phys. 2002; 203(15):2196–204.

37. Liu S, Tian J, Wang L, Zhang Y, Qin X, Luo Y. *et al*. Hydrothermal treatment of grass: a low-cost, green route to nitrogen-doped, carbon-rich, photoluminescent polymer nanodots as an effective fluorescent sensing platform for label-free detection of Cu (II) ions. Adv Mater. 2012; 24: 2037–41.

38. Shoujun Z, Yubin S, Xiaohuan Z, Jieren S, Junhu Z, Bai Y. The photoluminescence mechanism in carbon dots (graphene quantum dots, carbon nanodots and polymer dots): current state and future perspective. Nano Res. 2014; 8: 355–381.

39. Lyu Y, Fang Y, Miao Q, Zhen X, Ding D, Pu K. Intraparticle molecular orbital engineering of semiconducting polymer nanoparticles as amplified theranostics for in vivo photoacoustic imaging and photothermal therapy. ACS Nano. 2016; 10:4472 –81.

40. Feng L, Liu L, Lv F, Bazan GC, Wang S. Preparation and biofunctionalization of multicolor conjugated polymer nanoparticles for imaging and detection of tumor cells. Adv Mater. 2014; 26:3926–30.

41. Liu SG, Luo D, Li N, Zhang W, Lei JL, Li NB, *et al*. Water-soluble nonconjugated polymer nanoparticles with strong fluorescence emission for selective and sensitive detection of nitro-explosive picric acid in aqueous medium. ACS Appl Mater Interfaces. 2016; 8(33):21700–9.

42. Qiao ZA, Huo Q, Chi M, Veith GM, Binder AJ, Dai S. A "ship-in-a-bottle" approach to synthesis of polymer dots@silica or polymer dots@carbon core-shell nanospheres. Adv Mater. 2012; 24:6017–21.

43. Zhu S, Zhang J, Wang L, Song Y, Zhang G, Wang H. *et al*. A general route to make non-conjugated linear polymers luminescent. Chem Commun. 2012; 48:10889–91.

44. BelBruno JJ. Molecularly imprinted polymers. Chem Rev. 2019; 119:94–119.

45. Aghaei A, Milani Hosseini MR, Najafi M. A novel capacitive biosensor for cholesterol assay that uses an electropolymerized molecularly imprinted polymer. Electrochim Acta. 2010; 55:1503–8.

46. Shekarchizadeh H, Ensafi AA, Kadivar M. Selective determination of sucrose based on electropolymerized molecularly imprinted polymer modified multiwall carbon nanotubes/glassy carbon electrode. Mater Sci Eng C. 2013; 33:3553–61.

47. Poma A, Guerreiro A, Whitcombe MJ, Piletska EV, Turner APF, Piletsky SA. Solid-phase synthesis of molecularly imprinted polymer nanoparticles with a reusable template-plastic antibodies. Adv Funct Mater. 2013; 23:2821–7.

48. Svenson J, Nicholls IA. On the thermal and chemical stability of Molecularly imprinted polymers. Anal Chim Acta. 2001; 435:19–24.

49. Pedroso MM, Foguel MV, Silva DHS, Sotomayor MPT, Yamanaka H. Electrochemical sensor for dodecyl gallate determination based on electropolymerized molecularly imprinted polymer. Sens Actuators B Chem. 2017; 253:180–6.

50. Jafari S, Dehghani M, Nasirizadeh N, Azimzadeh M. An azithromycin electrochemical sensor based on an aniline MIP film electropolymerized on a gold nano urchins/graphene oxide modified glassy carbon electrode. J Electoanal Chem. 2018; 829:27–34.

51. Weber P, Riegger BR, Niedergall K, Tovarb GEM, Bach M, Gauglitz G. Nano-MIP based sensor for penicillin G: sensitive layer and analytical validation. Sens Actuators B Chem. 2018; 267:26–33.

52. Alizadeh T, Atashi F, Ganjali MR. Molecularly imprinted polymer nano-sphere/multi-walled carbon nanotube coated glassy carbon electrode as an ultra-sensitive voltammetric sensor for picomolar level determination of RDX. Talanta. 2019; 194:415–21.

53. Guney S, Guney O. Development of an electrochemical sensor based on covalent molecular imprinting for selective determination of bisphenol-A. Electroanalysis. 2017; 29:2579–90.

54. Akhoundian M, Alizadeh T, Ganjali MR, Norouzi P. Ultra-trace detection of methamphetamine in biological samples using FFT-square wave voltammetry and nano-sized imprinted polymer/MWCNTs -modified electrode. Talanta. 2019; 200:115–23.

55. Akhoundian M, Alizadeh T, Ganjali MR, Rafiei F. A new carbon paste electrode modified with MWCNTs and nano-structured molecularly imprinted polymer for ultratrace determination of trimipramine: the crucial effect of electrode components mixing on its performance. Biosens Bioelectron. 2018;11: 27–33.

56. Chen L, Wang X, Lu W, Wu X, Li J. Molecular imprinting: perspectives and applications. Chem Soc Rev. 2016; 45:2137–211.

57. He F, Jiang Y, Ren C, Dong G, Gan Y, Lee MJ. *et al*. Generalized electrical conductivity relaxation approach to determine electrochemical kinetic properties for MIECs. Solid State Ion. 2016; 297:82–92.

58. Nezhadali A, Mojarrab M. Fabrication of an electrochemical Molecularly imprinted polymer triamterene sensor based on multivariate optimization using multi-walled carbon nanotubes. J Electroanal Chem. 2015; 744:85–94.

59. Pan Y, Shang L, Zhao F, Zeng B. A novel electrochemical 4-nonyl-phenolsensor based on molecularly imprinted poly(o-phenylenediamine-co-o-toluidine)-nitrogendoped grapheme nano ribbons-ionic liquid composite film. Electrochim Acta. 2015; 151:423–8.

60. Rezaei B, Boroujeni MK, Ensafi AA. Development of Sudan II sensor based on modified treated pencil graphite electrode with DNA, o-phenylenediamine, and gold nanoparticle bioimprinted polymer. Sens Actuators B Chem. 2016; 222:849–56.

61. Ansari S. Combination of molecularly imprinted polymers and carbon nanomaterials as a versatile biosensing tool in sample analysis: recent applications and challenges. TrAC Trends Ana. Chem. 2017; 93:134–51.

62. Mayes AG, Mosbach K. Molecularly imprinted polymer beads: Suspension polymerization using a liquid perfluorocarbon as the dispersing phase. Anal Chem. 1996; 68(21):3769–74.
63. Hosoya K, Yoshizako K, Tanaka N, Kimata K, Anaki T, Haginaka J. Uniform-size macroporous polymer-based stationary phase for HPLC prepared through molecular imprinting technique. Chem Lett. 1994; 1437.
64. Mosbach K, Ramstrom O. The emerging technique of molecular iImprinting and its future impact on biotechnology. Biotechnol. 1996; 14:163.
65. Perez N, Whitcombe MJ, Vulfson EN. Molecularly imprinted nanoparticles prepared by core-shell emulsion polymerization. J Appl Polym. Sci. 2000; 77:1851.
66. Soysal M, Muti M, Esen C, Gencdag K, Aslan A, Erdem A. *et al.* A novel and selective methylene blue imprinted polymer modified carbon paste electrode. Electroanalysis. 2013; 25:1278–85.
67. Novak P, Muller K, Santhanam KSV, Haas O. Electrochemically active polymers for rechargeable batteries. Chem Rev. 1997; 97:207–82.
68. Manuel J, Kim JK, Matic A, Jacobsson P, Chauhan GS, Ha JK. Electrochemical properties of lithium polymer batteries with doped polyaniline as cathode material. Mater Res Bull 2012; 47:2815–818.
69. Manuel J, Kim M, Fapyane D, Chang IS, Ahn HJ, Ahn JH. Preparation and electrochemical properties of polyaniline nanofibers using ultrasonication. Mater Res Bull. 2014; 58:213–17.
70. Manuel J, Salguero T, Ramasamy R. Synthesis and characterization of polyaniline nanofibers as cathode active material for sodium-ion battery. J Appl Electrochem. 2019; 49:529–37.
71. Heinze J, Frontana-Uribe BA, Ludwigs S. Electrochemistry of conducting polymers - persistent models and new concepts. Chem Rev. 2010; 110(8):4724–71.
72. Zhou M, Heinze J. Electropolymerization of pyrrole and electrochemical study of polypyrrole. 3. Nature of Bwater effect in acetonitrile. J Phys Chem B. 1999; 103(40):8451–7.
73. Heinze J, Rasche A, Pagels M, Geschke B. On the origin of the so-called nucleation loop during electropolymerization of conducting polymers. J Phys Chem B. 2007; 111(5):989–97.
74. Janata J, Josowicz M. Conducting polymers in electronic chemical sensors. Nature Materials . 2003; 2(1):19–24.
75. Madden JD, Vandesteeg NA, Anquetil PA, Madden PG, Takshi A, Pytel AZ. *et al.* Artificial muscle technology: physical principles and naval prospects. IEEE J Oceanic Eng. 2004; 29:706.
76. Kornbluh R, Pelrine R, Pei Q, Heydt R, Stanford S, Oh S. *et al.* Electroelastomers: applications of dielectric elastomer transducers for actuation, generation, and smart structures. Proc SPIE EAPAD 2002; 4698:254.
77. Pons JL. Emerging actuator technologies: a micromechatronic approach. New Jersey: Wiley; 2005.
78. Kaneto K, Kaneko M, Min Y, MacDiarmid AG. Artificial muscle–electrochemical actuators using polyaniline films. Synth Met. 1995; 71: 211.
79. Yu Z, Yuan W, Brochu P, Chen B, Liu Z, Pei Q. Large-strain, rigid-to-rigid deformation of bistable electroactive polymers. Appl Phys Lett. 2009; 95:192904.
80. Yu Z, Niu X, Brochu P, Yuan W, Li H, Chen B. *et al.* Bistable electroactive polymers (BSEP): large-strain actuation of rigid polymers. Proc SPIE. 2010; 7642:76420.
81. Rasmussen L. Electroactivity in polymeric materials. New York: Springer; 2012.
82. Zihang P, Yu Q, Ziyang Z, Ye S, Adie A, Roshan P. *et al.* Bistable electroactive polymers for refreshable tactile displays. SPIE Digial Library.
83. Kagami Y, Gong JP, Osada Y. Shape memory behaviors of crosslinked copolymers containing stearyl acrylate. Macromol Rapid Commun. 1996; 17(8):539–43.
84. Matsuda A, Sato JI, Yasunaga H, Osada Y. Order-disorder transition of a hydrogel containing an n-alkyl acrylate. Macromolecules. 1994; 27(26):7695–8.
85. Plate NA, Shibaev VP, Petrukhin BS, Zubov YA, Kargin VA. Structure of crystalline polymers with unbranched long side chains. J Polym Sci Pol Chem. 1971; 9(8):2291–8.
86. Furukawa T. Ferroelectric properties of vinylidene fluoride copolymers. Phase Transit. 1989; 18:143–211.
87. Nalwa H. Ferroelectric Polymers. 9th ed. Boca Raton: CRC Press. 1995.
88. Takase Y, Lee JW, Scheinbeim JI, Newman BA. High-temperature characteristics of nylon-11 and nylon-7 piezoelectrics. Macromolecules. 1991; 24: 6644.
89. Su J, Ma ZY, Scheinbeim JI, Newman BA. Ferroelectric and piezoelectric properties of nylon 11/poly(vinylidene fluoride) bilaminate films. J Polym Sci B. 1995; 33:85.

90. Gao Q, Scheinbeim JI, Newman BA. Dipolar intermolecular interactions, structural development, and electromechanical properties in ferroelectric polymer blends of nylon-11 and poly(vinylidene fluoride). Macromolecules. 2000; 33:7564.

91. Furukawa T. Structure and functional properties of ferroelectric polymers. Adv Colloid Interface Sci. 1997; 71–72:183–208.

92. Bauer F. Ferroelectric polymers for high pressure and shock compression sensors. Mat Res Soc Symp Proc. 2002; 698.

93. Reynolds G. Piezoluminescence from a ferroelectric polymer and quartz. J Lumin. 1997; 75(4):295–9.

94. Pelrine RE, Kornbluh RD, Joseph JP. Electrostriction of polymer dielectrics with compliant electrodes as a means of actuation. Sens Actuators A. 1998; 64:77.

95. Krakovsky I, Romjin T, Posthuma de Boer A. A few remarks on the electrostriction of elastomers. J Appl Phys. 1999; 85:628.

96. Pelrine R, Kornbluh R, Joseph J, Chiba S. Electrostriction of polymer films for microactuators. IEEE Tenth Annual International Workshop on MEMS 238. 1997.

97. Pelrine R, Kornbluh R, Kofod G. High-strain actuator materials based on dielectric elastomers. Adv Mater. 2000; 12:1223.

98. Pelrine R, Kornbluh R, Joseph J, Heydt R, Pei Q, Chiba S. High-field deformation of elastomeric dielectrics for actuators. Mater Sci Eng C. 2000; 11:89.

99. Rogers JA. A clear advance in soft actuators. Science. 2013; 341(6149):968–9.

100. Liu Y, Gao M, Mei S, Han Y, Liu J. Ultra-compliant liquid metal electrodes with in-plane self-healing capability for dielectric elastomer actuators. Appl Phys Lett. 2013; 103(6):064101.

101. Frederikke BM, Daugaard AE, Hvilsted S, Skov AL. The current state of silicone-based dielectric elastomer transducers. Macromol Rapid Commun. 2016; 37(5):378–413.

102. Koh SJA, Keplinger C, Li T, Bauer S, Suo Z. Dielectric elastomer generators: How much energy can be converted. Trans Mechatron. 2011;16(1):33–41.

103. Sessler GM. Electrets. 3rd ed. 1. Berlin: Laplacian Press. 1998.

104. Bauer S. Piezeo-, pyro- and ferroelectrets: soft transducer materials for electromechanical energy conversion. IEEE Trans Dielectr Electr Insul. 2006; 13:953.

105. Cheng Z, Zhang Q. Field-activated electroactive polymers. MRS Bull. 2008; 33:183.

106. Bauer S, Gerhard-Multhaupt R, Sessler G. Ferroelectrets: soft electroactive foams for transducers. Phys Today . 2004; 57:37.

107. Lovera D, Ruckdaschel H, Goldel A, Behrendt N, Frese T, Sandler JKW. *et al.* Tailored polymer electrets based on poly(2,6-dimethyl-1,4- phenylene ether) and its blends with polystyrene. Eur Polym J. 2007; 43:1195.

108. de Gennes PG. A semi-fast artificial muscle. CR Acad Sci Paris Se. II B. 1997; 324:343.

4 Intelligent Electroactive Polymers

Sapana Jadoun and Ufana Riaz

CONTENTS

4.1 INTRODUCTION

Before 1960, researchers prepared some conjugated polymers in the semiconductor field, which were used as insulators. Heeger *et al.* (1977) presented the first conducting polymer, iodine (I_2) doped poly(acetylene) (PAc). Its conductivity was 10^3 S/cm. The synthesis of PAc and the enhancement in its conductivity after I_2 doping developed a new area of research (Inzelt, 2012). Hideki Shirakawa, Alan J. Heeger, and Alan MacDiarmid received the Nobel prize in Chemistry in 2000 for discovering the conductivity in PAc (Wallace *et al.,* 2008). This discovery opened a new area of research for the commercialization of these polymers as electrical conductors. These new conducting polymers have numerous applications in the field of sensors, actuators, molecular electronics, supercapacitors, electrochromic windows, corrosion protection, photovoltaics, and optoelectronics (Namsheer and Rout, 2021).

Electroactive polymers (EAPs) are materials that have conjugated π bonds in which electrons are delocalized along their polymeric backbone to conduct electricity (Guo and Facchetti, 2020). Intelligent EAPs are the class of polymers that can be excited by the electrical field that causes them to change their shape and size. These have received recent attention from industry and academia in polymer science and are designed based on a response or stimulus behavior. They can transfer electrons in the electric field, and therefore, have applications in various fields, such as biosensors, actuators, artificial muscles, soft robotic, and solar materials. They have several advantages including they are easy to process, have low density and mechanical flexibility, and can be mass produced. In addition, they can sense mechanical strain and produce electrical energy (Khan and Alamry, 2020).

A lot of biological systems have the ability of sensing, evaluating, and responding and are considered to be intelligent. A variety of intelligent behaviors can be seen in some types of materials, which are known as stimuli-responsive polymers. Some electroactive polymers possess conductivity and can monitor and respond to the surrounding chemical environment, which are known as intelligent EAPs. These polymers show a mechanical response to electrical stimulation and can respond to any change in the environment of living or non-living things, which is known as called actuation.

DOI: 10.1201/9781003173502-4

These polymers can be categorized as ionic or electronic. Numerous EAPs demonstrated a fast electromechanical response under electrical stimulation and mimicked natural muscles, and therefore, were called artificial muscles. They can change their shape and size in response to appropriate stimuli that can be electrical, thermal, chemical, or magnetic (X. Feng *et al.*, 2019; Perković *et al.*, 2020; S. Feng *et al.*, 2019; Wells *et al.*, 2019).

In this chapter, intelligent EAPs are summarized based on their classification and applications. In addition, conducting EAPs are discussed.

4.2 INTELLIGENT ELECTROACTIVE POLYMERS

Recently, polymers have become one of the most promising and attractive materials for biomedical applications due to their unique features and biocompatibility. The flexibility in their functionalization and properties makes them the best candidate for biodegradable thermoplastic polymers (Song *et al.*, 2018). In addition, various features allow numerous applications of these polymers. Of these, EAPs are considered to be a new generation of intelligent polymers that can respond to an electric field and change their properties for biomedical applications (Figure 4.1). As a polymeric material, these polymers include an anionic conductive hydrogel, percolated nanocomposites, and intrinsically conducting polymers (Kirillova and Ionov, 2019).

EAPS have attracted attention because many electro-sensitive tissues are present in the human body, such as the skin, heart, bones, and blood vessels and they have been used in tissue engineering due to their advantageous properties. They can be prepared in numerous sizes and shapes with the desired morphology and have a wide range of chemical and physical properties. Many review articles have been published on conducting polymers (CPs) (Jadoun and Riaz, 2020a, b; Jangid, Jadoun, and Kaur, 2020; Jadoun *et al.*, 2017; Khokhar, Jadoun, Arif, and Jabin, 2020; Jangid *et al.*, 2020; Jadoun, Riaz, and Budhiraja, 2020; Bach-Toledo *et al.*, 2020; R Murad *et al.*, 2020; Dunlop

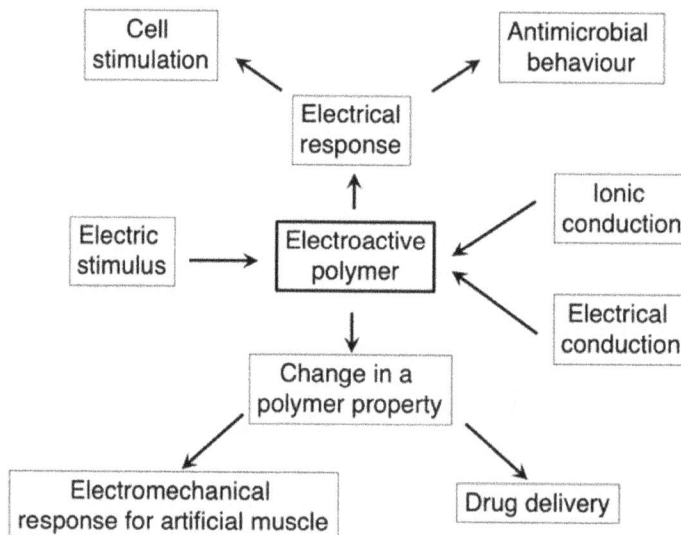

FIGURE 4.1 General representation of smart electroactive polymers: Its mechanism via ionic or electric conduction, mechanism proceed via electric current and producing cell stimulation or antimicrobial behavior or via the change in polymer property such as shape/size of polymer provided a response in electrochemical nature for artificial muscle and drug delivery.

(From Palza, Zapata, and Angulo-Pineda, 2019. With permission.)

and Bissessur, 2020) and piezoelectric polymers (Covaci and Gontean, 2020; Mishra *et al.*, 2019; Sappati and Bhadra, 2018). These intelligent materials can initiate suitable chemical responses when required and these chemical processes can be induced in CPs chemically or electrochemically.

To prepare these materials, the coating of non-conductive substance is carried out using a chemical polymerization method. During chemical polymerization, incorporation of counterions of oxidants occurs and there is the potential for counterion exchange; however, this type of synthesis restricted the chemical properties of the designed materials and electropolymerization offered a wide range of incorporations for counterion into the polymer. The advantage of electropolymerization is that it does not require an oxidant. Using anodic oxidation, a conducting film can be directly synthesized on the surface and these EAPs can be converted from an insulating to conducting state by various functionalization and doping techniques, which include charge injection at the polymeric interface, electrochemical doping, photodoping, chemical doping by charge transfer, and doping by acid–base chemistry (Pathiranage *et al.*, 2017; Guo and Ma, 2018; Zarren, Nisar, and Sher, 2019).

4.3 CLASSIFICATION OF ELECTROACTIVE POLYMERS

These polymers can be classified into two different groups:

1. Electronic electroactive polymers: these polymers are driven by Coulomb forces or electric and can be further classified into various types
   a. Ferroelectric polymers
   b. Dielectric electroactive polymers
   c. Electrostrictive graft elastomers
   d. Electrostrictive paper
   e. Electro viscoelastic elastomers
   f. Liquid crystal elastomer materials
2. Ionic electroactive polymers: these are driven by mobility or diffusion of ions and can be further classified into
   a. Ionic polymer gels
   b. Ionomeric polymer–metal composites
   c. Conductive Polymers
   d. Carbon Nanotubes
   e. Electrorheological Fluids

These polymers experience dramatic changes in some of the properties when placed in an electric field. A general representation of the electrical and ionic conduction electroactive polymers are shown in Figure 4.1 (Palza, Zapata, and Angulo-Pineda, 2019). In this chapter, conductive EAPs that behave intelligently when subjected to an electric field are focused on.

4.4 CONDUCTIVE ELECTROACTIVE POLYMERS

Conductive EAPs work on the principle of counter ion exchange during the redox process. The exchange of ions with an electrolyte induces a volume change when reduction and oxidation occur at the electrodes. CPs have the desirable electric and optical properties of semiconductors to metals and are easy to synthesize (Naarmann, 2004). The initial conducting polymer was discovered as polyacetylene but now polyaniline (PANI) (Stejskal and Gilbert 2002; Boeva and Sergeyev, 2014) and their derivativesm such as poly(o-phenylenediamine) (POPD) (Khokhar *et al.*, 2020; Khokhar *et al.*, 2021), polypyrrole (PPY) (Jain, Jadon, and Pawaiya, 2017), polythiophene (PTH) (Kaloni *et al.*, 2017), and poly(ethylenedioxy thiophene) (PEDOT) (Gueye *et al.*, 2020) have been

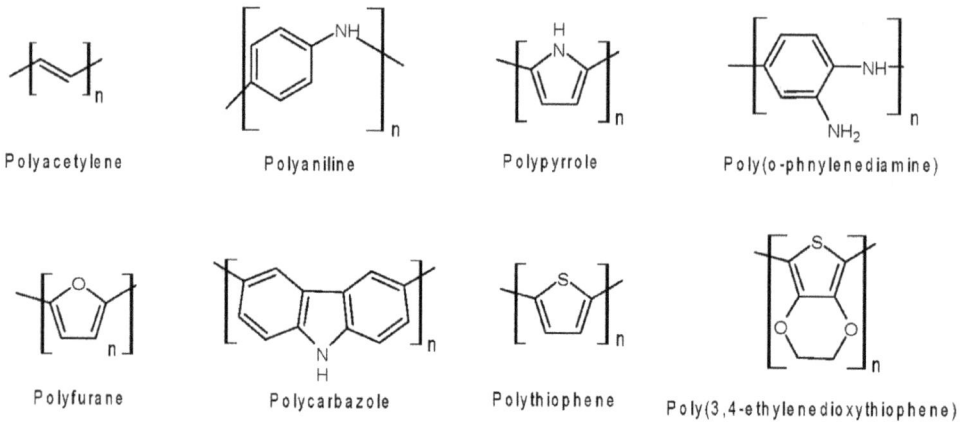

| Polyacetylene | Polyaniline | Polypyrrole | Poly(o-phnylenediamine) |

| Polyfurane | Polycarbazole | Polythiophene | Poly(3,4-ethylenedioxythiophene) |

FIGURE 4.2 Structures of some CPs and their derivatives.

(From Khokhar, Jadoun, Arif, and Jabin, 2020. With permission.)

extensively studied for their electrical conducting. The structures of some CPs and their derivatives are shown in Figure 4.2.

4.4.1 POLYANILINE

Polyaniline (PANI) is one of the extensively studied polymers that exists in three forms: (1) fully oxidized (pernigraniline base); (2) half-oxidized (emeraldine base); and (3) completely reduced (leucoemeraldine base). The half-oxidized form of PANI emeraldine base exhibits the highest level of stability and conductivity and is inexpensive and has high processibility (Jangid, Jadoun, and Kaur, 2020). In addition, PANI exhibits the charge transport phenomena via doping and dedoping and is biocompatible for in vivo and in vitro applications. This polymer could be used in commercial applications, because it has good processibility, is stable and is inexpensive. However, chemical and electrochemical synthesis limit its solubility due to some structural defects that limit its practical applications (Ameen, Shaheer Akhtar, and Husain, 2010). Some researchers developed different approaches to overcome this problem and incorporated the conjugated and well-defined oligoaniline in the backbone of copolymers that resulted in the combined properties of an oligomer and polymer, such as film-forming ability and mechanical strength (Zhou *et al.*, 2006; Martin and Diederich, 1999). Huang *et al.* (2009) prepared electroactive polymers of polyimide via inserting amine-capped aniline trimers in between the backbone (Figure 4.3), which had higher electroactivity and could potentially be applied in anticorrosive coatings by forming a passive layer of metal oxide.

4.4.2 POLYPYRROLE

Polypyrrole (PPY) is the most explored CPs due to its easy synthesis (Jadoun, Biswal, and Riaz 2018), high electrical conductivity, and surface modification properties and it has shown high stability and stimuli-responsive properties. Due to its chemical stability, biocompatibility, and conductivity under physiological conditions, and high response in the electrical field, it is referred to as an intelligent electroactive material. Various studies showed its biocompatibility with osteoblasts and mesenchymal stem cells, nerve cells, myocardial cells, and endothelial cells. When repairing and regenerating damaged tissues, PPY was a promising material (Forciniti *et al.*, 2014). In the fabrication of a bending electroactive actuator, a sandwich of PPY and other conducting polymer electrodes with an electrolyte was used (Nezakati *et al.*, 2018; Pei and Inganäs, 1992).

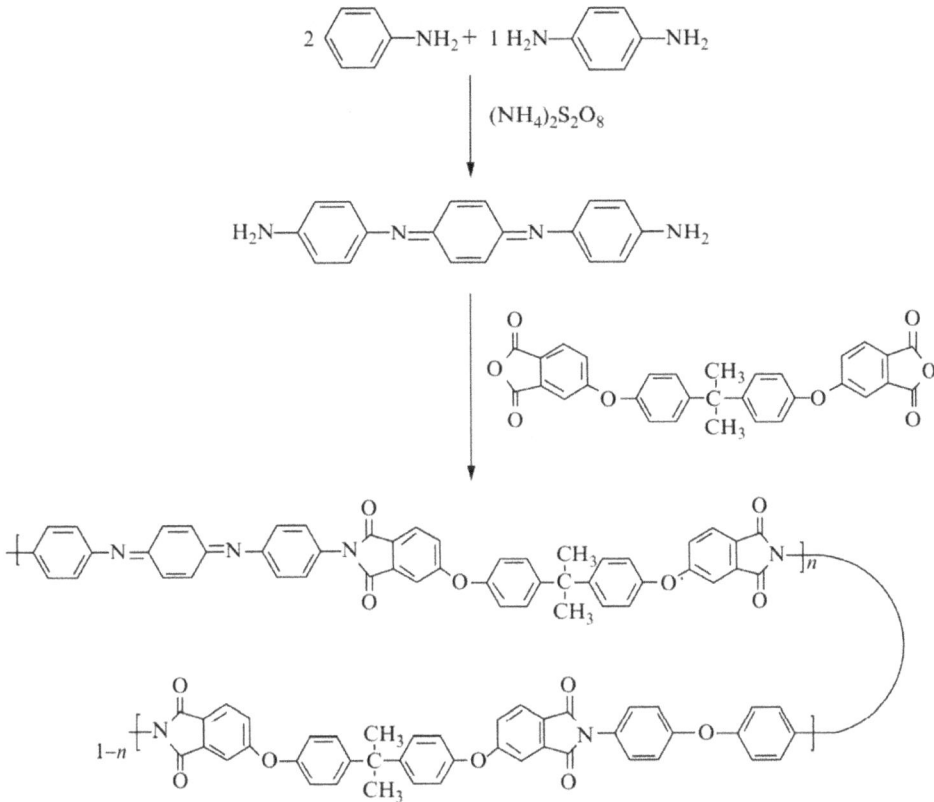

FIGURE 4.3 Representation for the synthesis of aniline trimers and polyimides.

(From Huang *et al.*, 2009. With permission.)

4.4.3 POLY(3,4-ETHYLENEDIOXYTHIOPHENE)

Poly(3,4-ethylenedioxythiophene) (PEDOT) is a very interesting CP that can be synthesized by chemical or electrochemical polymerization that uses an EDOT monomer. It is a derivative of polythiophene (PTH) that has superior properties to PTH, such as its redox potential, lower bandgap and it is more conducive and thermally established, which provides it with environmental, chemical, and electrical stability (Higgins *et al.*, 2012; Nambiar and Yeow, 2011). PEDOT has been used for preparing electrochemical transistors for biosensing (Mabeck and Malliaras, 2006), and it is a useful regenerative material due to the development of a neurotransmitter delivery system (Simon *et al.*, 2009). PEDOT has been employed to prepare biomaterials that have electrical conductivity and can make an electrical interface with tissues that are electrically responsive, such as the nervous system, heart, and skeletal muscles. In addition, PEDOT was used to coat neural cells and resulted in the fabrication of PEDOT-live cell electrodes and microelectrodes (Ghasemi-Mobarakeh *et al.*, 2011).

4.4.4 FUNCTIONALIZED CONDUCTING POLYMERS

The advantages of CPs can be enhanced by modifications or functionalization of them by composite formation, copolymerization, or by doping to make them more biocompatible for biomedical applications (Jadoun, Ashraf, and Riaz, 2018; Jadoun, Verma, and Riaz, 2018; Jadoun, Sharma, *et al.*, 2017; Riaz, Ashraf, *et al.*, 2017). Functionalization has enhanced the solubility, stability, electrical conductivity, mechanical stability, and the power to dispense biological moiety (Jadoun,

Ashraf, and Riaz, 2017; Riaz, Jadoun, *et al.*, 2017; Riaz *et al.*, 2016). For example, Riaz *et al.* (2019) modified POPD by doping it with dyes to make the materials water-soluble and they could be applied for the detection of BSA and bioimaging. Some researchers used PVDF as fillers with PANI and PPY enhancement in processability (Merlini *et al.*, 2014; Saïdi *et al.*, 2013). A lot of in vitro studies have been performed on these functionalized conducting materials, which showed responses after interacting with cells and tissues, which made them suitable for applications in the biomedical field (Riaz et al. 2018). Where electrical conductivity is needed for cell growth, these are promising for use as medical implant materials, such as in sensing, neural stimulation, and regenerating tissues.

4.5 APPLICATIONS

Intelligent EAPs are widely applied in the biomedical fields and other fields, such as corrosion protection, and aerospace applications. The applications of intelligent EAPs in numerous biomedical fields, such as sensors, drug release, robotics, artificial muscles, tissue engineering, and actuators, are shown in Figure 4.4.

The intelligent EAPs provide various types of electrical stimulation to cells, and therefore, are potential candidates for tissue engineering and the healing of bones and nerve regeneration. These polymers recently attracted a lot of attention due to their use in many sensitive tissues in the human body, such as nerves, skin, vessels, heart, and bones (Guo and Ma, 2018). PPY has shown stimuli-responsive properties and high stability under environmental conditions, which proved that the polymer was a biocompatible material that could be used in tissue engineering applications (Ravichandran *et al.*, 2012). In neural and cardiac tissue engineering, PPY was suitable due to its electrical conductivity. In addition, some two-dimensional substrates, and fibrous or porous scaffolds were promoted for cardiac differentiation using PANI or PPY (Kai et al. 2011; Qazi et al. 2014). PANI is suitable for in vivo applications (Mattioli-Belmonte *et al.*, 2003). Nanofibers of PANI and gelatin were synthesized by blending them to favor H9c2 cell adhesion and growth (Guo *et al.*, 2012), which demonstrated the use of PANI for nerve and cardiac tissue regeneration. Therefore, using CPs or their composites are good alternatives for materials in tissue engineering applications. Some potential applications of intelligent EAPs in tissue regeneration are shown in Figure 4.5.

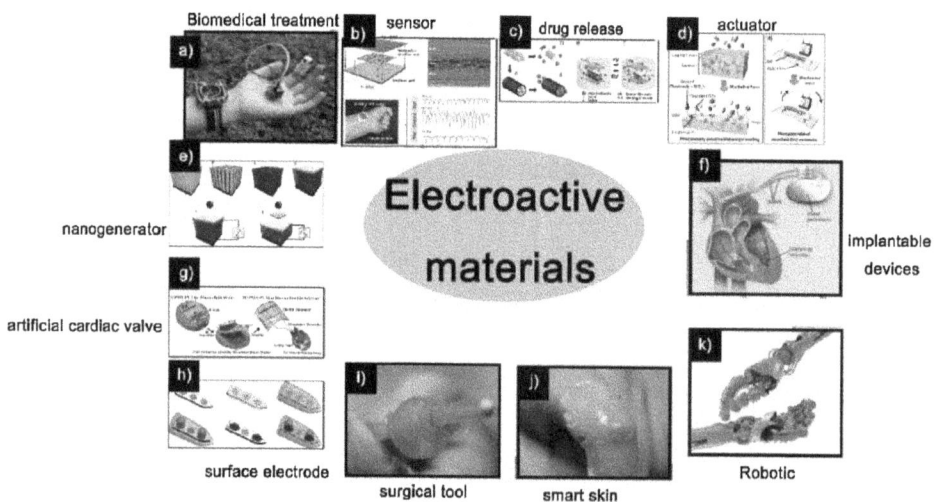

FIGURE 4.4 Applications of intelligent electroactive polymers in various biomedical fields.

(From Ning *et al.*, 2018. With permission.)

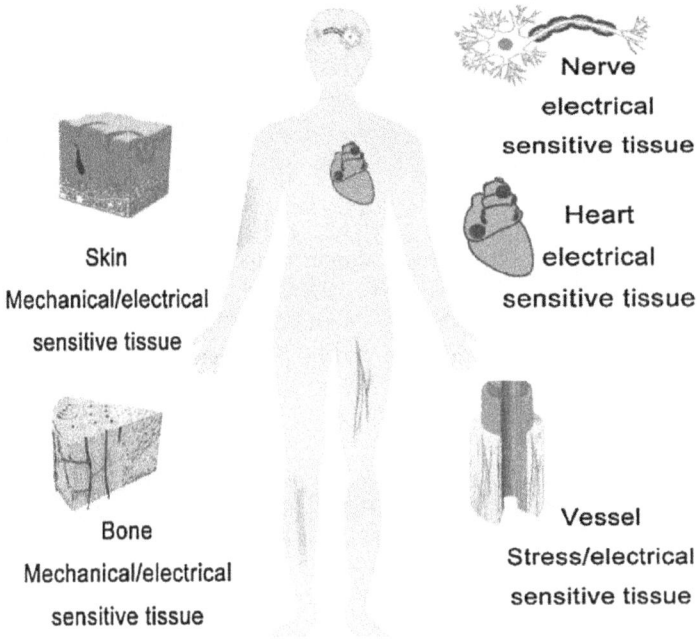

FIGURE 4.5 Potential applications of intelligent electroactive polymers in tissue engineering. **(From Ning *et al.*, 2018. With permission.)**

When chemical interactions between materials occurs with the surrounding atmosphere, the deterioration of the materials takes place, which is known as corrosion. The protection of materials can be carried out by various coatings and intelligent EAPS are one of these best substitutes (Chang and Yeh, 2015). In these polymers, an amine-capped aniline trimer was first evaluated for corrosion protection. It was prepared by an oxidative coupling reaction using the one-step method with aniline and p-phenylenediamine in an acidic aqueous solution (Huang *et al.*, 2009; Peng *et al.*, 2011).

Intelligent EAPs are very convenient and practical for actuation using electricity, because they can easily emulate biological muscles. These polymers provide approximately 20 J/cm³ of mechanical energy densities, which is quite high, and therefore, counted as effective actuation materials. Some researchers fabricated an effective actuator by sandwiching solid polymer electrolytes between two PEDOT electrodes. In addition, Ribeiro *et al.* (2018) studied solid-state CPs actuators. During the redox cycle, these actuators work via the insertion and expulsion of reversible counter ions (Otero *et al.*, 1993). Oxidation and reduction at the anode and cathode occurred because of the applied voltage between the electrodes, which resulted in the change in volume due to ion and electrolytes exchange (Madden, 2018) and by ions migration between the electrodes and electrolytes, and the electric charge was balanced. These ions swelled the polymer and after removing the ions, swelling reversed and resulted in the sandwich bending. Cui *et al.* (2020) reviewed the polymers-based bioinspired actuators, which included their design strategies, capabilities, and applications in robotic and biomedical areas.

4.6 CONCLUSION AND FUTURE PERSPECTIVES

Intelligent EAPs have shown significant potential for the development of biologically inspired unique devices. In addition, they have shown many potential advantages in tissue regeneration, actuating, and corrosion protection coatings. This chapter provides the reader with an understanding of general intelligent EAPs and their classification. It focused on conducting EAPs and summarized their applications in various fields. In the future, research should focus on the long-term stability and performance of EAPs by designing a water-resistance surface, which will prevent water loss from the polymers and decrease the positive counter ion loss during the functioning of them in an aqueous environment.

ACKNOWLEDGMENTS

The author Sapana Jadoun is grateful for the support of the National Research and Development Agency of Chile (ANID) and the projects FONDECYT Postdoctoral 3200850, FONDECYT 1191572 and ANID/FONDAP/15110019. The authors are also thankful to Elsevier, Springer, American Chemical Society, Taylor & Francis, and MDPI for copyright permissions.

REFERENCES

Ameen, S., Akhtar, M.S., and Husain, M. (2010) 'A review on synthesis processing, chemical and conduction properties of polyaniline and its nanocomposites', *Science of Advanced Materials*, 2(4), pp. 441–462.

Bach-Toledo, L. *et al.* (2020) 'Conducting polymers and composites nanowires for energy devices: A brief review', *Materials Science for Energy Technologies*, 3, pp. 78–90.

Boeva, Z.A, and Sergeyev V.G. (2014) 'Polyaniline: Synthesis, properties, and application', *Polymer Science Series C*, 56(1), pp. 144–153.

Chang, K.C., and Yeh J.M. (2015) 'Electroactive polymer-based anticorrosive coatings', in Tiwari, A., Rawlins, J., and Lloyd, H.B.T. (eds) *Intelligent Ccoating (for corrosion control Hihara)*. Boston: Butterworth-Heinemann, pp. 557–583.

Covaci, C., and Gontean A. (2020) 'Piezoelectric energy harvesting solutions: A review', *Sensors*, 20(12). 3512.

Cui, H., Zhao Q., Zhang L., and Du X. (2020) 'Intelligent polymer-based bioinspired actuators: From monofunction to multifunction', *Advanced Intelligent Systems*, 2(11), pp. 2000138. doi.10.1002/aisy.202000138.

Dunlop, M.J., and Bissessur R. (2020) 'Nanocomposites based on graphene analogous materials and conducting polymers: A review', *Journal of Materials Science*, 55(16), pp. 6721–6753.

Feng, S. *et al.* (2019) 'Review on smart gas sensing technology', *Sensors*, 19(17) p. 3760.

Feng, X., Li M., Li Y., and Ding J. (2019) '*Smart and functional Polymers*', Basel: Multidisciplinary Digital Publishing Institute.

Forciniti, L., Ybarra J., Zaman M.H., and Schmidt, C.E. (2014) 'Schwann cell response on polypyrrole substrates upon electrical stimulation', *Acta Biomaterialia*, 10(6), pp. 2423–2433. doi.10.1016/j.actbio.2014.01.030.

Ghasemi-Mobarakeh, L. *et al.* (2011) 'Application of conductive polymers, and electrical stimulation for nerve tissue engineering', *Journal of Tissue Engineering and Regenerative Medicine*, 5(4), e17–35. doi.10.1002/term.383.

Gueye, M.N. *et al.* (2020) 'Progress in understanding structure and transport properties of PEDOT-based materials: A critical review', *Progress in Materials Science*, 108., p. 100616.

Guo, B., and Ma, P.X. (2018) 'Conducting polymers for tissue engineering', *Biomacromolecules*, 19(6), pp. 1764–1782.

Guo, B. *et al.* (2012) 'Electroactive porous tubular scaffolds with degradability and non-cytotoxicity for neural tissue regeneration', *Acta Biomaterialia*, 8(1), pp. 144–153. doi.10.1016/j.actbio.2011.09.027.

Guo, X., and Facchetti, A. (2020) 'The journey of conducting polymers from discovery to application', *Nature Materials*, 19(9), pp. 922–928.

Higgins, M.J., Molino, P.J., Yue, Z., and Wallace, G.G. (2012) 'Organic conducting polymer–protein interactions', *Chemistry of Materials*, 24(5), pp. 828–839. doi:10.1021/cm203138j.

Huang, K-Y. *et al.* (2009) 'Electrochemical studies for the eElectroactivity of amine-capped aniline trimer on the anticorrosion effect of as-prepared polyimide coatings', *European Polymer Journal*, 45(2), pp. 485–493. doi.10.1016/j.eurpolymj.2008.10.033.

Inzelt, G. (2012) 'Historical background (or: there is nothing new under the sun)', *Conducting Polymers*, pp. 295–297.

Jadoun, S., Ashraf, S.M., and Riaz, U. (2017) 'Tuning the spectral, thermal and fluorescent properties of conjugated polymers: Via random copolymerization of hole transporting monomers', *RSC Advances*. doi:10.1039/c7ra04662f.

Jadoun, S., Ashraf, S.M., and Riaz, U. (2018) 'Microwave-assisted synthesis of copolymers of luminol with anisidine: Effect on spectral, thermal and fluorescence characteristics', *Polymers for Advanced Technologies*, 29(2), pp. 1007–1017.

Jadoun, S., Biswal L., and Riaz, U. (2018) 'Tuning the optical properties of poly(o-phenylenediamine-co-pyrrole) via template mediated copolymerization', *Designed Monomers and Polymers*, 21(1), pp.75–81. doi:10.1080/15685551.2018.1459078.

Jadoun, S., and Riaz, U. (2020a) 'Conjugated polymer light-emitting diodes', in *Polymers for Light-Emitting Devices and Displays*. doi:10.1002/9781119654643.ch4.

Jadoun, S., and Riaz, U. (2020b) 'A review on the chemical and electrochemical copolymerization of conducting monomers: Recent advancements and future prospects', *Polymer-Plastics Technology and Materials*, 59(5), pp. 484–504. doi:10.1080/25740881.2019.1669647.

Jadoun, S., Riaz U., and Budhiraja V. (2020) 'Biodegradable conducting polymeric materials for biomedical applications: A review', *Medical Devices & Sensors*. doi:10.1002/mds3.10141.

Jadoun, S., Sharma, V., Ashraf, S.M., and Riaz, U. (2017) 'Ponolytic doping of poly(1-naphthylamine) with luminol: Influence on spectral, morphological and fluorescent characteristics', *Colloid and Polymer Science*. doi:10.1007/s00396-017-4055-3.

Jadoun, S., Verma, A., Ashraf, S.M., and Riaz, U. (2017) 'A short review on the synthesis, characterization, and application studies of poly(1-naphthylamine): A seldom explored polyaniline derivative', *Colloid and Polymer Science*. doi:10.1007/s00396-017-4129-2.

Jadoun, S., Verma, A., and Riaz, U. (2018) 'Luminol modified polycarbazole and poly(o-anisidine): Theoretical insights compared with experimental data', *Spectrochimica Acta - Part A: Molecular and Biomolecular Spectroscopy*. doi:10.1016/j.saa.2018.06.025.

Jain, R., Jadon, N., and Pawaiya, A. (2017) 'Polypyrrole based next generation electrochemical sensors and biosensors: A review', *TrAC Trends in Analytical Chemistry*, 97, pp. 363–373.

Jangid, N.K., Jadoun, S., and Kaur, N. (2020) 'A review on high-throughput synthesis, deposition of thin films and properties of polyaniline', *European Polymer Journal*, 125, p. 109485.

Jangid, N. *et al.* (2020) 'Polyaniline-TiO2-based photocatalysts for dyes degradation', *Polymer Bulletin*. doi:10.1007/s00289-020-03318-w.

Kai, D., Prabhakaran, M.P., Jin, G., and Ramakrishna, S. (2011) 'Polypyrrole-contained electrospun conductive nanofibrous membranes for cardiac tissue engineering', *Journal of Biomedical Materials Research Part A*, 99(3), pp. 376–385. doi.10.1002/jbm.a.33200.

Kaloni, T.P., Giesbrecht, P.K., Schreckenbach, G., and Freund, M.S. (2017) 'Polythiophene: From fundamental perspectives to applications', *Chemistry of Materials*, 29(24), pp. 10248–10283.

Khan, A., and Alamry, K.A. (2020) 'Stimuli-responsive conducting polymer composites: Recent progress and future prospects', *Actuators: Fundamentals, Principles, Materials and Applications*. pp. 159–186.

Khokhar, D., Jadoun, S., Arif, R., and Jabin, S. (2020) 'Functionalization of conducting polymers and their applications in optoelectronics', *Polymer-Plastics Technology and Materials*. doi:10.1080/25740881.2020.1819312.

Khokhar, D., Jadoun, S., Arif, R., and Jabin, S. (2021) 'Tuning the spectral, thermal and, morphological properties of poly(o-phenylenediamine-co-vaniline)', *Materials Research Innovations*. doi:10.1080/14328917.2020.1870330.

Khokhar, D., *et al.* (2020) 'Copolymerization of o-phenylenediamine and 3-aAmino-5-methylthio-1H-1,2,4-triazole for tuned optoelectronic properties and its antioxidant studies', *Journal of Molecular Structure*, p. 129738. doi.10.1016/j.molstruc.2020.129738.

Kirillova, A., and Ionov, L. (2019) 'Shape-changing polymers for biomedical applications', *Journal of Materials Chemistry B*, 7(10), pp. 1597–1624.

Mabeck, and J.T, Malliaras, G.G. (2006) 'Chemical and biological sensors based on organic thin-film transistors', *Analytical and Bioanalytical Chemistry*, 384(2), pp.343–353.

Madden, J.D.W. (2018) '25 years of conducting polymer actuators: History, mechanisms, applications, and prospects', University of British Columbia, Canada.

Martin, R.E., and Diederich, F. (1999) 'Linear monodisperse Π-conjugated oligomers: Model compounds for polymers and more', *Angewandte Chemie International Edition*, 38(10), pp. 1350–1377.

Mattioli-Belmonte, M. *et al.* (2003) 'Tailoring biomaterial compatibility: In vivo tissue response versus in vitro cell behavior', *The International Journal of Artificial Organs*, 26(12), pp. 1077–1085. doi:10.1177/039139880302601205.

Merlini, C., Barra, G.M.O., Medeiros Araujo, T., and Pegoretti, A. (2014) 'Electrically pressure sensitive poly (vinylidene fluoride)/polypyrrole electrospun mats', *RSC Advances*, 4(30), pp. 15749–15758.

Mishra, S., Unnikrishnan, L., Nayak, S.K., and Mohanty, S. (2019) 'Advances in piezoelectric polymer composites for energy harvesting applications: A systematic review', *Macromolecular Materials and Engineering*, 304(1), pp. 1800463.

Murad, R. *et al.* (2020) 'Conducting polymers for optoelectronic devices and organic solar cells: A review', *Polymers*, 12(11), p. 2627.

Naarmann, H. (2004) 'Conducting polymers', in Handbook of Polymer Synthesis, 2nd edn. doi:10.1201/9781315101217-2.

Nambiar, S., and Yeow, J.T.W. (2011) 'Conductive polymer-based sensors for biomedical applications', *Biosensors and Bioelectronics*, 26(5), pp. 825–832. doi.10.1016/j.bios.2010.09.046.

Namsheer, K., and Rout, C.S (2021) 'Conducting polymers: A comprehensive review on recent advances in synthesis, properties and applications', *RSC Advances*, 11(10), pp. 5659–5697.

Nezakati, T., Seifalian, A., Tan, A., and Seifalian, A.M. (2018) 'Conductive polymers: Opportunities and challenges in biomedical applications', *Chemical Reviews*, 118(14), pp. 6766–6843.

Ning, C. *et al.* (2018) 'Electroactive polymers for tissue regeneration: Developments and perspectives', *Progress in Polymer Science*, 81, pp. 144–162.

Otero, T.F., Rodriguez, J., Angulo, E., and Santamaria, C. (1993) 'Artificial muscles from bilayer structures', *Synthetic Metals*, 57(1), pp. 3713–3717.

Palza, H., Zapata, P.A., and Angulo-Pineda, C. (2019) 'Electroactive smart polymers for biomedical applications', *Materials*. doi:10.3390/ma12020277.

Pathiranage, T.M.S.K. *et al.* (2017) 'Role of polythiophenes as electroactive materials', *Journal of Polymer Science Part A: Polymer Chemistry*, 55(20), pp. 3327–3346.

Pei, Q., and Inganas, O. (1992) 'Electrochemical applications of the bending beam method. 1. Mass transport and volume changes in polypyrrole during redox', *The Journal of Physical Chemistry*, 96(25), pp. 10507–10514.

Peng, C-W. *et al.* (2011) 'Electrochemical corrosion protection studies of aniline-capped aniline trimer-based electroactive polyurethane coatings', *Electrochimica Acta*, 58, pp. 614–620. doi.10.1016/j.electacta.2011.10.002.

Perković, T. *et al.* (2020) 'Smart parking sensors: State of the art and performance evaluation', *Journal of Cleaner Production*, 262, pp. 121181.

Qazi, T.H. *et al.* (2014) 'Development and characterization of novel electrically conductive PANI–PGS composites for cardiac tissue engineering applications', *Acta Biomaterialia*, 10(6), pp. 2434–2445. doi.10.1016/j.actbio.2014.02.023.

Ravichandran, R. *et al.* (2012) 'Advances in polymeric systems for tissue engineering and biomedical applications', *Macromolecular Bioscience*, 12(3), pp. 286–311.

Riaz, U. *et al.* (2016) 'Microwave-assisted green synthesis of some nanoconjugated copolymers: Characterisation and Ffluorescence quenching studies with bovine serum albumin', *New Journal of Chemistry*, 40(5), pp. 4643–4653. doi:10.1039/C5NJ02513C.

Riaz, U., Ashraf, S.M., Fatima, T., and Jadoun, S. (2017) 'Tuning the spectral, morphological and photophysical properties of sonochemically synthesized poly(carbazole) using acid orange, fluorescein and rhodamine 6G', *Spectrochimica Acta - Part A: Molecular and Biomolecular Spectroscopy*. doi:10.1016/j.saa.2016.11.003.

Riaz, U. *et al.* (2017) 'Influence of luminol doping of poly(o-phenylenediamine) on the spectral, morphological, and fluorescent properties: A potential fluorescent marker for early detection and diagnosis of *Leishmania donovani*', *ACS Applied Materials and Interfaces*, 9(38), pp. 33159–33168. doi:10.1021/acsami.7b10325.

Riaz, U. *et al.* (2018) 'Microwave-assisted facile synthesis of poly(luminol-co-phenylenediamine) copolymers and their potential application in biomedical imaging', *RSC Advances*. doi:10.1039/c8ra08373h.

Riaz, U. *et al.* (2019) 'Spectroscopic and biophysical interaction studies of water-soluble dye modified poly(o-phenylenediamine) for its potential application in BSA detection and bioimaging', *Scientific Reports*, 9(1), p. 8544.

Ribeiro, F.B. *et al.* (2018) 'All-solid state ionic actuators based on polymeric ionic liquids and electronic conducting polymers', *Electroactive Polymer Actuators and Devices (EAPAD) XX*, 10594:105941H.

Saidi, S. *et al.* (2013) 'Effect of PANI rate percentage on morphology, structure and charge transport mechanism in PANI–PVDF composites above percolation threshold', *Journal of Physics D: Applied Physics*, 46(35), p. 355101.

Sappati, K.K., and Bhadra, S. (2018) 'Piezoelectric polymer and paper substrates: A review', *Sensors*, 18(11), p. 3605.

Simon, D.T. *et al.* (2009) 'Organic electronics for precise delivery of neurotransmitters to modulate mammalian sensory function', *Nature Materials*, 8(9), pp. 742–746. doi:10.1038/nmat2494.

Song, R. *et al.* (2018) 'Current development of biodegradable polymeric materials for biomedical applications', *Drug Design, Development and Therapy*, 12, p. 3117.

Stejskal, J., and Gilbert, R.G. (2002) 'Polyaniline. Preparation of a conducting polymer (IUPAC Technical Report)', *Pure and Applied Chemistry*. doi:10.1351/pac200274050857.

Wallace, G.G., Teasdale, P.R.,. Spinks, G.M., and Kane-Maguire, L.A.P. (2008) *Conductive Electroactive Polymers: Intelligent Polymer Systems*. Boca Raton: CRC Press.

Wells, C.M. *et al.* (2019) 'Stimuli-responsive drug release from smart polymers', *Journal of Functional Biomaterials*, 10(3), p. 34.

Zarren, G., Nisar, B., and Sher, F. (2019) 'Synthesis of anthraquinone-based electroactive polymers: A critical review', *Materials Today Sustainability*, 5, p. 100019.

Zhou, Y. *et al.* (2006) 'Crystal structure and morphology of phenyl-capped tetraaniline in the leucoemeraldine oxidation state', *Journal of Polymer Science Part B: Polymer Physics*, 44(4), pp. 764–759. doi.org/10.1002/polb.20700.

5 History and Progress of Electroactive Polymers

Aniruddha Jaiswal and Kashish Gupta

CONTENTS

5.1 INTRODUCTION

In the modern technological world, polymers are replacing conventional materials that are made from metals and alloys in various areas, such as electronics, automobiles, and biomedical industries due to their light weight, easy manufacturing processes, less expensive, pliable, tolerant to fractures, and configurable into desirable shapes. Electroactive polymers (EAPs) are those that transform an electrical field into mechanical energy and have attracted interest in science and technology recently (Nassab et al., 2017; Jou et al., 2017; Zeng et al., 2017). EAPs are polymers that show changes in shape, or dimensions, or both when under the influence of an electric field; therefore, they are responsive to electrical stimulation. EAPs have advantages over conventional materials for automation, programmed control, and remote control.

EAPs are polymers that are discussed in this chapter in actuators, artificial muscles, electromagnetic systems, and energy generators. Some classes of EAPs can respond to mechanical strain and harvest electrical energy from it.

5.2 HISTORICAL BACKGROUND

The Origin of EAPs can be traced back to an experiment performed by Wilhelm Röntgen in 1880, when a natural rubber strip that was attached to a body weight was subjected to an electrostatic field. Under the applied electric field, the length of the rubber strip changed by a few centimeters; this

DOI: 10.1201/9781003173502-5

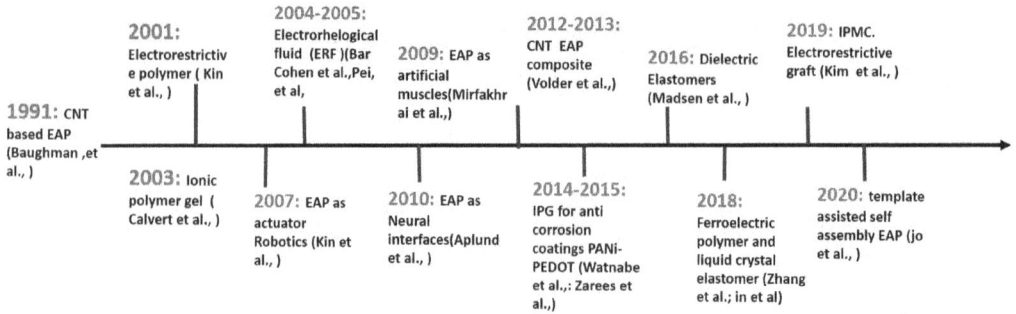

FIGURE 5.1 Historical overview in technological developments of EAPs.

is the first such report on the electroactivity of a polymeric material (Roentgen, 1880). Sacerdote (1899) put forward a theory on the strain response of a polymer under an applied electric field.

Electret, which was discovered in 1925, was the first such polymer that exhibited piezoelectric effects and produced an electric field when deformed and vice versa. Eguchi (1925)prepared it by cooling the mixture of rosin, carnauba wax, and beeswax together under an applied DC bias field.

In addition, EAPs cover polymers that respond to the natural stimulus other than the applied field and change their size or shape. Collagen fibers are such a polymer that is responsive to these conditions and show expansion or contraction under acidic and alkaline environments (Katchalsky et al., 1955). Osada (1991) and Osada et al. (1989) demonstrated that responsive gels were significant in this field.

Kawai (1969) discovered the large piezoelectric effect in polyvinylidene fluoride (PVDF) and attempted to discover similar polymeric systems that exhibit a greater piezoelectric effect. The applications of PVDF are limited as sensors and ultrasonic wave transducers due to the limited strains produced (Bauer and Bauer, 2008). The discovery of electrical conductivity in polyacetylene by Shirakawa et al. (1977) was another milestone. They demonstrated that doping of polyacetylene with iodine vapor causes enhancement in its conductivity up to eight times, which was close to the metal counterparts.

Research work in the early 1990s, led to the development of ionic polymer–metal composites (IPMCs), which are a synthetic composite nanomaterial, and opened avenues for the development of various other classes of EAPs, because IPMCs demonstrated electroactive properties that were superior to all the previously reported EAPs. IPMCs showed larger deformation at lower voltages (1 or 2 V) than previously reported EAPs. It requires low activation energy and exhibits strain ≤ 380% (Oguro et al., 1992; Nemat-Nasser et al., 2004).

The invention by employees of SRI International (SRI) led to the development of dielectric elastomer (DE) actuators. Pelrine et al. (2000) working at SRI demonstrated strain >100% with a fast response speed (<0.1 s).

To further advance the development of EAPs and to exchange the information collected, the first SPIE annual EAP Actuators and Devices (EAPAD) Conference was held in March 1999 under the aegis of the SPIE Smart Structures and Materials Symposium (Bar-Cohen, 1999). An EAP activated robotic arms wrestling challenge against humans proposed by Dr. Yoseph Bar-Cohen was part of this event to realize the potential of EAPs. The first commercial device that used an EAP was an artificial muscle developed by Eamex Corporation, Japan 2002. A detailed historical overview is shown in Figure 5.1.

5.3 TYPES OF ELECTROACTIVE POLYMER

Based on their mechanism of activation, EAPs can be classified into two major classes: (1) ionic; and (2) electronic. Ionic EAPs exhibit electrical activity due to the migration of ions, and electronic

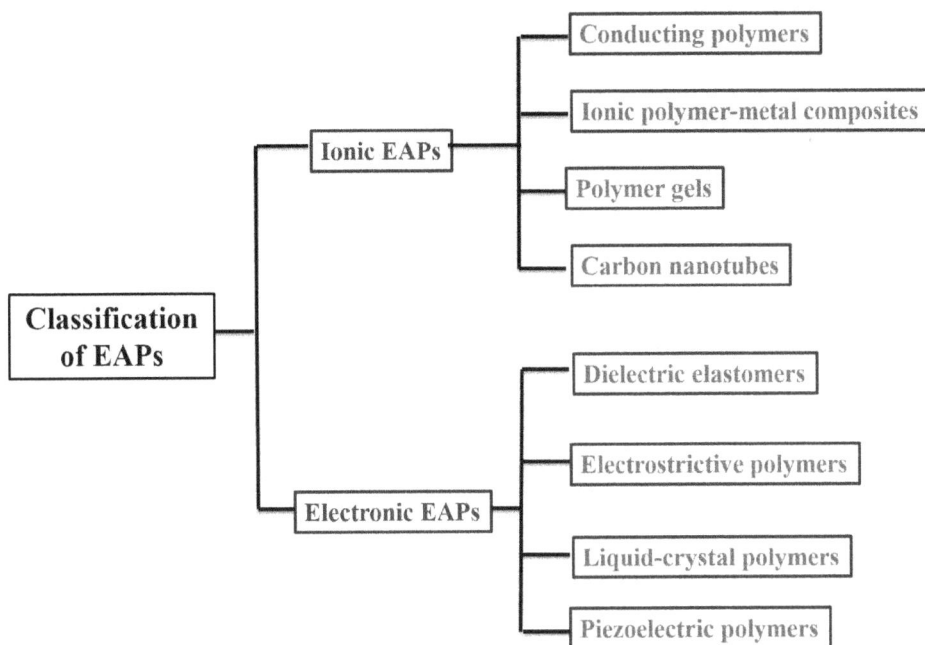

FIGURE 5.2   Classification of EAPs.

(field activated) EAPs have electrical activation because of applied electric fields and coulomb forces (Bar-Cohen, 2004). Details of both types of EAPs with further classification are shown in Figure 5.2.

5.3.1   IONIC ELECTROACTIVE POLYMERS

In general, ionic EAPs require low voltage for activation (1–2 V) and generate greater deformation or displacement (Jean-Mistral et al., 2010). Since activation is by the migration of ions or molecules (solvent), an ion reservoir is needed through which ions or molecules are transported (Park et al., 2008). To maintain the ionic flow and actuator at a given position, higher electrical energy is needed. Materials used for ionic EAPs are biocompatible, such as polypyrrole (PPy) and poly(3,4-ethylene dioxythiophene) (PEDOT), which can be applied in biological systems (Carpi et al., 2009). Ionic EAPs are further categorized into conducting polymers (CPs), ionic polymer–metal composites (IPMCs), and ionic polymer gels (Bar-Cohen, 1999, 2004; Jean-Mistral et al., 2010). Carbon nanotubes (CNTs) show features of ionic EAPs due to their excellent conductivity and charge transfer property, and were added to this group later (Baughman et al., 1999; Carpi, 2010).

5.3.1.1   Conducting Polymers

CPs are referred to as intrinsic conducting polymers (ICPs), which are organic polymers. CPs/ICPs have high electrical conductivity but do not have the mechanical features that are characteristics of commercial polymers. CPs are widely known due to research into the conducting features of poly-acetylene, for which the Nobel prize for Chemistry 2000 was awarded to Prof. H. Shirakawa, A.G. MacDiarmid, and A.J. Heeger (Shirakawa et al., 1977). CP based EAPs have mechanical energy densities of 20 J/cm³, and therefore, have the potential to be used as highly effective actuators (Madden, 2018).

Although CPs are semiconductors (or insulators), but on oxidation by some oxidizing agent (e.g., bromine or iodine gas) or oxidation in electrochemical cells, the conductivity of oxidized CPs are similar or slightly higher than metals. During oxidation, there is intercalation of heavy ions into the polymer network, which makes it swell. In electrochemical cell reactions, the conductivity of CPs can be controlled by the applied potential in the cell and there is uniform oxidation. These CPs are highly conductive, soft, flexible, and possess mechanical strength; therefore, they are suitable for electrically controllable soft actuators. This actuation by electrochemical cell reaction is called electro–chemo–mechanical deformation (ECMD). Trilayer structures based on CPs have emerged as substitutes for piezoelectric and electrostatic actuators (Wu et al., 2007; Festin et al., 2014; Otero et al., 2003 & 2003a).

Since the discovery of CPs, numerous nanomaterials, such as CNTs, graphene, inorganic, and polymeric materials have been incorporated to enhance their properties. Efforts have been made to use CPs in biomimetic applications. Ribeiro et al. (2018) reported recent progress on all solid-state CP actuators.

5.3.1.2 Ionic Polymer–Metal Composites

In 1993, a research group from Japan and USA independently reported the development of IPMCs as EAPs (Oguro et al., 1992; Sadeghipour et al., 1992; Shahinpoor et al., 2000). IPMCs show large bending deformation by charge migration and redistribution under the application of a small, applied voltage (1–5V). The response of the IPMC is <10Hz, because the ions take time to travel through the polymer.

IPMCs consist of a thin ionic polymer membrane (e.g., Nafion or Flemion) with a surface chemically coated or physically plated with a Nobel metal (e.g., gold or platinum). In contrast, if these deformations are externally applied to the strip, an output voltage signal of a few mV is generated. IPMCs have low impedance and have been applied in self-powered strain or deformation sensors, soft actuation applications, and for biomimetic robotic soft artificial muscles (Shahinpoor, 2016; Stalbaum et al., 2018).

5.3.1.3 Ionic Polymer Gels

Polymer gels are those stimuli-responsive gels that reversibly change their volume, optical, mechanical, and other properties in response to change to an external electric field, as reported by Katchalsky et al. (1955). It shows the reversible conversion of chemical energy into mechanical energy (chemo–mechanical behavior) in response to stimulation. Among them, the hydrogel has been studied for sensing applications (Dragan, 2014; Ullah et al., 2015). Swelling or shrinking of hydrogels by diffusion phenomenon led to changes in the volume. Some researchers prepared the combination of an ionic liquid with macromolecules to make actuators that were less sensitive to air(Watanabe et al., 2014). Ionic liquid-based polymer electrolytes consisting of block copolymers and polyimides are being developed to enhance the performance and durability of these actuators.

5.3.1.4 Carbon Nanotubes

CNTs are thermally conductive, one-dimensional, high surface area materials that possess high stiffness and strength and can also store charge and molecules. The first report of CNTs as an EAP actuator was reported in 1999 (Baughman et al., 1999; Qu et al., 2008; Spinks et al., 2004). CNT actuators are made to operate as bending devices in a three-layer strip, with two narrow CNT sheets on either side of an electrically insulating adhesive layer. Potential applied to CNT electrodes in electrolytes causes either insertion or expulsion of ions by the electrode; therefore, generating positive or negative strains. In contrast, the application of tensile stress in megapascals (MPa) causes voltage generation in charged multi-wall carbon nanotube (MWNT) yarns (Mirfakhrai et al., 2008).

5.3.1.5 Electrorheological Fluid

ERFs are ionic EAPs that change the viscosity of a solution under an applied electric field. It is a suspension of a polymer in a low dielectric constant liquid, and under an applied field the viscosity of the suspension increases. ERFs have found applications in shock absorbers, engine mounts, and acoustic dampers (Glass et al., 1991).

5.3.2 ELECTRONIC ELECTROACTIVE POLYMERS

In electronic EAPs, actuation is caused by the electrostatic force that is generated in between two electrodes and it requires higher input voltage (>10 V/μm), unlike the ion migration and low voltage, such as in ionic EAPs (Bar-Cohen, 2004; Cheng et al., 2008). Therefore, electronic EAPs take less time for response or relaxation (<1 ms) than ionic EAPs and can work at higher frequencies for dynamic sensing (Jean-Mistral et al., 2010).

There are two principal causes for electric- activation in electronic EAPs: (1) intrinsic field-induced molecular conformational changes, such as in ferroelectric polymers; and (2) extrinsic electronic charge attraction–repulsion phenomenon at the surface of the actuator electrode. Electronic EAPs are further divided into four subgroups: DEs, piezoelectric polymers, liquid crystal polymers, and electrostrictive polymers. (Carpi et al., 2015; Zhang et al., 1998; Lehmann et al., 2001; Ji et al., 2012; Nalwa et al., 1995).

5.3.2.1 Dielectric Elastomers

DEs are capacitors whose capacitance changes with an applied voltage. DE-based sensors have two compliant electrodes with a dielectric layer in between them. In the presence of an applied electric field, the polymer compresses, and the area increases, which leads to a change in the capacitance of the material. Dielectric EAPs do not require an external power source to sustain the actuator at the desired position. Modern methods of DE actuator fabrication involve spin coating, screen printing of the electrode, and use of a commercially available dielectric membrane material; however, these manufacturing techniques are still under development (Lotz et al., 2011; Maas et al., 2015; Fasolt et al., 2017).

5.3.2.2 Liquid Crystal Polymers

Liquid crystal (LC) polymers are an intermediate phase polymer of a crystalline state (i.e., highly ordered) and liquid (i.e., fully disordered) state, and therefore, they are partially crystalline and retain the property to flow and display crystal-like optical properties (i.e., birefringence) (Collings et al., 1997). Molecules of LC polymers are long-rigid rods and have anisotropic properties at the molecular level. By adding some conducting filler material (e.g., CNTs and magnetic nanoparticles) to liquid crystal elastomers (LCEs), they become responsive to an electric field. The applied electric field causes the flow of current and heats the materials. Shrinkage of the material happens along the LC director due to this increased disorder (i.e., thermal-induced actuation) and is reversible (Clarke et al., 2002; Chambers et al., 2009).

LCEs have great potential as mechanical actuators due to their ability to sustain large strain changes and tunability. However, costly, and complex manufacturing limits their commercialization.

5.3.2.3 Piezoelectric Polymers

The majority of the piezoelectric materials are inorganic, since they have a high piezoelectric strain constant (d). Piezoelectric polymeric materials have advantages over these conventional materials for their soft elasticity and cost effectiveness; (Feng et al., 2010) therefore, they are mainly applied in organic field effect transistors, flexible sensors, and energy harvesting devices(Lin et al., 2006; Kim et al., 2015; Li et al., 2010; Koc et al. 2013). Piezoelectric polymers have lower piezoelectric strain constants, but good piezoelectric voltage constants due to the low dielectric permittivity of the polymer. These polymeric materials are most suited for mechanical force and temperature gradients (Murat et al., 2013).

PVDF and its copolymers are some of the most successfully commercialized piezoelectric polymers (i.e., high electromechanical coupling of approximately 28 pCN^{-1}) (Crossley et al., 2014; Crossley et al., 2015; Seminara et al., 2011) and have wide bandwidth, fast electromechanical response, low acoustic and mechanical impedance, high voltage sensitivity, high strength, and high impact resistance.

5.3.2.4 Electrostrictive Graft Polymers

Electrostriction is defined mathematically as the quadratic coupling between strain (S_{ij}) and polarization (P_m). Electrostrictive graft polymers consist of a flexible backbone of a macromolecule (i.e., chain structure) and grafted polymer that is made up of a polarizable molecular or nanocrystalline structure (Su et al., 1999; Su et al., 2018). The polarizable section can be aligned under the influence of an applied field. An electrostrictive polymer uses two methods for energy harvesting via cycling and pseudo piezoelectric behavior. Electrostrictive polymers polyurethane and polyvinylidene fluoride-tri fluoroethylene (PVDF-TrFE) generate strain >10% under a moderate electric field of 20 V/μm, and therefore, can be used as potential actuators (Guiffard et al., 2006, 2009; Cottinet et al., 2010).

Piezoelectric polymers and electrostrictive graft polymers are classified as ferroelectric due to their permanent electrical polarization that changes under an applied electric field.

5.4 COMPARATIVE STUDY OF IONIC AND DIELECTRIC ELECTROACTIVE POLYMERS

Ionic EAPs require a low driving voltage (1–2 V) and have low electromechanical coupling. They operate in a wet environments or in solid electrolytes. Deformation in ionic EAPs is larger than electronic EAPs and this mechanism resembles to biological muscle deformation. These virtues favor the biomimetic applications of ionic EAPs. Ionic EAPs need constant wetting by the electrolyte to operate in air and require a constant DC voltage (i.e., high energy) to maintain the displacement (except for CPs and CNTs).

In contrast, the activation field required for dielectric EAPs is larger (>150V/μm for approximately 10% strain) at which the materials might break. These materials hold the induced displacement under the applied DC voltage, and therefore, are potential candidates for use in robotic applications and can be operated in the air without any major challenges. Dielectric EAPs have high mechanical energy density and a rapid response time (i.e., in m/sec) (Bar-Cohen et al., 2007).

5.5 APPLICATION AREAS FOR ELECTROACTIVE POLYMERS

EAP materials can be formed into different shapes, such as fibers, films, fabrics, and strips due to ease of processing. The integration of EAPs into micro electromechanical system (MEMS) produces smart actuators. They are applied most in artificial muscles, tactile displays (i.e., using the EAP actuator configured in an array form), and as micro fluids (i.e., a drug delivery system) (Kim et al., 2007; Bar-Cohen, 2009; Richter et al., 2009; Yu et al., 2003). EAP actuators have a low modulus and mechanical impedance compared with common optical membranes. Therefore, these actuators can be used in optical membrane technology.

5.6 ELECTROACTIVE POLYMERS FOR BIOMEDICAL APPLICATIONS

For biomedical application purposes, the EAPs used are CPs that are derived from electrochemical and interfacial polymerization. Polyaniline, polypyrrole, polythiophene and poly (3,4-ethylene dioxythiophene) are a few examples of CPs that are used in biomedical industries. These EAPs are further divided into biodegradable EAPs and non-biodegradable EAPs. In biomedical industries, biodegradable EAPs are used in devices that have a short duration, such as tissue scaffolds or drug delivery devices, and non-biodegradable EAPs are used for long-term applications, such as electrodes for neural activity recording or stimulation. The first conceptually biodegradable EAPs were reported in 1995 (Hong and Miller, 1995). Some recent research has attempted to develop EAP-based sensors that mimic a neuron interface and are compatible with neuron systems to convey information from sensor to recipient (Balint et al., 2014; Zhu et al., 2014; Cullen et al., 2008).

5.7 CONCLUSIONS AND FUTURE SCOPE

EAPs are a long way from wide-scale deployment due to some major constraints that hinder their capability of full-scale commercialization. The performance and stability of EAPs deployed in aqueous environments are compromised by water loss from the surface, which results in a change in potential. Therefore, a water impermeable surface needs to be designed to address this issue. In addition, enhanced surface conductivity is required to obtain a defect-free conductive surface. EAPs that function at higher voltages suffer damage due to heat generation; therefore, heat-resistant EAPs need to be explored.

Manufacturing human-friendly robots that use artificial muscles or soft actuators have not been fully developed and commercialized despite the knowledge of soft actuators for more than three decades. Increased research activity is required to explore the full potential of EAPs in sensor-based technology and to commercialize it similar to other technologies.

REFERENCES

Balint, R., Cassidy, N.J., & Cartmell, S.H. (2014). Conductive polymers: towards a smart biomaterial for tissue engineering. *Acta Biomaterials*, *10*:2341–2353. doi.10.1016/j.actbio.2014.02.015.

Bar-Cohen, Y. (Ed.) (1999). *Proceedings of the first SPIE's Electroactive Polymer Actuators and Devices (EAPAD) Conference, Smart Structures and Materials Symposium*, *3669*:1–414.

Bar-Cohen, Y. (Ed.) (2004). *Electroactive Polymer (EAP) Actuators as Artificial Muscles - Reality, Potential and Challenges* (2nd ed) (pp. 1–765) Bellingham: SPIE Press.

Bar-Cohen, Y. (2009). *Electroactive polymers for refreshable Braille displays*. Retrieved from www.spie.org/news/1738-electroactive-polymers-for-refreshable-braille-displaysSSO.

Bar-Cohen, Y., Kim, J.K., Choi, H.R. & Madden, J.D.W. (2007). Electroactive polymer materials. *Smart Materials and Structures*, *16*(2). doi.10.1088/0964-1726/16/2/E01.

Bauer, S., & Bauer, F. (2008). Piezoelectric polymers and their applications. In W. Heywang, K., Lubitz, W. Wersing (eds.) *Piezoelectricity Evolution and Future of a Technology Series; Springer Series in Materials Science* (pp.157–177) *114*.

Baughman, R.H., Cui C., Zakhidov, A.A., Iqbal, Z., Barisci, J.N., Spinks, G.M., & Wallace, G.G. (1999). Carbon nanotube actuators. *Science*, *284*(5418):1340–1344. doi:10.1126/science.284.5418.1340.

Carpi, F. (2010). Electromechanically active polymers. *Polymers International*, *59*:277–278. doi./10.1002/pi.2790.

Carpi, F., Anderson, I., Bauer, S., Frediani, G., Gallone, G., Gei, M., & Graaf, C. (2015). Standards for dielectric elastomers transducers. *Smart Materials and Structures*, 24:105025. doi.10.1088/0964-1726/24/10/105025.

Carpi, F., & Smela, E. (eds) (2009). *Biomedical applications of electroactive polymer actuators*. Chippenham: John Wiley & Sons.

Chambers, M., Finkelmann, H., Remsˇkar M., Saˊnchez-Ferrer, A., Zalar, B., & Zˇumer, S. (2009). Liquid crystal elastomer–nanoparticle systems for actuation. *Journal of Materials Chemistry*, *19*:1524–1531. doi.10.1039/B812423J.

Cheng, Z., & and Zhang, Q. (2008). Field-activated electroactive polymers. Special issue dedicated to EAP. *Materials Research Society (MRS) Bulletin*, *33*:183–187. doi.10.1557/mrs2008.43.

Clarke, S.M., Hotta, A., Tajbakhsh, A.R. & Terentjev, E.M. (2002). Effect of cross-linker geometry on dynamic mechanical properties of nematic elastomers. *Physical Review E*, *65*:021804. doi.10.1103/PhysRevE.65.021804.

Collings, P.J., & Hird, M. (1997). *Introduction to liquid crystals: chemistry and physics.* Boca Raton: CRC Press. doi.10.1201/9781315272801.

Cottinet, P.J., Guyomar, D., Guiffard B., Lebrun, L., & Putson, C. (2010). Electrostrictive polymers as high-performance electroactive polymers for energy harvesting, *Piezoelectric Ceramics.* doi:10.5772/9946.

Cottinet, P.-J., Guyomar, D., Guiffard, B., Putson, C., & Lebrun, L. (2010). Modeling and experimentation on an electrostrictive polymer composite for energy harvesting. *IEEE Transactions on ultrasonics, ferroelectrics, and frequency control, 57*(4), 885–3010. doi:10.5772/9946.

Crossley, S., & Kar-Narayan, S. (2015). Energy harvesting performance of piezoelectric ceramic and polymer nanowires. *Nanotechnology, 26*:344001. doi:10.1088/0957–4484/26/34/344001.

Crossley, S., Whiter, R..A., & Kar-Narayan, S. (2014). Polymer-based nanopiezoelectric generators for energy harvesting applications. *Materials Science and Technology, 30*:1613–1624. doi:10.1179/1743284714Y.0000000605.

Cullen, D.K., Patel, A.R., Doorish, J.F., Smith, D.H. & Pfister, B.J. (2008). Developing a tissue-engineered neuralelectrical relay using encapsulated neuronal constructs on conducting polymer fibers. *Journal of Neural Engineering, 5*:374–384. doi:10.1088/1741-2560/5/4/002.

Dragan, E.S. (2014). Design and applications of interpenetrating polymer network hydrogels: A review. *Chemical Engineering Journal,* 243:572–590. doi.10.1016/j.cej.2014.01.065.

Eguchi, M. (1925). On the permanent electret. *Philosophical Magazine, 49*:178. doi.10.1080/14786442508634594.

Fasolt, B., Hodgins, M., Rizzello, G., & Seelecke, S. (2017). Effect of screen printing parameters on sensor and actuator performance of dielectric elastomer (DE) membranes. *Sensors Actuators A: Physics, 265*:10–19. doi.10.1016/j.sna.2017.08.028.

Feng, G.H., & Tsai, M.Y. (2010). Acoustic emission sensor with structure-enhanced sensing mechanism based on micro-embossed piezoelectric polymer. *Sensors Actuators A: Physics, 162*:100–106. doi:10.1016/j.sna.2010.06.019.

Festin, N., Plesse, C., Pirim, P., Chevrot, C., & Vidal, F. (2014). Electro-active interpenetrating polymer networks actuators and strain sensors: fabrication, position control and sensing properties. *Sensors Actuators B, 193*:82–88. doi.10.1016/j.snb.2013.11.050.

Glass, J.E., Schulz, D.N., & Zukosi C.F. (1991). Polymers as rheology modifiers. *ACS Symposium Series, 462*:2–17. doi:10.1021/bk-1991-0462.

Guiffard, B., Guyomar, D., Seveyrat, L., Chowanek, Y., Bechelany, M., Cornu, D., & Miele, P. (2009). Enhanced electroactive properties of polyurethane films: Loaded with carbon-coated SiC nanowires. *Journal of Physics D: Applied Physics, 42*(5):055503.

Guiffard, B., Seveyrat, L., Sebald, G., & Guyomar, D. (2006). Enhanced electric field induced strain in non-percolative carbon nanopowder/polyurethane composites. *Journal of Physics D: Applied Physics, 39*(14):3053–3057. doi:10.1088/0022-3727/39/14/027.

Hong, Y.L., & Miller, L.L. (1995). An electrically conducting polyester that has isolated quatrathiophene units in the main chain. *Chemistry of Materials, 7*(11):1999–2000.

Jean-Mistral, C., Basrour, S., & Chaillout, J.-J. (2010). Comparison of electroactive polymers for energy scavenging applications. *Smart Materials and Structures, 19*(8):085012. doi:10.1088/0964-1726/19/8/085012.

Ji, Y., Marshall J.E., & Terentjev, E.M. (2012). Nanoparticle liquid crystalline elastomer composites. *Polymers, 4*(1):316–340. doi:10.3390/polym4010316.

Jou, A.P., Micheletti, P., Estrany, F., del Valle, L.J., & Alemán, C. (2017). Electrostimulated release of neutral drugs from polythiophene nanoparticles: Smart regulation of drug–polymer interactions. *Advanced Healthcare Materials, 6*:1700453. doi.10.1002/adhm.201700453.

Katchalsky, A., & Zwick, M. (1955). Mechanochemistry and ion exchange. *Journal of Polymer Science, 16*(82):221–234. doi.10.1002/pol.1955.120168212.

Kawai, H. (1969). Piezoelectricity of poly(vinylidene fluoride). *Japan Journal of Applied Physics, 8*:975–976. doi.10.1143/JJAP.8.975/pdf.

Kim, K.J., & Tadokoro, S. (Eds.) (2007). *Electroactive polymers for robotic applications, artificial muscles and sensors.* London: Springer.

Kim, K.N., Chun, J., Kim, J.W. Lee, K.Y., Park, J.-U., Kim, S.-W., ...Baik, J.M. (2015). Highly stretchable 2D fabrics for wearable triboelectric nanogenerator under harsh environments. *ACS Nano, 9*:6394–6400. doi.10.1021/acsnano.5b02010.

Koç, I.M. & Akça, M. (2013). Design of a piezoelectric based tactile sensor with bio-inspired micro/nanopillars. *Tribology International*, 59:321–331. doi.10.1016/j.triboint.2012.06.003.

Lehmann, W., Skupin, H., Tolksdorf, C., Gebhard, E., Zentel, R., Krüger, P., & Lösche, M., (2001). Giant lateral electrostriction in ferroelectric liquidcrystalline elastomers. *Nature*, *410*:447–450. doi.10.1038/35068522.

Li, C., Wu, P.M., & Shutter, L.A. (2010). Dual mode operation of flexible piezoelectric polymer diaphragm for intracranial pressure measurement. *Applied Physics Letters*, 96:5–8. doi:10.1063/1.3299003.

Lin, B., & Giurgiutiu, V. (2006). Modeling and testing of PZT and PVDF piezoelectric wafer active sensors. *Smart Materials and Structures*, *15*:1085–1093. doi:10.1088/0964-1726/15/4/022.

Lotz, P., Matysek, M., & Schlaak, H.F. (2011). Fabrication and application of miniaturized dielectric elastomer stack actuators. *IEEE/ASME Transactions on Mechatronics*, *16*(1):58–66. doi 10.1109/TMECH.2010.2090164.

Maas, J., Tepel, D., & Hoffstadt, T. (2015). Actuator design and automated manufacturing process for DEAP-based multilayer stack-actuators. *Meccanica*, *50*(11):2839–2854. doi.10.1007/s11012-015-0273-2.

Madden, J.D.W. (2018). 25 years of conducting polymer actuators: history, mechanisms, applications, and prospects, in Anderson, I. (ed.) *Proceedings of the EAPAD Conference, SPIE Smart Structures and Materials Symp.* Denver, Colorado.

Mirfakhrai, T., Oh, J.Y., & Kozlov, M. (2008). Carbon nanotube yarns as high load actuators and sensors. *Advances in Science and Technology*, *61*:65–74. doi.10.4028/www.scientific.net/AST.61.65.

Murat Koc, I., & Akc, A.E. (2013). Design of a piezoelectric based tactile sensor with bio-inspired micro/nanopillars. *Tribology International*. 59, 321–331. doi:10.1016/j.triboint.2012.06.003.

Nalwa, H.S. (Ed.). (1995). *Ferroelectric polymers: chemistry, physics, and applications*. New York: Marcel Dekker.

Nassab, N.H., Samanta, D., Abdolazimi, Y., Annes, J.P. & Zare, R.N. (2017). Electrically controlled release of insulin using polypyrrole nanoparticles. *Nanoscale*, 9:143–149. doi.10.1039/C6NR08288B.

Nemat-Nasser, S. & Thomas, C.W. (2004). Ionomeric polymer-metal composites. In Y. Bar-Cohen (Ed.) *Electroactive polymer (EAP) actuators as artificial muscles: Reality, potential and challenges* (2nd ed.). SPIE Press. Retrieved from http://ceam.ucsd.edu/documents/papers/ch6.pdf.

Oguro, K., Kawami, Y., & Takenaka, H. (1992). Bending of an ion-conducting polymer film-electrode composite by an electric stimulus at low voltage. *Materials Science*. Retrieved from www.semanticscholar.org/paper/Bending-of-an-Ion-Conducting-polymer-Film-Electrode-Oguro/4e2a860a40cd053f4721010cc9ec04e98fd2d4d7.

Osada, Y. (1991). Chemical valves and gel actuators. *Advanced Materials*, *3*(2):107–108. doi.10.1002/adma.19910030209.

Osada, Y. & Kishi, R. (1989). Reversible volume change of microparticles in an electric field. *Journal of the Chemistry Society*, *85*(3):655–662. doi.10.1039/F19898500655.

Otero, T.F. & Corte´s, M.T. (2003). Artificial muscles with tactile sensitivity. *Advanced Materials*, *15*:279–282. doi.10.1002/adma.200390066.

Otero, T.F. & Corte´s, M.T. (2003a). A sensing muscle. *Sensors Actuators B:* *96*:152–156. doi.10.1016/S0925-4005(03)00518-5.

Park, I.-S., Jung, K., Kim, D., Kim, S.M., & Kim, K.J. (2008). Physical principles of ionic polymer-metal composites as electroactive actuators and sensors. Special Issue dedicated to EAP, *Materials Research Society (MRS) Bulletin*, *33*(3):190–195. doi.10.1557/mrs2008.44.

Pelrine, R., Kornbluh, R., Pei, Q., & Joseph, J. (2000). High-speed electrically actuated elastomers with strain greater than 100%. *Science*, *287*(5454):836–839. doi:10.1126/science.287.5454.836.

Qu, L., Peng, Q., & Dai L. (2008). Carbon nanotube electroactive polymers: opportunities and challenges. Special Issue dedicated to EAP, *Materials Research Society (MRS) Bulletin*, *33*(3):215–234. doi.10.1557/mrs2008.47.

Ribeiro, F.B., Plesse, C., Nguyen, G.T.M., Morozova, S.M., Drockenmuller, E., Shaplov, A.S., & Vidal, F. (2018). All-solid-state ionic actuator based on polymeric ionic liquids and electronic conducting polymer. In I. Anderson. (Ed.) *Proceedings of the EAPAD Conference, SPIE Smart Structures and Materials Symposium*. Denver, Colorado, U.S. doi.10.1117/12.2300774.

Richter A., Klatt, S., Paschew, G., & Klenke, C. (2009). Micropumps operated by swelling and shrinking of temperature-sensitive hydrogels. *Lab on a Chip*, *9*(4):613–618. doi.10.1039/B810256B.

Röntgen, W.C. (1880). About the changes in shape and volume of dielectrics caused by electricity. *Annual Review of Physical Chemistry*, *11*:771–786. doi.10.1002/andp.18802471304.

Sacerdote, M.P. (1899). On the electrical deformation of isotropic dielectric solids. *Journal of Physics: Conference Series*, *VIII*,3:282–285.

Sadeghipour, K., Salomon, R., & Neogi, S. (1992). Development- of a novel electrochemically active membrane and 'smart' material based vibration sensor/damper. *Journal of Smart Materials and Structures*, *1*(1):172–179. doi.10.1088/0964-1726/1/2/012/pdf.

Seminara, L., Capurro, M., Cirillo, P., Cannata, G., & Valle, M. (2011). Electromechanical characterization of piezoelectric PVDF polymer films for tactile sensors in robotics applications. *Sensors Actuators A*, *169*:49–58. doi:10.1016/j.sna.2011.05.004.

Shahinpoor, M. (2000). Elastically-activated artificial muscles made with liquid crystal elastomers. Paper presented at the Proceedings of the SPIE's 7th Annual International Symposium on Smart Structures and Materials, EAPAD Conference. Newport Beach, CA, USA. *3987*: 187–192. doi:10.1117/12.387777.

Shahinpoor, M. (ed.) (2016). *Ionic Polymer Metal Composites (IPMCs): Smart Multi-Functional Materials and Artificial Muscles*. Cambridge: Royal Society of Chemistry.

Shirakawa, H., Louis, E.J., MacDiarmid, A.G., Chiang, C.K., & Heeger, A.J. (1977). Synthesis of electrically conducting organic polymers: halogen derivatives of polyacetylene, $(CH)_x$. *Journal of the Chemical Society, Chemical Communications*, 578–580. doi.10.1039/C39770000578.

Spinks, G.M., Wallace, G.G., & Baughman, R.H. (2004). Carbon nanotube actuators: synthesis, properties and performance. In Y. Bar-Cohen (Ed). *Electroactive polymer (EAP) actuators as artificial muscles: Reality, potential, and challenges*. 261–295. doi.org/10.1117/3.547465.ch8.

Stalbaum, T., Trabia, S., Hwang, T., Olsen, Z., Nelson, S., Shen, Q., & Lee, D.-C. (2018). Guidelines for making ionic polymer-metal composite (IPMC) materials as artificial muscles by advanced manufacturing methods: state-of-the-art. In Y. Bar-Cohen. (Ed.) *Advances in manufacturing and processing of materials and structures* (pp. 377–394). Boca Raton: CRC Press, Taylor & Francis Group.

Su, J. (2018). A review of electrostrictive graft elastomers: structures, properties, and applications. In Y. Bar-Cohen & I. Anderson (Eds.), *SPIE Smart Structures and Materials*. Paper presented at the Symposium Proceedings of the EAPAD Conference Denver, Colorado, USA: SPIE.

Su, J., Harrison, J.S., Clair, T.L., Bar-Cohen, Y., & Leary, S. (1999). Electrostrictive graft elastomers and applications. *MRS Symposium Proceedings*, *600*:131–136. doi.10.1557/PROC-600-131.

Ullah, F., Othman, M.B.H., Javed, F., Ahmad, Z., & Akil, H.M. (2015). Classification, processing and application of hydrogels: a review. *Materials Science and Engineering: C. 57*:414–433. doi.10.1016/j.msec.2015.07.053.

Watanabe, M., Imaizumi, S., Yasuda, T. & Kokubo, H. (2014). Ion gels for ionic polymer actuators. In K. Asaka, & H. Okuzaki (Eds) *Soft actuators: materials, modeling, applications, and future perspectives* (pp. 141–156).Tokyo: Springer. doi.10.1007/978-4-431-54767-9.

Wu, Y., Alici, G., Madden, J.D.W., Spinks, G.M., & Wallace, G.G. (2007). Soft mechanical sensors through reverse actuation in polypyrrole. *Advances in Functional Materials*, *17*:3216–3222. doi.10.1002/adfm.200700060.

Yu, C., Mutlu, S., Selvaganapathy, P., Mastrangelo, C.H., Svec, F., & Fréchet J.M.J. (2003). Flow control valves for analytical microfluidic chips without mechanical parts based on thermally responsive monolithic polymers. *Analytical Chemistry*, *75*(8):1958–1961. doi.10.1021/ac026455j.

Zeng, H., Zhang, Y., Mao, S., Nakajima, H., & Uchiyama K. (2017). A reversibly electro-controllable polymer brush for electro-switchable friction. *Journal of Material Chemistry*, *5*:5877–5881. doi.10.1039/C7TC01624G.

Zhang, Q.M., Bharti, V., & Zhao, X. (1998). Giant electrostriction and relaxor ferroelectric behavior in electronirradiated Poly(vinylidene fluoride-trifluoroethylene) copolymer. *Science*, *280*:2101–2104. doi:10.1126/science.280.5372.2101.

Zhu, B., Luo, S.-C., Zhao, H. Lin, H.A., Sekine, J., Nakao, A., & Chen, C. (2014). Large enhancement in neurite outgrowth on a cell membrane-mimicking conducting polymer. *Nature Communications*, *5*:4523. doi:10.1038/ncomms5523.

6 Electroactive Polymers for Smart Window Technology

Marcero R. Romero and Santiago P. Fernandez Bordín

CONTENTS

6.1 INTRODUCTION

Windows are architectural elements that have been developed for hundreds of years and play a key role in house functionality and other environments where the humans have developed their lives, transport and work. According to the Oxford Dictionary, a window is "an opening in the wall or roof of a building or vehicle, fitted with glass in a frame to admit light or air and allow people to see out." As the definition indicates, windows were created to meet the ventilation and lighting needs in homes, and for people to look through them. Etymologically the word "window" comes from the language of Northern Europe, Old Norse *vindauga*, from *vindr* "wind" and *auga* "eye". The initial implementation of windows consisted of an opening in the walls of rudimentary stone houses. Therefore, ventilation was easily and successfully achieved by making an opening in the walls; however, the entry of light was always accompanied by deficiencies in thermal insulation. The opening-type window had to be covered with animal skins or other available materials to prevent the entry of cold or heat due to unfavorable environmental conditions; however, the simultaneously entry of light into the environment was affected.

During development, semi- or transparent materials were gathered and formed into sheets to allow the passage of light when achieving some isolation or at least preventing air circulation when it was necessary. Among the different materials, minerals, such as lapis specularis were used in ancient Rome, which is a type of translucent plaster; and sheets of this were assembled using small pieces. Semi-transparent laminas of paper, which were sometimes greased, to improve the passage of light, were used by ancient people of Asia. Later, a hydrated phyllosilicate mineral of aluminum and potassium (muscovite) and commonly called mica was used in Russia and then in England in the Middle Ages, harnessing the ability of this material as an excellent thermal insulator.

It is necessary to discuss glass in a separate paragraph, because it has its own history of discovery and development. In the stone age, it was collected in small pieces of obsidian, a natural volcanic mineral for which a broad diversity of tools were made. Then, in Egypt small rudimentary pieces were obtained and later in the Roman Empire the production techniques were perfected. This body of knowledge generated one of the most important centers of glass production in the world in Murano, Italy from the twelfth century. The most important activity was the production of containers, ornamental products, and the construction of stained glass windows for cathedrals and churches, in which religious figures were obtained by incorporating impurities from different minerals to achieve the final coloring of different pieces that were later assembled with lead. However, its cost was very high and was restricted to public buildings. In the twentieth century, with the mass production of laminates using the float method, thin, transparent, and regular sheets were achieved. Similary, the cost of production decreased and the use of glass in windows became widespread.

In 1902, Benedictus invented the first laminated glass with cellulose acetate, which gave the glass excellent mechanical properties and resistance to breaking. This allowed it to be used safely as automotive window glass. Then, in 1907 Baekeland invented the first polymeric synthetic material, Bakelite and in 1922 Staudinger received the Nobel Prize for explaining the chemical nature of polymeric materials. Polymers quickly proved to be very versatile materials that could be laminated, and sheets that were equal to or even more transparent than glass, such as polymethylmethacrylate (Plexiglas), which was discovered in 1930, were achieved.

Currently, buildings have a large number of walls and glazed ceilings, and the cost of the latter is approximately 25% the construction cost. The energy used by buildings is one-third of the total consumption in developed countries, which surpasses other areas, such as industries. Of this third, half of the energy in buildings is used for heating, ventilation, or cooling. Considering all the elements that make up a building, windows are the least energy-efficient elements. The thermal performance of glazed units is very rudimentary, because 50% of the energy is lost or gained through the windows, especially recently with the increased interest in building structures with large glazing. The thermal conductivity of glass is similar to that of bricks or concrete. However, the glass sheets are much thinner than the rest of the materials of the walls and the temperature gradient is large across the windows, and therefore, so is the heat transfer. Therefore, the windows alter the thermal demand of buildings, which means that it is necessary to optimize these characteristics.

To prevent sunlight from entering excessively, different strategies have been designed, which includes the implementation of fenestration devices. Currently, these are integrated with control devices that can be operated manually or are automated, which covers windows with constant optical properties. All these systems do not incorporate transmittance adjustments in the glazing and in many cases treatment of the glasses gives them opaque characteristics to avoid glare. This makes rooms dark and they require artificial lighting during the day and limit visibility to the outside. Faced with these drawbacks, current research is focused on the development of glazing that can dynamically change their optical properties, which are called smart windows. Figure 6.1 shows the evolution of windows and as an analogy, a sequence of human evolution.

The driving force beyond scientific and academic interest is the large market for smart windows, because they function as a tool to improve energy efficiency, and can reduce approximately 10% of the energy used by a building.

6.2   RELEVANT PHYSICAL PARAMETERS

For the characterization of electrochromic (EC) devices, several terms of recurrent use appear. Many times, these are usually mentioned without giving a specific definition; therefore, leading to confusion. To prevent this, some terms have been defined.

The visible transmittance (T_{vis}) is defined by the following:

$$T_{vis} = \frac{\sum\limits_{380\,nm}^{780\,nm} T(\lambda) D_\lambda V(\lambda) \Delta\lambda}{\sum\limits_{380\,nm}^{780\,nm} D_\lambda V(\lambda) \Delta\lambda} \tag{6.1}$$

FIGURE 6.1 Evolution of windows technology in the building construction industry.

(From the authors.)

where:

D = relative spectral distribution of illuminant D65 (i.e., mean illumination distribution defined in intervals = 10 nm)

$V(\lambda)$ = spectral luminous efficiency for the photopic vision of a standard observer (i.e., a function that refers to the perception of light by an observer) $T(\lambda)$ = spectral transmittance of the glass at the given wavelength

$\Delta\lambda$ = wavelength interval.

To account for the entire solar spectrum, solar transmittance (T_{sol}) is used:

$$T_{sol} = \frac{\sum_{300\,nm}^{2500\,nm} T(\lambda) S_\lambda \Delta\lambda}{\sum_{380\,nm}^{780\,nm} S_\lambda \Delta\lambda} \tag{6.2}$$

where:

S = relative spectral distribution of solar radiation.

Which broadens the spectrum of visible wavelengths, adding infrared radiation and a portion of UV.

Another very commonly used parameter is the contrast ratio (*CR*), defined by:

$$CR = T_b/T_c \qquad\qquad (6.3)$$

where:

$T_b = T_{vis}$ in the bleached form
T_c = transmittance in the colored state.

As *CR* increases, the ability of the device to alter its optical transmission properties improves.

6.3   SMART WINDOWS

Smart windows are named because they are devices that, according to requirements, can dynamically modulate the transmittance of light. They are a promising technology for saving energy in buildings by controlling indoor solar radiation. Therefore, smart windows are being seriously considered as indispensable devices to reduce the consumption of cooling energy, adjusting the transport of sunlight that allows it to pass through on cold days, partially adjusting the entrance on warm days, and preventing environmental heat on hot days. Important efforts have been made in recently for the development of smart windows, among which the most developed are electro, thermo, mechano, and photoresponsive. The most common arrangement consists of a sandwich-type structure with a functional material between two transparent electrodes in which thermochromic and photochromic substances are introduced, this simple arrangement has received rapid development recently. A thermochromic window uses thermo-responsive materials that can be placed as films on the surface of the glass or encapsulated between two sheets of glass. Similar to thermochromic windows, photochromics are commonly based on a photoresponsive material. The latest smart mechano–chromic windows are materials that show changes in structure, such as morphology in response to external deformation. Among all the types of smart windows, EC windows have the best chance of reaching the market since they can be adjusted and integrated into the electrical systems of a building, which allows some products to reach and become commercially available. In these windows, the functional material plays a dominant role in the modulation of light and can show different states, and therefore, different optical properties due to the composition or structural changes. EC smart windows are formed by a multilayer of electrochemical cells that respond to electric fields by reversibly changing their optical transmittance. These types of devices could allow energy savings, improvements in natural lighting, and decrease the requirements for cooling equipment in buildings. In addition, they have great potential to be used in automotive glazing, sunglasses, mirrors, sunroofs, and aviation (Figure 6.2).

For these windows to meet their objectives, they are expected to operate at a low electrical potential, respond reasonably quickly to switching, they need to be sufficiently transparent, have low thermal conductivity, be stable against temperature variations, and have a long service life. The main EC materials are based on conjugated polymers, which include polythiophenes (PT), polypyrrole (PPy), polyanilines (PANI), polyfurans, and polycarbazoles (PCz).

6.4   CONDUCTIVE AND CONJUGATED POLYMERS

The first studies on conductive polymers were in 1977 when Drs. H. Shirakawa, O.A.G. MacDiarmid, and A.J. Heeger studied the electrical conductivity of organic polymers based on halogen derivatives of polyacetylene (Shirakawa,1977). The study of this phenomenon earned them the Nobel Prize in 2000 for discovering that partial oxidation with iodine or other reagents to transform the polyacetylene film, which was 10^9 times more conductive than the original (Figure 6.3).

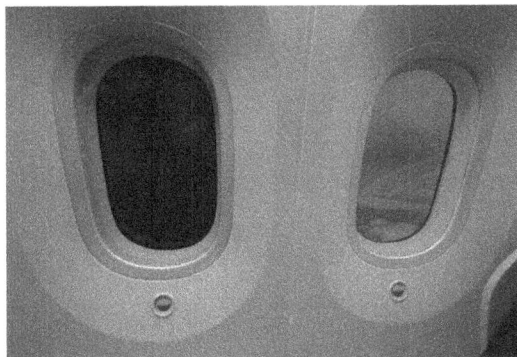

FIGURE 6.2 Smart windows of Boeing 787–8.

(Image of the public domain by Spaceaero2 (CC BY-SA 3.0.)

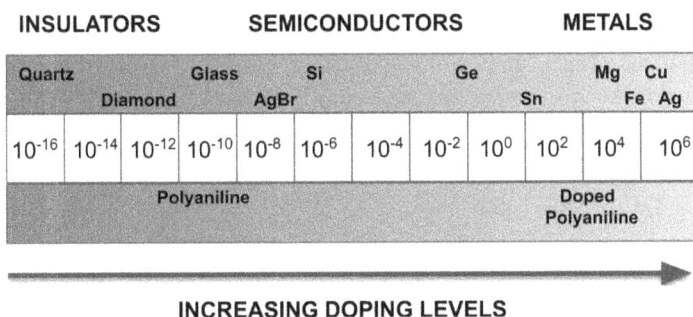

FIGURE 6.3 Effect of doping in polymers on the conductivity of conductive polymers and comparison with conventional materials.

(Based on MacDiarmid, 2001.)

From these discoveries, the name of intrinsically conductive polymer (ICP) or synthetic metal arose, which is an organic polymer that possesses the electromagnetic, electronic, and optical properties of a metal combined with the properties that are commonly associated with a conventional polymer. Its properties are intrinsic to a doped form of the polymer (MacDiarmid, 2001).

Within the ICPs the conjugated polymers (CP) are the most remarkable constituents with chains that are formed by an extended π-conjugated system and a backbone made up of double and single bonds alternately arranged (Figure 6.4). The superposition of p-orbitals occupied by an unpaired electron of the carbon atoms from the polymer chain causes the delocalization of electrons along the backbone. This overlap produces a band structure (Chiang, 1977) in which there is a large energy interval between the valence band and the conduction band. Thus, in their neutral state, the CPs are considered as semiconductors (MacDiarmid, 2001). Through a doping process, CPs can increase their conductivity by several orders of magnitude, leaving behind their semiconductor regime to become conductors (MacDiarmid, 2001). The alteration in the number of π electrons as a consequence of doping results in the modification of their electronic, magnetic, and optical properties, which has been of particular interest in the use of these materials in technological applications, such as biosensors (Travas-Sejdic, 2014), electromagnetic radiation shielding (Joo, 1994), organic light-emitting diodes (OLEDs) (Sekine, 2014), and smart windows.

As mentioned previously, one of the main points in the development of smart windows is their ability to control the flow of sunlight and heat according to the external climate conditions and the

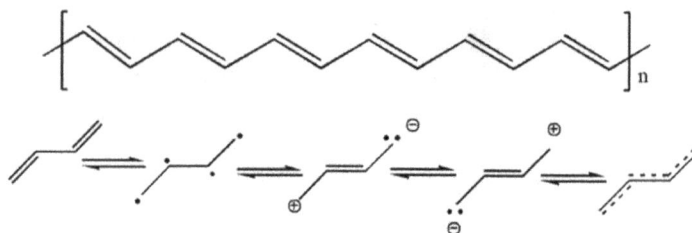

FIGURE 6.4 Conjugated polymer and resonant structures.

(Based on image of the public domain by Iqmanuelnavarro.)

desired comfort requirements in the building. Therefore, the property of being able to produce a reversible and fast response electrochemical–optical changes added to its ability to precisely adjust its electronic properties, a low operating voltage, and the processability of the solution (Kim, 2011), positions CP as a central material in this type of device (Kim, 2020, Wang, 2016).

In general, CPs have a color tone in their neutral state, in which they absorb visible light. But after oxidation, the absorbance shifts to longer wavelengths in the near infrared, which results in a loss of hue toward a whitish or transparent color. Furthermore, for their EC performance, CPs exhibit high coloring efficiency, color versatility, high contrast, and fast response times.

6.5 POLYMER DOPING

Polymer doping is a transformation in which a semiconductor polymer increases its conductivity by several orders. The processes have the same name as semiconductor metal doping, in which impurities are added to a very pure semiconductor metal to change its electrical properties, the process for polymers is different and only shares the result. The doping process is the reaction of polymers with a chemical reagent that oxidizes (or reduces) the system, which makes the electrons pass from the valence band to the conduction band, and modifies their electrical properties. There are two main methods for polymer doping, chemical and electrochemical, which are based on an oxidation–reduction (redox) process. In addition, there are non-redox doping methods.

The conductivity mechanism in these polymers within the conjugated structure is based on the movement of charge carriers (e.g., oxidation or reduction products of the polymer) either of the p-type (positive) or the n-type (negative). Therefore, the concept of polarons and bipolarons could describe this phenomenon (Figure 6.5). The polaron is a radical ion that is produced by an excitation that leads to the loss (p-doping) or gain (n-doping) of an electron by the π structure of the conjugated polymer backbone (MacDiarmid, 2001). If another electron is lost or gained in a different segment of the polymer chain or another hole is created, another electron or positive charge, double-charge (bipolaron) is produced. This process results in polarons or bipolarons that can move along the polymer chain through the rearrangement of the double bonds of the conjugated system, which occurs in the presence of an electric field. Therefore, the structural deformation caused leads to the appearance of bonding orbitals that are destabilized (e.g., donor or acceptor states) between the valence band and the conduction band that is energetically accessible to the π electrons. Consequently, the bandgap reduces its separation and produces a significant increase in the polymer conductivity.

The conductivity in polymeric semiconductors depends on many factors, such as nature, the concentration of the dopants, the homogeneity of the doping, the carrier mobility, and the crystallinity and morphology of the polymers (Amb, 2011; Wang, 2016).

Because of doping, a change in the conductivity of CPs is accompanied by a variation in color. For p-doping, the polymer presents a state of coloration in its neutral form, which is lost when

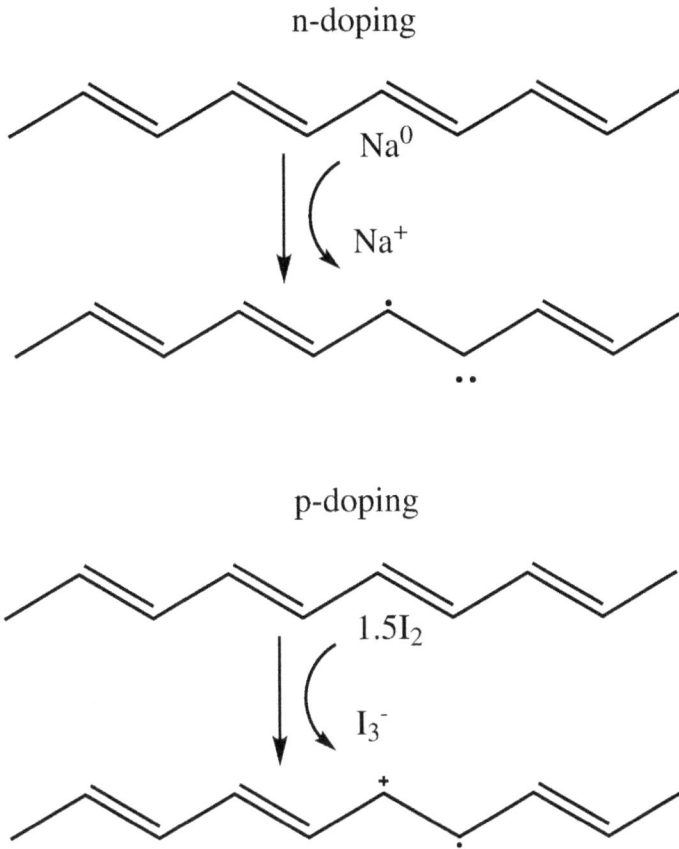

FIGURE 6.5 Conductivity mechanisms of p- and n-doped CPs.

(Based on image of the public domain by Iqmanuelnavarro (CC BY-SA 2.5).)

doped, and its conductivity increases. The neutral and colored polymer forms are reversible states and are regenerated with a further reduction. The cathodically coloring EC is a process of reversible color change through doping or de-doping, and is summarized in the following equation (Kim, 2020; Shin et al., 2016):

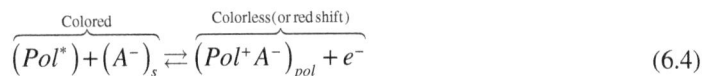

$$\overbrace{\left(Pol^*\right)+\left(A^-\right)_s}^{\text{Colored}} \rightleftarrows \overbrace{\left(Pol^+A^-\right)_{pol}+e^-}^{\text{Colorless(or red shift)}} \tag{6.4}$$

where:

(Pol*) = active centers capable of absorbing light in neutral CP films
$(A^-)_s$ = counterion of the electrolyte injected or ejected to maintain the electroneutrality of the film
$(Pol^+A^-)_{pol}$ = transparent CP films containing the counter ions
e^- = electrons transported to the surface of indium tin oxide (Shin et al., 2016).

In addition, anodically coloring EC placement is possible (Christiansen, 2019). In this example, the polymer is colorless in its neutral state and acquires color when n-doped. This process can be summarized in the following equation:

$$\underbrace{\left(Pol^*\right)+\left(A^-\right)_s}_{\text{Colorless}} \rightleftharpoons \underbrace{\left(Pol^+A^-\right)_{pol}+e^-}_{\text{Colored}} \qquad (6.5)$$

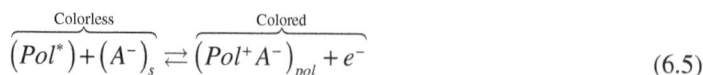

Since the color of a material is determined by its bandgap, for instance, the energy difference between the highest-occupied molecular orbital (HOMO) and the lowest unoccupied molecular orbital (LUMO) of a CP, this optical bandgap (Eg) is a critical parameter for color switching. Any modification that impacts the charge transport, the MO levels, the Eg, the electrical potential that is associated with the EC reaction will change the EC properties (Kim, 2020)

The color properties depend on the energy difference between the HOMO and LUMO of a CP that correspond to the optical bandgap. Therefore, the electric potential associated with the EC reaction (Eg) will affect the color switching since any modification in charge transport between MOs will produce changes in the EC properties of the material.

6.6 MAIN TYPES OF ELECTROCHROMIC CONJUGATED POLYMERS

6.6.1 ELECTROCHROMIC POLYTHIOPHENES

PT and its derivatives are an important and widely studied class of linear CPs that are often used as a model for the study of charge transport in conductive polymers (Roncali, 1992). PTs have become materials of interest in electronic and optoelectronic applications, mainly due to a range of possibilities for synthetic approaches, which are chemical and electrochemical. In addition, PTs functionalization produces highly environmentally stable doped and undoped states (Heeney, 2005) with unique electrochemical and optical properties (Kaneto, 1983) that are extensively modulated.

The chemical structure of PT (Figure 6.6) is derived from the homocoupling of the thiophene ring. Sulfur is an electron donor heteroatom, which contributes two electrons to the π system of the ring. It has another solitary electron pair in a hybrid sp² orbital in the ring plane, which is an aromatic heterocycle rich in electrons.

Unsubstituted PT in the neutral state exhibits a red color (λ_{max}=470 nm) and turns blue (λ_{max}=730 nm) when oxidized by p-doping (Beaujuge, 2010, Mortimer, 2006). The bandgap for this polymer can be between 2.0 and 2.2 eV and depends on the molecular weights present, with a value characteristic of a π–π * interband transition (Beaujuge, 2010). When doping occurs, the interband transition decreases, and two new optical low energy transitions are produced (at approximately 1.25 and 0.80 eV). The characteristic absorption pattern with free charge carriers is a metallic-like state and appears when the valence and conduction bands merge after various doping stages (Mortimer, 2006).

One of the characteristics that have drawn attention to PTs is the simplicity of controlling the optical properties of the material from subtle modifications in the backbone (Figure 6.7). These changes significantly alter the spectral properties of the material, which allows the different color states to be adjusted.

These substitutions allow the control of the optical bandgap from values significantly lower than the unsubstituted PT value of approximately 0.9 eV, to higher values of approximately 2.9 eV.

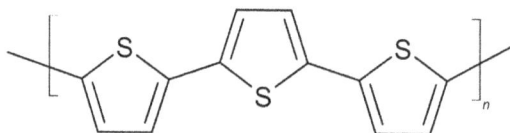

FIGURE 6.6 Repetitive units of polythiophene.

(From the authors.)

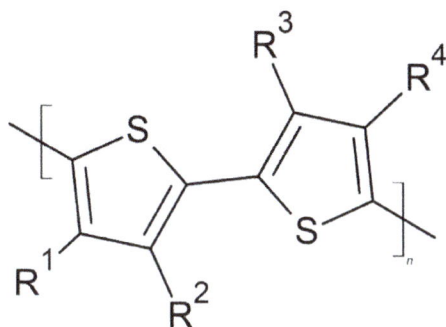

FIGURE 6.7 Different positions for substitutions in PT.

(From the authors.)

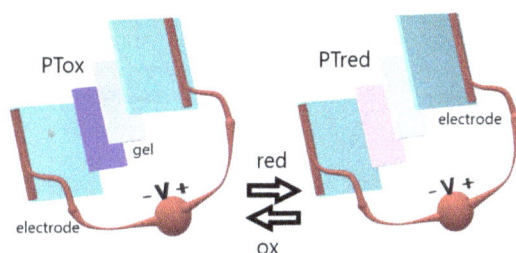

FIGURE 6.8 Typical electrode configuration for electrochromic window experiments. Scheme shows PT reduced and oxidized states.

(From the authors.)

In applications, such as smart windows, one of the most important aspects is the ability to obtain a transparent or translucent state with low loss when penetrated by light. These properties have been achieved by PT fused with aromatic systems that generate high-energy resonance systems. For example, with the introduction of alkoxy substituents that donate electrons, a bandgap of approximately 0.95 eV is obtained with blue-black coloration in its neutral state and transparent light yellow in its oxidized state (Figure 6.8) (Hung, 1999). Furthermore, benzene modified with methyl, fluorine, or chlorine groups generate polymers with lower reduction potentials (i.e., higher electron affinity) but maintain similar bandgap and electrochromic properties (King, 1995).

6.6.2 ELECTROCHROMIC POLYPYRROLE

PPy and its derivatives, have been the subject of many scientific studies for applications that require electrochromic materials, because they are relatively easy to synthesize chemically and electrochemically, have adequate bandgap for this type of application, and have low redox potential (Mortimer, 2006; Wang, 2016). However, one of the disadvantages of PPy is the short life cycle, which makes it difficult to use them in real applications (Carpi, 2006).

PPys and their derivatives are obtained by the oxidation of pyrrole monomers or substituted pyrrole. In general, these oxidations are carried out by electropolymerization on a conductive substrate (i.e., electrode) and applying an external electric potential or by chemical polymerization in a solution with a chemical oxidant (Wallace, 2002). Different synthesis routes result in PPy with varied

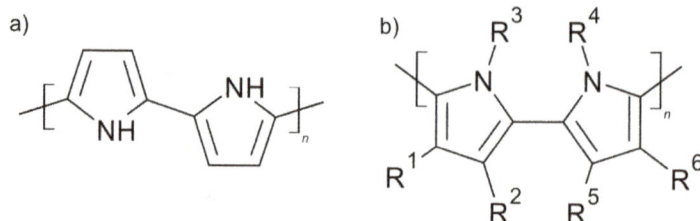

FIGURE 6.9 Chemical structure of the repeating unit of PPy in: (a) absence; and (b) presence of substituents. **(From the authors.)**

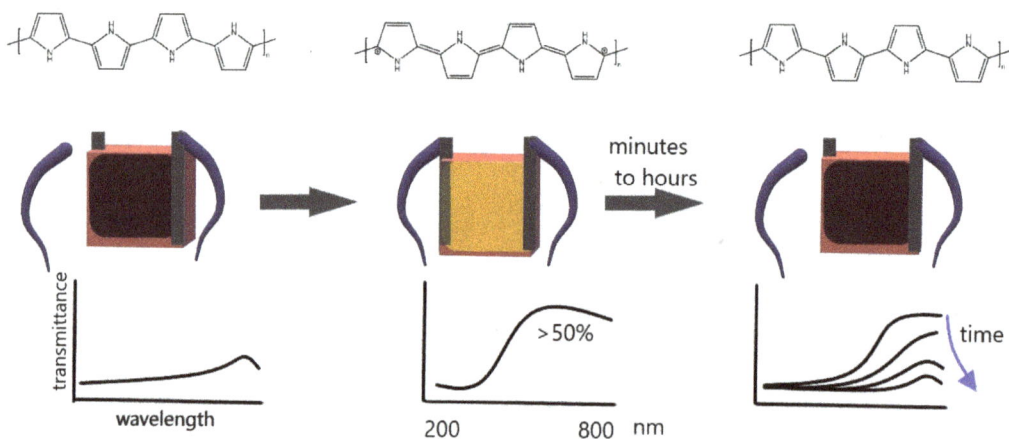

FIGURE 6.10 PPy redox states and spectral changes from these transitions as function of time. **(Based on Yang, 2019.)**

features: chemical oxidations generally produce powders, and electrochemical synthesis produces films. In addition, these polymeric products have different chemical and electrical properties.

The chemical structure of PPy is shown in Figure 6.9. PPy films exhibit a bandgap of approximately 2.7 eV and have a yellow/green color in their neutral state and turn a brown-black color in their conductive doped state (Beaujuge, 2010; Genies, 1983). Similarly, some derivatives of PPy do not change these properties. For example, poly(N-alkyl pyrrole), which contains various methyl, butyl, or phenyl substituents present similar properties to the original polymer (Diaz, 1982). However, the N-substituted benzene diamine groups in pyrrole exhibit different chromic shifts in the visible range from red to blue. However, some substitutions show limited or no EC contrasts, such as N-benzyl-PPy and N-phenyl-PPy. (Bjorklund, 1985)

Some substituted PPys have attracted attention for presenting different states with multiple colors that depend on the doping level. This is the case for poly[3,4-(propylene dioxy) pyrrole] (PProDOP), MePPProDOP, and Me2PProDOP, which have a bandgap of approximately 2.2 eV (Schottland, 2000). These polymers change from an orange color to a red-brown and finally to light blue when they go from a neutral state to a partially oxidized state and finally a completely oxidized state. In addition, Me2PProDOP shows an optical contrast of 76% and can maintain 90% of its electroactivity after thousands of commutation cycles (Camurlu, 2014). The response time observed was from 5 to 1 h to recover 50%–90% of the original color, respectively (Figure 6.10). However, one of the biggest problems that faces these materials for smart windows applications is the lack of a clear or very transparent state (Wang, 2016). This drawback occurs in a variety of organic electrochromic materials.

6.6.3 Electrochromic Polyaniline

PANI is one of the most famous and oldest electrochromic polymers. It is a promising material for electrode development in supercapacitors due to its low cost, simplicity of synthesis, and flexibility. However, due to its ability to transition electrochromically in a reversible way, good cyclic stability in non-aqueous electrolytes, and its intense color contrast in thin films, PANI could be used in multiple EC devices (Baetens, 2010, Wang, 2012).

PANI is obtained in an acidic solution by the oxidative polymerization of an aniline monomer. However, this polymer can be synthesized by chemical, electrochemical, and photochemical methods (Wallace, 2002). The main difference between PANI and PTs and PPys is that the N heteroatom participates directly in the polymerization process and integrates into the backbone and the conjugation of the conductive form of the polymer.

The base PANI formula consists of repeating units that alternate between reduced (y) and oxidized (1-y) forms (Figure 6.11). The average oxidation (1-y) can range from y=0, with a fully reduced polymer, y=0.5 for a half-oxidized state, to y=1, in which the material is in a fully reduced form. (MacDiarmid, 2001).

Due to its variety of redox states, PANI is multi-chromic. In the fully reduced form (leucoemeraldine), PANI has a yellow color that turns green (the emeraldine salt) when the polymer chain is partially oxidized. Subsequent redox mechanisms in PANI achieve other oxidation and coloration states, such as a partially oxidized state (i.e., emeraldine base) with a blue color or the fully oxidized form (i.e., pernigraniline), which is black (Kim, 2020, Mortimer,2006).

Substitutions improve the EC properties of PANI by modifying the structure of aniline. For example, those with alkyl groups are a good alternative. Poly(o-toluidine) and poly(m-toluidine) films exhibit a better multi-EC response stability (Mortimer, 1995). In addition, the functionalization of nitrogen with alkyl sulfonates leads to a self-doped PANI with better redox stability. Furthermore,

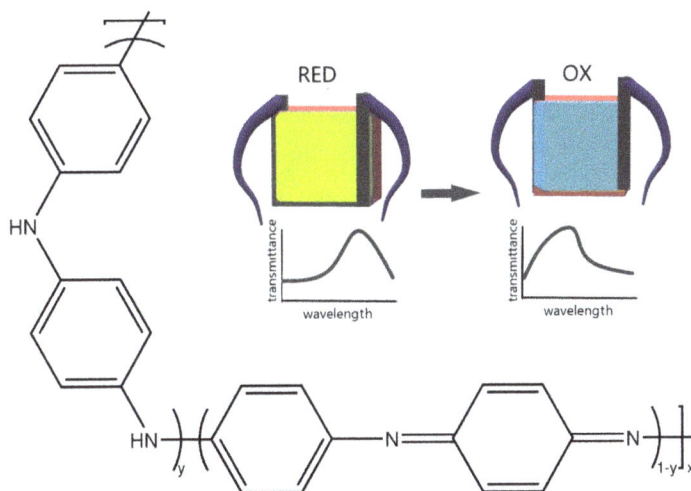

FIGURE 6.11 Repetitive units of PANI. Inset: reduced emeraldine (green) and oxidized pernigraniline (blue) states. These redox materials show shift in the absorption curve but similar transmittance.

(From the authors.)

this material has a lyotropic structure and is polyionic, which favors the assembly by layers, and improves the EC properties (Kim, 2006, Kim, 2020).

6.6.4 ELECTROCHROMIC POLYCARBAZOLES

PCz are based on repeating units of carbazole, which consists of a compound containing a nitrogen heterocycle. Its basic structure (Figure 6.12(a)) consists of two benzene rings coupled with a central pyrrole ring. The interest in these materials lies in their stability and high redox potential compared to other conductive polymers. Furthermore, the base structure can be conjugated by associating the nitrogen atom (Figure 6.12(b)) with a wide variety of substituents (Yasutani, 2012), adjusting the material properties to the specific application requirements.

Similarly, in most conductive polymers, there are different polymerization methods to obtain these materials. The most common synthesis processes are chemical and electrochemical polymerization.

EC, unsubstituted PCz have yellow coloration in their neutral state that changes green after being p-doped, and this oxidation process can occur at relatively low potentials (Figure 6.13) (Beaujuge, 2010). However, in most electrochromic systems the use of derivatives with substituents in the N-position is usually preferred. In general, the functionalization of the N-group is generally accompanied by a hypsochromic shift. This process can also bring other advantages such as transparency of their neutral states and have intermediate oxidation steps with different colorations. For example, substitution with an alkyl group results in a material without color in its neutral state to transition from green to blue for partially and fully oxidized states, respectively. (Chevrot. 1996).

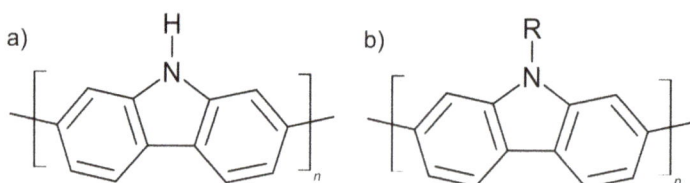

FIGURE 6.12 Basic repeat of: (a) PCz; and (b) derivative modified in the heterocyclic N atom.

(From the authors.)

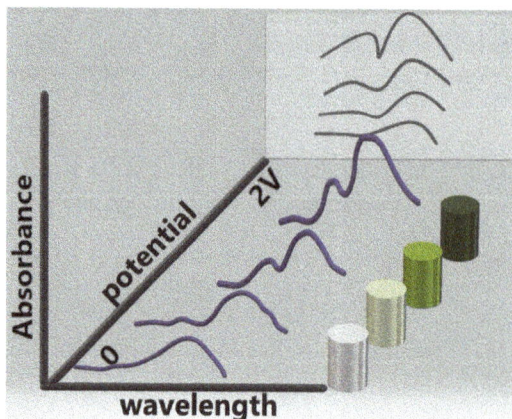

FIGURE 6.13 PCz change color responses from white to green at potentials <2 V.

Besides, poly (N-vinyl carbazole) (PVK) also appears as a relevant derivative of PCz due to its good thermal stability and good behavior against doping. However, the processability is tedious since the π - π electron system reduces the stability of the oxidized state producing a conductivity decrease (Bekkar, 2020).

6.6.5 ELECTROCHROMIC COPOLYMERS

There are a large number of EC polymers available for smart window technology. In each of the main groups, a range of possibilities open up when making substitutions in the base chain.

Another interesting strategy to expand the variety of electrochromic polymers is through the copolymerization of multiple monomeric units. By following this strategy, the copolymer properties can be tuned to expand the color palette and synergistically improve the commutation response time (Kim, 2020).

One of the most outstanding achievements obtained through copolymerization is the development of a material capable of passing from a state of black coloration to a light transmissive state. Studies to achieve these properties were made by copolymerizing propylene dioxythiophene (ProDOT) and 2,1,3-benzothiadiazole (BTD), in different proportions or using oligomers (Beaujuge, 2008, Shi, 2010) (Figure 6.14). In the first studies, a contrast of 52% at 592 nm was reached (Beaujuge, 2008). Subsequently, Shi (2010) developed copolymers with better and more uniform visible absorption. Furthermore, these combined monomers showed high optical contrast, fast redox switching, and redox species with long-term stability.

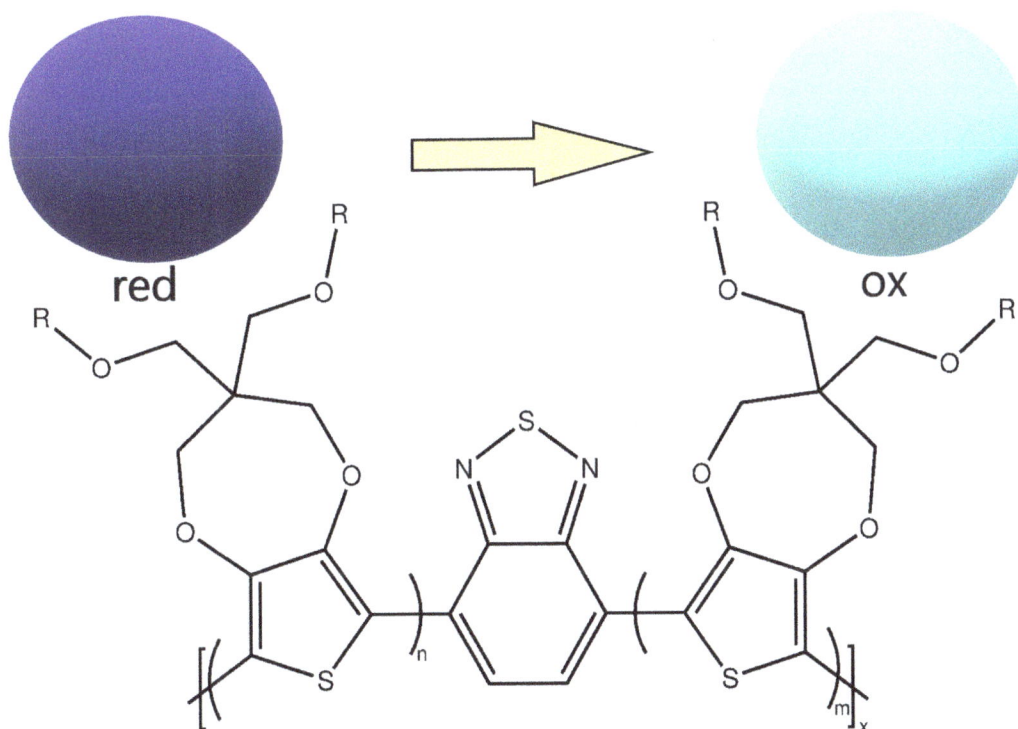

FIGURE 6.14 Copolymer of ProDOT with BTD in reduced form (dark blue). The oxidized form of this copolymer is light blue.

(From the authors.)

6.7 CONCLUSIONS AND PROSPECTS

During this chapter, the main groups of CPs that possess electroactive properties have been presented, which are very promising for the development of smart windows and other EC devices. Numerous scientific studies have shown that CPs have good coloring efficiency, fast response times, and high optical contrast. In addition, their performance and affordable production costs place them in a competitive position compared with other materials with similar characteristics, such as metal oxides.

The main advantage that CPs present is the adaptability of their properties through modifications in their main chain or the side chains. The introduction of branches, functional groups, or the copolymerization of different monomers allows the control of the response time, color, contrast, and stability of the material, which are properties that are essential for specific applications. Furthermore, it is possible to increase the possibilities and combination of properties using the hybridization and multilayer combination strategy in different CPs.

All these advantages account for the coming advances that highlight the integration of multiple devices with different functionalities, either through the introduction of an external circuit or the development of a multi-responsive polymer. Smart window creations could integrate solar cells, some energy storage device, and thermal or light sensors to provide the user experience with an autonomous and self-sufficient device and might be relevant. However, there are still challenges to overcome, such as having devices at an affordable cost, which would allow them to increase the market and remove the premium status that they currently have. In addition, cost reduction must be accompanied by greater durability and energy efficiency to achieve wide distribution and public acceptance.

ACKNOWLEDGMENTS

We would like to thank Agustina A. Romero for her assistance with the edition of the illustrations of this chapter. We would also like to thank Prof. Inamuddin for his invitation to join us in his book project. Financial support from FAMAF –IFEG and FCQ-UNC and IPQA – CONICET (Argentina) are gratefully acknowledged.

REFERENCES

Amb, C.M., Dyer, A.L., & Reynolds, J.R. (2011). Navigating the color palette of solution-processable electrochromic polymers. *Chemistry of Materials*, *23*(3), 397–415.

Baetens, R., Jelle, B.P., & Gustavsen, A. (2010). Properties, requirements and possibilities of smart windows for dynamic daylight and solar energy control in buildings: A state-of-the-art review. *Solar energy materials and solar cells*, *94*(2), 87–105.

Beaujuge, P.M., Ellinger, S., & Reynolds, J.R. (2008). The donor–acceptor approach allows a black-to-transmissive switching polymeric electrochrome. *Nature materials*, *7*(10), 795–799.

Beaujuge, P.M., & Reynolds, J.R. (2010). Color control in π-conjugated organic polymers for use in electrochromic devices. *Chemical Reviews*, *110*(1), 268–320.

Bekkar, F., Bettahar, F., Moreno, I., Meghabar, R., Hamadouche, M., Hernáez, E. ... Ruiz-Rubio, L. (2020). Polycarbazole and its derivatives: Synthesis and applications. A review of the last 10 years. *Polymers*, *12*(10), 2227.

Bjorklund, R., Andersson, S., Allenmark, S., & Lundstrøm, I. (1985). Electrochromic effects of conducting polymers in water and acetonitrile. *Molecular Crystals and Liquid Crystals*, *121*(1–4), 263–270.

Camurlu, P. (2014). Polypyrrole derivatives for electrochromic applications. *RSC Advances*, *4*(99), 55832–55845.

Carpi, F., & De Rossi, D. (2006). Colours from electroactive polymers: Electrochromic, electroluminescent and laser devices based on organic materials. *Optics & Laser Technology*, *38*(4–6), 292–305.

Chevrot, C., Ngbilo, E., Kham, K., & Sadki, S. (1996). Optical and electronic properties of undoped and doped poly (N-alkylcarbazole) thin layers. *Synthetic metals*, *81*(2–3), 201–204.

Chiang, C.K., Fincher, Jr., C.R., Park, Y.W., Heeger, A.J., Shirakawa, H., Louis, E.J. ... MacDiarmid, A.G. (1977). Electrical conductivity in doped polyacetylene. *Physical review letters*, *39*(17), 1098.

Christiansen, D.T., Tomlinson, A.L., & Reynolds, J.R. (2019). New design paradigm for color control in anodically coloring electrochromic molecules. *Journal of the American Chemical Society*, *141*(9), 3859–3862.

Diaz, A.F., Castillo, J., Kanazawa, K.K., Logan, J.A., Salmon, M., & Fajardo, O. (1982). Conducting poly-N-alkylpyrrole polymer films. *Journal of Electroanalytical Chemistry and Interfacial Electrochemistry*, *133*(2), 233–239.

Genies, E.M., Bidan, G., & Diaz, A.F. (1983). Spectroelectrochemical study of polypyrrole films. *Journal of Electroanalytical Chemistry and Interfacial Electrochemistry*, *149*(1–2), 101–113.

Heeney, M., Bailey, C., Genevicius, K., Shkunov, M., Sparrowe, D., Tierney, S., & McCulloch, I. (2005). Stable polythiophene semiconductors incorporating thieno [2, 3-b] thiophene. *Journal of the American Chemical Society*, *127*(4), 1078–1079.

Hung, T.T., & Chen, S.A. (1999). The synthesis and characterization of soluble poly (isothianaphthene) derivative: poly (5, 6-dihexoxyisothianaphthene). *Polymer*, *40*(13), 3881–3884.

Joo, J., & Epstein, A.J. (1994). Electromagnetic radiation shielding by intrinsically conducting polymers. *Applied Physics Letters*, *65*(18), 2278–2280.

Kaneto, K., Yoshino, K., & Inuishi, Y. (1983). Electrical and optical properties of polythiophene prepared by electrochemical polymerization. *Solid state communications*, *46*(5), 389–391.

Kim, E., & Kim, Y. (2006). Layer-by-layer assembled electrochromic film based on an alkylsulfonated polyaniline. *Molecular Crystals and Liquid Crystals*, *447*(1), 173–491.

Kim, J., Rémond, M., Kim, D., Jang, H., & Kim, E. (2020). Electrochromic conjugated polymers for multifunctional smart windows with integrative functionalities. *Advanced Materials Technologies*, *5*(6), 1900890.

Kim, J., You, J., Kim, B., Park, T., & Kim, E. (2011). Solution processable and patternable poly (3, 4-alkylenedioxythiophene) s for large-area electrochromic films. *Advanced Materials*, *23*(36), 4168–4173.

King, G., & Higgins, S.J. (1995). Synthesis and characterization of novel substituted benzo [c] thiophenes and polybenzo [c] thiophenes: Tuning the potentials for n-and p-doping in transparent conducting polymers. *Journal of Materials Chemistry*, *5*(3), 447–455.

MacDiarmid, A.G. (2001). Nobel Lecture: "Synthetic metals": A novel role for organic polymers. *Reviews of Modern Physics*, *73*(3), 701.

Mortimer, R.J. (1995). Spectroelectrochemistry of electrochromic poly (o–toluidine) and poly (m-toluidine) films. *Journal of Materials Chemistry*, *5*(7), 969–973.

Mortimer, R.J., Dyer, A.L., & Reynolds, J.R. (2006). Electrochromic organic and polymeric materials for display applications. *Displays*, *27*(1), 2–18.

Roncali, J. (1992). Conjugated poly (thiophenes): synthesis, functionalization, and applications. *Chemical Reviews*, *92*(4), 711–738.

Schottland, P., Zong, K., Gaupp, C.L., Thompson, B.C., Thomas, C.A., Giurgiu, I., & Reynolds, J.R. (2000). Poly (3, 4-alkylenedioxypyrrole) s: Highly stable electronically conducting and electrochromic polymers. *Macromolecules*, *33*(19), 7051–7061.

Sekine, C., Tsubata, Y., Yamada, T., Kitano, M., & Doi, S. (2014). Recent progress of high performance polymer OLED and OPV materials for organic printed electronics. *Science and Technology of Advanced Materials*, *15*(3).

Shi, P., Amb, C.M., Knott, E.P., Thompson, E.J., Liu, D.Y., Mei, J. … Reynolds, J.R. (2010). Broadly absorbing black to transmissive switching electrochromic polymers. *Advanced Materials*, *22*(44), 4949–4953.

Shin, H., Seo, S., Park, C., Na, J., Han, M., & Kim, E. (2016). Energy saving electrochromic windows from bistable low-HOMO level conjugated polymers. *Energy & Environmental Science*, *9*(1), 117–122.

Shirakawa, H., Louis, E.J., MacDiarmid, A.G., Chiang, C.K., & Heeger, A.J. (1977). Synthesis of electrically conducting organic polymers: halogen derivatives of polyacetylene, (CH) x. *Journal of the Chemical Society, Chemical Communications*, 16, 578–580.

Travas-Sejdic, J., Aydemir, N., Kannan, B., Williams, D.E., & Malmström, J. (2014). Intrinsically conducting polymer nanowires for biosensing. *Journal of Materials Chemistry B*, *2*(29), 4593–4609.

Wallace, G.G., Teasdale, P.R., Spinks, G.M., & Kane-Maguire, L.A. (2002). *Conductive electroactive polymers: intelligent materials systems*. Boca Ratón: CRC Press.

Wang, K., Wu, H., Meng, Y., Zhang, Y., & Wei, Z. (2012). Integrated energy storage and electrochromic function in one flexible device: an energy storage smart window. *Energy & Environmental Science*, *5*(8), 8384–8389.

Wang, Y., Runnerstrom, E.L., & Milliron, D.J. (2016). Switchable materials for smart windows. *Annual review of chemical and biomolecular engineering, 7*, 283–304.

Yang, B., Ma, D., Zheng, E., & Wang, J. (2019). A self-rechargeable electrochromic battery based on electrodeposited polypyrrole film. *Solar Energy Materials and Solar Cells, 192*, 1–7.

Yasutani, Y., Honsho, Y., Saeki, A., & Seki, S. (2012). Polycarbazoles: Relationship between intra-and intermolecular charge carrier transports. *Synthetic metals, 162*(17–18), 1713–1721.

7 Systematic Investigation of the Revolutionary Role of Electroactive Polymers in Modifying Microelectromechanical Systems

Shaan Bibi Jaffri and Khuram Shahzad Ahmad

CONTENTS

7.1 INTRODUCTION

A myriad of materials have been synthesized and others are emerging with time, which are aimed at the welfare of humans by revolutionizing several domains in human life. The scientific community has been actively engaged in the development of novel materials that have varied compositions, for example, oxides, sulfides, tellurides, selenides, polymeric substances, and composite materials (Afsheen et al., 2020; Ahmad and Jaffri, 2018; Ijaz et al., 2020a). They are employed in a variety

DOI: 10.1201/9781003173502-7

of applications that range from biomedical to photovoltaic applications, which depend on the characteristics required. Among different materials, polymeric substances have gained special attention for many decades due to their characteristics (Bonneaud et al., 2021; Nenna et al., 2021). Today, there are no areas of life where the utilization of polymers and polymeric composites has not been implemented. The field of polymers has grown considerably over time and a significant amount of research has been conducted to resolve the concerns raised against polymers for their persistence and stability in environmental matrixes that negatively affect life (Forouzanfar et al., 2021; Hafner et al., 2021). Polymers can be broadly categorized in different classes depending upon their composition, nature, type of application, and other characteristics. Electroactive polymers (EAPs) are the major class of polymers and polymeric substances that are known for their feature of undergoing change in their size ranges, morphology, and volume when they come into contact with an electrical field of higher strength (Peng et al., 2021; Marín et al., 2021). They are one of the active polymeric materials, such as magnetostrictive materials, piezoelectrics, alloys, and polymers that have shape memory, and thermoelastic polymers These active materials are mainly marked by their remarkable active disfiguration capacity, higher speed of response, lower density, and upgraded resilience. In addition, EAPs are remarkably lightweight materials that have good fracture tolerance and compliance. Furthermore, they are economically viable (Bar-Cohen, 2004). This category has grown into a large number of polymeric materials that are highly responsive toward electric field application.

Synthetic EAPs emerged in the 1990s and since then, they have been inspiring the scientific community and engineering, which has led to their application in wider spectrums that encompass biomedical and other microelectromechanical systems (MEMs). For instance, they have similar aspects with the muscles, and therefore, they have been seen as artificial muscles based on their mode of action (Sarikaya et al., 2021; Rohtlaid et al., 2021). In addition, the softer and robust EAPs are associated with the provision of the larger strains compared with traditionally employed piezoceramics. Such features advocate their utilization in a wide range of applications. The literature is indicative of the larger number of publications that report the utilization of EAPs in actuators and sensors (Takagi et al., 2021; Yang et al., 2021; Rivkin et al., 2021). Of note, most actuators were developed previously using ceramic piezoelectric substances. Ceramic piezoelectric materials can withstand larger forces that are exerted on them and the consequent deformation in them is very minimal. However, with the development of EAPs, they can withstand 380% strain. Therefore, they have exceeded the ceramic piezoelectric materials to be considered in actuators. Of interest, these EAPs undergo greater deformation although they support large forces (Surmenev et al., 2021).

EAPs have undergone rapid transformation from their previous forms to their present form. Electromechanical coupling is usually carried out in EAPs to use them in actuation in addition to other applications of chemical sensors and mechanical stimuli. EAPs represents a unique category of materials that are known for their potential to conform to the surface regions of varied morphologies. These characteristics renders them suitable for applications in sensors and actuators (Minaian et al., 2021). EAPs are usually categorized into two classes that consider the major types of charge carriers in them (Bar-Cohen 2004) and this system of categorization is often referred as the Yoseph Bar-Cohen classification of EAPs. The classes of EAPs are ionic and electronic. Ionic EAPs include polymeric gels, carbon nanotubes (CNTs), conducting polymers (CPs), and ionic polymer–metal composites (IPMC) and the electronic EAPs include dielectric elastomers (DEs), liquid-crystal polymers, and piezoelectric polymers (PE) (Wang et al., 2016). EAPs is a generally used term and includes lightweight, flexible, and organic materials that are highly responsive to electrical stimuli. Ionic polymers usually function based on the migration of ions in the polymer matrix, and electronic polymers are functional after activation by an extrinsic electrical field. Electronic polymers usually required a higher voltage of >100 V, which exhibit shorter responsive durations.

MEMS signify the advanced systems that consist of smaller scale electrical and mechanical constituents and are used for specialized purposes. The translation of MEMS into specialized systems was carried out by the incorporation of electromechanical components and since then the boundaries of MEMS has been extended to include various devices, for example, nano, optical, and radio–frequency. Therefore, MEMS-based devices are known by various names that depend on the type of application and device where they are employed. For instance, optical applications make use of micro–optoelectromechanical systems (MOEMS) (Hillmer et al., 2021; Busurin et al., 2021), radio–frequency constituents and applications make use of radio–frequency MEMS (RF–MEMS) (Iannacci et al., 2021; Shanthi et al., 2021), and nano systems and devices have one dimension <1 μm that make use of nanoelectromechanical systems (NEMS) (Auciello & Aslam, 2021; Song et al., 2021). In addition, MEMS are used in biological domains for the detection of a desired target or for the manipulation of the cells. These biological systems that incorporate MEMS are often referred to as bioMEMS (Espinosa-Hernandez et al., 2021; Garcia-Ramirez & Hosseini, 2021). MEMS in a general sense are known by different names in different parts of the globe. For instance, they are referred to as microsystems technology (MST) in European countries and in Japan they are known as micromachines. In brief, MEMS can be used to describe smaller machines that are composed of a set of electrical and mechanical constituents that aim to achieve a specific purpose. According to the requirements of specific devices, other components can be added to the MEMS device, for instance, reflective surficial components used in micrometers. The components of MEMS are between 1 and 200 μm with an overall size approximately <1 mm.

The features of polymers that have been used in MEMS are given in Table 7.1. Different types of EAPs have been utilized in different kinds of microdevices. For instance, CPs have expressed

TABLE 7.1
Utilization of different types of EAPs for development into MEMS

Name of EAP (pristine or modified)	Role	Type of MEMS	References
Poly(2,2-dimethyl-3,4-propylene-dioxythiphohene) (PProDOT-Me2)	Polymeric electrodes	Supercapacitor modules	Liu & Reynolds (2010)
Polyhydroquinone–graphene hydrogel composites	Electrode material	Supercapacitor	Chen et al. (2015)
Polyaniline and carbon fiber graphene oxide	Electrode material	Supercapacitor	Gandara & Gonçalves (2020)
ligninsulfonate/single-wall carbon nanotube film/holey reduced graphene oxide (Lig/SWCNT/HrGO)	Film for energy storage	Wearable supercapacitors	Peng & Zhong (2020)
Poly(vinyl alcohol)-PANI nanofiber/graphene hydrogel	Energy storage	Coin Cell Supercapacitor	Joo et al. (2020)
Polyaniline/Ti$_3$C$_2$T$_x$ composites (i-PANI@Ti$_3$C$_2$T$_x$)	Wearable energy storage	Flexible all-solid-state supercapacitors	Zhou et al. (2020)
Nitrogen-doped ordered mesoporous carbon, in which g-C3N4	Mesoporosity provisioning	Volumetric supercapacitor	Xie et al. (2021)
Palladium oxide-polypyrrole (PdP)	Active electrode material	Supercapacitors	Jose (2021)
PEDOT:PSS/PPy	Electrode material	Flexible fiber-shaped supercapacitor	Teng et al. (2020)
Silsesquioxane -containing graphene oxide (SSQ-GO)	Energy storage	Supercapacitor	Ajdari et al. (2020)
PEDOT nanotubes	Energy storage	Supercapacitor	Hryniewicz et al. (2020)
Pyrolyzed polyacrylonitrile particles	Energy storage	Supercapacitor	Abalyaeva et al. (2020)

(continued)

TABLE 7.1 (Continued)
Utilization of different types of EAPs for development into MEMS

Name of EAP (pristine or modified)	Role	Type of MEMS	References
PVDF–PZT nanohybrid	Nanogenerator	Energy harvesting applications	Wankhade et al. (2020)
Contact-type (CT)	Dielectric elastomer generator	CT energy harvesting	Zhang et al. (2020)
Poly(ε-caprolactone) (PCL) membranes	Nano harvester	Energy harvesting	Sencadas (2020)
PVDF	Effective energy harvesting	Light detection	Si et al. (2020)
VHB elastomer	Dielectric elastomer generator	Energy harvesting	Jiang et al. (2020)
Poly(vinylidene fluoride-co-hexafluoroproplyene) (PVDF-HFP)	Energy harvesting	Energy harvesting	Ponnamma et al. (2020)
PVDF and trifluoroethylene (P(VDF-TrFE))	Piezoelectric medium	Energy conversion and sensing	Jiang et al. (2020a)
Ionic polymer–metal composite (IPMC)	Ionic medium	Wearable sensors	Patel & Mukherjee (2020)
Nafion IPMC	Porous media	Energy harvesting	Kweon et al. (2020)
β-PVDF/rGO	Piezoelectric nanogenerators	Energy harvesting	Ongun et al. (2020)
Terpolymer P(VDF-TrFE-CFE)/diisononyl phthalate (DINP)	Actuation	Optical applications inside Live Mirror	Thetpraphi et al. (2020)
HNTs/P(VDF-CTFE) nanocomposite	Capacitance	Dielectric capacitor	Ye et al. 2020
Ferroelectric copolymer P(VDF-TrFE)	Actuation	Dielectric material and actuators	Thuau et al. 2020
Poly(3,4-ethylenedioxythiophene): poly(styrene sulfonate) (PEDOT:PSS)	Electrode material	Linear actuators	Nguyen 2020
IPMC	Actuation	Actuators	Chang et al. 2020
Poly(ethylene glycol) diacrylate-poly(acrylic acid)	Electroactive scaffold	Excitable tissues	Gupta et al. 2021

remarkable performances in different for their potential for energy storage, sensing ability, and actuation. In addition, they are preferred because they are compatible with biological environments that have higher levels of moisture. MEMS-based devices made using EAPs can be further engineered for their speed by their enhancement at macroscale, because the ionic transfer paths are smaller, and the surface areas are relatively larger. Furthermore, a myriad of patterning methods are available and suitable for EAPs, which ensures their utilization in microfabrication (Spinks et al., 2007). The micro robots prepared with EAPs are based on a number of actuators that are marked by their movements that mimic the motion of elbows, wrists, and fingers.

The utilization of EAPs in different types of MEMS are in development and considers the suitability and end products prepared. In different types of MEMS, EAPs are used as integral components and some studies have suggested the advantages and disadvantages associated with them. A lot of data has been published that describes the use of EAPs in different types of MEMS; however, the utilization of these polymers in energy system is gaining considerable attention due to the present status of energy demand on a global scale. Research has been carried out on single EAPs and MEMS in addition to the utilization of these polymers in MEMS. However, to the best of our knowledge, no work has provided a comprehensive overview of the utilization of EAPs in MEMS.

In particular, there is limited literature on comprehensive reviews that cover energy systems based on EAPs. Therefore, this chapter will, for the first time, study the utilization of EAPs in MEMS, which specifically targets EAPs in energy systems. This chapter will focus on EAPs, their various types in ionic and electronic polymers, MEMS, the commercially significant uses of MEMS, and EAP-based energy MEMS will be discussed. In addition, this chapter will present some significant future considerations and challenges that are associated with the utilization of EAPs in different types of MEMS.

7.2 METHODS

7.2.1 SEARCH STRATEGY

This chapter has adopted a literature survey on previous years to compile a brief review that considers the favorability of EAPs in MEMS. Therefore, the literature studied was from published items that included research articles, book chapters, short communications, reviews, perspectives, and reports. The published items were selected from authentic publishers, such as Taylor & Francis, Elsevier, Springer, Springer Nature, JStore, Science direct, and IEEE. In general, the literature studied included articles from 2005 to 2021; however, considering the relevance of the topics and the space requirements, the selected papers were from 2010 to 2021. Different types of search engines were used, specifically Google scholar and the terms used to search different databases were varied, such as: ["polymers", "electroactive polymers", "microelectromechanical systems", "MEMs", "EAPs classes", "MEMS types", "electroactive polymers and MEMS", "electroactive polymers and energy harvesting", "Issues of electroactive polymers"].

7.3 BROAD CATEGORIZATION OF ELECTROACTIVE POLYMERS

EAPs have a significant position as polymers, and they are employed in a wide range of devices. EAPs are a very large family of polymers that have been further divided into the major groups of ionic and electronic EAPs according to Yoseph Bar-Cohen classification as shown in Figure 7.1. The categorization of EAPs is carried out based on their actuation. Specifically, the ionic EAPs are include materials that function as a result of the motion or diffusion of ions. Electronic materials are the EAPs that provide a mechanical response toward the Coulombic forces or other types of electrical fields. Recently, EAPs have emerged due to the favorable results and prompt reaction toward electrical stimulation with the following transformations in their morphology. These features have prompted the scientific community since the 1990s to explore EAPs and their use in different types of devices, especially MEMS (Defaz et al., 2021). Therefore, they have been used in increasing applications with time. Robust polymers are known for providing larger strains compared with other conventional materials.

7.3.1 IONIC ELECTROACTIVE POLYMERS

Ionic EAPs include a wide range of materials, which differ based on different factors. In addition, they are further classified into polymeric gels, CPs, CNTs, and IPMC. In all these ionic EAPs, the major mechanism of functionality is via electrically triggered ionic diffusion that occurs inside the material in bulk. The following section will discuss these types of ionic EAPs.

7.3.1.1 Polymeric Gels

Polymeric gels are a widely investigated type of EAP. Among the different types of polymeric gels, a highly advanced form of polymeric gel has been studied, which has been modified by metallic cross-linkers to improve the characteristics. Researchers have been using metal cross-linkers to modify

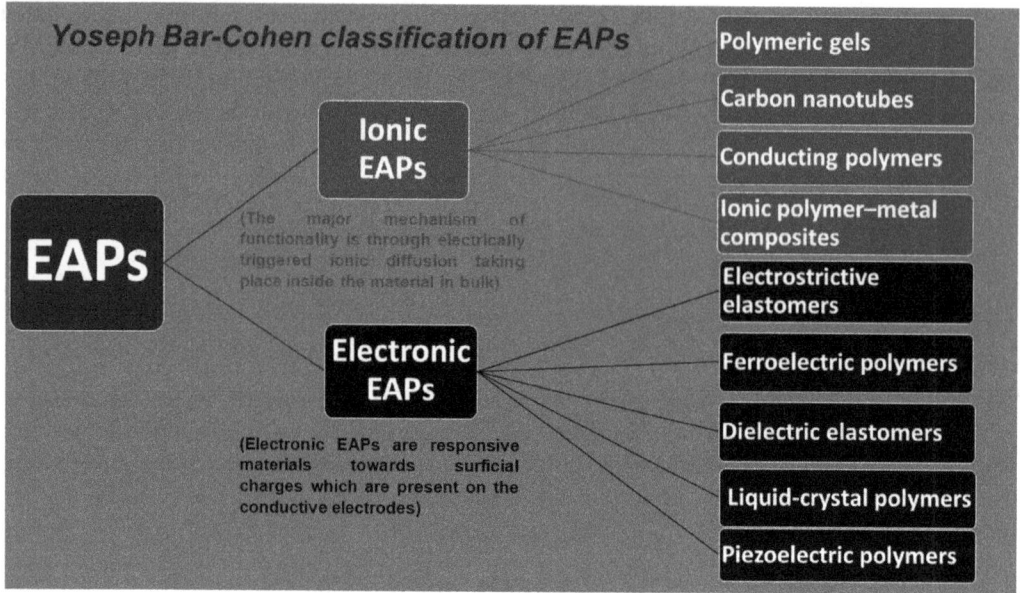

FIGURE 7.1 Yoseph Bar-Cohen classification of EAPs in sub-classes.

polymeric gels to overcome different limitations. For instance, such modification has been carried out to resolve in situ monomeric units and polymeric gels for stability in varying thermal conditions, plugging potential, and viscoelasticity (Boakye & Mahto, 2021). In 1990, the first polymeric gel was fabricated, investigated, and modified by the used of metallic cross-linkers. It was referred to as HPAM/Cr(III) (Karimi et al., 2016). Based on the cross-linkers used, polymeric gels can be further categorized into organically and inorganically cross linked polymeric gels. Organic cross-linkers include phenolics and polyethyleneimine (PEI) (Zhu et al., 2016). An organic cross linked polymeric gel was formed using hydroquinone (HQ) and hexamethylenetetramine (HMTA) that had amazing strength at 120°C (Sengupta et al., 2017). Inorganically cross linked polymeric gels include metallic ions, for example, Zr, Al, and Cr (Jia & Chen, 2018). Zhao et al. (2013) prepared an inorganic polymeric gel using HPAM and a zirconium salt (i.e., acetate). Rheological measurements in this inorganic polymeric gel indicated the division of the gelation process in three stages: induction, quicker mode cross-linking, and stabilization (Zhao et al., 2013).

7.3.1.2 Conducting Polymers

CPs are the fourth generation of polymeric materials (Ke et al., 2021). This class of EAPs is important for energy, because the final physicochemical characteristics of CPs can be altered positively compared with conventional polymeric substances through the combination of the remarkable electrical conductivity of the metals with polymers (Ibanez et al., 2018; Swager et al., 2017). In addition to the utilization of CPs in energy associated MEMS, which is due to their favorable electrical and electronic features, CPs have been employed in biomedical fields. Their use in biomedical applications is attributed to their responsive chemistry toward electrical fields (Llerena Zambrano et al., 2021). Since 1980, polymer and synthetic scientists have been synthesizing novel combinations of CPs that possesses favorable characteristics. Since then, a large number of CPs have been synthesized, in addition to their derivatives. However, some of the CPs have been developed based on their favorable features (e.g., polyaniline, polyacetylene, polypyrrole, polythiophene, and poly[3–ethylenedioxy)thiophene]) (Alipour et al., 2021; Boswel et al., 2021; Synodis et al., 2021)

7.3.1.3   Carbon Nanotubes

CNTs are the stiffest anthropogenic material with elevated strength. In addition to this, CNTs have significant electrical conductivity, and therefore, they are gaining a significant place in different types of electrical devices and applications in of communications. However, the favorable features of CNTs can only be employed in practical applications if care is taken due to their smaller size and embedding is carried out meticulously inside light weight matrixes in a homogenous manner (Spitalsky et al., 2010). CNTs are a type of ionic EAPs that has great flexibility, lower mass density, and a larger aspect ratio that is typically >1,000. In addition to this, experiments carried out with CNTs indicated their higher strengths and tensile moduli. CNTs are further categorized into different types, for example, single-walled carbon nanotubes (SWCNTs) and multi-walled CNTs (MWCNTs) (Yoo et al., 2021). These types can be metallic or semiconducting. In semiconducting CNTs, they have potential for the transport of electronic species over a range of longer lengths. There is no interruption in this electronic transferal, which makes them highly conductive compared with copper.

7.3.1.4   Ionic Polymer–Metal Composites

IPMCs are another significant type of EAPs that are available in composite forms of some conductive media, for instance, metals. IPMCs expresses a larger dynamic disfiguration during its development into electrode material in an electric field that varied as a function of time (Washington et al., 2021; Olsen et al., 2021). Typical sensors and actuators that use IPMCs are composed of a thinner layer of polyelectrolyte membrane, for example, Nafion, Flemion, or Aciplex, which is plated over the surface of the noble metal (Jamil et al., 2021). The noble metal is often platinum (Pt) or gold (Au) or Pt with top finishing layer of Au to improve conductivity. They are neutralized using different counter ions, and the balancing occurs for the anions that have been covalently bonded onto the membrane backbone. Sulfonates are typical anions in Aciplex and Nafion, and carboxylates are the major anions in Flemion (Sanginov et al., 2021; Taufiq Musa et al., 2021).

7.3.2   Electronic Electroactive Polymers

Electronic EAPs are responsive toward the surface charges that are present on the conductive electrodes. They are often bulk insulators. The surface charges over the conductive electrodes are known for the application of Coulombic forces over the materials that causes stress and strain in them. Electronic EAPs have further been categorized in different types. The following section gives a brief description of these types.

7.3.2.1   Electrostrictive Elastomers

Electrostrictive elastomers are electronic EAPs that are characterized by their light weight, flexibility, and economic viability and they are especially considered for their potential toward morphological molding into the desired shapes (Diguet et al., 2021). Furthermore, their mechanical energy density is remarkable compared with piezoelectric single crystals. Electrostrictive elastomers and piezoelectric EAPs are still emerging, and research is continuing in new areas in polymers. Compared with electroactive ceramics, electrostrictive elastomers are gaining attention, which could be attributed to their characteristics of responsiveness to electrical fields. Research in the field of electrostrictive elastomers led to the development of poly(vinylidene fluoride) (PVDF or PVF2) in 1969 (Su & Tajitsu, 2016).

7.3.2.2   Ferroelectric Polymers

Ferroelectric polymers and ferroelectret polymers are important types of electronic EAPs that are used in a wide range of applications (Mi et al., 2021). However, there is a fundamental difference between ferroelectric and ferroelectret EAPs due to the larger elastic anisotropic features of ferroelectret EAPs. They are prone to deformation due to compression and expansion. Ferroelectric EAPs are often incompressible that leads to stronger coupling. This type of coupling often develops between longitudinal piezoelectricity that occurs in the thick direction and transverse piezoelectricity. The latter piezoelectricity is often known for coupling the crosswise elastic disfiguration with transformation of the perpendicular electric polarization. Ferroelectric EAPs are used in a wide range of commercial applications, for example, piezoelectric ignition systems, hydrophones, and clamp-on transducers (Bauer & Bauer, 2008).

7.3.2.3 Dielectric Elastomers

DEs are a unique class of electronic EAPs that consist of longer chains of monomeric units, and they have the potential for larger disfiguration ≤300%. When subjected to higher voltage ranges, DEs undergo deformation and convert electrical energy into mechanical energy. Similar to other electronic EAPs, DE have been used in a variety of appliances, for example, valves, robotics, pumps, prosthetic devices, and medical implants. These applications mainly includes DE actuators with various configurations (Ni et al., 2021). On exposure to a constant voltage, DEs produce an augmented electrical field that leads to a further thinning of the film and instability in DE actuation. Therefore, further research is needed to resolve the issue of electromechanical instability that is encountered in DE-based actuation.

7.4 MICROELECTROMECHANICAL SYSTEMS AS REVOLUTIONIZERS

Currenlty, there has been in increase in the fabrication of new materials in different applications (Ijaz et al., 2002b; Iqbal et al., 2019; Jaffri & Ahmad 2018, 2018a). A lot of materials have been synthesized (Ahmad & Jaffri, 2002a; Ijaz et al., 2020a), which were aimed at utilization in MEMS. MEMS-based devices and structures that have smaller mechanical and electromechanical constituents are adding value to human life due to their development into different types of applications, which have been generated via microfabrication techniques. The revolutionary role of MEMS could be understood from the large number of polymer and other chemical constituent-based MEMS-based devices that are available on a commercial scale. In addition, research is being carried out to optimize the results for different types of MEMS (Gupta et al., 2021; Nguyen et al., 2020). One typical example of MEMS is a micromirror, which is used for the transmission and display of desired information. Other examples of MEMS include pressure sensors, gyroscopes, digital micromirror devices, and accelerometers that have industrial significance (Ye et al., 2020; Thuau et al., 2020; Ren et al., 2012). These devices have been commercialized and are being utilized in a wide range of applications that benefit humans. Despite the benefits that MEMS offer to humans, there is an ongoing debate about MEMS on the consolidation of the softer, cheaper, and efficient materials in operationally active MEMS that does not impact their functionality. The synergistic mixture of microelectronics and mechanical structures of micron sizes leads to the development of MEMS, which has gained interest due to the favorable results (Tiliakos, 2013).

7.5 MICROELECTROMECHANICAL SYSTEMS: COMMERCIALLY SIGNIFICANT APPLICATIONS

7.5.1 MICROCOOLING

Microcooling is a type of fluid acceleration. MEMS-based micro heat pipes (MHPs) are an emerging technology aimed at cooling heat pipes and they have acquired a significant place in thermal managerial strategies. The best use of this MEMS-based technology is carried out in microelectronic circuits packing, laser diodes, and concentrating photovoltaic cells. MHPs have often been employed as thermal spreaders that are the best technology the removal of heat directly from devices made from semiconductors. The fabrication and integration of the MHPs can be achieved by the electronic or optoelectronic chips over the MEMS surface. Research has been carried out to improve thermal conductivity and to augment heat transferal (Qu et al., 2017). In a recent report by Sun et al. (2017), an investigation was carried out on a MEMS-based micro oscillating heat pipe (micro-OHP) to determine the characteristics of the flow and thermal functionality using the working fluid, for instance, a dielectric liquid HFE-7100. The integration of the micro-OHP was carried out on a silicon wafer that had trapezoidal channels 357 μm of the hydraulic diameter (Sun et al., 2017). In another report, a feasibility study was conducted for a liquid microthruster that was based on channels with regenerative microcooling. The microthruster was based on the materials that were thermally fragile (Huh et al., 2017).

7.5.2 MICROELECTROMECHANICAL SYSTEM-BASED MICROSCOPY

MEMS have had an impact on all fields of human life including science, technology, and engineering. Microscopic techniques have acquired an important place in science for the detection of morphology and dimension-based measurements. Microscopy depends on the use of MEMS for proper functionality. Recently, in a microscopy-based report, an array of a single-chip atomic force microscope (sc-AFM) was reported for the first time, which could capture a lot of images of the target sample simultaneously (Olfat et al., 2017). Another report based on an AFM that was modified by MEMS demonstrated the use of a MEMS-based probe scanner. The device was composed of a stage (in place) that was equipped with electrostatic actuators and sensors for electrothermal displacement. For implementation, a microfabrication procedure was used, which was based on a reference silicon-on-insulator (Maroufi et al., 2019). Sun et al. (2020), used a MEMS-based multiprobe scanning probe microscope (SPM) that was developed to improve the efficiency of the images of target samples. The developed SPM was based on seven MEMS probes that had identical features. Each MEMS probe was integrated over the displacement sensor and actuator's z-axis (Sun et al., 2020).

7.6 ELECTROACTIVE POLYMER-BASED MICROELECTROMECHANICAL SYSTEMS AND ENERGY HARVESTING

Energy is an integral factor for the progress of any nation. Therefore, energy is considered to be one of the Sustainable Development Goals (SDGs) (Jaffri &Ahmad 2020, 2020a). The achievement of a cheap and clean energy resource would fulfill our energy needs and would protect future generations from insufficient energy. Fossil fuels are exhaustible energy resources; therefore, the development of novel energy resources highlighted by sustainability and SDGs is urgent. Therefore, a large number of materials have been synthesized and used and further research is ongoing to achieve the most suitable energy sources that have the capacity for replenishment through anthropogenic means. A number of researchers have reported the use of various types of EAPs in MEMS, which are aimed at development into energy systems (Sunithamani et al., 2020; Yang et al., 2017; Wang et al., 2018). For instance, a recent report used microturbines to extract wind energy, which was different to traditionally used batteries and other techniques at the microscale and provided the power to microsystems that operated without any batteries. In brief, the experiment designed, developed, and tested microturbines composed of 160 μm blade lengths. This type of the microturbine-based MEMS were developed using EAPs that is known as polysilicon surficial micromachining silicon technology (PolyMUMPs), which is known for the effective transformation of wind kinetic energy into torque to drive electric generators. Furthermore, PolyMUMPs are associated with the provision of direct power. This system could be a future candidature for commercialization, since compared with the traditionally employed batteries that cannot be scaled down effectively to a microscale, these MEMS-based polymeric systems are characterized by the potential to scale down in addition to the extension of the operational range of the appliances that run on conventional batteries (Visconti et al., 2020). There is increased demand to provide power for sensors and other types of sensing devices that are present at distant locations. Therefore, research is being carried out on harvesting vibrational energy as an effective substitute for batteries. This research into harvesting vibrational energies at the microscale has carried out over the last two decades. In particular, significant recognition has been given to the use of mechanical nonlinearity for the dynamic response of the structure that has piezoelectric power generation. This is considered to be one of the promising solutions for the problem of vibrations with lower frequency that are used in ecospheric applications. To combat these issues, a vibrational energy harvester based on a nonlinear MEMS-scale was developed and tested for efficiency (Derakhshani et al., 2018).

EAPs have emerged as a promising component of MEMS applications that are aimed at energy systems, which could be attributed to the favorable features of these EAPs for strain rate, rapid responsiveness, compliance, and higher mechanical flexibility, although they are manufactured through harder synthetic routes (Bashir & Rajendran, 2018; Guo et al., 2019). In a previous study carried out with poly(vinylidenefluoride-co-trifluoroethylene) (P(VDF-TrFE)), which are piezo-electric polymers, the focus was on harvesting the mechanical energy. The study highlighted the possible use of electrostrictive materials for energy harvesting since they are operational at lower voltages and have promising electromechanical characteristics (Liu et al., 2005). Therefore, studies carried out with electrostrictive materials produced interesting results for functionality in harvesting vibrational energy by generating 1.5 μW/cm^3 of the power density. The material specifically used was polyaniline (PANI), which has significant conduciveness articulated in a polyurethane (PU) matrix at a considerably lower percolation threshold (Jaaoh et al., 2016). Other research investigated the use of an electrostrictive terpolymer, for instance, P(VDF-TrFE-CTFE). Here, the power generation capacity was higher and produced 7.2 μW/cm^3 on the application of a 5 V/μm of polarization field (Cottinet et al., 2011).

Design modifications have been carried out for the incorporation of the EAPs into MEMS. For instance, in a recent report, researchers integrated electrostrictive nanocomposites inside the matrix of microcantilever resonators with an organic composition, which was designed to harvest the mechanical energy produced from environmental vibrations. Specifically, the dispersion of the nanocomposite material that had considerable strain sensitivity, for instance, reduced graphene oxide (rGO) was carried out inside polydimethylsiloxane (PDMS). This type of EAP embedded MEMS is a cost-effective and eco-friendly procedure that combined printing techniques and xurography (Nesser et al., 2018). In another study on the use of nanoscale materials in MEMS, ternary composites were synthesized for higher density high density polyethylene (HDPE)/boron nitride (BN)/CNTs via melt blending and hotter sheets were obtained. This studied signified the development of polymeric composites that possess super thermal conductivity (Che et al., 2018). The demonstration of the synergistic impacts of the use of CNTs and graphene nanoplatelets (GNPs) has been studied for energy harvesting via the construction of a hybrid filler inside PVDF composites inside a thermal conduction network (Xiao et al., 2016). In a recent report, PDMS-rGO/C hybrid membranes were used for self-charging supercapacitors (PSCS) that had flexible piezoelectric characteristics (Lu et al., 2020).

A significant amount of work has been carried out using different types of EAPs in MEMS for the harvesting of the different types of energies (Cao et al., 2019; He et al., 2021; Mariappan et al., 2019). In a report on the harvesting of electrochemical energy, a powder of carbyne and N-doped carbyne polysulfide was synthesized by a mechanochemical procedure (Liu et al., 2017). Similarly, the preparation of carbyne polysulfide was carried out by combining carbyne (i.e., dehydrochlorination of polyvinylidene chloride) and sulfur in elemental form using coheating. The final product was then used as a cathodic material for use is a lithium-sulfur (Li/S) battery (Duan et al., 2013). Furthermore, a carbyne-rich carbon film has been utilized as an electrode material in supercapacitor MEMS (Bettini et al., 2016). In a recent report, the energy storage capacity of the chemically synthesized carbyne was carried out for electrochemical energy. A carbon film enriched with carbyne was used as a MEMS for harvesting the energy, and it produced 72 nW/cm^2 energy density instantly in addition to exhibiting promising electromechanical stability (Krishnamoorthy et al., 2019). A polymeric composition-based composite matrix based on PVDF was shown to be a self-charging system that was designed by embedding PVDF/sodium niobate, which behaved as an energy harvester, and PVDF/rGO SSC behaved as an energy storage unit (Pazhamalai et al., 2020). Other advances have improved the functionality of EAP-based MEMS aimed at harvesting different types of energy (Sunithamani & Lakshmi, 2015; Deng et al., 2018; Mangaiyarkarasi et al., 2019).

7.7 CHALLENGES AND CONCLUSIONS

The promising characteristics associated with different types of EAPs means that they are suitable candidates for use in different types of MEMS, which are aimed at applications in different domains of life. In particular, they are appreciated for their remarkable actuation strains and comparatively elevated energy densities compared with other materials. In addition, they have other favorable features, such as lighter weight, noiseless, economic viability, and remarkable tolerance toward damage. The first interest in this field occurred in 2000 and, since then, a significant amount of research has been carried out by altering the types of EAPs through rigorous experiments and optimization. In addition to this, their use in different types of MEMS has been attempted from different perspectives. EAPs were finally commercialized in 2011. Since then, many market-based products have been released to ease human life in different ways.

The favorable features associated with the use of EAPs in MEMS has revolutionized human life. However, some of the challenges associated with these polymeric substances cannot be overlooked. First, the persistence of all polymeric substances including EAPs and their derivatives in the ecosphere pose a threat to the environment and the inhabitants from microbial species that reside inside hydrospheric bodies to the top predators in the food chain that are the residents of the lithosphere. Therefore, the complete commercialization, synthesis, and design of these polymeric substances that does not consider the environment and biotic health could be devastating. Although the world has benefited from EAPs; however, some ares have suffered because of their persistence, which has led to harming the precious biota in the oceans. In addition, EAPs can exist in the soil for millions of years. Therefore, these challenges need to be resolved by the development of more eco-friendly plastics, especially those that can be easily biodegraded by microbial consortia. In addition to this ecospheric perspective, there are some technical problems associated with the use of EAPs in MEMS. For instance, the electrical responsiveness of EAPs is remarkable; however, on a practical scale, this responsiveness can only be obtained by the application of higher operational voltages, especially if a larger response is required. Therefore, their practical implementation in different types of MEMS requires further research and development to obtain maximum benefit.

REFERENCES

Abalyaeva, V. V., Efimov, M. N., Efimov, O. N., Karpacheva, G. P., Dremova, N. N., Kabachkov, E. N., & Muratov, D. G. (2020). Electrochemical synthesis of composite based on polyaniline and activated IR pyrolyzed polyacrylonitrile on graphite foil electrode for enhanced supercapacitor properties. *Electrochimica Acta, 354*:136671.

Afsheen, S., Naseer, H., Iqbal, T., Abrar, M., Bashir, A., & Ijaz, M. (2020). Synthesis and characterization of metal sulphide nanoparticles to investigate the effect of nanoparticles on germination of soybean and wheat seeds. *Material Chemistry and Physics* 252:123216.

Ahmad, K. S., and Jaffri, S. B. (2018). Carpogenic ZnO nanoparticles: amplified nanophotocatalytic and antimicrobial action. *IET nanobiotechnology, 13*:150–159.

Ahmad, K. S., & Jaffri, S. B. (2018a). Photosynthetic Ag doped ZnO nanoparticles: Semiconducting green remediators: Photocatalytic and antimicrobial potential of green nanoparticles. *Open Chemistry, 16*:556–570.

Ajdari, F. B., Kowsari, E., Nadri, H. R., Maghsoodi, M., Ehsani, A., Mahmoudi, H., & Ramakrishna, S. (2020). Electrochemical performance of Silsesquioxane-GO loaded with alkoxy substituted ammonium-based ionic liquid and POAP for supercapacitor. *Electrochimica Acta, 354*:136663.

Alipour, A., Lakouraj, M. M., & Tashakkorian, H. (2021). Study of the effect of band gap and photoluminescence on biological properties of polyaniline/CdS QD nanocomposites based on natural polymer. *Scientific Reports, 11*:1–15.

Auciello, O., & Aslam, D. M. (2021). Review on advances in microcrystalline, nanocrystalline and ultrananocrystalline diamond films-based micro/nano-electromechanical systems technologies. *Journal of Materials Science, 1*:1–60.

Bar-Cohen, Y. (2004). *Electroactive polymer (EAP) actuators as artificial muscles: Reality, potential, and challenges*. Bellingham: SPIE Press.

Bashir, M., & Rajendran, P. (2018). A review on electroactive polymers development for aerospace applications. *Journal of Intelligent Material Systems and Structures, 29*:3681–3695.

Bauer, S., & Bauer, F. (2008). Piezoelectric polymers and their applications. In W. Heywang, K. Lubitz, & W. Wersing (Eds.) *Piezoelectricity: evolution and future of a technology* (Vol. 114) Springer series in materials science. (pp. 157–177)Berlin/Heidelberg: Springer.

Bettini, L. G., Della Foglia, F., Piseri, P., & Milani, P. (2016). Interfacial properties of a carbyne-rich nanostructured carbon thin film in ionic liquid. *Nanotechnology, 27*:115403.

Boakye, C., & Mahto, V. (2021). Investigating the performance of an organically cross-linked grafted copolymer gel under reservoir conditions for profile modifications in injection wells. *Journal of Sol-Gel Science and Technology, 97*:71–91.

Bonneaud, C., Howell, J., Bongiovanni, R., Joly-Duhamel, C., & Friesen, C. M. (2021). Diversity of synthetic approaches to functionalized perfluoropolyalkylether polymers. *Macromolecules, 54*:521–550.

Boswell, B. R., Mansson, C. M., Cox, J. M., Jin, Z., Romaniuk, J. A., Lindquist, K. P., & Burns, N. Z. (2021). Mechanochemical synthesis of an elusive fluorinated polyacetylene. *Nature Chemistry, 13*:41–46.

Busurin, V. I., Korobkov, V. V., Korobkov, K. A., & Koshevarova, N. A. (2021). Micro-opto-electro-mechanical system accelerometer based on coarse-fine processing of fabry–perot interferometer signals. *Measurement Techniques, 1*:1–8.

Cao, L. M., Li, Z. X., Guo, C., Li, P. P., Meng, X. Q., & Wang, T. M. (2019). Design and test of the MEMS coupled piezoelectric–electromagnetic energy harvester. *International Journal of Precision Engineering and Manufacturing, 20*:673–686.

Chang, X. L., Chee, P. S., & Lim, E. H. (2020). Ionic polymer actuator with crenellated structures for MEMs application. 2020 IEEE International Conference on Semiconductor Electronics (ICSE) (pp. 160–163). IEEE.

Che, J., Jing, M., Liu, D., Wang, K., & Fu, Q. (2018). Largely enhanced thermal conductivity of HDPE/boron nitride/carbon nanotubes ternary composites via filler network-network synergy and orientation. *Composites Part A: Applied Science and Manufacturing, 112*:32–39.

Chen, L., Wu, J., Zhang, A., Zhou, A., Huang, Z., Bai, H., & Li, L. (2015). One-step synthesis of polyhydroquinone–graphene hydrogel composites for high performance supercapacitors. *Journal of Materials Chemistry A, 3*:16033–16039.

Cottinet, P. J., Lallart, M., Guyomar, D., Guiffard, B., Lebrun, L., Sebald, G., & Putson, C. (2011). Analysis of AC-DC conversion for energy harvesting using an electrostrictive polymer P (VDF-TrFE-CFE). *IEEE Transactions on Ultrasonics, Ferroelectrics, and Frequency Control, 58*:30–42.

Defaz, R. I., Epstein, M., & Federico, S. (2021). Analysis of solitary waves in fluid-filled thin-walled electroactive tubes. *Mechanics Research Communications, 113*:103654.

Deng, L., Fang, Y., Wang, D., & Wen, Z. (2018). A MEMS based piezoelectric vibration energy harvester for fault monitoring system. *Microsystem Technologies, 24*:3637–3644.

Derakhshani, M., Allgeier, B. E., & Berfield, T. A. (2018). A MEMS-scale vibration energy harvester based on coupled component structure and bi-stable states. *arXiv preprint*.

Diguet, G., Cavaille, J. Y., Sebald, G., Takagi, T., Yabu, H., Suzuki, A., & Miura, R. (2021). Physical behavior of electrostrictive polymers. Part 1: Polarization forces. *Computational Materials Science, 190*:110294.

Duan, B., Wang, W., Wang, A., Yuan, K., Yu, Z., Zhao, H., & Yang, Y. (2013). Carbyne polysulfide as a novel cathode material for lithium/sulfur batteries. *Journal of materials chemistry A*, *1*:13261–13267.

Espinosa-Hernandez, M. A., Reveles-Huizar, S., & Hosseini, S. (2021). Bio-microelectromechanical systems (BioMEMS) in bio-sensing applications-colorimetric detection strategies. Biomedical MEMS: 21–67.

Forouzanfar, S., Pala, N., Madou, M., &Wang, C. (2021). Perspectives on C-MEMS and C-NEMS biotech applications. *Biosensors and Bioelectronics*, *1*:113119.

Gandara, M., & Gonçalves, E. S. (2020). Polyaniline supercapacitor electrode and carbon fiber graphene oxide: Electroactive properties at the charging limit. *Electrochimica Acta*, *345*:136197.

Garcia-Ramirez, R., & Hosseini, S. (2021). History of bio-microelectromechanical systems (BioMEMS). *Biomedical MEMS*: 1–20.

Guo, Y., Cao, C., Luo, F., Huang, B., Xiao, L., Qian, Q., & Chen, Q. (2019). Largely enhanced thermal conductivity and thermal stability of ultra high molecular weight polyethylene composites via BN/CNT synergy. *RSC Advances*, *9*:40800–40809.

Gupta, K., Patel, R., Dias, M., Ishaque, H., White, K., & Olabisi, R. (2021). Development of an Electroactive Hydrogel as a Scaffold for Excitable Tissues. *International Journal of Biomaterials*, *1*:1–10.

Hafner, J., Teuschel, M., Disnan, D., Schneider, M., & Schmid, U. (2021). Large bias-induced piezoelectric response in the ferroelectric polymer P (VDF-TrFE) for MEMS resonators. *Materials Research Letters*, *9*:195–203.

He, T., & Yuan, J. (2021). A microbeam based piezoelectric energy harvester with LASMP. *Ferroelectrics*, *570*:1–14.

Hillmer, H., Iskhandar, M. S. Q., Hasan, M. K., Akhundzada, S., Al-Qargholi, B., & Tatzel, A. (2021). MOEMS micromirror arrays in smart windows for daylight steering. *Journal of Optical Microsystems*, *1*:014502.

Hryniewicz, B. M., Bach-Toledo, L., & Vidotti, M. (2020). Harnessing energy from micropollutants electrocatalysis in a high-performance supercapacitor based on PEDOT nanotubes. *Applied Materials Today*, *18*:100538.

Huh, J., Seo, D., & Kwon, S. (2017). Fabrication of a liquid monopropellant microthruster with built-in regenerative micro-cooling channels. *Sensors and Actuators A: Physical*, *263*:332–340.

Iannacci, J. (2021). Study of the Radio Frequency (RF) performance of a Wafer-Level Package (WLP) with Through Silicon Vias (TSVs) for the integration of RF-MEMS and micromachined waveguides in the context of 5G and Internet of Things (IoT) applications. Part 2: parameterised 3D model and optimisation. *Microsystem Technologies*, *27*:223–234.

Ibanez, J. G., Rincón, M. E., Gutierrez-Granados, S., Chahma, M., Jaramillo-Quintero, O. A., & Frontana-Uribe, B. A. (2018). Conducting polymers in the fields of energy, environmental remediation, and chemical–chiral sensors. *Chemical Reviews*, *118*:4731–4816.

Ijaz, M., Aftab, M., Afsheen, S., & Iqbal, T. (2020). Novel Au nano-grating for detection of water in various electrolytes. *Applied Nanoscience*, *10*:4029–4036.

Ijaz, M., Zafar, M., & Iqbal, T. (2020a). Green synthesis of silver nanoparticles by using various extracts: a review. *Inorganic Nano-Metal Chemistry*: 1–12.

Ijaz, M., Zafar, M., Afsheen, S., & Iqbal, T. (2020b). A review on Ag-nanostructures for enhancement in shelf time of fruits. *Journal of Inorganic and Organometallic Polymeric Materials*, *30*:1475–1482.

Iqbal, T., Farooq, M., Afsheen, S., Abrar, M., Yousaf, M., & Ijaz, M. (2019). Cold plasma treatment and laser irradiation of *Triticum* spp. seeds for sterilization and germination. *Journal of Laser Applications*, *31*:042013.

Jaaoh, D., Putson, C., & Muensit, N. (2016). Enhanced strain response and energy harvesting capabilities of electrostrictive polyurethane composites filled with conducting polyaniline. *Composites Science and Technology*, *122*:97–103.

Jaffri, S. B., & Ahmad, K. S. (2018). Augmented photocatalytic, antibacterial and antifungal activity of prunosynthetic silver nanoparticles. *Artificial Cells and Nanomedicine and Biotechnology*, *46*:127–137.

Jaffri, S. B., & Ahmad, K. S. (2018a). Neoteric environmental detoxification of organic pollutants and pathogenic microbes via green synthesized ZnO nanoparticles. *Environmental Technology*, *1*:1–10.

Jaffri, S. B., & Ahmad, K. S. (2020). Interfacial engineering revolutionizers: perovskite nanocrystals and quantum dots accentuated performance enhancement in perovskite solar cells. *Critical Reviewsin Solid State and Material Science*, *1*:1–29.

Jaffri, S. B., Ahmad, K. S., Thebo, K. H., & Rehman, F. (2020a). Sustainability consolidation via employment of biomimetic ecomaterials with an accentuated photo-catalytic potential: emerging progressions. *Reviews in Inorganic Chemistry*: 1–11.

Jamil, S. M., Abd Rahman, M., Shabri, H. A., & Othman, M. H. D. (2021). Solid Electrolyte Membranes for Low-and High-Temperature Fuel Cells. In Z. Zhang, W. Zhang, & M. Chehimi (eds), *Membrane Technology Enhancement for Environmental Protection and Sustainable Industrial Growth* (pp. 109–125). Cham: Springer.

Jia, H., & Chen, H. (2018). Using DSC technique to investigate the non-isothermal gelation kinetics of the multi-crosslinked chromium acetate (Cr^{3+}s)-Polyethyleneimine (PEI)-Polymer gel sealant. *Journal of Petroleum Science and Engineering, 165*:105–113.

Jiang, H., Yang, J., Xu, F., Wang, Q., Liu, W., Chen, Q., & Zhu, G. (2020a). VDF-content-guided selection of piezoelectric P (VDF-TrFE) films in sensing and energy harvesting applications. *Energy Conversion and Management, 211*:112771.

Jiang, Y., Liu, S., Zhong, M., Zhang, L., Ning, N., & Tian, M. (2020). Optimizing energy harvesting performance of cone dielectric elastomer generator based on VHB elastomer. *Nano Energy*.

Joo, H., Han, H., & Cho, S. (2020). Fabrication of poly (vinyl alcohol)-polyaniline nanofiber/graphene hydrogel for high-performance coin cell supercapacitor. *Polymers, 12*:928.

Jose, J., Jose, S. P., Prasankumar, T., Shaji, S., & Pillai, S. (2021). Emerging ternary nanocomposite of rGO draped palladium oxide/polypyrrole for high performance supercapacitors. *Journal of Alloys and Compounds, 855*:157481.

Karimi, S., Kazemi, S., & Kazemi, N. (2016). Syneresis measurement of the HPAM-Cr (III) gel polymer at different conditions: An experimental investigation. *Journal of Natural Gas Science and Engineering, 34*:1027–1033.

Ke, Z., You, L., Tran, D. T., He, J., Perera, K., Gumyusenge, A., & Mei, J. (2021). Thermally stable and solvent-resistant conductive polymer composites with cross-linked siloxane network. *ACS Applied Polymer Materials, 1*:1–10.

Krishnamoorthy, K., Mariappan, V. K., Pazhamalai, P., Sahoo, S., & Kim, S. J. (2019). Mechanical energy harvesting properties of free-standing carbyne enriched carbon film derived from dehydrohalogenation of polyvinylidene fluoride. *Nano Energy, 59*:453–463.

Kweon, B. C., Sohn, J. S., Ryu, Y., & Cha, S. W. (2020). Energy harvesting of ionic polymer-metal composites based on microcellular foamed Nafion in aqueous environment. *Actuators, 9*(3):71.

Liu, D. Y., & Reynolds, J. R. (2010). Dioxythiophene-based polymer electrodes for supercapacitor modules. *ACS Applied Materials & Interfaces, 2*:3586–3593.

Liu, Y., Ren, K. L., Hofmann, H. F., & Zhang, Q. (2005). Investigation of electrostrictive polymers for energy harvesting. *IEEE Transactions on Ultrasonics, Ferroelectrics, and Frequency Control, 52*:2411–2417.

Liu, Y., Wang, W., Wang, A., Jin, Z., Zhao, H., & Yang, Y. (2017). N-doped carbyne polysulfide as cathode material for lithium/sulfur batteries. *Electrochimica Acta, 232*:142–149.

Llerena Zambrano, B., Renz, A. F., Ruff, T., Lienemann, S., Tybrandt, K., Vörös, J., & Lee, J. (2021). Soft electronics based on stretchable and conductive nanocomposites for biomedical applications. *Advanced Healthcare Materials, 10*:2001397.

Lu, Y., Jiang, Y., Lou, Z., Shi, R., Chen, D., & Shen, G. (2020). Wearable supercapacitor self-charged by P (VDF-TrFE) piezoelectric separator. *Progress in Natural Science: Materials International, 30*:174–179.

Mangaiyarkarasi, P., Lakshmi, P., & Sasrika, V. (2019). Enhancement of vibration based piezoelectric energy harvester using hybrid optimization techniques. *Microsystem Technologies, 25*:3791–3800.

Mariappan, V. K., Krishnamoorthy, K., Pazhamalai, P., Sahoo, S., & Kim, S. J. (2019). Carbyne-enriched carbon anchored on nickel foam: A novel binder-free electrode for supercapacitor application. *Journal of Colloid and Interface Science, 556*:411–419.

Marín, F., Martínez-Frutos, J., Ortigosa, R., & Gil, A. J. (2021). A convex multi-variable based computational framework for multilayered electro-active polymers. *Computer Methods in Applied Mechanics and Engineering, 374*:113567.

Maroufi, M. A. Alipour, H. Alemansour, H., & Moheimani, S. O. R. (2019). Design and characterization of a MEMS probe scanner for on-chip atomic force microscopy, *International Conference on Manipulation, Automation and Robotics at Small Scales (MARSS)*, Helsinki, Finland, pp. 1–6.

Mi, Z., Zhang, Y., Hou, X., & Wang, J. (2021). Phase field modeling of dielectric breakdown of ferroelectric polymers subjected to mechanical and electrical loadings. *International Journal of Solids and Structures, 1*:1–10.

Minaian, N., Olsen, Z. J., & Kim, K. J. (2021). Ionic polymer-metal composite (IPMC) artificial muscles in underwater environments: Review of actuation, sensing, controls, and applications to soft robotics. *Bioinspired Sensing, Actuation, and Control in Underwater Soft Robotic Systems*, *1*:117–139.

Nenna, A., Nappi, F., Larobina, D., Verghi, E., Chello, M., & Ambrosio, L. (2021). Polymers and nanoparticles for statin delivery: Current use and future perspectives in cardiovascular disease. *Polymers*, *13*:711.

Nesser, H., Debeda, H., Yuan, J., Colin, A., Poulin, P., Dufour, I., & Ayela, C. (2018). All-organic microelectromechanical systems integrating electrostrictive nanocomposite for mechanical energy harvesting. *Nano Energy*, *44*:1–6.

Nguyen, T. M. G. (2020). Soft microelectromechanical systems and artificial muscles based on electronically conducting polymers (Conference Presentation). In *Electroactive Polymer Actuators and Devices (EAPAD) XXII* (Vol. 11375, p. 1137512). International Society for Optics and Photonics.

Ni, L., Pocratsky, R. M., & de Boer, M. P. (2021). Demonstration of tantalum as a structural material for MEMS thermal actuators. *Microsystems & Nanoengineering*, *7*:1–13.

Olfat, M., Strathearn, D., Lee, G., Sarkar, N., Hung, S. C., & Mansour, R. R. (2017). A single-chip scanning probe microscope array. In *2017 IEEE 30th International Conference on Micro Electro Mechanical Systems (MEMS)* (pp. 1212–1215). IEEE.

Olsen, Z. J., Kim, K. J., & Oh, I. K. (2021). Developing next generation ionic polymer–metal composite materials: perspectives for enabling robotics and biomimetics. *Polymer International*, *70*:7–9.

Ongun, M. Z., Oguzlar, S., Doluel, E. C., Kartal, U., & Yurddaskal, M. (2020). Enhancement of piezoelectric energy-harvesting capacity of electrospun β-PVDF nanogenerators by adding GO and rGO. *Journal of Materials Science: Materials in Electronics*, *31*:1960–1968.

Patel, S. N., & Mukherjee, S. (2020). Modeling and analysis of a taper ionic polymer metal composite energy harvester. *ISSS Journal of Micro and Smart Systems*, *9*:143–150.

Pazhamalai, P., Mariappan, V. K., Sahoo, S., Kim, W. Y., Mok, Y. S., & Kim, S. J. (2020). Free-standing pvdf/reduced graphene oxide film for all-solid-state flexible supercapacitors towards self-powered systems. *Micromachines*, *11*:198.

Peng, C. J., Seurre, L., Cattan, É., Nguyen, G. T. M., Plesse, C., Chassagne, L., & Cagneau, B. (2021). Toward an electroactive polymer-based soft microgripper. *IEEE Access*, *9*:32188–32195.

Peng, Z., & Zhong, W. (2020). Facile preparation of an excellent mechanical property electroactive biopolymer-based conductive composite film and self-enhancing cellulose hydrogel to construct a high-performance wearable supercapacitor. *ACS Sustainable Chemistry & Engineering*, *8*:7879–7891.

Ponnamma, D., Aljarod, O., Parangusan, H., & Al-Maadeed, M. A. A. (2020). Electrospun nanofibers of PVDF-HFP composites containing magnetic nickel ferrite for energy harvesting application. *Materials Chemistry and Physics*, *239*:122257.

Qu, J., Wu, H., Cheng, P., Wang, Q., & Sun, Q. (2017). Recent advances in MEMS-based micro heat pipes. *International Journal of Heat and Mass Transfer*, *110*:294–313.

Ren, H., Lee, H. S., & Chae, J. (2012). Miniaturizing microbial fuel cells for potential portable power sources: Promises and challenges. *Microfluidics and Nanofluidics*, *13*:353–381.

Rivkin, B., Becker, C., Akbar, F., Ravishankar, R., Karnaushenko, D. D., Naumann, R., & Schmidt, O. G. (2021). Shape-controlled flexible microelectronics facilitated by integrated sensors and conductive polymer actuators. *Advanced Intelligent Systems*, *1*:2000238.

Rohtlaid, K., Nguyen, G. T., Ebrahimi-Takalloo, S., Nguyen, T. N., Madden, J. D., Vidal, F., & Plesse, C. (2021). Asymmetric PEDOT: PSS trilayers as actuating and sensing linear artificial muscles. *Advanced Materials Technologies*, *1*:2001063.

Sanginov, E. A., Borisevich, S. S., Kayumov, R. R., Istomina, A. S., Evshchik, E. Y., Reznitskikh, O. G., & Bushkova, O. V. (2021). Lithiated Nafion plasticised by a mixture of ethylene carbonate and sulfolane. *Electrochimica Acta*, *1*:137914.

Sarikaya, S., Gardea, F., Auletta, J. T., Kavosi, J., Langrock, A., Mackie, D. M., & Naraghi, M. (2021). Athermal artificial muscles with drastically improved work capacity from pH-responsive coiled polymer fibers. *Sensors and Actuators B: Chemical*, *1*:129703.

Sencadas, V. (2020). Energy harvesting applications from poly (ε-caprolactone) electrospun membranes. *ACS Applied Polymer Materials*, *2*:2105–2110.

Sengupta, B., Sharma, V. P., & Udayabhanu, G. (2017). Gelation studies of an organically cross-linked polyacrylamide water shut-off gel system at different temperatures and pH. *Journal of Petroleum Science and Engineering*, *81*:145–150.

Shanthi, G., Rao, K. S., & Sravani, K. G. (2021). Performance analysis of EBG bandstop filter using U: shaped meander type electrostatically actuated RF MEMS switch. *Microsystem Technologies, 1*:1–8.

Si, S. K., Paria, S., Karan, S. K., Ojha, S., Das, A. K., Maitra, A., & Khatua, B. B. (2020). In situ-grown organo-lead bromide perovskite-induced electroactive γ-phase in aerogel PVDF films: an efficient photoactive material for piezoelectric energy harvesting and photodetector applications. *Nanoscale, 12*:7214–7230.

Song, J. H., Raza, S., van de Groep, J., Kang, J. H., Li, Q., Kik, P. G., & Brongersma, M. L. (2021). Nanoelectromechanical modulation of a strongly-coupled plasmonic dimer. *Nature Communications, 12*:1–7.

Spinks, G. M. & Smela, E. (200). *Handbook of Conducting Polymers: Conjugated Polymers Processing and Applications*, T. Skotheim, & J. R. Reynolds (Eds) (3rd ed.) (pp. 14–1).Boca Raton: CRC Press.

Spitalsky, Z., Tasis, D., Papagelis, K., & Galiotis, C. (2010). Carbon nanotube–polymer composites: chemistry, processing, mechanical and electrical properties. *Progress in polymer science, 35*:357–401.

Su, J., & Tajitsu, Y. (2016). Piezoelectric and electrostrictive polymers as EAPs: Materials. *Electromechanically Active Polymers: A Concise Reference, 509*–531.

Sun, F., Zhu, Z., & Ma, L. (2020). Multiprobe scanning probe microscope using a probe-array head. *Review of Scientific Instruments, 91*:123702.

Sun, Q., Qu, J., Yuan, J., & Wang, Q. (2017). Operational characteristics of an MEMS-based micro oscillating heat pipe. *Applied Thermal Engineering, 124*:1269–1278.

Sunithamani, S., & Lakshmi, P. (2015). Simulation study on performance of MEMS piezoelectric energy harvester with optimized substrate to piezoelectric thickness ratio. *Microsystem Technologies, 21*:733–738.

Sunithamani, S., Rooban, S., Nalinashini, G., & Rajasekhar, K. (2020). A review on mems based vibration energy harvester cantilever geometry (2015–2020). *Materials Today: Proceedings, 1*:1–10.

Surmenev, R. A., Chernozem, R. V., Pariy, I. O., & Surmeneva, M. A. (2021). A review on piezo-and pyroelectric responses of flexible nano-and micropatterned polymer surfaces for biomedical sensing and energy harvesting applications. *Nano Energy, 79*:105442.

Swager, T. M. (2017). 50th anniversary perspective: Conducting/semiconducting conjugated polymers. a personal perspective on the past and the future. *Macromolecules, 50*:4867–4886.

Synodis, M., Pyo, J. B., Kim, M., Wang, X., & Allen, M. G. (2021). Lithographically patterned polypyrrole multilayer microstructures via sidewall-controlled electropolymerization. *Journal of Micromechanics and Microengineering, 31*:025008.

Takagi, K., Kitazaki, Y., & Kondo, K. (2021). A simple dynamic characterization method for thin stacked dielectric elastomer actuators by suspending a weight in air and electrical excitation. *Actuators, 10*(3):40.

Taufiq Musa, M., Shaari, N., & Kamarudin, S. K. (2021). Carbon nanotube, graphene oxide and montmorillonite as conductive fillers in polymer electrolyte membrane for fuel cell: an overview. *International Journal of Energy Research, 45*:1309–1346.

Teng, W., Zhou, Q., Wang, X., Che, H., Hu, P., Li, H., & Wang, J. (2020). Hierarchically interconnected conducting polymer hybrid fiber with high specific capacitance for flexible fiber-shaped supercapacitor. *Chemical Engineering Journal, 390*:124569.

Thetpraphi, K., Chaipo, S., Kanlayakan, W., Cottinet, P. J., Le, M. Q., Petit, L., & Capsal, J. F. (2020). Advanced plasticized electroactive polymers actuators for active optical applications: Live mirror. *Advanced Engineering Materials, 22*:1901540.

Thuau, D., Sun, Q., Dufour, I., & Ayela, C. (2020). PVDF-based electroactive polymers for flexible electronics (Conference Presentation). In *Electroactive Polymer Actuators and Devices (EAPAD) XXII* (Vol. 11375, p. 1137514). International Society for Optics and Photonics.

Tiliakos, N. (2013). MEMS for harsh environment sensors in aerospace applications: selected case studies. *MEMS for Automotive and Aerospace Applications*: 245–282.

Visconti, P., Varona, J., Vel, R., Giannocaro, N. I., De Fazio, R., & Carrasco, M. (2020). MEMS-based Micro-scale Wind Turbines as Energy Harvesters of the Convective Airflows in Microelectronic Circuits. *International Journal of Renewable Energy Research, 10*:1213–1225.

Wang, T., Farajollahi, M., Choi, Y. S., Lin, I. T., Marshall, J. E., Thompson, N. M., & Smoukov, S. K. (2016). Electroactive polymers for sensing. *Interface focus, 6*:20160026.

Wang, Z. G., Gong, F., Yu, W. C., Huang, Y. F., Zhu, L., Lei, J., & Li, Z. M. (2018). Synergetic enhancement of thermal conductivity by constructing hybrid conductive network in the segregated polymer composites. *Composites Science and Technology, 162*:7–13.

Wankhade, S. H., Tiwari, S., Gaur, A., & Maiti, P. (2020). PVDF–PZT nanohybrid based nanogenerator for energy harvesting applications. *Energy Reports*, *6*:358–364.

Washington, A., Neubauer, J., & Kim, K. J. (2021). Soft actuators and their potential applications in rehabilitative devices. *Soft Robotics in Rehabilitation*: 89–110.

Xiao, Y. J., Wang, W. Y., Chen, X. J., Lin, T., Zhang, Y. T., Yang, J. H., & Zhou, Z. W. (2016). Hybrid network structure and thermal conductive properties in poly (vinylidene fluoride) composites based on carbon nanotubes and graphene nanoplatelets. *Composites Part A: Applied Science and Manufacturing*, *90*:614–625.

Xie, M., Meng, H., Chen, J., Zhang, Y., Du, C., Wan, L., & Chen, Y. (2021). High-volumetric supercapacitor performance of ordered mesoporous carbon electrodes enabled by the faradaic-active nitrogen doping and decrease of microporosity. *ACS Applied Energy Materials*, *1*:1–10.

Yang, J., Tang, L. S., Bao, R. Y., Bai, L., Liu, Z. Y., Yang, W., & Yang, M. B. (2017). Largely enhanced thermal conductivity of poly (ethylene glycol)/boron nitride composite phase change materials for solar-thermal-electric energy conversion and storage with very low content of graphene nanoplatelets. *Chemical Engineering Journal*, *315*:481–490.

Yang, J., Yao, J., & Ma, Y. (2021). A highly flexible, renewable and green alginate polymer for electroactive biological gel paper actuators reinforced with a double-side casting approach. *Cellulose*: 1–16.

Ye, H., Wang, Q., Sun, Q., & Xu, L. (2020). High energy density and interfacial polarization in poly (vinylidene fluoride-chlorotrifluoroethylene) nanocomposite incorporated with halloysite nanotube architecture. *Colloids and Surfaces A: Physicochemical and Engineering Aspects*, *606*:125495.

Yoo, B., Xu, Z., & Ding, F. (2021). How single-walled carbon nanotubes are transformed into multiwalled carbon nanotubes during heat treatment. *ACS omega*, *1*:1–10.

Zhang, C. L., Lai, Z. H., Rao, X. X., Zhang, J. W., & Yurchenko, D. (2020). Energy harvesting from a novel contact-type dielectric elastomer generator. *Energy Conversion and Management*, *205*:112351.

Zhao, G., Dai, C., You, Q., Zhao, M., & Zhao, J. (2013). Study on formation of gels formed by polymer and zirconium acetate. *Journal of sol-gel science and technology*, *65*:392–398.

Zhou, Y., Zou, Y., Peng, Z., Yu, C., & Zhong, W. (2020). Arbitrary deformable and high-strength electroactive polymer/MXene anti-exfoliative composite films assembled into high performance, flexible all-solid-state supercapacitors. *Nanoscale*, *12*:20797–20810.

Zhu, D., Hou, J., Meng, X., Zheng, Z., Wei, Q., Chen, Y., & Bai, B. (2016). Effect of different phenolic compounds on performance of organically cross-linked terpolymer gel systems at extremely high temperatures. *Energy & Fuels*, *31*:8120–8130.

8 Electroactive Polymers for Sensors

Xiupeng Sun, Chaoyue Li, Dan Ling, Wentao Zheng, and Yanchao Mao

CONTENTS

8.1 INTRODUCTION

Electroactive polymers (EAPs) are materials that exhibit mechanical changes when stimulated by an electrical stimulus. EAPs are classified as ionic and electronic EAPs, according to the mechanism. In general, ionic EAPs are driven by the mobility or diffusion of ions, and electronic EAPs are driven by an electric field (Kim and Tadokoro, 2007). Some EAPs can convert mechanical motion into electrical signals, which means that EAPs can be used as generators and sensors (Bar-Cohen and Zhang, 2008). With proper design, EAP-based sensors can carry out haptic or pressure sensing, human motion monitoring, health monitoring, and even chemical sensing (Wang *et al.*, 2016). In addition, because of the softness of EAPs, they are more suitable for applications in flexible electronics than traditional sensors, which agrees with the future development trend in electronics. According to this, EAPs are promising materials for sensing and will play an increasingly significant role in flexible electronics and the Internet of Things (IoT). This chapter deals with EAP-based sensors using several examples. The sensing mechanism of different EAP-based sensors will also be discussed.

8.2 ELECTRONIC ELECTROACTIVE POLYMERS FOR SENSORS

For electronic EAPs, dielectric elastomers were used to assemble capacitive sensors and triboelectric sensors, and the piezoelectric polymers were applied as piezoelectric sensors. These sensors are suitable for health monitoring, human motion monitoring, and pressure detection.

DOI: 10.1201/9781003173502-8

8.2.1 Dielectric Elastomers for Sensors

8.2.1.1 Capacitive Sensors

Dielectric elastomers, which are a sub-group of electronic EAPs, are extensively used in capacitive sensing applications. The EAP-based capacitive sensors show low hysteresis, high stability, low cost, and easy fabrication, and are suitable for wearable electronics, soft robotics, and artificial intelligence devices (Atalay *et al.*, 2018; Zhang *et al.*, 2019).

Taking advantage of the EAPs-based capacitive sensors, Kim et al. reported a stretchable and transparent contact lens sensor to record intraocular pressure wirelessly with high sensitivity, as shown in Figure 8.1 (Kim *et al.*, 2017). The increase in intraocular pressure is the main risk factor for glaucoma, which probably leads to human blindness. This sensor can provide early diagnosis and therapy for glaucoma, and therefore, helps glaucoma patients. It is fabricated by sandwiching silicone elastomer (ecoflex) as a dielectric layer between two inductive spiral antennas that are made from graphene-silver nanowires hybrid electrodes. The sensor's circuit contains the antenna coil's inductance (L), the internal resistance (R), and the capacitance (C) that depend on the dielectric layer, which constitutes an electrical RLC resonant circuit. Figure 8.1(A) shows the principle of the contact lens sensor that responds to elevated intraocular pressure. When the intraocular pressure

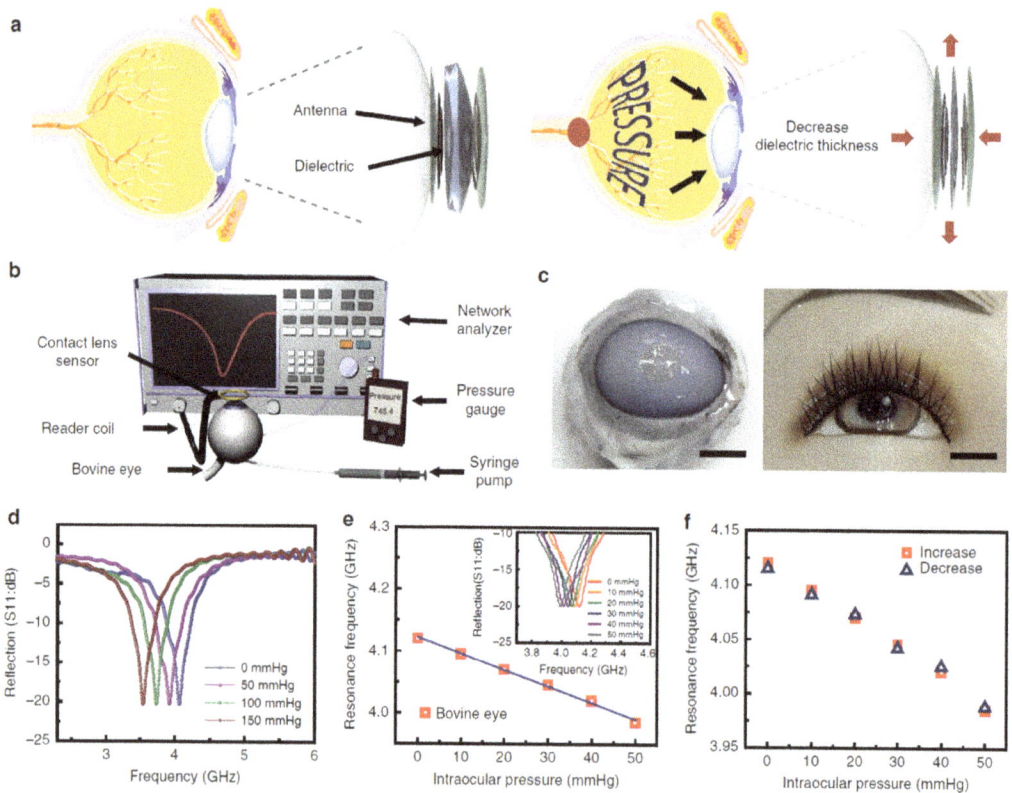

FIGURE 8.1 EAP-based capacitive sensor: (A) Sketch map presenting the principle of sensing intraocular pressure; (B) Arrangement schematic of in vitro testing of this sensor on a bovine eyeball; (C) photographs of the contact lenses sensor worn on the bovine eyeball and the model human eyeball (scale 1 cm); (D) Reflection spectra gathered from a bovine-eyeball-implanted sensor; (E) sensor frequency response when worn on bovine eye; and (F) sensor's frequency response during a pressing cycle.

(From Kim *et al.* 2017. With permission.)

increases, the curvature's corneal radius becomes larger, and therefore, the C and L of the sensor will become larger by reducing the dielectric layer and expanding the spiral coils, respectively. Therefore, the reflection spectra of the spiral antenna collected from the sensor will be moved to a lower frequency by ocular hypertension (Chen *et al.*, 2014). As shown in Figure 8.1(b), aligning the reader coil over the internal sensor, the intraocular pressure was in situ wireless sensed. The resonance frequency, which is a function of L and C, is the detected variable by the readout system. A bovine eyeball, having a similar structure to the human eyeball, was used to in vitro test the contact lens sensor. As shown in Figure 8.1(c) the sensor on the eye does not obstruct the field of vision because of its adequate transparency. The reflection spectra shown in Figure 8.1(d) was gathered from a bovine-eyeball-implanted sensor. The decline in the resonance frequency resulted in the raised C at a higher intraocular pressure. As shown in Figure 8.1(e), the resonance frequency decreased linearly with the sensitivity of 2.64 MHz/mm Hg. The frequency that responded to the intraocular pressure was tested using the pressure sensor inserted into an eyeball, which proved the reproducibility and relevance, as shown in Figure 8.1(F). Due to the excellent ability to monitor intraocular pressure, this contact lens sensor is promising for next-generation ocular diagnostics.

8.2.1.2 Triboelectric Sensors

Using dielectric elastomers as a triboelectric layer to assemble a triboelectric nanogenerator, some sensitive sensors can be produced (Pu *et al.*, 2017; Jin *et al.*, 2020; Zhou *et al.*, 2020). Zhou et al. reported yarn-based extensible triboelectric sensor arrays (YSSAs) for a wearable system of sign-to-speech translating (Zhou *et al.*, 2020). The YSSA's fundamental sensing unit is a strip-type composite structure, which consists of a central rubber microfiber and surrounding conductive yarn coil that are all encapsulated by a polydimethylsiloxane sleeve. When the strip-type composite structure was stretched or released, the contact area between the polydimethylsiloxane sleeve and the conductive yarn changed and the charge transfer occurred due to the triboelectrification and electrostatic induction. Therefore, the hand gesture movements and other body movements could be converted by YSSAs into electrical signals, when installing these sensor arrays conformally on the human body. Combined with a wireless printed circuit board that has the functions of signal-regulating, processing and wireless transmitting, real-time sign language acquisition equipment was produced, as shown in Figure 8.2(a). To evaluate the capability of the YSSAs to respond to finger motions, a sensing unit's output signals on the index finger at five bending points was measured (Figure 8.2(b)). The larger bending of the finger triggered a larger electrical signal, which proved the YSSAs' reliability of monitoring motion status. As shown in Figure 8.2(c), the multiple finger motions were measured using the YSSAs attached to a glove, which made it possible to monitor the motion of the whole hand. Moreover, the YSSAs could detect changes in facial expression (Figure 8.2(d)). When frowning or opening the mouth, the YSSAs will generate the corresponding feature signals, as shown in Figures 8.2(e and f), respectively. The signal flow in this system is shown in Figure 8.2(g), which shows that this system wirelessly transmits signals to terminal equipment after converting the YSSAs generated analog signals into digital signals. Using machine learning, this system could translate the sign language into speech in a short time. This wearable system of sign-to-speech translating could promote communicating efficiently between non-signers and signers (Zhou *et al.*, 2020).

8.2.2 Piezoelectric Polymers for Sensors

Recently, piezoelectric polymers have been considered to be promising materials for pressure sensors, because they possess high sensitivity, have a self-powered ability, have fast response time, and have great mechanical properties (Han *et al.*, 2019; Gao *et al.*, 2019; Jiang *et al.*, 2020). Polyvinylidene fluoride (PVDF) is one of the most common piezoelectric polymer materials (Cha *et al.*, 2011; Ribeiro *et al.*, 2018).

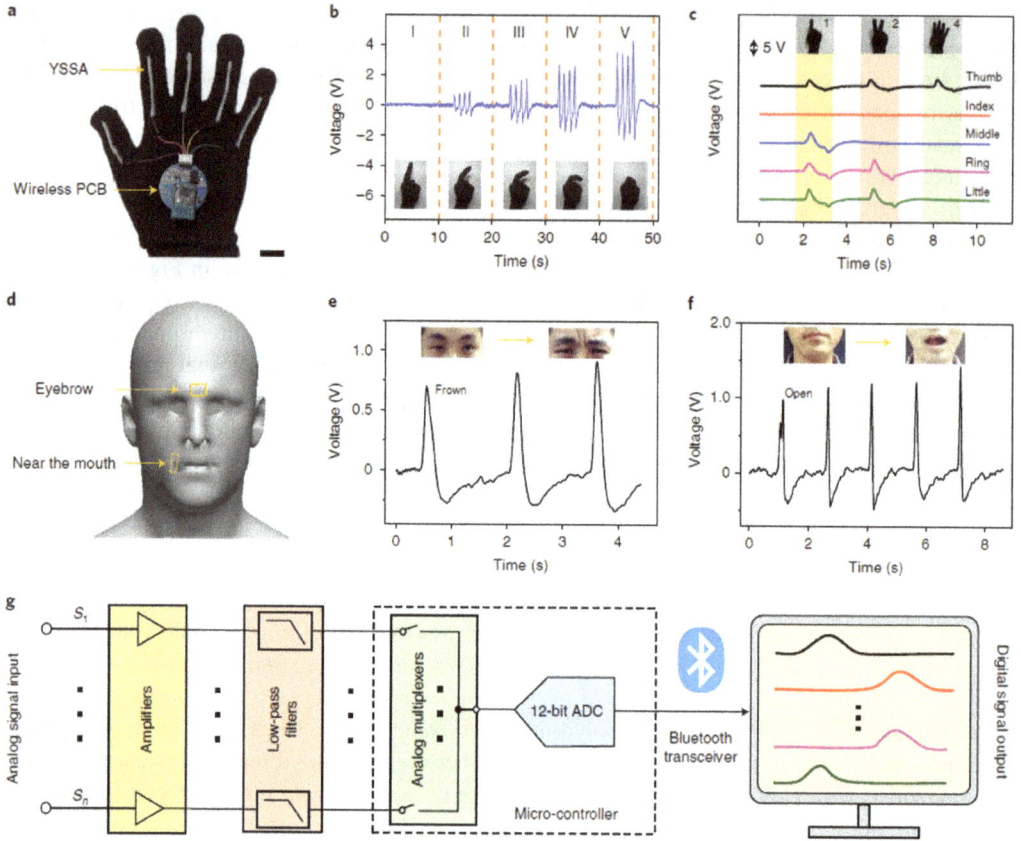

FIGURE 8.2 EAP-based triboelectric sensor: (a) photograph of a glove with the YSSAs and wireless PCB (2 cm scale); (b) One sensing unit's output signals on the finger at five bending degrees; (c) measured signals of YSSA when expressing numbers 1, 2, and 4; (d) location of YSSA when detecting changes in facial expression; (e) corresponding feature signals of frowning; (f) corresponding feature signals of opening the mouth; and (g) signal flow of this system.

(From Zhou _et al._ 2020. With permission.)

Tian et al. fabricated one type of zirconate titanate (PZT)/PVDF piezoelectric sensor that was integrated into a smart racket to monitor the movement from table tennis balls in training, as shown in Figure 8.3(a) (Tian _et al._, 2019). This device is composed of a signal processing module, wireless transmission module, and PVDF/PZT composites. The sensor arrays manufactured from PVDF/PZT composites were spread evenly on the racket, and the signal processing module and wireless transmission module were mounted on the racket's handle. In addition, as shown in Figure 8.3(b), the piezoelectric sensor unit's detailed structure is a layered sandwich structure. The functional layer, which is a baklava-structured piezo composite, contains PVDF lamellar crystal and PZT particles. Meanwhile, Kapton films and aluminum foils act as encapsulation and electrodes, respectively, which make the sensor flexible. The layered baklava structure responded to external forces very quickly and was sensitive to stress changes. Of note, the piezoelectric performance was improved by the accumulative effect of the potential and the inorganic particles synergistic effect. The schematic diagram and physical appearance of the process to manufacture this type of sensor is shown in Figure 8.3(c). To show the application potential of this device, the sensor arrays were integrated on a racket to monitor the signals from ping pong (Figure 8.3(d)), then the contact force and hit location were obtained in detail, which provided unique guidance for an individual.

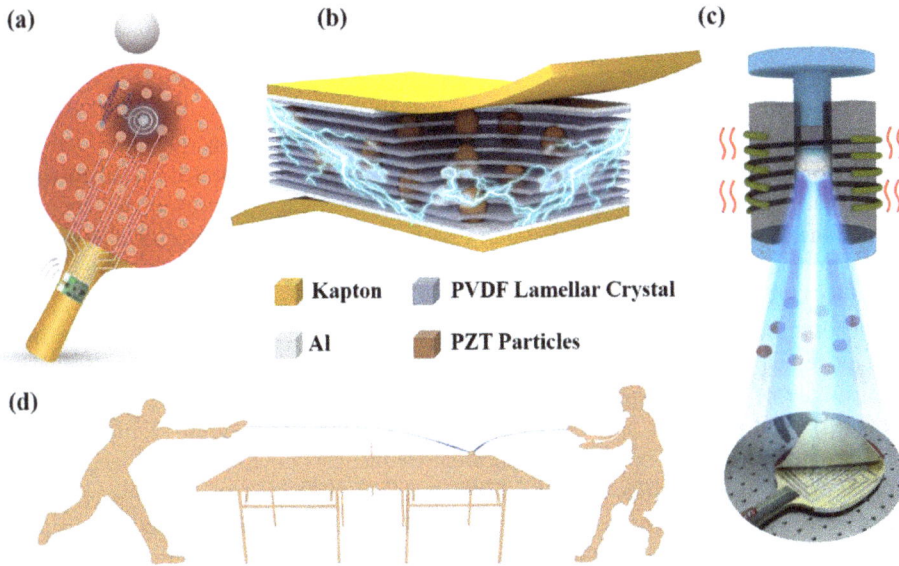

FIGURE 8.3 PZT/PVDF piezoelectric sensor integrated into a smart racket for individual table tennis training; (a) Schematic of the sensor device; (b) sensor unit's detailed structure; (c) manufacturing method of polymer piezoelectric sensor and optical photograph of it; and (d) application of the device integrated piezoelectric polymer sensor.

(From Tian *et al.* 2019. With permission.)

Liquid crystal elastomers (LCEs), which are a special type of EAP, are prepared by mixing ferroelectric liquid crystals into a cross-linked matrix (White and Broer, 2015; Kato *et al.*, 2018). The mechanism of LCE-based sensors is similar to piezoelectric sensors, which result from the piezoelectric effect of ferroelectric liquid crystals. Therefore, LCE-based sensors will not be discussed separately.

8.3 IONIC ELECTROACTIVE POLYMERS FOR SENSORS

Ionic EAPs, including ionic polymer-metal-composites (IPMCs), conducting ionic polymer gels, and carbon nanotubes (CNTs), can be used as vibration sensors, haptic sensors, and chemical sensors. They are discussed in the following sections.

8.3.1 CONDUCTING IONIC POLYMER GELS FOR SENSORS

Ionic polymer gels (Wang *et al.*, 2019; Ge *et al.*, 2020; Yeom *et al.*, 2020), which are an important class of ionic EAPs, have been widely used in skin-like strain sensors for their excellent electrical signal transduction properties, stretchability, and biocompatibility. Wang et al. developed a gel-based strain sensor with poly(3,4-ethylene dioxythiophene) polystyrene sulfonate PEDOT–sulfonated lignin (SL) as the conductive material, which can precisely detect human physiological signals (Wang *et al.*, 2019). For the preparation of the hydrogel, lignin was separated from the black liquor of radiated pine Kraft pulping was sulfonated to turn into SL first. The next step is combing SL with PEDOT to form the PEDOT:SL solution. Then, to prepare a PEDOT:SL-PAA-X' hydrogel, MBA, APS and AA were mixed into the previous solution to obtain a conductive hydrogel with a cross-linked structure. Finally, the previous hydrogel was immersed in glycerol to replace the solvent to obtain the PEDOT:SL-PAA-X organohydrogel. The hydrogel was connected to a LED bulb in an energized circuit.

As shown in Figure 8.4(a), the LED bulb's brightness gradually decreased as the hydrogel stretched from 0% to 200%. This indicated that the strain affected the resistance of the hydrogel. The change in the hydrogel's resistance with strain was due to the change in the conductive network's distribution, because of the hydrogel's deformation that was caused by an external force (Figure 8.4(b)). Therefore, the conductivity of hydrogel changed in response. The gauge factor, which is the ratio of relative resistance change to strain, is used to express the strain sensitivity of resistance. The hydrogel's gauge factor improved when the PEDOT:SL content increased, which could reach seven when the strain was 100%, as shown in Figure 8.4(c). The previous characterizations meant that the hydrogels could be used to develop sensors for strain sensing. By attaching the hydrogel to a volunteer's wrist and finger, the corresponding signals were generated due to the wrist and finger's bending movements, which are shown in Figures 8.4 (d and e). In addition, monitoring the large movements of a human, the hydrogel sensor could detect weak physiological signals due to the sensitivity of the hydrogel sensor. Figure 8.4(f) shows that a regular pulse from the throat (Figure 8.4(g)), and a dicrotic notch were observed, as shown in the inset. Vibrations in the throat could be detected; therefore, breathing at different frequencies could be recognized by the sensor, which is shown in Figure 8.4(h). Because different throat vibrations are generated when speaking different English phrases, the sensor could distinguish the phrases, which is shown in Figures 8.4(I and J). There are many differences between the figures, which correspond to "Hello" and "Thank you", respectively. These findings demonstrate that hydrogel is an extremely promising material for precise strain sensing, which has the potential for use in human motion monitoring.

8.3.2 Ionic Polymer–Metal Composites for Sensors

IPMCs have excellent electromechanical conversion performance and act as a type of piezoelectric sensor, which has the potential for monitoring various human activities (Biddiss and Chau, 2006; Ming et al., 2018). When they are bent by an external force, the movable ions of the ionic polymer layer will move in a specific direction, which results in a potential difference between the electrodes.

Ming et al. fabricated an IPMC sensor that was easy to manufacture to diagnose a subtle pulse, perceive accurate braille and control a machine hand precisely (Ming et al., 2018). When an IPMC sensor deforms due to external stress, it causes ions anisotropic transmission and accumulation; therefore, the sensing signal is generated, which is shown in Figure 8.5(a). As shown in Figure 8.5(b), the sensor was given six forward bends of 3 mm bending displacement and six reverse bends for identical bending displacement by the displacement platform. The output signals revealed that the direction of the sensor bending by the signal from the voltage from the corresponding signal can be judged. To investigate the repeatability of the IPMC sensor and the electrical output signal for different stretches, a sequence of displacement deformations from 1 to 7 mm were applied 100 times, and the corresponding signal is shown in Figure 8.5(c). Figure 8.5(d) shows the change in voltage increased linearly with the strain, which reflected a high degree of linearity and strain coefficient. As shown in Figure 8.5(e), the IPMC sensor was given 3 mm displacement for 8,000 cycles in approximately 9 h and the output signals were stable during the process. All of these demonstrate the electrical properties of this IPMC sensor. The IPMC sensors ensured excellent performance to realize the application for human–computer interaction and portable wearable medical devices.

8.3.3 Carbon Nanotubes for Sensors

Compared with other EAPs, CNTs have some unique properties, such as their one-dimensional nature and large surface area (Zaporotskovaa et al., 2016; Rasheed et al., 2019). Due to their ability to store charge and molecules, CNTs exhibit actuation and sensing behavior (Shaalan et al., 2020; Doshi et al., 2018) and have received much attention. A CNTs-based ethylene gas sensor was developed

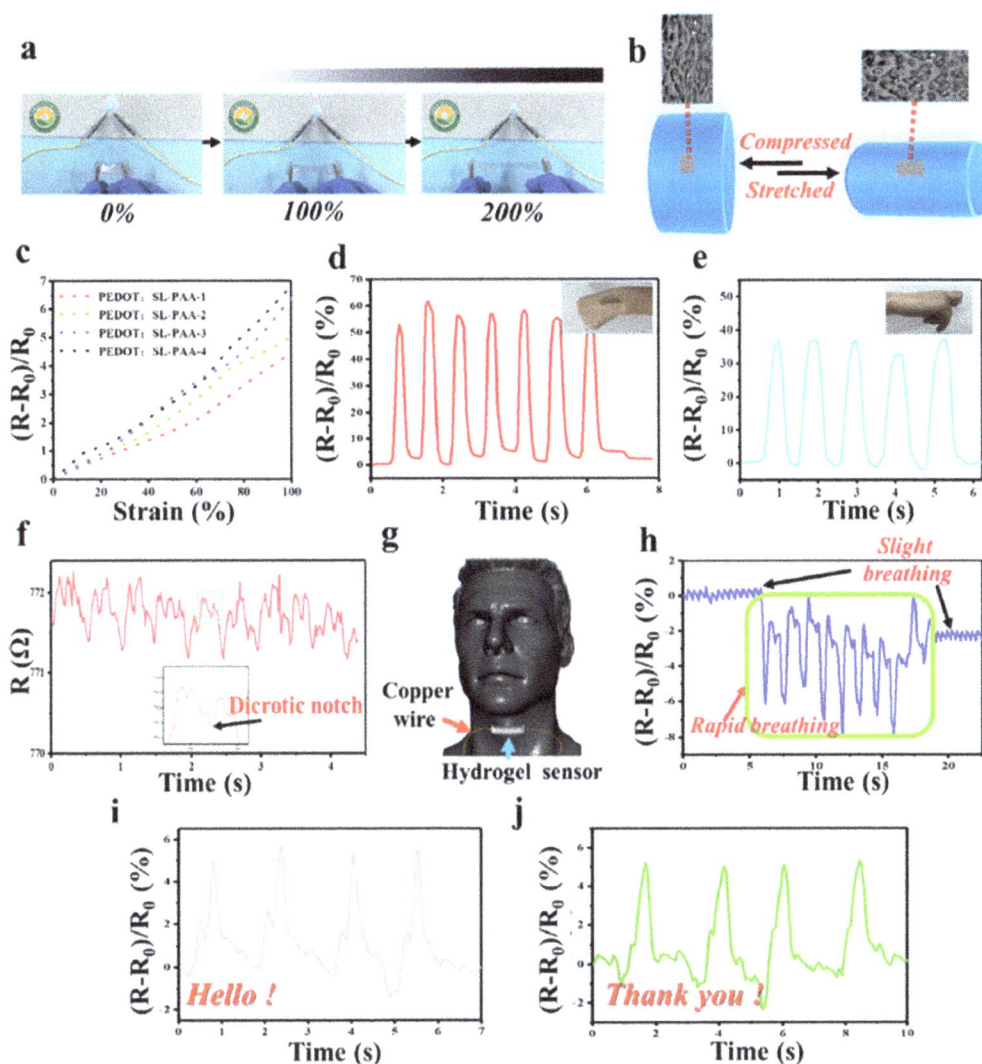

FIGURE 8.4 Hydrogel-based sensor for monitoring human motion: (a) variations in the LED bulb's brightness with different strains applied to the hydrogel; (b) change in internal conductive network when stretching the sensor; (c) hydrogel sensor's relative resistance changes; (d)relative resistance changes when bending wrist; (e) bending finger; (f) relative resistance changes in response to human pulse; (g) sensing on the throat; (h) resistance changes corresponding to two frequencies of respiration; and (i) different English phrases can be distinguished due to the different resistance changes of two kinds of soundings, "Hello" (i) and "Thank you" .

(From Wang *et al.* 2019. With permission.)

for monitoring food spoilage by Shaalan *et al.* (2020). The sensor was based on multi-walled carbon nanotubes, and the reaction of the ethylene with its surface was the mechanism for sensing. The reaction caused a charge transfer in the CNTs and then the resistance of the CNTs changed. To carry out the sensing measurements, a gas senor testing platform was set up, as shown in Figure 8.6. As shown in this figure, the measurement system consisted of microbial fuel cells, a temperature-controlled chamber, and electrical measurement which included the CNTs-based gas sensor. The gas sensor is shown in this figure. The first step to prepare the gas sensor was depositing two gold

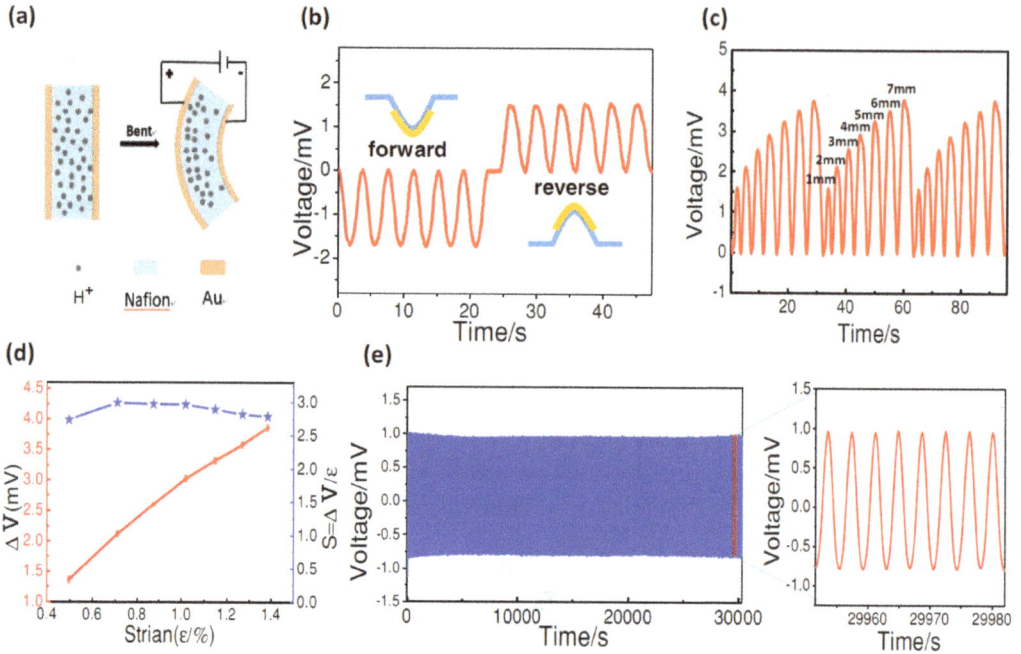

FIGURE 8.5　IPMC sensor: (a) IPMC sensor's working principle; (b) Difference between forward and reverse bending signals; (c) Potential measuring from 1 to 7 mm bending displacement; (d) sensitivity and relative potential changes of the sensor at different strains; and (e) repeatability and stability test of the sensor.

(From Ming *et al.* 2018. With permission.)

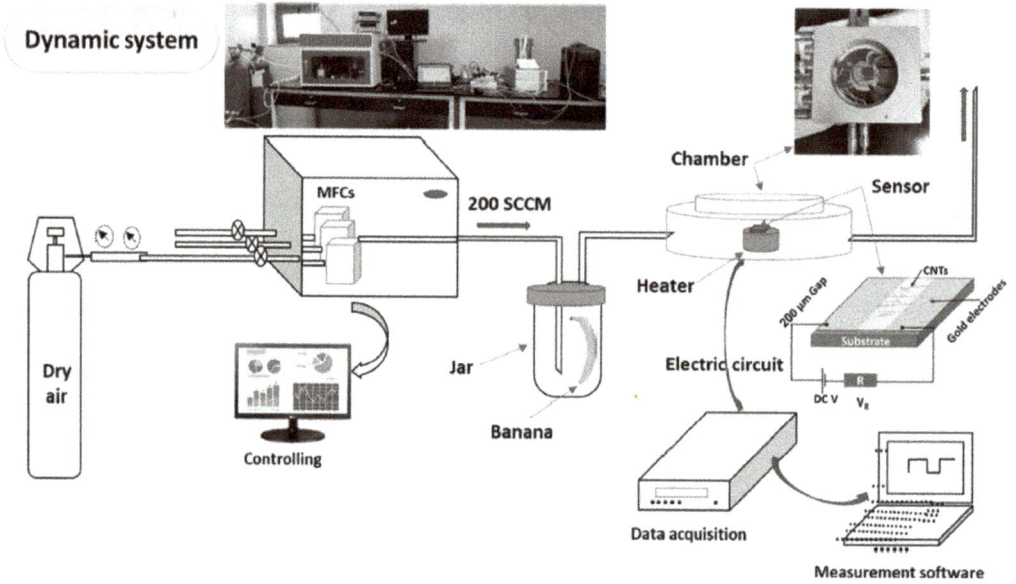

FIGURE 8.6　CNT-based measurement system for gas sensing.

(From Shaalan *et al.* 2020. With Permssion.)

electrodes onto a glass substrate by DC-sputtering with approximately a 200 μm gap. Then, a solution mixture in ethanol was used to line the CNT powder in the gap. Following drying in ambient air, the gas sensor was inserted into a Linkam chamber whose temperature was controlled by the heating system. A sensor characterization instrument system, which was attached to the chamber, was built to carry out all measurements. For testing sensing characterizations, the resistance for the baseline in the air was attained. In the test, the sensor was exposed to the gas for 120 s. Then, it was exposed to dry air for 300 s to recover. The cycles of exposure to gas and dry air were examined. Before measuring the sensor performance response to ethylene, a banana was placed in a closed jar to obtain ethylene. To pass the ethylene in the jar into the chamber, the microbial fuel cells were used to enable the dry air to pass into the jar at a 200 mL/min rate. Then, the electrical measurement was carried out by data acquisition and software, which was used to calculate the response of the sensor. The results of the test showed that the sensor had a reliable response and short response-recovery times. Therefore, it could be applied for fast food ripening or spoilage monitoring.

8.4 SUMMARY

In this chapter, a brief introduction of the EAPs-based sensors was presented. Due to their unique electrical and mechanical properties, EAPs have been successfully applied to assemble multiple sensors for haptic and pressure sensing, human motion monitoring, health monitoring, and chemical sensing. The EAP-based sensors were divided into two categories for discussion because there are two types of EAPs. Due to different sensing principles and various design methods, EAPs-based sensors have great potential in flexible electronics and IoT. With the development of electronic manufacturing technology, EAP-based sensors could be more significant and prevalent in the near future.

REFERENCES

Atalay, O., A. Atalay, J. Gafford, & C. Walsh. (2018). A highly sensitive capacitive-based soft pressure sensor based on a conductive fabric and a microporous dielectric layer. *Advanced Materials Technologies*, 3:1700237.

Bar-Cohen, Y., & Q. Zhang. (2008). Electroactive polymer actuators and sensors. *MRS Bulletin*, 33(3):173–181.

Biddiss, E., & T. Chau. (2006). Electroactive polymeric sensors in hand prostheses: Bending response of an ionic polymer-metal composite. *Medical Engineering & Physics*, 28(6):568–578.

Cha, S., Kim S. M., Kim, H., Ku, J., Sohn, J. I., Park, Y. J. ... Kim, K. (2011). Porous PVDF as Effective Sonic Wave Driven Nanogenerators. *Nano Letters*, 11(12):5142–5147.

Chen, L., Tee, B. K. Chortos, A., Schwartz, G., Tse, V. Lipomi, D. ... Bao Z. (2014). Continuous wireless pressure monitoring and mapping with ultra-small passive sensors for health monitoring and critical care. *Nature Communications*, 5:5028.

Doshi, S. M., & Thostenson, E. T. (2018). Thin and flexible carbon nanotube-based pressure sensors with ultrawide sensing range. *ACS Sensors*. 3 (7): 1276–1282.

Gao, X., Zheng, M., Yan, X., Fu, J., Zhu, M., & Hou, Y. (2019). The alignment of BCZT particles in PDMS boosts the sensitivity and cycling reliability of a flexible piezoelectric touch sensor. *Journal of Materials Chemistry C*, 7:961–967.

Ge, G., Lu, Y., Qu, X., Zhao, W., Ren, Y., Wang, W.... Dong, X. (2020). Muscle-inspired self-healing hydrogels for strain and temperature sensor. *ACS Nano*, 14(1)218–228.

Han, J., Li, D., Zhao, C., Wang, X., Li, J., & Wu, X. (2019). Highly sensitive impact sensor based on PVDF-TrFE/Nano-ZnO composite thin film. *Sensors*, 19(4):830.

Jiang, J., Tu, S., Fu, R., Li, J., Hu, F., Yan, B....Chen, S. (2020). Flexible piezoelectric pressure tactile sensor based on electrospun BaTiO3/Poly(vinylidene fluoride) nanocomposite membrane. *ACS Applied Materials & Interfaces*, 12(30):33989–33998.

Jin, T., Sun, Z., Li, L., Zhang, Q., Zhu, M., Zhang, Z. ... Lee, C. (2020). Triboelectric nanogenerator sensors for soft robotics aiming at digital twin applications. *Nature Communications*, 11:5381.

Kato, T., Uchida, J., Ichikawa, T., & Sakamoto, T. (2018). Functional liquid crystals towards the next generation of materials. *Angewandte Chemie International Edition*, *57*(16):4355–4371.

Kim, J., Kim, M., Lee, M. S., Kim, K., Ji, S., Kim, Y. … Park, J. U. (2017). Wearable smart sensor systems integrated on soft contact lenses for wireless ocular diagnostics. *Nature Communications*, 8:14997.

Kim, K. J. & Tadokoro, S. (2007). Electroactive Polymers for Robotic Application. (pp. 1–36) *Artificial Muscles and Sensors*. London: Springer.

Ming, Y., Yang, Y., Fu, R. P., Lu, C., Zhao, L., Hu, Y. M. … Chen, W. (2018). IPMC sensor integrated smart glove for pulse diagnosis, braille recognition, and human–computer interaction. *Advanced Materials Technology*, 3:1800257.

Pu, X., Guo, H., Chen, J., Wang, X., Xi, Y., Hu, C., & Wang, Z. L. (2017). Eye motion triggered self-powered mechnosensational communication system using triboelectric nanogenerator. *Science Advances*, *3*(7):e1700694.

Rasheed, T., Nabeel, F., Adeel, M., Rizwan, K., Bilal, M., & Iqbal, H. M. N. (2019). Carbon nanotubes-based cues: A pathway to future sensing and detection of hazardous pollutants. *Journal of Molecular Liquids*, *292*:111425.

Ribeiro, C., Costa, C., Correia, D., Nunes-Pereira, J., Oliveira, J., Martins, P. … Lanceros-Méndez, S. (2018). Electroactive poly(vinylidene fluoride)-based structures for advanced applications. *Nature Protocols*, *13*:681–704.

Shaalan, N. M., Ahmed, F., Kumar, S., Melaibari, A., Hasan, P. M. Z. & Aljaafari, A. (2020). Monitoring food spoilage based on a defect-induced multiwall carbon nanotube sensor at room temperature: Preventing food waste. *ACS Omega, 5* (47):30531–30537.

Tian, G., Deng, W., Gao, Y., Xiong, D., Yan, C., He, X. … Yang, W. (2019). Rich lamellar crystal baklava-structured PZT/PVDF piezoelectric sensor toward individual table tennis training. *Nano Energy*, *59*:574–581.

Wang, Q., Pan, X., Lin, C., Lin, D., Ni, Y., Chen, L. … Ma, X. (2019). Biocompatible, self-wrinkled, anti-freezing and stretchable hydrogel-based wearable sensor with PEDOT: Sulfonated lignin as conductive materials. *Chemical Engineering Journal*, *370*:1039–1047.

Wang, T., Farajollahi, M., Choi, Y. S., Lin, I. T., Marshall, J. E., Thompson, N. M. …Smoukov, S. K. (2016). Electroactive polymers for sensing. *Interface Focus*, *6*:20160026.

White, T., & Broer, D. (2015). Programmable and adaptive mechanics with liquid crystal polymer networks and elastomers. *Nature Materials*, *14*:1087–1098.

Yeom, J., Choe, A., Lim, S., Lee, Y., Na, S., & Ko, H. (2020). Soft and ion-conducting hydrogel artificial tongue for astringency perception. *Science Advances*, *6*(23):eaba5785. doi:10.1126/sciadv.aba5785.

Zaporotskovaa, I., Borozninaa, N., Parkhomenkob, Y., & Kozhitovb, L. (2016). Carbon nanotubes: Sensor properties. A peview. *Modern Electronic Materials*, *2*(4):95–105.

Zhang, J., Wan, L., Gao, Y., Fang, X., Lu, T., Pan, L., & Xuan, F. (2019). Highly stretchable and self-healable MXene/polyvinyl alcohol hydrogel electrode for wearable capacitive electronic skin. *Advanced Electronic Materials*, *5*:1900285.

Zhou, Z., Chen, K., Li, X., Zhang, S., Wu, Y., Zhou, Y., & Chen, J. (2020). Sign-to-speech translation using machine-learning-assisted stretchable sensor arrays. *Nature Electronics*, *3*:571–578.

9 Conductive Electroactive Polymers in Electrocatalysis and Sensing Applications

Achi Fethi, Benmoussa Fateh, Henni Abdellah,
Zembouai Idris, and Kaci Mustapha

CONTENTS

9.1 INTRODUCTION

Most polymers are good electrical insulators and any electrical conductivity in a polymer was initially considered to be an undesirable phenomenon. By the 1970s, excellent conductivity properties in certain polymers had been discovered, which allowed the development of a new class of materials called intrinsically electronically conductive polymers (PCEIs) (Heeger, 2001). This class of conductive polymers has become increasingly important, thanks to the awarding of the Nobel Prize for chemistry in 2000 to A. Heeger, A. MacDiarmid, and H. Shirakawa, following the discovery of the first conductive polymer: polyacetylene (PA) in 1977 (Shirakawa *et al.*, 1977). Therefore, many conjugated polymers that combine excellent conductivity and good stability have been developed (Li *et al.*, 2021; Balint, Cassidy, and Cartmell, 2014; Wang, Y. *et al.* 2020). Biosensors based on conducting polymers are miniaturizable and have specific electroactive and mechanical characteristics. The potential benefits of these instruments is their low-cost and detection at trace levels, which potentially leads to improved therapy access (Tandon *et al.*, 2020). Cancer, which has

DOI: 10.1201/9781003173502-9

a high death rate, has become more common worldwide. Early detection, prognosis, and recovery management with robust and non-invasive techniques will potentially be the focus in the future (Wang, L. *et al.* 2017).

Electrochemical biosensors could be a good candidate for cancer theranostics due to their advantage of ultra-sensitivity, high selectivity, low-cost, fast readability mand simplicity. In addition, electrochemical biosensors are simpler to miniaturize and mass produce, which makes them more suitable for point-of-care applications (Zhang, W. *et al.,* 2020). The use of polymers as immunosensor materials has been encouraging, especially for cancer biomarkers detection, such as carcinoembryonic antigen (CEA) (Tothill, 2009, Tahalyani, Rahangdale, and Khushbu, 2016). Nanostructured conducting polymers, such as polyaniline (PANI), polythiophene (PTh), and polypyrrole (Ppy) are strongly favored to improve cancer biomarkers immunosensor properties.

This chapter describes the utilization of conducting polymers in sensing applications and highlights their benefits for trace-level monitoring of organic compounds and for monitoring cancer biomarkers. The electrodeposition methods for conductive polymers to obtain thin films for sensing interfaces will be reviewed. In addition, the use of biopolymers and nanocomposites including carbon materials is discussed.

9.2 CONDUCTING POLYMERS FOR ELECTROCHEMICAL SENSING APPLICATIONS

The determination of chemical compounds by electrochemical sensing platforms offers real-time monitoring and on-site analysis. The analytical performance of electrochemical sensors strongly depends on the nature of the added nanomaterials (Bensana and Achi, 2020). In addition to their high conductivity, which provides the sensitivity of the electrochemical sensors, conducting polymers offer a rigid, highly stable, and suitable microstructure for immobilizing other biomaterials (Lange, Roznyatovskaya, and Mirsky, 2008) (Janata and Josowicz, 2003).

This section will discuss the effect of this type of nanomaterials on the analytical characteristics of electrochemical sensors for monitoring phenols and cancer biomarkers.

9.2.1 Conducting Polymers: Synthesis and Applications

9.2.1.1 Polyaniline

PANI is a low-cost conductive polymer, which is easily prepared, has tunable properties, high capacitance values, and enhanced chemical stability (Bhadra *et al.,* 2009). PANI's electric conductivity varies from 10 to 100 S/cm, which makes it a suitable material to construct immunosensors (Epstein, 2007). For instance, a novel conductive and protein-resistant redox PANI-polythionine hydrogel (PANI-PThi gel) that successful detected tumor markers, such as carcinoma-125 (CA125) with LOD= 0.00125 U/mL (Zhao, L. and Ma, 2018). In addition, electro-polymerized PANI on the surface of a glassy carbon electrode (GCE) forms thin layer modified with functionalized carbon nanotubes (CNTs) for the determination of prostate-specific antigen (PSA) (0.5 pg/mL). This strategy is fast and promotes high electrical conductivity and demonstrates good mechanical properties of thin film-based immunosensing platforms (Assari *et al.,* 2020). Of interest, mixing PANI nanowires with antifouling materials, such as zwitterionic poly(carboxybetaine methacrylate) is stable and displays an ultra-sensitive response for CEA detection in blood serum (LOD=3.05 fg/mL) (Wang, J. and Hui, 2019).

9.2.1.2 Polypyrrole

PPy is a highly conductive polymer with good stability in oxidized states, its high water solubility and excellent catalytic activity mean that is can be used in constructing chemical sensors using one-step electrodeposition methods. However, compared with PANI, the cost of pyrrole monomers is higher than that of aniline, which means it is less desirable for certain applications (MacDiarmid,

1997). Employing PPy as a conductive polymer significantly improves the electron transfer rate in the recognition process and promotes the integration of high quantities of antibodies. Therefore, Moon et al. (2014) developed an immunosensor platform based on direct integration of PSA into PPy electropolymerized three-dimensional nanowire. This method enhanced the molecular interaction with trace-level PSA detection (0.3 fg/mL) and a wide linear range (from 10 fg/mL to 10ng/mL) (Moon et al., 2014). Similarly, Pei et al. (2019) obtained a limit of detection for alpha-fetoprotein (AFP) of 17 fg/m using PPy nanotubes (PPyNTs) with platinum nano dendrites (PtNDs) functionalized with molybdenum disulfide (MOS_2) (Pei et al., 2019).

Furthermore, Truong et al. (2011) assembled a sensitive immunosensor by screen printing carbon ink based PPy carboxylic acid copolymer to allow HCG antibody immobilization through the COOH groups. The polymer film showed high conductivity and strong biocompatibility with a limit of detection of 2.3 pg/mL (Truong et al., 2011).

A sandwich-type electrochemical immunosensor using functionalized PPy microspheres that could simultaneously monitor two tumor markers (e.g., LOD= 0.40 pg/mL and 0.33 pg/mL) for CEA and AFP, respectively has been developed (Zhao J. et al., 2016a).

9.2.1.3 Polythiophene

PTh polymer has abundant carboxyl groups, which improves the electroanalytical performance of immunosensors. For instance, detecting Interleukin 1β in human serum and saliva by ITO modified PTh displays a wide linear range from 0.01–3.00 pg/mL with LOD= 3 fg/mL (Aydın, Aydın, and Sezgintürk, 2018). In addition, PTh derivatives, such as epoxy-substituted-PTh polymer (PThiEpx) has several epoxy groups that facilitate its binding to the NH_2 antibody groups, and therefore, effectively immobilizes Interleukin-1 alpha antibodies (Table 9.1) (Aydın, 2019).

9.2.1.4 Poly-amidoamine

This type of conducting polymer has multiple chain ends and is a highly branched dendritic macromolecule. The unique surface properties make it a suitable material for electrochemical sensing interfaces. Therefore, PAMAM can be easily combined with ferric oxide (Fe_3O_4) magnetic nanoparticles (Yin et al., 2011), or with CoTe quantum dots (CoTe QDs) composite (LOD=1 nM) in milk (Yin et al., 2010). Similarly, a BPA sensor based on PAMAM dendrimers displays a low detection limit toward BPA detection (LOD=0.5 nM) (Yin et al., 2010).

9.2.1.5 Polymerized Ionic Liquids and Other Conducting Polymers

Unlike conductive polymers, where mobility is usually due to the π-conjugated electrons along the polymer chain, this type of polymer has high ionic conductivity without ion doping. Polymerized ionic liquids provide excellent adaptability with various forms of ionicity in polymer chains (Eftekhari and Saito, 2017). Therefore, very high stability of 95% for 2 months for the detection of Bisphenol A was attained using a polymerized ionic liquid (Ma et al., 2014). Moreover, Wang et al. (2018) constructed a BSA sensor using a polymerized ionic liquid functionalized with graphene oxide (Go) that displays a high sensitivity of 0.2629 µA/µM (Wang, Y. et al., 2018). In addition, a new strategy based on the combination of cetyltrimethylammonium bromide with multi-walled carbon nanotubes (MWCNTs) using an electropolymerization technique onto the surface of pencil graphite electrode. The obtained sensor was highly sensitive toward BPA detection (84.6 µA/µM) and provides a lower detection limit (LOD=134 pM) (Bolat, Yaman, and Abaci, 2018).

9.2.2 Sensors Based on Conducting Polymers for the Detection of Phenolic Compounds

As previously described, PANI is a low-cost and biocompatible polymer with excellent thermal and mechanical properties. The use of PANI as a material modifier with a silver nanowire (AgNWs)

TABLE 9.1
Electroanalytical performance of conducting polymer-based sensors for cancer detection

Conductive polymer	Immunosensor composition	Cancer biomarker	Technology	Linear range	LOD	Reference
PANI	GCE/COOH-MWCNTs/PANI/AuNPs/anti-PSA/BSA	PSA	DPV	1.66 ag/mL– 1.3 ng/mL	0.5 pg/mL	(Zhao, J. et al., 2016b)
	PtlPANI-Au/N,S-GQDs/anti-CEA	CEA	EIS	0.5–1000 ng/mL	0.01 ng/mL	(Ganganboina and Doong, 2019)
	GCE/rGO/MoS$_2$@PANI BSA/anti-CEA	CEA	CV	0.001–80 ng/mL	0.3 pg/L	(Song et al., 2020)
	anti-CEA/polyCBMA/PANI/GCE	CEA	DPV	1×10^{-14} g/mL– 1×10^{-10} g/mL	3.05 fg/mL	(Wang J. and Hui, 2019)
PPy	ITO AB/EpxS-PPyr Composite/IL6 receptor/BSA	IL6	EIS	0.01–50 pg/mL	3.2 fg/mL	(Aydın, Aydın, and Sezgintürk, 2021)
	BSA/anti-PSA doped Ppy NWs	PSA	DPV	10 fg/mL–10 ng/mL	0.3 fg/mL	(Moon et al., 2014)
	BSA/Anti AFP/Pt NDs/PDDA/MoS$_2$@PPyNTs/GCE	AFP	Amp.	50 fg/mL–50 ng/mL	17 fg/mL	(Pei et al., 2019)
	CE/PPy–PPa copolymer/anti-HCG	HCG	EIS	0–1000 pg/mL	2.3 pg/mL	(Truong et al., 2011)
	BSA/anti-PSA/AgPt@PtHNs/PPyNS/GCE	PSA	Amp.	0.0005–50 ng/mL	120.3 fg/mL	(Wang, P. et al., 2020)
	PPy@signaltags@Au NPs/Ab2/BSA/Ab1/AuNPs/rGO/GCE	AFP CEA	DPV	1 pg ml^{-1} – 50 ng ml^{-1}	0.33 pg m L^{-1} 0.40 pg mL^{-1}	(Zhao, J. et al., 2016a)
PTh	ITO/PThiEpx/anti-IL 1α	IL–1α	EIS	0.01 pg/mL– 5.5 pg/mL	3.4 fg mL^{-1}	(Aydın. 2019)
	ITO/Polymer P3-TMA /anti-IL-1β/BSA	IL-1β	EIS	0.01-3 pg/mL	3 fg mL^{-1}	(Aydın, Aydın, and Sezgintürk, 2018)

composite can detect p-nitrophenol at a nano level scale (LOD=52 nM) (Zhang, C. *et al.*, 2017). PANI can be used to modify the surface of an ITO electrode for the sensing of p-nitrophenol (Roy *et al.*, 2013). Poly(vinyl ferrocenium) perchlorate (PVF⁺) is a conducting polymer that acts as an electron transfer mediator. As shown by Kavanoz and Pekmez (2012), its combination with PANI provided a thin film that promoted wide linear concentration ranges for the detection of hydroquinone from 0.16 µM to 115 mM (Table 9.2).

Wan *et al.* (2016) demonstrated that the sensitivity of a Bisphenol A sensor could be enhanced (1.14 µA/µM) using electrochemical deposition of Pt nanoclusters on a GCE surface modified with a PTh-MWCNT composite (Wan *et al.*, 2016). Bianchini *et al.* (2014) electrodeposited poly(3,4-ethylenedioxythiophene) on the surface of a Pt electrode using sodium poly(styrene-4-sulfonate) as a surfactant. The method formed a thin film and the sensing interface could determine caffeic acid in wine from 10.00 nM to 6.50 mM (Bianchini *et al.*, 2014). Similarly, the direct electropolymerization of poly(3,4-ethylenedioxythiophene) on the surface of a GCE could detect BPA at a micro level (Mazzotta, Malitesta, and Margapoti, 2013).

9.2.3 Conducting Polymers as Sensor Modifiers for Cancer Detection

The incorporation of PANI decorated gold nanowires via thiol bonding with graphene QDs can anchor anti-CEA and acts as a probe for amplifying the electrochemical current signal. This label-free immunosensor exhibits a wide linear range from 0.5 to 1000.0 ng/mL with a LOD of 0.01 ng/mL (Ganganboina and Doong, 2019). Functionalizing PANI with transition metal dichalcogenides is a method to obtain nanocomposites with a high specific surface area and a good loading ability for antibodies. For instance, an immunosensor for detecting CEA at pico level concentration (0.3 pg/mL) was constructed using PANI functionalized molybdenum disulfide (MoS₂) mixed with reduced graphene oxide (rGO) (Ganganboina and Doong, 2019). Another electrochemical immunosensor for detecting Interleukin 6 with acetylene black and epoxy-substituted-PPy was constructed (Table 9.2). The nanocomposite deposited on an ITO electrode displayed good biocompatibility and a stable response with a low detection limit (3.2 fg/mL) (Aydın, Aydın, and Sezgintürk, 2021).

9.2.4 Conducting Polymer-Based Carbon Nanocomposites

The use of polymer nanocomposites is a suitable approach to improve the performance of polymers, and graphene is possibly the most promising nanofiller. Currently, the chemical synthesis of Go from graphite (GR) is a common method to obtain a layered material with more functional groups (Paredes *et al.*, 2008, Bourlinos *et al.*, 2003, Niyogi *et al.*, 2006). The groups are hydrophilic and good interfacial linkers, which improves polymer incorporation (Madhad and Vasava, 2019).

The use of conducting polymers to construct an efficient sensing platform are essential. Especially, when using molecular imprinting polymers (MIPs) techniques to assemble various graphene materials, such as GR, Go, or rGo (Liang *et al.*, 2017). The use of a graphene nanosheet and poly(4–vinylpyridine) material as electrode modifiers offers sensitive detection of catechol and provides good stability (99% 2 months) with high sensitivity (580 µA/µM) (Tehrani, Ghadimi, and Ab Ghani, 2013).

To enhance the electron transfer in the electrochemical oxidation of hydroquinone, Kavanoz and Pekmez (2012) synthesized a polydopamine-coated graphene sheets decorated with Ag. The catechol sensor displayed a wide linear range (from 0.5 to 240.0 µM) and a detection limit of 0.1 µM. Tan *et al.* (2016) prepared graphene quantum dots (GQDs) dropped onto the GCE surface followed by electropolymerization of pyrrole to obtain a PPy film. This method enhanced water solubility and minimized the interface resistance between aqueous solutions and the graphene interface.

TABLE 9.2

Electrochemical sensors based conductive polymers for the detection of phenolic compounds

E-Matrix/Electrode	Sensitivity	L.R.	L.O.D.OD	pH	Stability (%)	Reference
NiTPPS/MWCNTs-Nafion/GCE	0.142 µA/µM	0.05–50 µM	15 nM	7.2	N.R.	(Liu et al., 2011)
Poly(CTAB)/MWCNTs/PGE	84.6 µA/µM	2 nM–0.808 µM	0.134 nM	5.0	N.R.	(Bolat, Yaman, and Abaci, 2018)
MIPPy/GQDs/GCE	1,1716 µA/µM	0.1–50 µM	40 nM	7.0	95% 15 days	(Tan et al., 2016)
PBPIDS/GCE	0.2008 µA/µM	10 nM–10 µM	8 nM	8.0	95% 2 months	(Ma et al., 2014)
Pyrogallol red/CPE	0.623 µA/µM	10–120 µM	18 nM	7.4	N.R.	(Ganesh et al., 2018)
Gs-P4VP/GCE	660 µA/µM	0.1–10 µM	8.1 nM	2.5	99% 2 months	(Tehrani, Ghadimi, and Ab Ghani, 2013)
PVF$^+$-PANI/Pt	0.83 µA/mM	0.16 µM–115mM	49,4 nM	4.0	65% 40 days	(Kavanoz and Pekmez, 2012)
GO-poly (NPBimBr)/GCE	0.2629 µA/µM	0.2–10 µM	17 nM	7.0	N.R.	(Yanying Wang et al., 2018)
PDNPH/AGCE	1.19 µA/µM/cm^2	20–250 µM	0.76 µM	7.0	95% 2 weeks	(Lopa et al., 2017)
MIPs/Go/GCE	1.295 µA/µM	4 nM–10 µM	0.5 nM	6.0	98,6% 10 days	(Liang et al., 2017)
AgNWs-PANI/GCE	1.032 µA/µM	0.6–32 µM	52 nM	7.0	87% 20 days	(Zhang et al., 2017)
PAMAM/Fe$_3$O$_4$/GCE	N.R.	0.01–3.07 µM	5 nM	7.0	86% 30 days	(Yin et al., 2011)
PAMAM/CoTeQDs/GCE	59.27 nA/µM	0.013–9.89 µM	1 nM	8.0	72% 35 days	(Yin et al., 2010)
PAMAM-AuNPs-SF/GCE	0.4455 µA/µM	1 nM – 1.33 µM	0.5 nM	8.0	91,4% 2 weeks	(Yin et al., 2010)
PANI-PVSA/ITO	1.5 mA/mM	N.R.	1µM	7.0	45 days (shelf-life)	(Roy et al., 2013)

9.3 ELECTRODEPOSITION METHODS FOR CONDUCTIVE POLYMERS

Electropolymerization can be defined as an electrochemical process for manufacturing a polymer film on a substrate, which is composed of a working electrode, from a solution that contains the monomer, the solvent, and the supporting electrolyte. These will be incorporated into the polymer during the process as a dopant ion. Electrochemical syntheses are carried out in aqueous or organic solvents, using assemblies with three electrodes: a working electrode that oxidizes the polymer; a reference electrode to control; and a counter electrode that allows the passage of current (Figure 9.1(a)). The electropolymerization process involves the transfer of electrons, in either direction, between the substrate and the monomer in solution (Berkes, Bandarenka, and Inzelt, 2015). It is the charged monomer that then allows the polymerization reaction to take place.

Many electrochemical techniques, such as cyclic voltammetry (Choo *et al.*, 2020, Samukaite-Bubniene *et al.*, 2021), photocurrent spectroscopy (Walsh *et al.*, 2013), electrochemical impedance spectroscopy (Olean-Oliveira, Oliveira Brito, and Teixeira, 2020), or electrochemical quartz–crystal microbalance (Zhao, M. *et al.* 2021) have been applied for the deposition of the polymer onto the surface of the anode. The quality of an electrochemically prepared polymeric film depends on many factors (Mello and Mulato, 2018). The most commonly employed electrochemical methods for forming polymer films from a monomer solution are cyclic voltammetry, chronopotentiometry, or chronoamperometry. As previously described, PTh and its derivatives (Contreras-Herrera *et al.*, 2018; Rajendran *et al.*, 2021), PANI (Korent *et al.*, 2020), and PPy (Rakhrour *et al.*, 2021) are among the most conductive polymers used in electrochemical polymerization.

9.3.1 POTENTIODYNAMIC ELECTROPOLYMERIZATION

These methods allow the very precise control of the morphology of the polymer, and the mass and thickness that is deposited. Cyclic voltammetry is useful for observing the progression of the electrochemical reaction and often displays useful information about the polymerization method and for the development and characterization of polymers (Babaiee, Pakshir, and Hashemi, 2015) (Figure 9.1(a)). It consists of a continuous potential sweep that varies with time. The result is the appearance of the oxidation or reduction reactions of the electroactive species in solution, possibly

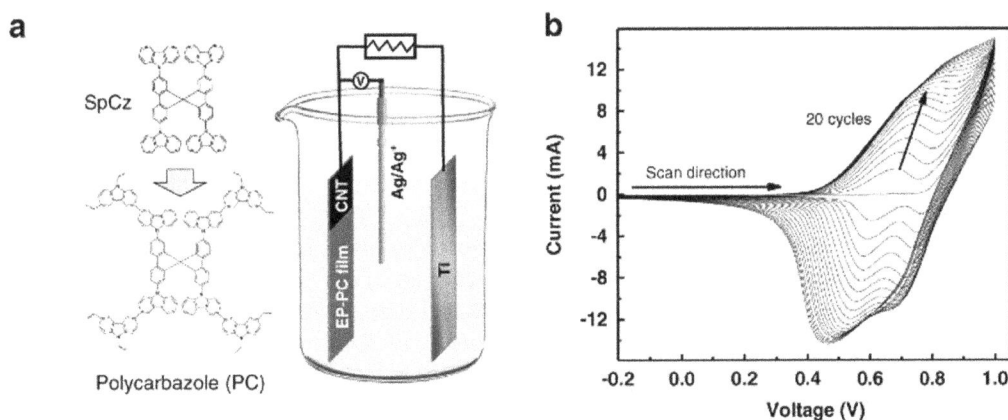

FIGURE 9.1 Showing: (a) setup of a three-electrode electrochemical cell for the electropolymerization process of polycarbazole; (b) CV profiles of the electropolymerization process pf polycarbazole recorded for 20 scan cycles.

(From Zhou *et al.* 2020. With permission.)

the adsorption of the species that depends on the potential, and a capacitive current due to the charge of the double layer. In general, polymers are characterized by large waves of oxidation and reduction. During polymerization, oxidation is followed by chemical coupling rather than reduction. Therefore, each oxidation peak is not systematically coupled with a reduction peak.

Polymerization and deposition of polymer films are characterized by the increase in peak currents of oxidation and reduction of the monomer during successive sweeping (Figure 9.1(b)) and the development of redox waves. Polymers have a lower potential than the oxidation of the monomer (Zhou *et al.*, 2020). The PANI film obtained by potential cycling is very adherent to the surface of the electrode (Holze, 2017); this method makes it possible to monitor the redox activity of the deposited polymer because the first redox couple of PANI is constantly monitored during cycling. Therefore, polymerization can be stopped when the voltammetric characteristics of the polymer formed are optimal, the disadvantage of the cycling method is that a large part of the deposition time corresponds to potentials where there is no polymerization; this explains why the yield of this method is lower than that of the other two.

9.3.2 POTENTIOSTATIC ELECTROPOLYMERIZATION

Synthesis in the potentiostatic mode can be carried out at a single potential or in successive stages at different potentials and allows a thin and homogeneous film to be obtained (Patois *et al.*, 2011). This method consists of applying a constant potential (E) to a working electrode and measuring the variation in the current as a function of time (Ruiz *et al.*, 2004). The applied potential is suitable for the oxidation of the monomer used, which generates oxidized monomer species that can be coupled to the surface of the working electrode. It is generally accepted that the potentiostatic method avoids the effects of overoxidation, because the oxidation potential is strictly controlled; in addition, it is very effective when preparing thick films over a short time. However, the reduction due solely to the binding of the monomer to the PANI chain during potentiostatic deposition is incomplete (Cui, Su, and Lee, 1993). This leads to the buildup of residual oxidized PANI and hydrolysis products in the film.

9.3.3 GALVANOSTATIC ELECTROPOLYMERIZATION

The choice of the applied current (i) in chronopotentiometry makes it possible to obtain either thin and homogeneous films (i.e., low current densities), or nodular structures (i.e., high current densities). The galvanostatic method consists of applying a fixed current to a working electrode and the potential is recorded as a function of time. The direct relationship between the time of electrosynthesis and the thickness of the polymer that is produced on the electrode surface is an advantage of galvanostatic polymerization (Jiang *et al.*, 2017). The application of a constant current allows a linear increase in the load over time when the current losses in the cell and the phenomena at the interfaces are neglected.

The flexibility of the used potential over time to adapt to variations in solution concentrations or to the passivity of the electrode is considered to be the second advantage of this polymerization method. Therefore, unlike the potentiostatic method, the potential drop at the electrode (i.e., with a large thicknesses of the polymer) is controlled by galvanometry to achieve the required current density. Therefore, galvanostatic polymerization is more suitable than potentiostatic polymerization for the preparation of thick films and especially with materials of low conductivities (Uang and Chou, 2002).

9.4 BIOPOLYMER-BASED CONDUCTING NANOCOMPOSITES

Biopolymers are in increasing demands due to their biodegradability, low-cost and versatility, especially due to the increasing harmful effects of non-biodegradable plastics (Touati *et al.*, 2011,

Zembouai *et al.*, 2013). Therefore, the increased use of green polymers (Zembouai *et al.*, 2016), such polylactic acid (PLA) or poly(3-hydroxybutyrate-co-3-hydroxyvalerate) (PHBV) and they are produced at an industrial scale (Zembouai *et al.*, 2018).

9.4.1 POLYLACTIDE

PLA is an aliphatic polyester obtained via the direct polycondensation of lactic acid monomers or via the ring opening polymerization of cyclic lactide dimers using a metal catalyst (Zaidi *et al.*, 2013). There are several different types of PLA, which have slightly different characteristics but are similar in that they are produced from a renewable resource. PLA is transparent with a high gloss (60%–110%) and displays good water and oxygen barrier properties and high oil resistance. The melting temperature (T_m) of PLA is generally in the range of 150°C–190°C, depending on PLA grades and their molecular weight (Zaidi *et al.*, 2010).

The mechanical proprieties of PLA are similar to polystyrene (PS) and polyethylene terephthalate (PET). The tensile modulus and tensile strength of PLA are in the range of 3,000–4,000 MPa and 50–60 MPa, respectively. However, PLA is a brittle material with an elongation at break <10% according to ISO 527 conditions (Zembouai *et al.*, 2014).

9.4.2 POLY(3-HYDROXYBUTYRATE-CO-3-HYDROXYVALERATE)

PHBV can be synthesized and accumulated intracellularly by a number of microorganisms(Gerard and Budtova, 2012). The properties of PHBV depend on the structure of the copolymer (Corre *et al.*, 2012). PHBV is semi-crystalline thermoplastic, with a degree of crystallinity from 40% to 60%. PHBV has very high oxygen and water barriers properties.

The melting temperature of PHBV is between 160°C and 180°C. The mechanical and thermal properties of PHBV are similar to those of poly(propylene) (PP) (Bledzki and Jaszkiewicz ,2010) (Hassaini *et al.*, 2017). The density of PHBV is similar to that of PLA (1.25 g/cm³). The thermal degradation of PHBV and PLA produces polymeric chains terminated with carboxyl and vinyl groups and carboxyl end groups of polyester catalyze hydrolysis reaction (Figure 9.2).

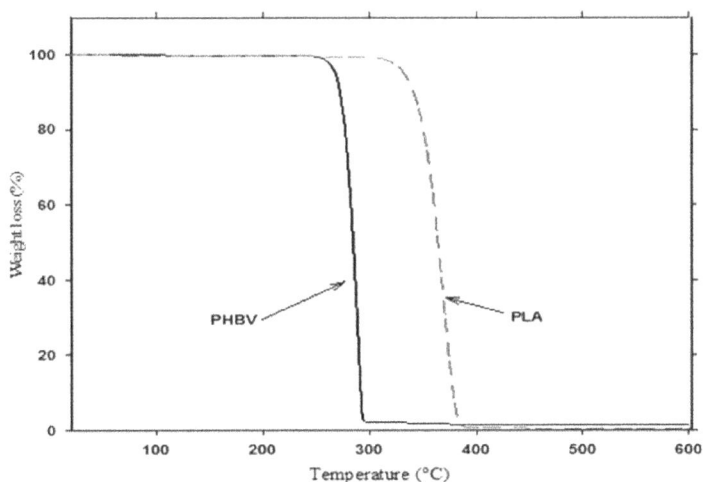

FIGURE 9.2 Thermal degradation of PLA and PHBV.

(From Zembouai *et al.* 2018.)

9.5 CONCLUSIONS

This chapter described the effect of conducting polymers on the analytical performance of electrochemical sensing platforms. It described the relationship between the use of conducting polymers and the analytical characteristics of electrochemical sensors, such as stability and sensitivity, linear concentration range, and the limit of detection. This provided advances in the use of conducting polymer functionalized nanocomposites with a focus on the application of electrodeposition methods of conducting polymers for electrochemical sensing applications. These methods are a good route to control the film thickness of conductive polymers. In addition, they are suitable to produce thin films with high conductivity directly rather than using chemical synthesized films. The application of electrodeposition potentiostatic methods minimize and even eliminate the passivation phenomena of electrochemical sensing interfaces. The use of conducting biopolymers is now widespread and offers flexible and stretchable platforms with high conductivity. Innovative experimental strategies applied for the preparation of sensors based conducting polymers will offer various robust and sensitive analytical tools.

REFERENCES

Assari, P. *et al.* (2020) 'Fabrication of a sensitive label free electrochemical immunosensor for detection of prostate specific antigen using functionalized multi-walled carbon nanotubes/polyaniline/AuNPs', *Materials Science and Engineering: C*, 115(October), p. 111066. doi:10.1016/j.msec.2020.111066.

Aydın, E.B., Aydın, M, and Sezgintürk, M.K. (2018) 'Highly sensitive electrochemical immunosensor based on polythiophene polymer with densely populated carboxyl groups as immobilization matrix for detection of Interleukin 1β in human serum and saliva', *Sensors and Actuators B: Chemical*, 270(October), pp. 18–27. doi:10.1016/j.snb.2018.05.014.

Aydın, E.B., Aydın, M, and Sezgintürk, M.K. (2021) 'A novel electrochemical immunosensor based on acetylene black/epoxy-substituted-polypyrrole polymer composite for the highly sensitive and selective detection of Interleukin 6', *Talanta*, 222(January), p.121596. doi:10.1016/j.talanta.2020.121596.

Aydın, M. (2019) 'A sensitive and selective approach for detection of IL 1α cancer biomarker using disposable ITO electrode modified with epoxy-substituted polythiophene polymer', *Biosensors and Bioelectronics*, 144(November), p. 111675. doi:10.1016/j.bios.2019.111675.

Babaiee, M., Pakshir, M., and Hashemi, B. (2015) 'Effects of potentiodynamic electropolymerization parameters on electrochemical properties and morphology of fabricated PANI nanofiber/graphite electrode', *Synthetic Metals*, 199(January), p.110–20. doi:10.1016/j.synthmet.2014.11.012.

Balint, R., Cassidy, N.J., and Cartmell, S.H. (2014) 'Conductive Polymers: Towards a Smart Biomaterial for Tissue Engineering', *Acta Biomaterialia*, 10 (6), pp. 2341–53. doi:10.1016/j.actbio.2014.02.015.

Bensana, A., and Achi, F. (2020) 'Analytical performance of functional nanostructured biointerfaces for sensing phenolic compounds', *Colloids and Surfaces B: Biointerfaces*, 196(December): 111344. doi:10.1016/j.colsurfb.2020.111344.

Berkes, B.B., Bandarenka, A.S., and Inzelt ,G. (2015) 'Electropolymerization: Further insight into the formation of conducting polyindole thin films', *The Journal of Physical Chemistry C*, 119(4), pp. 1996–2003. doi:10.1021/jp512208s.

Bhadra, S. *et al.* (2009) 'Progress in preparation, processing and applications of polyaniline', *Progress in Polymer Science*, 34(8), pp. 783–810. doi:10.1016/j.progpolymsci.2009.04.003.

Bianchini, C. *et al.* (2014) 'Determination of caffeic acid in wine using PEDOT film modified electrode', *Food Chemistry*, 156(August), pp. 81–86. doi:10.1016/j.foodchem.2014.01.074.

Bledzki, A.K., and Jaszkiewicz, A. (2010) 'Mechanical performance of biocomposites based on PLA and PHBV reinforced with natural fibres – A comparative study to PP', *Composites Science and Technology*, 70(12), pp. 1687–96. doi:10.1016/j.compscitech.2010.06.005.

Bolat, G., Yaman, Y.T., and Abaci S. (2018) 'Highly sensitive electrochemical assay for Bisphenol A detection based on poly(CTAB)/MWCNTs modified pencil graphite electrodes', *Sensors and Actuators B: Chemical*, 255(February), pp. 140–48. doi:10.1016/j.snb.2017.08.001.

Bourlinos, A.B. *et al.* (2003) 'Graphite oxide: Chemical reduction to graphite and surface modification with primary aliphatic amines and amino acids', *Langmuir*, 19(15), pp. 6050–55. doi:10.1021/la026525h.

Choo, Y. *et al.* (2020) 'Diffusion and migration in polymer electrolytes', *Progress in Polymer Science*, 103(April), p. 101220. doi:10.1016/j.progpolymsci.2020.101220.

Contreras-Herrera, K.M. *et al.* (2018) 'Influence of the electropolymerization parameters on the doping level of polybithiophene films grown in acetonitrile and water', *ECS Transactions*, 84(1), p. 35–39. doi:10.1149/08401.0035ecst.

Corre, Y-M. *et al.* (2012) 'Morphology and functional properties of commercial polyhydroxyalkanoates: A comprehensive and comparative Study', *Polymer Testing*, 31(2), p. 226–35. doi:10.1016/j.polymertesting.2011.11.002.

Cui, C.Q., Su, X.H., and Lee, J.Y. (1993) 'Measurement and evaluation of polyaniline degradation', *Polymer Degradation and Stability*, 41(1), pp. 69–76. doi:10.1016/0141-3910(93)90063-O.

Eftekhari, A., and Saito, T. (2017) 'Synthesis and properties of polymerized ionic liquids', *European Polymer Journal*, 90(May), pp.: 245–72. doi:10.1016/j.eurpolymj.2017.03.033.

Epstein, A.J. (2007) 'Conducting polymers: Electrical conductivity', in Mark, J.E. *(ed.) Physical properties of polymers handbook*. New York: Springer. doi:10.1007/978-0-387-69002-5_46. pp. 725–755.

Ganesh, P.S., *et al.* (2018) 'Interference free detection of dihydroxy benzene isomers at pyrogallol film coated electrode: A voltammetric method', *Journal of Electroanalytical Chemistry*, 813(March), pp. 193–99. doi:10.1016/j.jelechem.2018.02.018.

Ganganboina, A.B., and Doong, R-A. (2019) 'Graphene quantum dots decorated gold-polyaniline nanowire for impedimetric detection of carcinoembryonic antigen', *Scientific Reports*, 9(1), p. 7214. doi:10.1038/s41598-019-43740-3.

Gerard, T., and Budtova, T. (2012) 'Morphology and molten-state rheology of polylactide and polyhydroxyalkanoate blends', *European Polymer Journal* 48 (6): 1110–17. doi:10.1016/j.eurpolymj.2012.03.015.

Hassaini, L. *et al.* (2017) 'Valorization of olive husk flour as a filler for biocomposites based on poly(3-Hydroxybutyrate-co-3-Hydroxyvalerate): Effects of silane treatment', *Polymer Testing*, 59(May), pp. 430–440. doi:10.1016/j.polymertesting.2017.03.004.

Heeger, A.J. (2001) 'Semiconducting and metallic polymers: The fourth generation of polymeric materials', *The Journal of Physical Chemistry B*, 105(36), pp. 8475–8491. doi:10.1021/jp011611w.

Holze, R. (2017) Metal oxide/conducting polymer hybrids for application in supercapacitors', *Metal Oxides in Supercapacitors*. doi:10.1016/B978-0-12-810464-4.00009-7.

Janata, J., and Josowicz, M. (2003) 'Conducting polymers in electronic chemical sensors', *Nature Materials*, 2(1), pp. 19–24. doi:10.1038/nmat768.

Jiang, L. *et al.* (2017) 'Electropolymerization of camphorsulfonic acid doped conductive polypyrrole anti-corrosive coating for 304SS bipolar plates', *Applied Surface Science*, 426(December), pp. 87–98. doi:10.1016/j.apsusc.2017.07.077.

Kavanoz, M., and Pekmez, N.O. (2012) 'Poly(vinylferrocenium) perchlorate–polyaniline composite film-coated electrode for amperometric determination of hydroquinone', *Journal of Solid State Electrochemistry*, 16(3), pp. 1175–1186. doi:10.1007/s10008-011-1505-6.

Korent, A., Soderžnik, K.Z., Šturm, S., and Rožman, K.Z. (2020) 'A correlative study of polyaniline electropolymerization and its electrochromic behavior', *Journal of The Electrochemical Society*, 167(10), p. 106504. doi:10.1149/1945-7111/ab9929.

Lange, U., Roznyatovskaya, N.V, and Mirsky, V.M. (2008) 'Conducting Polymers in Chemical Sensors and Arrays', *Analytica Chimica Acta*, 614(1), PP. 1–26. doi:10.1016/j.aca.2008.02.068.

Li, X. *et al.* (2021) 'Recent progress in conductive polymers for advanced fiber-shaped electrochemical energy storage devices', *Materials Chemistry Frontiers*, 5(3), pp. 1140–1163. doi:10.1039/D0QM00745E.

Liang, Y. *et al.* (2017) 'High sensitive and selective graphene oxide/molecularly imprinted polymer electrochemical sensor for 2,4-dichlorophenol in water', *Sensors and Actuators B: Chemical*, 240(March), pp. 1330–1335. doi:10.1016/j.snb.2016.08.137.

Liu, X. *et al.* (2011) 'Electrocatalytic detection of phenolic estrogenic compounds at NiTPPS|carbon nanotube composite electrodes|carbon nanotube composite electrodes', *Analytica Chimica Acta*, 689(2), pp. 212–18. doi:10.1016/j.aca.2011.01.037.

Lopa, N.S. *et al.* (2017) 'A glassy carbon electrode modified with poly(2,4-dinitrophenylhydrazine) for simultaneous detection of dihydroxybenzene isomers', *Microchimica Acta*, 185(1), p. 23. doi:10.1007/s00604-017-2567-7.

Ma, M. *et al.* (2014) 'Electrochemical sensor for Bisphenol A based on a nanoporous polymerized ionic liquid interface', *Microchimica Acta*, 181(5), pp. 565–572. doi:10.1007/s00604-013-1151-z.

MacDiarmid, A.G. (1997) 'Polyaniline and polypyrrole: Where are we headed?' *Synthetic Metals*, 84(1), pp. 27–34. doi:10.1016/S0379-6779(97)80658-3.

Madhad, H.V., and Vasava, D.V. (2019) 'Review on recent progress in synthesis of graphene–polyamide nanocomposites', *Journal of Thermoplastic Composite Materials*, October. doi:10.1177/0892705719880942.

Mazzotta, E., Malitesta, C., and Margapoti, E. (2013) 'Direct electrochemical detection of Bisphenol A at PEDOT-modified glassy carbon electrodes', *Analytical and Bioanalytical Chemistry*, 405(11), pp. 3587–3592. doi:10.1007/s00216-013-6723-6.

Mello, H.J., Dias, N.P., and Mulato, M. (2018) 'Effect of aniline monomer concentration on PANI electropolymerization process and its influence for applications in chemical sensors', *Synthetic Metals*, 239(May), pp. 66–70. doi:10.1016/j.synthmet.2018.02.008.

Moon, J-M., Kim, Y.H., and Cho, Y. (2014) 'A nanowire-based label-free immunosensor: Direct incorporation of a PSA antibody in electropolymerized polypyrrole', *Biosensors and Bioelectronics*, 57(July), pp.157–161. doi:10.1016/j.bios.2014.02.016.

Niyogi, S. *et al.* (2006) 'Solution properties of graphite and graphene', *Journal of the American Chemical Society*, 128(24), pp. 7720–7721. doi:10.1021/ja060680r.

Olean-Oliveira, A. *et al.* (2020) 'Mechanism of nanocomposite formation in the layer-by-layer single-step electropolymerization of π-conjugated azopolymers and reduced graphene oxide: An electrochemical impedance spectroscopy study', *ACS Omega*, 5(40), pp. 25954–25957. doi:10.1021/acsomega.0c03391.

Paredes, J. I, *et al.* (2008) 'Graphene oxide dispersions in organic solvents', *Langmuir*, 24(19), pp.10560–10564. doi:10.1021/la801744a.

Patois, T. *et al.* (2011) 'Characterization of the surface properties of polypyrrole films: Influence of electrodeposition parameters', *Synthetic Metals*, 161(21), pp. 2498–2505. doi:10.1016/j.synthmet.2011.10.003.

Pei, F. *et al.* (2019) 'Sensitive label-free immunosensor for alpha fetoprotein detection using platinum nanodendrites loaded on functional MoS_2 hybridized polypyrrole nanotubes as signal amplifier', *Journal of Electroanalytical Chemistry*, 835(February), pp.197–204. doi:10.1016/j.jelechem.2019.01.037.

Rajendran, R. *et al.* (2021) 'A study on polythiophene modified carbon cloth as anode in microbial fuel cell for lead removal', *Arabian Journal for Science and Engineering*. doi:10.1007/s13369-021-05402-3.

Rakhrour, W. *et al.* (2021) 'Electrochemical synthesis of an organometallic material based on polypyrrole/MnO_2 as high-performance cathode', *Journal of Inorganic and Organometallic Polymers and Materials*, 31(1), pp. 62–69. doi:10.1007/s10904-020-01664-w.

Roy, A.C. *et al.* (2013) 'Molecularly imprinted polyaniline-polyvinyl sulphonic acid composite based sensor for para-nitrophenol detection', *Analytica Chimica Acta*, 777(May), pp. 63–71. doi:10.1016/j.aca.2013.03.014.

Ruiz, V. *et al.* (2004) 'Electropolymerization under potentiodynamic and potentiostatic conditions', *Electrochimica Acta*, 50(1), pp. 59–67. doi:10.1016/j.electacta.2004.07.013.

Samukaite-Bubniene, U. *et al.* (2021) 'Toward supercapacitors: Cyclic voltammetry and fast fourier transform electrochemical impedance spectroscopy based evaluation of polypyrrole electrochemically deposited on the pencil graphite electrode', *Colloids and Surfaces A: Physicochemical and Engineering Aspects*, 610(February), p.125750. doi:10.1016/j.colsurfa.2020.125750.

Shirakawa, H. *et al.* (1977) 'Synthesis of electrically conducting organic polymers: Halogen derivatives of polyacetylene, (CH)x', *Journal of the Chemical Society, Chemical Communications*, 16(January), pp. 578–580. doi:10.1039/C39770000578.

Song, Y. *et al.* (2020) 'Optimal film thickness of RGO/MoS_2 @ polyaniline nanosheets of 3D arrays for carcinoembryonic antigen high sensitivity detection', *Microchemical Journal*, 155(June), p. 104694. doi:10.1016/j.microc.2020.104694.

Tahalyani, J. *et al.* (2016) 'The dielectric properties and charge transport mechanism of π-conjugated segments decorated with intrinsic conducting polymer', *RSC Advances*, 6(74), pp. 69733–69742. doi:10.1039/C6RA09554B.

Tan, F. *et al.* (2016) 'An electrochemical sensor based on molecularly imprinted polypyrrole/graphene quantum dots composite for detection of Bisphenol A in water samples', *Sensors and Actuators B: Chemical*, 233(October), pp. 599–606. doi:10.1016/j.snb.2016.04.146.

Tandon, S., *et al.* (2020) 'Polymeric immunosensors for yumor detection', *Biomedical Physics & Engineering Express*, 6(3), p. 032001. doi:10.1088/2057-1976/ab8a75.

Tehrani M.A., *et al.* (2013) 'Electrochemical studies of two diphenols isomers at graphene nanosheet–poly(4-vinyl pyridine) composite modified electrode', *Sensors and Actuators B: Chemical*, 177(February), pp. 612–619. doi:10.1016/j.snb.2012.11.047.

Tothill, I.E. (2009) 'Biosensors for cancer markers diagnosis', *Seminars in Cell & Developmental Biology*, 20(1), pp. 55–62. doi:10.1016/j.semcdb.2009.01.015.

Touati, N. *et al.* (2011) 'The effects of reprocessing cycles on the structure and properties of isotactic polypropylene/cloisite 15A nanocomposites', *Polymer Degradation and Stability*, 96(6), pp. 1064–1073. doi:10.1016/j.polymdegradstab.2011.03.015.

Truong, L.T.N. *et al.* (2011) 'Labelless impedance immunosensor based on polypyrrole–pyrrolecarboxylic acid copolymer for HCG detection', *Talanta*, 85(5), pp.2576–2580. doi:10.1016/j.talanta.2011.08.018.

Uang, Y.-M., and Chou, T.-S. (2002) 'Criteria for designing a polypyrrole glucose biosensor by galvanostatic electropolymerization', *Electroanalysis*, 14(22), pp. 1564–1570. doi.10.1002/1521-4109(200211)14:22<1564:AID-ELAN1564>3.0.CO;2-H.

Walsh, J.J. *et al.* (2013) 'Visible light sensitized photocurrent generation from electrostatically assembled thin films of [Ru(Bpy)3]2+ and the polyoxometalate Γ*-[W18O54(SO4)(2)](4–): Optimizing performance in a low electrolyte medium', *Journal of Electroanalytical Chemistry*, 706(October): 93–101. doi:10.1016/j.jelechem.2013.07.020.

Wan, J. *et al.* (2016) 'Bisphenol A electrochemical sensor based on multi-walled carbon nanotubes/polythiophene/Pt nanocomposites modified electrode', *Analytical Methods*, 8(16), pp. 3333–3338. doi:10.1039/C6AY00850J.

Wang, J. and Hui, N. (2019) 'Zwitterionic poly(carboxybetaine) functionalized conducting polymer polyaniline nanowires for the electrochemical detection of carcinoembryonic antigen in undiluted blood serum', *Bioelectrochemistry*, 125(February), pp. 90–96. doi:10.1016/j.bioelechem.2018.09.006.

Wang, L. *et al.* (2017) '2D nanomaterials based electrochemical biosensors for cancer diagnosis', *Biosensors and Bioelectronics*, 89(March), pp. 136–151. doi:10.1016/j.bios.2016.06.011.

Wang, P. *et al.* (2020) 'The preparation of hollow AgPt@Pt core-shell nanoparticles loaded on polypyrrole nanosheet modified electrode and its application in immunosensor', *Bioelectrochemistry*, 131(February), p. 107352. doi:10.1016/j.bioelechem.2019.107352.

Wang, Yanying, Chunya Li, Tsunghsueh Wu, and Xiaoxue Ye (2018). "Polymerized Ionic Liquid Functionalized Graphene Oxide Nanosheets as a Sensitive Platform for Bisphenol A Sensing." *Carbon* 129 (April): 21–28. doi:10.1016/j.carbon.2017.11.090.

Yin, H. *et al.* (2010) 'Electrochemical behavior of Bisphenol A at glassy carbon electrode modified with gold nanoparticles, silk fibroin, and PAMAM dendrimers', *Microchimica Acta*, 170(1), pp. 99–105. doi:10.1007/s00604-010-0396-z.

Yin, H. *et al.* (2010) 'Sensitivity and selectivity determination of BPA in real water samples using PAMAM dendrimer and CoTe quantum dots modified glassy carbon electrode', *Journal of Hazardous Materials*, 174(1), pp. 236–243. doi:10.1016/j.jhazmat.2009.09.041.

Yin, H. *et al.* (2011) 'Amperometric determination of Bisphenol A in milk Using PAMAM–Fe_3O_4 modified glassy carbon electrode', *Food Chemistry*, 125(3), pp.1097–1103. doi:10.1016/j.foodchem.2010.09.098.

Zaidi, L. *et al.* (2010) 'Effect of natural weather on the structure and properties of polylactide/cloisite 30B nanocomposites', *Polymer Degradation and Stability*, 95(9), pp. 1751–1758. doi:10.1016/j.polymdegradstab.2010.05.014.

Zaidi, L. *et al.* (2013) 'The effects of gamma irradiation on the morphology and properties of polylactide/cloisite 30B nanocomposites', *Polymer Degradation and Stability*, 98(1), pp. 348–355. doi:10.1016/j.polymdegradstab.2012.09.014.

Zembouai, I. *et al.* (2013) 'A study of morphological, thermal, rheological and barrier properties of poly(3-hydroxybutyrate-Co-3-hydroxyvalerate)/polylactide blends prepared by melt mixing', *Polymer Testing*, 32(5), pp. 842–851. doi:10.1016/j.polymertesting.2013.04.004.

Zembouai, I. *et al.* (2014) 'Poly(3-hydroxybutyrate-Co-3-hydroxyvalerate)/polylactide blends: Thermal stability, flammability and thermo-mechanical behavior', *Journal of Polymers and the Environment*, 22(1), pp. 131–139. doi:10.1007/s10924-013-0626-7.

Zembouai, I. *et al.* (2016) 'Electron beam radiation effects on properties and ecotoxicity of PHBV/PLA blends in presence of organo-modified montmorillonite', *Polymer Degradation and Stability*, 132(October), pp. 117–126. doi:10.1016/j.polymdegradstab.2016.03.019.

Zembouai, I. *et al.* (2018) 'Combined effects of sepiolite and cloisite 30B on morphology and properties of poly(3-hydroxybutyrate-Co-3-hydroxyvalerate)/polylactide blends', *Polymer Degradation and Stability*, 153(July), pp. 47–52. doi:10.1016/j.polymdegradstab.2018.04.017.

Zhang, C. *et al.* (2017) 'AgNWs-PANI nanocomposite based electrochemical sensor for detection of 4-nitrophenol', *Sensors and Actuators B: Chemical*, 252(November), pp. 616–623. doi:10.1016/j.snb.2017.06.039.

Zhang, W. *et al.* (2020) 'Miniaturized electrochemical sensors and their point-of-care applications', *Chinese Chemical Letters*, 31(3), pp. 589–600. doi:10.1016/j.cclet.2019.09.022.

Zhao, J. *et al.* (2016a) 'Electrochemical detection of two tumor markers based on functionalized polypyrrole microspheres as immunoprobes', *RSC Advances*, 6(37), pp. 31448–33153. doi:10.1039/C6RA01773H.

Zhao, J. *et al.* (2016b) 'Electrochemical detection of two tumor markers based on functionalized polypyrrole microspheres as immunoprobes', *RSC Advances*, 6(37), pp. 31448–31453. doi:10.1039/C6RA01773H.

Zhao, L. and Ma, Z. (2018) 'Facile synthesis of polyaniline-polythionine redox hydrogel: Conductive, anti-fouling and enzyme-linked material for ultrasensitive label-free amperometric immunosensor toward carcinoma antigen-125', *Analytica Chimica Acta*, 997(January), pp.60–66. doi:10.1016/j.aca.2017.10.017.

Zhao, M. *et al.* (2021) 'Characterization of complicated electropolymerization using UV–Vis spectroelectrochemistry and an electrochemical quartz-crystal microbalance with dissipation: A case study of tricarbazole derivatives', *Electrochemistry Communications*, 123(February), p.106913. doi:10.1016/j.elecom.2020.106913.

Zhou, Z. *et al.* (2020) 'Electropolymerization of robust conjugated microporous polymer membranes for rapid solvent transport and narrow molecular sieving', *Nature Communications*, 11(1), p. 5323. doi:10.1038/s41467-020-19182-1.

10 Electroactive Polymers for Artificial Muscles

Zhangpeng Li, Qiulong Gao, and Jinqing Wang

CONTENTS

10.1 INTRODUCTION

Artificial muscles are a class of materials or devices with similar characteristics to a biological muscle that can change in size, or shape, or both, and therefore, generate force and displacement under the activation of voltage, current, magnetic field, pressure, light, or temperature. They can contract, expand, or rotate reversibly, which produces some action outputs that are similar to their biological counterparts, and can bear large deformation and external loads (Wang, Gao, & Lee, 2021; Spinks, 2020). As society has developed, artificial muscles have become of great significance in many practical applications, such as outer space manipulators, micro reconnaissance devices, prostheses, robotics, and miniature rotating motors (Bashir & Rajendran, 2018). Therefore, it is significant and highly desirable to develop new materials to improve the performance of artificial muscles (Wang & Qu, 2016). Typical artificial muscle materials or devices include fluid-driven artificial muscles (e.g., pneumatic and hydraulic), shape-memory materials, and piezoelectric ceramics (Gu et al., 2017).

Since the 1950s, researchers have developed suitable artificial muscle materials or devices, such as McKibben actuators, shape-memory materials, and piezoelectric ceramic materials (Bashir & Rajendran, 2018; Nickel et al., 1965). However, as artificial muscles, they have distinct limitations, such as the large size of the McKibben actuators, the unpredictability deformation, and slow response speed of the shape-memory materials (Chen et al., 2020), and the brittleness and small strain of the piezoelectric ceramic materials (Ma, 1995; Jiang, Ng, and Lam, 2000). Due to these limitations, the traditional artificial muscle materials and devices do not meet the requirements of current scientific development and technological innovations. This dilemma prompted research and development of new materials that can produce reversible changes in shape, size, and other mechanical properties

DOI: 10.1201/9781003173502-10

Type   Source of actuation

Type   Source of actuation

Artificial muscles

Fluid-driven artificial muscles — Pneumatic or Hydraulic

Shape memory materials — External stimulus such as light, heat, or voltage

Piezoelectric ceramics — Electric field

Electroactive polymers (EAPs) — Electric current or voltage

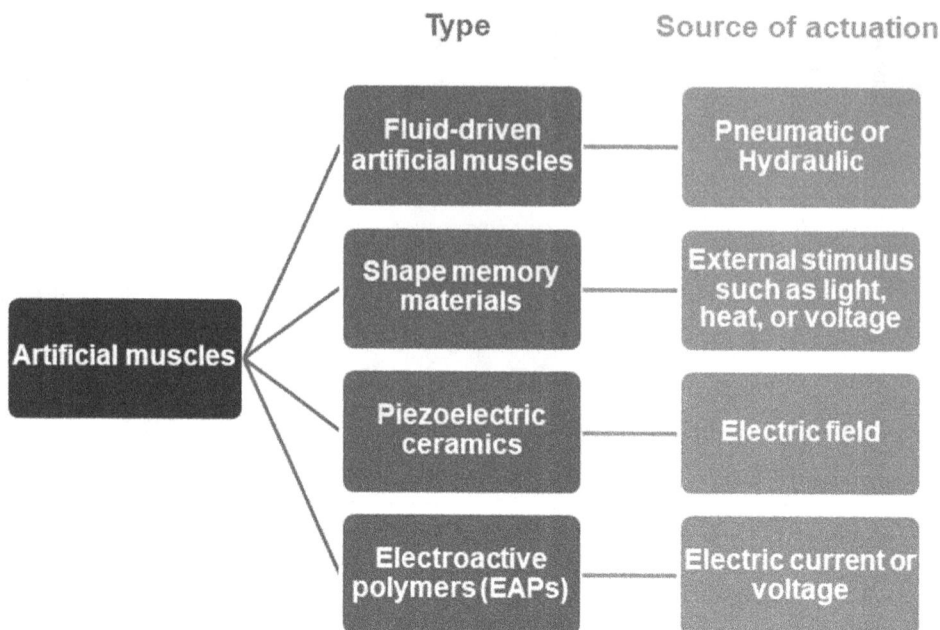

FIGURE 10.1   Types and source of actuation in different artificial muscle materials.

under external stimulation with improved performance (Bar-Cohen, 2000; Palza, Zapata, and Angulo-Pineda, 2019). Before suitable materials appeared, the research and development of artificial muscles were very slow. Until the emergence of a new type of material called electroactive polymers (EAPs), the artificial muscles increased rapidly (Pelrine et al., 2003, Bar-Cohen, 2002). Figure 10.1 shows the types and the corresponding source of actuation of typical artificial muscle materials.

10.2   ELECTROACTIVE POLYMERS

Various materials have been designed and used as promising materials for artificial muscle applications. Among them, EAPs have attracted attention due to their similarity to biological muscles when operating, especially elasticity, ability to induce large strain, and real-time actuation ability. They can efficiently generate mechanical motion in response to an electrical stimulus and could promote breakthroughs in miniaturized mechanical devices (Bar-Cohen, 2002; Baudis, Behl, and Lendlein, 2014). Due to their outstanding capabilities, such as large active strains, fast response, reliability, good flexibility, light weight, and no acoustic noise (Fried 2000; Carpi et al., 2011), accurately mimicking the functionalization of natural muscle (Bar-Cohen 2005), EAPs have been called artificial muscles (Chen and Pei 2017, Qiu et al. 2019).

EAP materials are responsive materials that can produce deformation through changes in their internal structure under the stimulation of an electric current, electric field, or voltage. The study of EAPs started in 1880 when Röntgen discovered that rubber can change length in an electric field. This discovery increased research on EAP materials (Bar-Cohen, 2004b). The emergence of EAPs has brought new research into artificial muscle. Therefore, since the early 1990s (Fannir et al. 2019), the application of EAPs in artificial muscles has been rapidly developed. Their distinguishing feature is that they can smoothly convert electrical energy into mechanical energy (Brochu and Pei, 2010). Compared with traditional artificial muscle materials (e.g., piezoelectric ceramics and shape-memory alloys), EAPs possess better deformation, faster response speed, lower density, and higher resilience (Pelrine et al., 2003). Considering these advantages, EAP materials have attracted

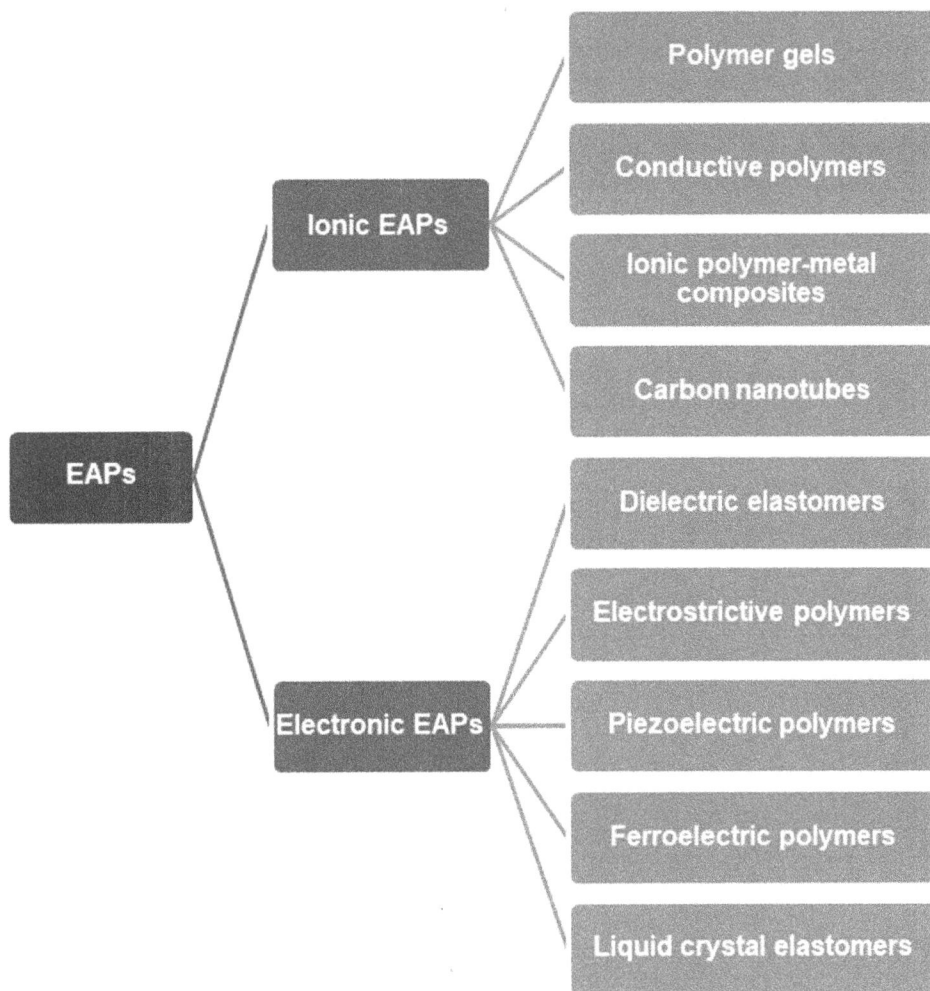

FIGURE 10.2   Categories of EAP materials.

widespread attention in many fields, such as implantable functional devices, microelectromechanical systems (MEMS), sensors, and artificial muscles. Based on their mechanism of deformation, EAP artificial muscle materials are mainly divided into two categories (Bar-Cohen, 2002), ionic and electronic EAPs. The former, which are known as wet EAPs, includes polymer gels, conductive polymers (CPs), ionic polymer–metal composites (IPMCs), and carbon nanotubes (CNTs). These materials usually contain two electrodes and an electrolyte, and their deformation is driven by the mobility or diffusion of ions or molecules under the stimulation of an external electrical field. The latter, which are classified as dry EAPs, include dielectric elastomers (DEs), electrostrictive polymers, ferroelectric polymers, piezoelectric polymers, and liquid crystal elastomers, whose actuation is based on an electrical field or Coulomb force. Figure 10.2 presents the classification of EAP materials. In the following sections, the basic features and research status of these EAPs will be introduced.

10.2.1   Ionic Electroactive Polymers

Ionic EAPs are materials that can generate displacement driven mainly by the transportation of ions or solvents, or both under electrical stimulation (Carpi et al., 2011). Typically, when electrical stimulation is applied to the system, it can accelerate the movement of ions and solvents and keep them in or out of the polymer matrix, and therefore, induce expansion and contraction of the materials. Ionic EAP materials must be in an ionization state to make the ions move; therefore, the material needs to operate in a wet environment.

For ionic EAPs, their primary advantage lies in their inherent ability to respond to extremely low driving voltages (e.g., from one to a few volts). In addition, they have large deformation capability and high flexibility. Their disadvantage is that, except for CPs, it is difficult to maintain a constant displacement when activated by a DC voltage. In addition, the disadvantages of maintaining wetness, long response time, and limited durability largely limit their applications (Carpi et al., 2011).

10.2.1.1 Polymer Gels

Polymer gels have been investigated for a long time as possible artificial muscles (Brock et al., 1994, Osada, Okuzaki, and Hori, 1992). They can swell or contract due to several environmental factors, including temperature (Kuckling, Richter, and Arndt, 2003), solvent composition (Arndt, Kuckling, and Richter, 2000), ionic strength (Liu et al., 2003), pH (Oktar, Caglar, and Seitz, 2005), and electric fields (Richter et al. 2003). For EAP gels, the change in gel volume is due to the ability of the gel network to absorb and remove the solvents under electrical stimulations, for instance, shape distortion of the gels. In general, the polymer gels are composed of anions, cations, and fixed charges, in which the fixed charges determine the displacement direction. For example, in a negative fixed charge system, when electrical stimulation is applied, cations will move to the cathode side, causing the bending of the polymer gel in the same direction (Jo, Naguib, and Kwon, 2011). In addition, the response speed of the gel is affected by the electrolyte, electrode material, the applied electrical force, and the shape of the polymer gels. For instance, the gel size needs to be controlled in a few microns if a fast response is desired in a specific application (Punning et al., 2014).

In 1992, a weakly crosslinked poly(2-acrylamide-2-methyl propane) sulfonic acid (PAMPS) gel with the electrically driven property was studied (Osada, Okuzaki, and Hori 1992). When a strip of gel was immersed in a solution of n-dodecyl pyridinium chloride (C_{12}PyCl) surfactant and an electric field was used, the strip could contract and curve. The underlying principle for the deformation of the gel and because the cooperative and reversible complex of C_{12}PyCl molecules on the PAMPS gel under the stimulation of an electric field caused the deformation of the gel. Specifically, the positively charged surfactant molecules migrated to the cathode by electrophoresis, and they bonded to the surface of the gel network (i.e., negatively charged) when DC voltage was used, which caused an isotropic contraction and induced the gel to bend toward the anode. In addition, by alternating the electric field's polarity repeatedly, the strip of gel underwent repeated stretching and bending; therefore, causing the gel artificial muscle actuators to achieve peristaltic behaviors.

Poly(vinyl chloride) (PVC) gel can be used for soft actuators (Hirai et al., 1999). The plasticized PVC gels can complete stretching and bending motion under electrical stimulation. Of note, it showed some outstanding characteristics, such as strain, strength, and rate that are similar to biological muscle even under a low voltage, which means that it has great potential for use as a soft actuator and artificial muscles (Li, Guo, and Li, 2019). Research has been carried out to develop various PVC gel-based materials and devices and there is increased research into artificial muscles.

An electrically driven artificial muscle constructed from a series of crosslinked polyacrylamide (PAAM)/polyacrylic acid (PAA) multilayer hydrogels was developed. The deformation depended on the ionic strength of the medium and the dissociation level of the PAA layer (Liu and Calvert 2000).

As an electroactive hydrogel, PAAM hydrogels can absorb and store a large amount of water and can be stimulated by an electric field to generate large volume changes. Bassil, Davenas, and Tahchi (2008) developed a method to characterize the pH gradient in a PAAM gel and investigated the actuator performance of PAAM under an electrical stimulus. They found that electricity was the driving force that induced changes in the volume of the gel, in which the applied electrical field imposed the electrochemical reaction kinetics and the volume variations. Therefore, the rate of bending movement of the gel was kinetically controlled by the driving electrical power.

Although various hydrogels have been studied for artificial muscle applications, the problems of slow response speed, low energy efficiency, and instability need to be addressed. Sun et al. (2020) reported a smart full-hydrogel artificial muscle that employed an ionic electrolyte membrane as the active material, which showed outstanding ion transportation efficiency, stability, ultrafast electromechanical response rate, deflection displacement (16.284 mm, 5 V DC power), and increased output force (4.153 mN).

10.2.1.2 Conductive Polymers

CPs were discovered by Shirakawa et al. (1977), and are one of the most important groups of the EAPs. Among them, polypyrrole (PPy), polyaniline (PANI), and polythiophene (PTh) are the most studied CPs. As artificial muscle materials, CPs exhibit unique properties, such as low operation voltage, large bending displacements, and strain-hold ability under DC voltage, which makes them a good choice for practical applications (Melling, Martinez, and Jager, 2019).

Currently, significant progress has been achieved in understanding the actuation mechanisms of these materials. In general, the mechanism is based on the migration of ions and solvent molecules between the electrolyte and polymer during electrochemical reactions. Therefore, CP-based artificial muscles are electro–chemo–mechanical devices (Melling, Martinez, and Jager, 2019).

Baughman (1991) proposed that CPs can be utilized as electromechanically active materials for driving actuators. Then, Pei and Inganläs (1992), and Otero et al. (1992) reported PPy-based artificial muscles, which triggered an upsurge in the research into CPs in the field of artificial muscles.

In general, the performance of actuation depends on lots of factors, such as the size of the ions, the thickness of the polymer film, the conductivity of the materials, and the applied voltage. For instance, the introduction of conductive nanomaterials, such as CNTs into a matrix of CPs promote the speed of charge injecting, which results in significantly improved response rate and stress intensity (100 MPa) (Spinks et al., 2006).

The influence of electrolytes on the electro–chemo–mechanical properties of active CPs has been well-studied (Tian and Wang, 2016; Bowers, 2004). Ebadi, Semnani, et al. (2020) found that in a polyurethane/PPy (PU/PPy) system, the nature of ions is the key parameter to the ion exchange mechanism within the polymer. They demonstrated that the larger anions were used in the greater bending movements in an anion exchange mechanism. Specifically, the bending displacement increased from 30° to 250° when the lithium chloride (LiCl) electrolyte was replaced with lithium bis(trifluoromethane sulfonyl)imide (LiTFSI) (Ebadi, Semnani, et al., 2020). To solve the problem of the potential drop along the length of conducting polymers, Ebadi, Fashandi, et al. (2020) fabricated a PU/copper/PPy (PU/Cu/PPy) nanofiber artificial muscle by a combination process of electrospinning, electroplating, and electrochemical polymerization. The resultant material showed an improved Young's modulus (≤62.32±5.42 MPa), electrical conductivity (10,103.06±14.28 S/cm), and enhanced bending actuation of 88° (scan rate 5 mV s^{-1}) between a voltage range of −0.8V–0.5 V. At the same time, due to the high specific surface area and porosity, the prepared nanofibers could be used to improve the flexibility and actuating speed of artificial muscle. The performance of the prepared artificial muscle could be improved by using Cu electroplating to form a conductive layer in the nanofiber actuator.

10.2.1.3 Ionic Polymer–Metal Composites

IPMCs are one of the most popular EAP-based artificial muscles due to their critical features of decent bending ability at low working voltages, ease of fabrication, light weight, and low cost. Typically, IPMCs arc composed of a laminated thin ionically conductive polymer film and two metal electrodes (e.g., percolated noble metals such as platinum (Pt) or gold (Au)) (Aabloo et al., 2020). The increase of IPMCs as artificial muscles dates back to Shahinpoor (1992), Shahinpoor

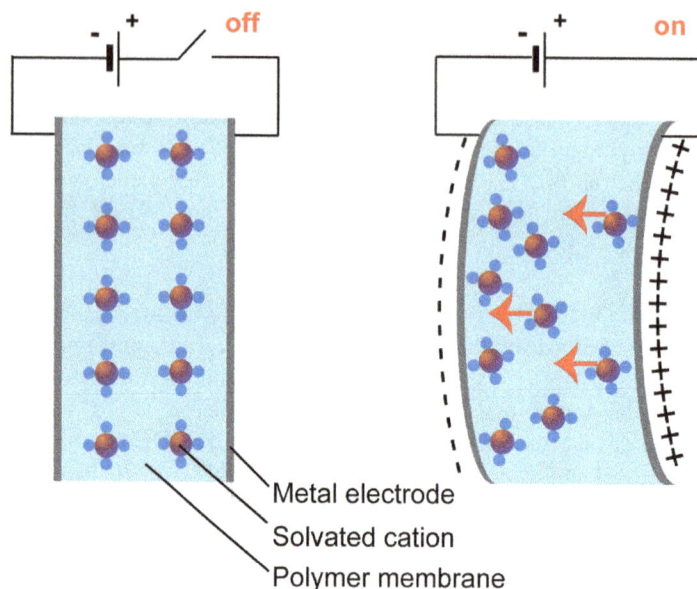

FIGURE 10.3 Working principle of an IPMC artificial muscle.

et al. (1998) and Oguro (1992). In traditional IPMCs, the membrane is made of polymers with covalently bonded and negatively charged groups on the polymer backbone, such as Nafion and Flemion with sulfonate and carboxylate groups, respectively. In the absence of electrical fields, the solvated cations are distributed uniformly in the IPMCs. When a small voltage is applied to the electrode, the solvated cations are forced to move to the cathode side, which results in a volume difference between the cathode and the anode, and therefore, the IPMC bends to the anode side (Figure 10.3).

The key parameters that determine the performance (e.g., amplitude and generated force) of IPMCs are ionic conductivity, liquid electrolyte uptake rate, ion exchange capacity, the mechanical stability of the ionic polymers, ionic conductivity, and electrochemical stability of the electrolytes, electrical conductivity, mechanical compliancy, morphology, and durability of the electrode. Recently, increased efforts have been made to improve the performance of IPMC-based artificial muscles by incorporating the polymers with nanoparticles (Kong and Chen, 2014), reducing surface resistance of electrodes, and applying alternative solvents (Mirvakili and Hunter, 2018; Carpi, 2016).

For example, Bian et al. (2016) fabricated a barium titanate ($BaTiO_3$)/Nafion-based IPMC with improved ionic conductivity and better actuation performance than the pure Nafion-IPMC counterpart. They demonstrated that $BaTiO_3$ nanoparticles significantly enhance the electrical and mechanical performance of Nafion. Specifically, the polymer film with 3% wt $BaTiO_3$ showed the optimal overall performance under a voltage (e.g., DC or AC, 3 V) excitation with deflection ≤101.4% and 250% under DC input and AC (1 Hz) input, respectively. Moreover, the blocking force was boosted by >375% under the DC input. In addition, Wang et al. (2017) used hot-pressing stacked Nafion films to fabricate ultra-thick polymer membranes. The resultant IPMC showed improved output forces. Nam et al. (2017) reported Nafion/polyimide blends based ion exchange membranes. Compared with traditional Nafion-based IPMC, the prepared materials showed enhanced thermal and mechanical properties.

Ma et al. (2020) employed an isopropanol-assisted electroless plating method to fabricate high-quality Pt electrodes in a Nafion-based IPMC. This method could improve the morphology and thickness of the electrode layers; therefore, the actuation performance of the Nafion-IPMC device was highly enhanced, which included a low driving voltage of <1 V, large displacement ≤35.3 mm, high bending speed of 224.2 mm/s, and ultrafast response (>10 Hz).

The long-term operation of the aqueous electrolyte-based IPMCs could not be achieved in the air due to their narrow electrochemically stable window and their poor stability under dry conditions, which limits their application. Significant efforts have been devoted to investigating new solvents to operate at a high voltage and in dry environments. For example, Bennett and Leo (2004) employed ionic liquids to displace the conventional aqueous solvent in Nafion. In addition, Nemat-Nasser, Zamani, and Tor (2006) investigated the effects of solvents (e.g., ethylene glycol and glycerol) on the properties of a series of Nafion- and Flemion-based IPMCs.

In addition, for electrode materials, beyond the traditional noble metal electrode materials, carbon electrodes, such as CNTs (Liu et al., 2010) and graphene (Xie et al., 2010), which have better flexibility and durability, has been explored as suitable electrode materials for IPMCs (Kim et al., 2014).

10.2.1.4   Carbon Nanotubes

CNT-based ionic EAP artificial muscles have been developed based on traditional IPMC to some extent. Due to the large strain resistance of metal electrodes in traditional IPMCs and the low combination force between the electrode and the electrolyte layer, the IPMC is prone to deformation that causes failure, which severely limits the practical applications. Therefore, the design and development of stable electromechanical artificial muscle materials or devices are of great significance today.

CNTs have attracted intense research interest due to their high conductivity, large specific surface area, light weight, and unique mechanical properties, which makes them excellent candidates for a lot of applications, including artificial muscles, biochemical sensors, tissue regeneration, energy storage, flexible or stretchable electronic devices, and electromagnetic shielding materials (Wang et al., 2021; De Volder et al., 2013). CNTs are generally classified into two categories: single-walled CNTs (SWCNTs) and multi-walled CNTs (MWCNTs). The former is generally considered to be cylindrical structures made of a single layer of graphite, and MWCNT can be considered as nested SWCNTs with different diameters.

The previously mentioned properties of CNTs are important factors for high performance electromechanical actuating. Baughman et al. (1999) reported the first electromechanical actuator based on SWCNTs, where SWCNT sheets laminated with Scotch tape served as the electrode and underwent deformation upon charge injection, which showed high stresses even compared with natural muscle and high strains in aqueous electrolyte solutions. Then, Fukushima et al. (2005) fabricated a fully plastic actuator by the layer-by-layer casting of SWCNT electrodes and an ionic liquid-based electrolyte in a gel mixture of 4-methyl-2-pentanone (MP) and poly(vinylidene fluoride-co-hexafluoropropylene) (PVdF(HFP)), which was the first actuator that could be operated in the air at low voltages. Under an applied electric potential of ±3.5 V and 0.01 Hz, the maximum displacement of the fabricated actuator strip bending to the anode side was 5 mm. Then, they studied the displacement performance of the strip under different frequencies and applied voltage and found that the displacement of the fabricated actuator strip was 0.36, 0.76, and 1.8 mm, respectively, with the applied voltage at ±1.0, ±2.0, and ±3.0 V at 0.1 Hz.

To improve the performance of SWCNTs in actuator applications, Chen et al. (2017) prepared SWCNT/ionic liquid-Nafion/SWCNT composite films using sulfonate groups in Nafion interacted with ionic liquid, and the fabricated bucky paper (BP) composite actuator showed an improved Young's modulus, good actuation stability (e.g., ≤43 h, >30,000 cycles) with a low alternating voltage under ambient conditions (Figure 10.4). In addition, the SWCNT BP/1-ethyl-3-methylimidazolium thiocyanate (EMI^+SCN^-)-Nafion/SWCNT BP actuator could achieve a large output blocking force on the Newton scale (1.5 N) operated at an applied voltage of 6 V, which was the first ionic electroactive polymer-based actuator with a Newton-level blocking force under a low applied voltage.

The previous research on SWCNT-based artificial muscle materials showed impressive actuation performance; however, the high price of SWCNTs hinders their practical application. MWCNTs

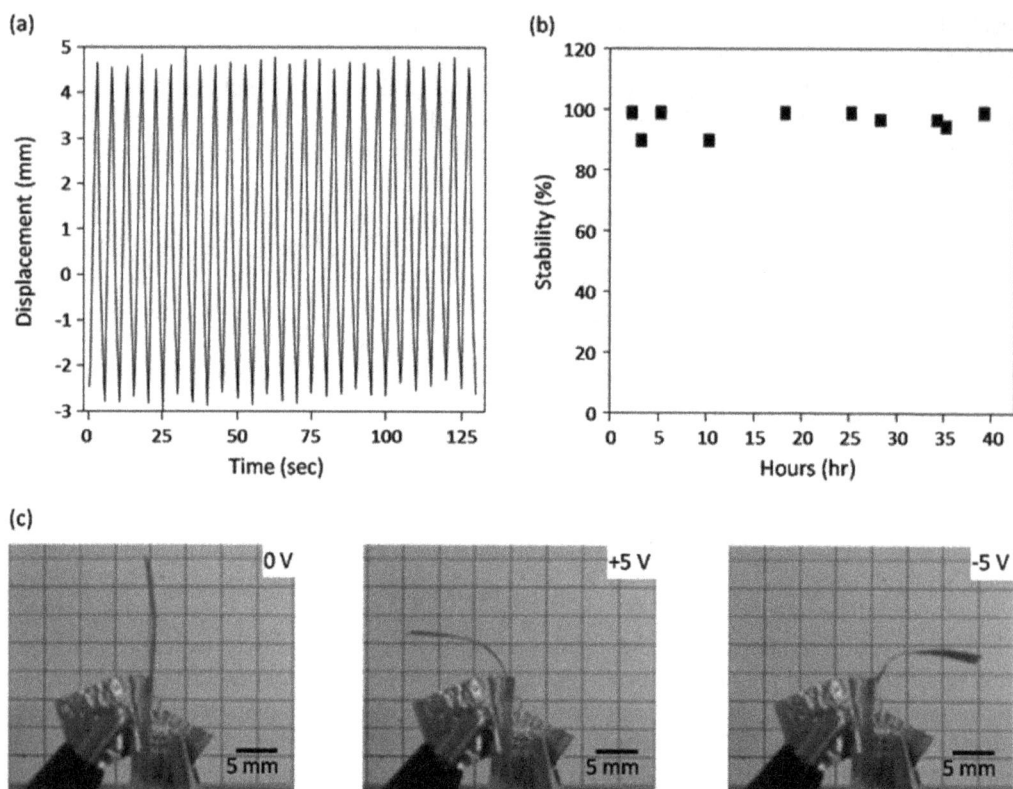

FIGURE 10.4 Actuation performances of the obtained SWCNT BP/EMI⁺SCN⁻-Nafion/SWCNT BP composite actuator: (a) initial 25 cycles and (b) operated for 43 h of actuator applying a voltage of ±2.5 V and 0.2 Hz; (c) bending displacement of the obtained actuator under a voltage of ±5 V at 0.15 Hz.

(From Chen et al. 2017. Copyright 2017 American Chemical Society. With permission.)

with similar sp^2 carbon structure and micromorphology to SWCNTs have attracted increased research interest in artificial muscles.

Aliev et al. (2009) found that CNT aerogel artificial muscles could achieve elongation≤220% and elongation rates as high as (3.7×10^4)%/s under a wide temperature range, and the strength in a specific direction was much higher than that of steel plate. They reported that the maximum achieved work per cycle was approximately 30 J/kg, which was comparable with natural muscle (approximately 40 J/kg), demonstrating potential applications. Then, various strategies were used to enhance the actuation performance and extend the applications for MWCNT-based devices, such as surface modification. Lu and Chen (2010) reported chitosan (CS)/MWCNTs based device, where the CS was employed as the matrix for the electrolyte and electrode due to its good compatibility with MWCNTs. The electrode and electrolyte layers were fabricated using CS/MWCNTs composite and 1-butyl-3-methylimidazolium tetrafluroborate ($BMIBF_4$)/CS, respectively. The prepared device showed enhanced response speed ≤2 mm/s.

10.2.2 ELECTRONIC ELECTROACTIVE POLYMERS

An electronic EAP material is an electrically field active material, due to the rearrangement of the electrostatic force (Coulomb force) on the polymer molecular chain to achieve the expansion and contraction of the volume in each dimension (Bar-Cohen, 2004a). The conversion process is a physical process, including an electrostrictive and Maxwell effect (Van Loocke, Lyons, and Simms, 2004). Electronic EAP materials mainly include DEs, electrostrictive polymers, piezoelectric polymers, ferroelectric polymers, and liquid crystal elastomers (Dang, Wang, and Wang, 2005). Among them, DEs and electrostrictive polymers have similar functional characteristics. The induced strain for the DEs and the electrostrictive polymers is principally a quadratic function of the applied voltage. The induced strain for piezoelectric polymers scales linearly as the electric field (Zhang, Bharti, and Zhao, 1998; Pelrine et al., 2000)

The advantages of electronic EAP materials include a high driving strain and stress, they respond at high frequency, have a long operational life and can hold strain under DC conditions. Meanwhile, the disadvantage is that it requires a high driving electric field (approximately 150 V/μm) during activation due to its electrostatic actuation mechanism. Therefore, to achieve low driving voltages, relatively thin-film materials should be used for electronic EAP artificial muscles.

10.2.2.1 Dielectric Elastomers

DEs are one of the most popular EAP artificial muscles that have been intensively studied due to their simple structure, large strain output, fast response speeds, high efficiency, and good reliability. DE-based artificial muscles are a class of deformable capacitors, which are composed of an elastomeric film with compliant electrodes on both sides of the film. When applying an electric field, pressure (i.e., Maxwell stress) is produced by the electrostatic attraction of opposite charges, which forces the film to expand in the lateral direction and contract in thickness, resulting in an electromechanical phenomenon. Figure 10.5 shows the deformation principle of a DE-based artificial muscle (Liu et al., 2009; Qiu et al., 2019). The thickness strains (S) can be described by assuming that the DE film is linear elastic in the following equation:

$$S = -\frac{p}{Y} = -\frac{\varepsilon_0 E^2}{Y}$$

10.1

where:

p = Maxwell stress
Y and ε_r = apparent elastic modulus and the dielectric constant of the DE, respectively
ε_0 = permittivity of vacuum
E = applied electric field.

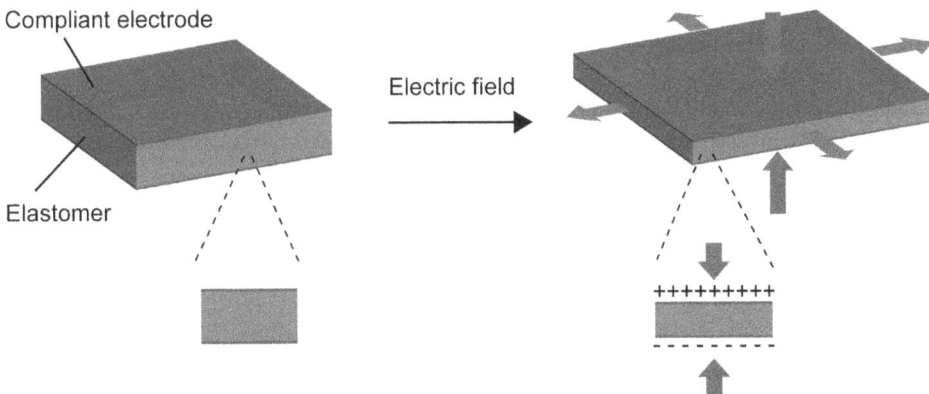

FIGURE 10.5 Deformation principle of a DE-based artificial muscle.

This equation demonstrates that the dielectric constant and electric field strength of the DEs are key factors since the Maxwell stress is proportional to them. In general, DEs are made from highly deformable elastomers with low Young's modulus. Polyacrylates (Shankar, Ghosh, and Spontak, 2007b), polysiloxanes (Pelrine et al., 2000), PU (Petit et al. 2008), and thermoplastic elastomer copolymers (Shankar, Ghosh, and Spontak, 2007a) are the most important active materials for DE artificial muscles that have large deformations under the Maxwell stress. In addition, the compliant electrodes are often made of conductive carbon materials (Carpi et al., 2003; Araromi et al., 2014) or silver pastes. Making the elastomer and the electrodes highly compliant without sacrificing conductivity and dielectric strength is the key to generating a large strain (Mirfakhrai, Madden, and Baughman, 2007).

Röntgen (1880) observed shape changes in a natural rubber strip during charging experiments for the first time. In the late 1990s, SRI International led the latest research on DE-based artificial muscles. Thereafter, efforts have been devoted to improving the intrinsic property of the DEs, such as the elastic stress–strain response and the dielectric constant. For example, an impressive mechanical deformation performance was demonstrated in a commercial polyacrylate with the trademark 3M VHB 4910, for which an actuated relative area strain >100% could be achieved under an electric field strength of 412 MV/m (Pelrine et al., 2000).

Niu et al. (2013) reported new acrylic DEs that showed large strain performance without prestrain. By changing the density of the cross-linker and the plasticizing agent percentage, the resultant elastomers achieved a large actuation strain (>100%), improved energy density (>1 J/g), and high electromechanical stability. To increase the dielectric constant of DEs and improve the artificial muscle performance, they synthesized an aluminum nanoparticle/acrylate copolymer nanocomposite, in which the surface of the aluminum nanoparticle was functionalized using methacrylate groups to form uniform composites (Figure 10.6). The obtained DEs showed a high dielectric constant, especially, the composite with 4% vol of aluminum nanoparticles exhibited a high dielectric constant of 8.4, a maximum area strain of 56% at a dielectric strength of 140 MV/m, and improved force output (Hu et al. 2014).

As mentioned previously, due to their high energy density and large strain, polyacrylate-based DEs have received increasing interest for many applications, such as tactile displays, biomimetic robotics, and microfluidics. Unfortunately, devices based on polyacrylate-based DEs have poor reliability and stability when operating under large strains due to the limited compliance of the electrode and the dielectric breakdown. To resolve this issue, Peng et al. (2021) reported a DE-based device with high strain and stable performance based on an interpenetrating bilayer electrode composed of a water-based PU (PU dispersion in water, WPU) thin layer covered on an ultrathin SWCNT film. This bilayer electrode showed a fair strain (225% under 3.8 kV) and a long durability (≥5.5 h operation at 150% actuation strain using a constant applied voltage). Moreover, the device achieved high stability of 1,000 cycles under a square-wave voltage at 0.05 Hz at 150% strain. These results demonstrated that the prepared bilayer electrode solved the problem of actuation instability of the bare polyacrylate electrodes.

In addition to experimental investigations, the dynamic performance of DE-based devices has been characterized theoretically. Recently, to optimize the power output performance, Cao et al. (2020) designed a double cone DE device (DCDEA) and developed a numerical model to simulate the corresponding dynamic response. Using this model, the power output was optimized for the prestretch rate and spacer length. In addition, a biologically inspired flapping wing mechanism was proposed to demonstrate the potential application of resonant actuation of DCDEA. Of interest, a peak flapping stroke of 31° at 30 Hz was obtained.

10.2.2.2 Electrostrictive Polymers

Electrostrictive polymers have similar functional characteristics to DEs. When the response is determined by the orientation of the electrically induced crystal or semi-crystal structure, the material

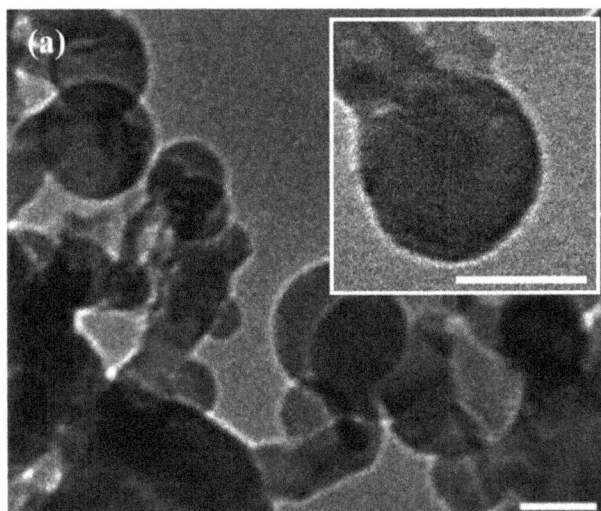

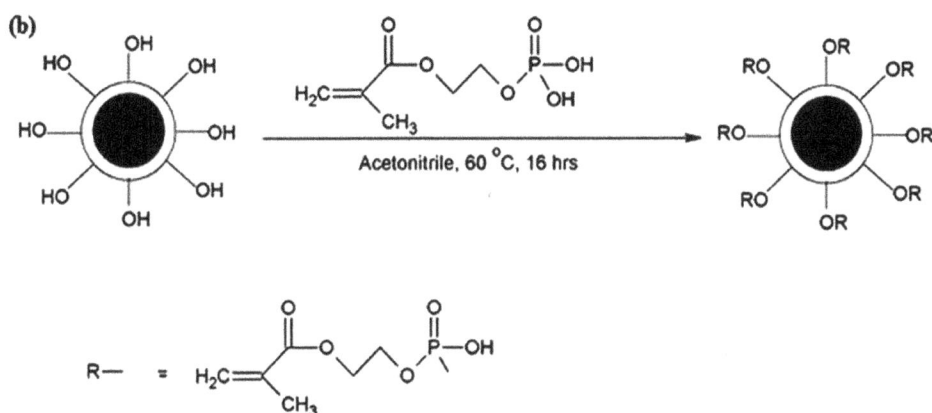

FIGURE 10.6 Showing: (a) transmission electron microscopic images of the used aluminum nanoparticles, scale bars 40 nm; (b) schematic of functionalizing the surfaces of aluminum nanoparticles with phosphoric acid 2-hydroxyethyl methacrylate ester.

(From Hu et al. 2014. With permission from The Royal Society of Chemistry.)

is a polymer electrostrictive material. Electrostrictive materials have been widely used in robots, artificial muscles, autofocus, and other fields due to their excellent physical, chemical, and mechanical properties (Qin and Yu, 1996; Jaaoh, Putson, and Muensit, 2015). In general, electrostrictive materials can be divided into relaxor ferroelectrics and graft copolymers. Poly(vinylidene fluoride) (PVDF) and PVDF copolymer or ternary polymer of PVDF are among the most well-studied electrostrictive polymers. These materials are of great interest in electrostriction because of their excellent properties, such as high strain, high energy density, and low hysteresis (Brochu and Pei, 2012).

Zhang, Bharti, and Zhao (1998)demonstrated a poly(vinylidene fluoride-trifluoroethylene) [P(VDF-TrFE)] copolymer with a good electrostrictive response. The results showed that the state of the crystalline region of the irradiated material was not a simple paraelectric state, but a phase that contained nanopolar regions. The changes in these nanopolar regions under the action of an applied external field lead to the observation of a slim polarization loop. Due to the large difference in lattice constants between the non-polar phase and polar phase of the P(VDF-TrFE) copolymer, the polarization produced an electrostrictive strain with high strain energy density as the electric field gradually increases in the relaxor P(VDF-TrFE) copolymer.

Huang et al. (2004) reported that using P(VDF-TrFE)-based terpolymers, such as poly(vinylidene fluoride-trifluoroethylene-chlorofluroethylene) terpolymer (P(VDF-TrFE-CFE), the thickness strains as high as 7% could be obtained. In addition, the influence of the defects on the micro and mesostructure of these polymers was investigated.

Khudiyev et al. (2017) constructed electrostrictive microelectromechanical fibers/textiles based on P(VDF-TrFE-CFE) ferrorelaxor terpolymer layer. A strain value of >8% was achieved for the fiber device. Under a DC voltage of 200 V, a maximum transverse deflection of 80 μm was established for a 3.5 cm long fiber by contact profilometry.

In addition to P(VDF-TrFE)-based electrostrictive polymers, Wongwirat et al. (2019) studied the electrostriction properties of nylon-12-based and poly(tetramethylene oxide) (PTMO)-based poly(ether-*b*-amide) multiblock copolymers (PEBAX). The soft PTMO block was used to modulate the stiffness of nylon-12 for enhancing electromechanical performance. They proposed the working mechanisms for the electrostriction of this alternative electrostrictive polymer.

However, electrostrictive materials have some disadvantages, such as low elastic coefficient, small dielectric constant, short service life, easy failure, and are fragile, which limit the application of this type of material.

10.2.2.3 Piezoelectric Polymers

Piezoelectric polymers are typical linear electromechanical materials whose stress and strain are linearly related to the electrical field and charge density, which play an important role in coupling the electrical and mechanical behaviors and are of great significance in modern science and technologies.

As a type of electronic EAP material, piezoelectric polymers that work in a similar way to the well-studied piezoelectric ceramics have been widely used in ultrasonic transducers, sensors, energy harvesters, and artificial muscles. As early as the 1920s, people had discovered piezoelectric polymers (Eguchi, 1925). However, due to their small strain and stress, they were not developed and used until the ferroelectric polymer material was discovered. Kawai (1969) demonstrated the tensile piezoelectricity in a stretched and polarized PVDF film for the first time. This work increased the research interests on the piezo, pyro, and ferroelectricity of PVDF. Then, researchers realized that a piezoelectric polymer was a typical linear electromechanical material, for example, the stress and strain of this material were linearly related to the electrical field and charge density. Then, research into piezoelectric polymer gradually increased. Most of the existing piezoelectric polymer materials are polarized bonded polymers, such as PVDF and its copolymers.

10.2.2.4 Ferroelectric Polymers

Ferroelectric polymers possess a non-centrosymmetric structure that shows permanent electric polarization, which can be aligned and maintains polarization under the action of electric fields. Valasek (1921) discovered and defined the ferroelectric property, for instance, materials have an electronic dipole moment, which can be reversed under an appropriate electrical field. As a subset of ferroelectric materials, ferroelectric polymers can be used as artificial muscles in air, vacuums, and underwater.

After the polarization of the material, a sufficiently large reverse electric field must be applied to reverse the polarization, which leads to a large amount of energy being consumed. PVDF and its copolymers, and odd-numbered polyamides are commonly used ferroelectric polymers. The polymer chain must be able to crystallize and maintain polarization. For instance, PVDF has four crystal structures, when an electric field stimulus is used on PVDF, the non-polar α phase state is transformed into a highly polar β phase state (Lovinger, 1983), resulting in contractions in the polarization direction and elongation in the chain direction. The β phase of PVDF is related to piezoelectricity, and the dipole moment that is generated by the orientation of the hydrogen and fluorine atoms contributes to polarization.

Ferroelectric polymers have the advantages of being lightweight, easy to fabricate, and good processability and they can adhere to various substrates. However, the disadvantage of low strain limits their practical applications (Kim and Tadokoro, 2007).

10.2.2.5 Liquid Crystal Elastomers

Liquid crystal elastomers are composed of polymer networks and monodomain nematic liquid crystal units. The working mechanism of these materials involves phase transition between nematic and isotropic phases in a short time (Bar-Cohen, 2002). The polymer network makes the matrix elastic, and under the applied electrical field, the rearrangement and orientation of the rigid liquid crystal cells will produce macroscopic deformations. Due to the shape and dielectric anisotropy of the material, the liquid crystal elastomer can exhibit significant electromechanical responses (Goodby et al., 2014). Long chains of molecules in liquid crystal elastomers can glide through each other, and therefore, elongate with very little driving strain. However, the rearrangement of the side-chain liquid crystal units can cause strain in the main polymeric chain, and then generates driving stress.

Liquid crystal elastomers have the dual characteristics of a liquid crystal and an elastomer, which retain the original properties of a non-crosslinked liquid crystal polymer and have excellent performance of orientation, piezoelectricity, ferroelectricity, and soft elastic under the action of a mechanical force field. Based on the excellent performance of liquid crystal elastomers, their application mainly covers sensors, actuators, and biomimetic machines. Therefore, it has great research values and application prospects in artificial muscles.

Urayama, Honda, and Takigawa (2005) investigated the electrically driven deformation of nematic networks. Their results showed that the maximum strain was 20% at a field strength of approximately 0.5 MV/m. Moreover, the nematic samples could be rapidly deformed within 1 s under electric fields. However, the shape recovery time after field removal was 10^3 s (Urayama, Honda, and Takigawa, 2005). Then, Hashimoto et al. (2008) found that a multifunctional main-chain liquid crystal elastomer (MCLCE) exhibited a large mechanical effect under electrical fields when it was in a nematic solvent. Of note, an electric-field-induced polydomain–monodomain transition in a swollen MCLCE was observed for the first time.

10.3 APPLICATIONS FOR ELECTROACTIVE POLYMER-BASED ARTIFICIAL MUSCLES

Until the emergence of EAP materials with large displacement responses, they received relatively little attention due to the limited availability of materials and satisfactory driving power (Bar-Cohen, 2000). Currently, due to their similar characteristics in operation to biological muscles, and their excellent elasticity and ability to induce large strains, EAP materials have inspired the exploration of unique robotic components and microdevices. In these explorations, EAPs as artificial muscles has gradually gained the interest of scientists and engineers to realize new functions. To date, EAP-based artificial muscles have been used in a wide range of fields, including wearable devices, soft robots (Carrico et al., 2019; Shintake et al., 2018), and biomedicine (Fang et al., 2007). Among the applications, soft robots are one of the most representative applications. Therefore, in this section, progress in the field of soft robots will be discussed.

Lau et al. (2017) fabricated a DE actuated finger with good mechanical strength and approximately 90° bending under voltage for object grasping and pinching (i.e., even pinching a highly deformable raw egg yolk). This finger had a roof shape membrane made from an acrylic elastomer (VHB 4910) on a frame that used polyimide and PVC, which exhibited a tension-induced moment 40 times higher than its flat-shaped DE counterpart. Furthermore, the gripper of these DE fingers showed good mechanical strength to life with a payload 8–9 times of its weight. Other strategies have been applied in DE-based grippers for soft robotics. For example, Shian, Bertoldi, and Clarke (2015) incorporated stiff fibers into DE beams to control their deformation. Specifically, a wrap-around gripper that contained vertically arranged fibers was used to pick and place different objects, such as metal cylinder and grapes (Figure 10.7(a)). A gripper with horizontally arranged fibers could grasp and lift a wooden frame (Figure 10.7(b)).

In addition to DE-based artificial muscles, IPMCs, CNTs (Hu et al., 2017), and polymer gels (Yang et al., 2017) have been used to mimic the function of fingers. Bar-Cohen et al. (1998) and Deole et al. (2008) constructed IPMC-based grippers to lift objects. However, due to their low induced stress and slow response, the application of this type of gripper is limited compared with those of DEs.

To achieve enhanced overall muscle-like performance, a new technology called hydraulically amplified self-healing electrostatic (HASEL) artificial muscle has been proposed, which couples electrostatic and hydraulic forces. Acome et al. (2018) developed a class of HASEL soft artificial muscles that was characterized by high performance, multi-function, and the ability to self-heal. In contrast to soft fluidic actuators, HASEL locally generates hydraulic pressure through an electrostatic force, which acts on the liquid dielectric distributed throughout the soft structure. Of note, the use of liquid dielectrics enabled HASEL to be self-healable from dielectric breakdown. Due to the excellent mechanical response and strain capacity of HASEL, the device can capture tiny objects. Two stacks of HASEL can be used as a soft gripper. When a DC voltage is applied to the stacked HASEL actuators, this device can gently grab fragile objects, such as raspberries and raw eggs. Rothemund et al. (2021) summarized recent advances in HASEL artificial muscles.

Aziz et al. (2020) proposed a yarn-based artificial muscle and applied it to a soft actuator. The material was made from CNT-coated polyethylene terephthalate (PET) yarn with electrochemically

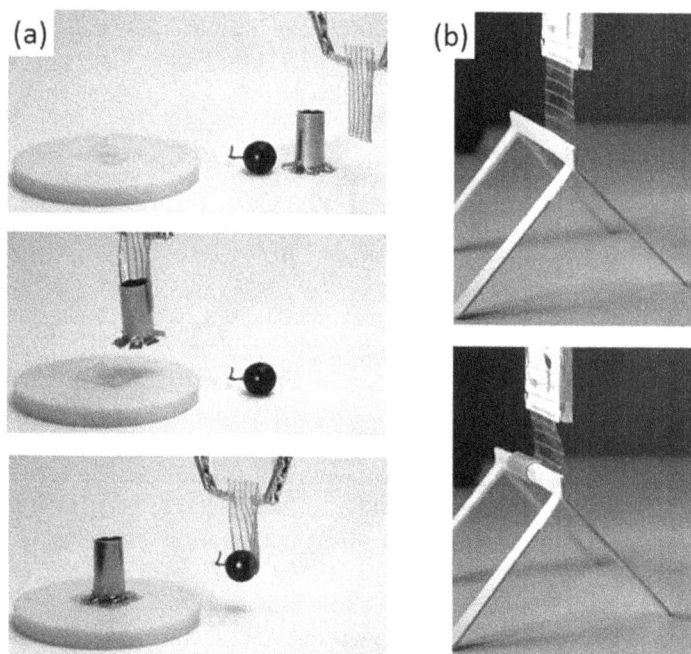

FIGURE 10.7   Examples of the DE-fibers based grippers.

(From Shian, Bertoldi, and Clarke 2015. Copyright 2015 John Wiley & Sons. With permission.)

deposited PPy coating. The twisted and coiled yarns delivered high tensile strength in wet and dry shearing conditions. It demonstrated that the constructed CNT-coated hybrid yarn actuators with desirable mechanical properties were of great interest as artificial muscles or soft actuators in textile exoskeletons and wearable devices.

A broad range of EAP materials has been investigated for soft robotic applications. Different types of EAP artificial muscles show distinct properties because these materials can produce specific responses and strains under applied conditions. Due to the progress in advanced materials and new design concepts, EAP materials have significant prospects in practical applications.

10.4 CONCLUSIONS AND OUTLOOK

The field of EAP-based artificial muscles has rapidly developed in the last 30 years due to the development of EAP materials. In this chapter, the important developments in EAP materials were reviewed, including responsive materials, working principles, and applications. Compared with the previous artificial muscle materials and devices, such as shape-memory materials and piezoelectric ceramic materials, EAP materials have the outstanding characteristics of high strain, fast response, good flexibility, light weight, and no noise, which means that they can simulate the functionalization of natural muscles.

Despite the previously mentioned advances, it remains challenging to design and assemble reliable EAP-based artificial muscles with excellent strain, response, and stability performance for different applications. One of the major limitations of the current EAP materials is that the advantages and disadvantages of each category are distinct; therefore, no individual material has a satisfactory comprehensive performance. For example, the development of new responsive materials with low activation voltage without losing much strain is of great importance for DE-based artificial muscles. To overcome these challenges, further innovation is needed, including the creation of new materials, and the design and fabrication of devices. In addition, the integration of multiple disciplines, such as computational chemistry, materials science, electrochemistry, and physics, will improve the application of these materials.

EAP-based artificial muscles face enormous challenges before realizing wide applications. However, EAPs are excellent candidates and could play an important role in developing future artificial muscle-related technologies.

REFERENCES

Aabloo, A., Belikov, J, Kaparin, V, & Kotta U. (2020). Challenges and perspectives in control of ionic polymer-metal composite (IPMC) actuators: A survey. *IEEE Access*, *8*, 121059–121073.

Acome, E, Mitchell, S. K., Morrissey, T. G., Emmett, M. B., Benjamin, C., King, M....Keplinger, C. (2018). Hydraulically amplified self-healing electrostatic actuators with muscle-like performance. *Science*, *359*(6371), 61–65.

Aliev, A. E., Oh, J., Kozlov, M. E., Kuznetsov, A. A., Fang, S., Fonseca, A. F. R. ... Gartstein, Y. N. (2009). Giant-stroke, superelastic carbon nanotube aerogel muscles. *Science*, *323*(5921), 1575–1578.

Araromi, O. A., Gavrilovich, I., Shintake, J., Rosset, S., Richard, M., Gass, V. ... Shea, H. R. (2014). Rollable multisegment dielectric elastomer minimum energy structures for a deployable microsatellite gripper. *IEEE-ASME Transactions on Mechatronics*, *20*(1), 438–446.

Arndt, K. F., Kuckling, D., and Richter, A. (2000). Application of sensitive hydrogels in glow control. *Polymers for Advanced Technologies*, *11*(8–12), 496–505.

Aziz, S., Martinez, J. G., Foroughi, J., Spinks, G. M., & Jager, E. W. H. (2020). Artificial muscles from hybrid carbon nanotube-polypyrrole-coated twisted and coiled yarns. *Macromolecular Materials and Engineering*, *305*(11), 2000421.

Bar-Cohen, Y., Xue, T., Shahinpoor, M., Simpson, J., & Smith J. (1998). *Proceedings of Robotics 98, American Society of Civil Engineers.* (pp.15–21) Albuquerque: ASCE.

Bar-Cohen, Y. (2000). Electroactive polymers as artificial muscles: Capabilities, potentials and challenges. *Robotics, 2000*, 188–196.

Bar-Cohen, Y. (2002). Electroactive polymers as artificial muscles: A review. *Journal of Spacecraft and Rockets, 39*(6), 822–827.

Bar-Cohen, Y. (2004a). *Electroactive Polymer (EAP) Actuators as Artificial Muscles: Reality, Potential, and Challenges*. (vol 136) Bellingham: SPIE Press.

Bar-Cohen, Y. (2004b). Electroactive polymers (EAP) as actuators for potential future planetary mechanisms. *6th NASA/DoD Conference on Evolvable Hardware 2004*, Proceedings, 309–317. Seattle, USA.

Bar-Cohen, Y. (2005). Current and future developments in artificial muscles using electroactive polymers. *Expert Review of Medical Devices, 2*(6), 731–740.

Bashir, M., and Rajendran, P. (2018). A review on electroactive polymers development for aerospace applications. *Journal of Intelligent Material Systems and Structures, 29*(19), 3681–3695.

Bassil, M., Davenas, J., & Tahchi, M. E. L. (2008). Electrochemical properties and actuation mechanisms of polyacrylamide hydrogel for artificial muscle application. *Sensors and Actuators B: Chemical, 134*(2), 496–501.

Baudis, S., Behl, M,. & Lendlein, A. (2014). Smart polymers for biomedical applications. *Macromolecular Chemistry and Physics, 215*(24), 2399–2402.

Baughman, R. H. (1991). Conducting polymers in redox devices and intelligent materials systems. *Makromolekulare Chemie-Macromolecular Symposia, 51*, 193–215.

Baughman, R. H., Cui, C., Zakhidov, A. A., Iqbal, Z., Barisci, J. N., Spinks, G. M., ... Rinzler, A. G. (1999). Carbon nanotube actuators. *Science, 284*(5418), 1340–1344.

Bennett, M. D., and Leo, D. J. (2004). Ionic liquids as novel solvents for ionic polymer transducers. *Smart Structures and Materials 2004: Electroactive Polymer Actuators and Devices (EAPAD), 5385*, 210–220.

Bian, K., Liu, H., Tai, G., Zhu, K., & Xiong. K. (2016). Enhanced actuation response of Nafion-based ionic polymer metal composites by doping $BaTiO_3$ nanoparticles. *The Journal of Physical Chemistry C, 120*(23), 12377–12384.

Bowers, T. A. (2004). *Modeling, Simulation, and Control of a Polypyrrole-Based Conducting Polymer Actuator*. (Unpublished MSME Dissertation). Massachusetts Institute of Technology. Cambridge, MA, USA.

Brochu, P., and Pei, Q. (2010). Advances in dielectric elastomers for actuators and artificial muscles. *Macromolecular Rapid Communications, 31*(1),10–36.

Brochu, P., and Pei, Q. (2012). Dielectric elastomers for actuators and artificial muscles. In L. Rasmussen (Ed.) *Electroactivity in Polymeric Materials* (pp. 1–56). New York: Springer.

Brock, D., Lee, W., Segalman, D., & Witkowski, W. (1994). A dynamic model of a linear actuator based on polymer hydrogel. *Journal of Intelligent Material Systems and Structures, 5*(6), 764–771.

Cao, C., Gao, X., Burgess, S., & Conn, A. T. (2020). Power optimization of a conical dielectric elastomer actuator for resonant robotic systems. *Extreme Mechanics Letters, 35*,100619.

Carpi, F. (2016). *Electromechanically Active Polymers: A Concise Reference*: Cham: Springer.

Carpi, F., Chiarelli, P., Mazzoldi, A., & De Rossi, D. (2003). Electromechanical characterisation of dielectric elastomer planar actuators: Comparative evaluation of different electrode materials and different counterloads. *Sensors and Actuators A: Physical, 107*(1), 85–95.

Carpi, F., Kornbluh, R., Sommer-Larsen, P., & Alici, G. (2011). Electroactive polymer actuators as artificial muscles: Are they ready for bioinspired applications? *Bioinspiration & Biomimetics, 6*(4), 045006.

Carrico, J. D., Hermans, T., Kim, K. J., & Leang, K. K. (2019). 3D-printing and machine learning control of soft ionic polymer-metal composite actuators. *Scientific Reports, 9*,17482.

Chen, D., & Pei, Q. (2017). Electronic muscles and skins: A review of soft sensors and actuators. *Chemical Reviews, 117*(17), 11239–11268.

Chen, I.-W. P., Yang, M.-C., Yang, C.-H., Zhong, D.-X., Hsu, M.-C., & Chen, Y. W. (2017). Newton output blocking force under low-voltage stimulation for carbon nanotube-electroactive polymer composite artificial muscles. *ACS Applied Materials & Interfaces, 9*(6), 5550–5555.

Chen, Y., Chen, C., Rehman, H. U., Zheng, X., Li, H., Liu, H., & Hedenqvist, M. S. (2020). Shape-memory polymeric artificial muscles: Mechanisms, applications and challenges. *Molecules, 25*(18), 4246.

Dang, Z.-M., Wang, L., & Wang, H.-Y. (2005). Novel smart materials: Progress in electroactive polymers. *Journal of Functional Materials, 36*(7), 981.

Deole, U., Lumia, R., Shahinpoor, M., & Bermudez, M. (2008). Design and test of IPMC artificial muscle microgripper. *Journal of Micro-Nano Mechanronics*, *4*, 95–102.

De Volder, M. F. L., Tawfick, S. H., Baughman, R. H., & Hart. A. J. (2013). Carbon nanotubes: Present and future commercial applications. *Science*, *339*(6119), 535–539.

Ebadi, S. V., Fashandi, H., Semnani, D., Rezaei, B., & Fakhrali, A. (2020). Overcoming the potential drop in conducting polymer artificial muscles through metallization of electrospun nanofibers by electroplating process. *Smart Materials and Structures*, *29*(8), 085036.

Ebadi, S. V., Semnani, D., Fashandi, H., Rezaei, B., & Fakhrali, A. (2020). Gaining insight into electrolyte solution effects on the electrochemomechanical behavior of electroactive PU/PPy nanofibers: Introducing a high-performance artificial muscle. *Sensors and Actuators B: Chemical*,*305*,127519.

Eguchi, M. (1925). On the permanent electret. *Philosophical Magazine*, *49*, 178–192.

Fang, B.-K., Ju, M.-S., & Lin, C.-C. K. (2007). A new approach to develop ionic polymer–metal composites (IPMC) actuator: Fabrication and control for active catheter systems. *Sensors and Actuators A: Physical*, *137*, 321–329.

Fannir, A., Temmer, R., Nguyen, G. T. M., Cadiergues, L., Laurent, E., Madden, J. D. W. …Plesse, C. (2019). Linear artificial muscle based on ionic electroactive polymer: A rational design for open-air and vacuum actuation. *Advanced Materials Technologies*, *4*(2), 1800519.

Fried, J. R. (2000). Polymers in aerospace applications. *Macromolecules*, *33*(6), 2171–83.

Fukushima, T., Asaka, K., Kosaka, A., & Aida. T. (2005). Fully plastic actuator through layer-by-layer casting with ionic-liquid-based bucky gel. *Angewandte Chemie International Edition*, *44*(16), 2410–2413.

Goodby, J. W., Collings, P. J., Kato, T., Tschierske, C., Gleeson, H., & Raynes, P. (2014). *Handbook of Liquid Crystals.* Vol. 1. Weinheim: Wiley-VCH.

Gu, G.-Y., Zhu, J., Zhu, L.-M., & Zhu, X. (2017). A survey on dielectric elastomer actuators for soft robots. *Bioinspiration & Biomimetics*, *12*(1), 011003.

Hashimoto, S., Yusuf, Y., Krause, S., Finkelmann, H., Cladis, P. E., Brand, H. R., & Kai S (2008). Multifunctional liquid crystal elastomers: Large electromechanical and electro-optical effects. *Applied Physics Letters*, *92*(18), 181902.

Hirai, T., Zheng, J., Watanabe, M., Shirai, H., & Yamaguchi M. (1999). Electroactive nonionic polymer gel-swift bending and crawling motion. *Electroactive Polymers (EAP)*, *600*(1), 267–272.

Hu, W., Zhang, S. N., Niu, X., Liu C., & Pei. Q. (2014). An aluminum nanoparticle-acrylate copolymer nanocomposite as a dielectric elastomer with a high dielectric constant. *Journal of Materials Chemistry C*, *2*(9), 1658–1666.

Hu, Y., Liu, J., Chang, L., Yang, L., Xu, A., Qi, K., … Wu, Y. (2017). Electrically and sunlight-driven actuator with versatile biomimetic motions based on rolled carbon nanotube bilayer composite. *Advanced Functional Materials*, *27*, 1704388.

Huang, C., Klein, R., Xia, F., Li, H., Zhang, Q. M., Bauer, F., & Cheng Z. Y. (2004). Poly (vinylidene fluoride-trifluoroethylene) based high performance electroactive polymers. *IEEE Transactions on Dielectrics and Electrical Insulation*, *11*(2), 299–311.

Jaaoh, D., Putson, C., & Muensit, N. (2015). Deformation on segment-structure of electrostrictive polyurethane/polyaniline blends. *Polymer*, *61*, 123–130.

Jiang, T. Y., Ng, T. Y., & Lam. K. Y. (2000). Optimization of a piezoelectric ceramic actuator. *Sensors and Actuators A: Physical*, *84*(1–2), 81–94.

Jo, C., Naguib, H. E., & Kwon, R. H. (2011). Fabrication, modeling and optimization of an ionic polymer gel actuator. *Smart Materials and Structures*, *20*(4), 045006.

Kawai, H. (1969). The piezoelectricity of poly (vinylidene fluoride). *Japanese Journal of Applied Physics*, *8*(7), 975.

Khudiyev, T., Clayton, J., Levy, E., Chocat, N., Gumennik, A., Stolyarov, A. M.…Fink Y. (2017). Electrostrictive microelectromechanical fibres and textiles. *Nature Communications*, *8*(1), 1435.

Kim, J., Jeonn, J.-H., Kim, H.-J., Lim, H., & Oh ,I.-K. (2014). Durable and water-floatable ionic polymer actuator with hydrophobic and asymmetrically laser-scribed reduced graphene oxide paper electrodes. *ACS Nano*, *8*(3), 2986–2997.

Kim, K. J., & Tadokoro, S. (2007). Electroactive polymers for robotic applications. *Artificial Muscles and Sensors*, *23*, 291.

Kong, L., & Chen, W. (2014). Carbon nanotube and graphene-based bioinspired electrochemical actuators. *Advanced Materials*, *26*(7, 1025–1043.

Kuckling, D., Richter, A., & Arndt, K.-F. (2003). Temperature and pH-dependent swelling behavior of poly (N-isopropyl acrylamide) copolymer hydrogels and their use in flow control. *Macromolecular Materials and Engineering, 288*(2), 144–151.

Lau, G. K., Heng, K. R., Ahmed, A. S., & Shrestha M. (2017). Dielectric elastomer fingers for versatile grasping and nimble pinching. *Applied Physics Letters, 110*, 182906.

Li, Y., Guo, M., & Li, Y. (2019). Recent advances in plasticized PVC cells for soft actuators and devices: A review. *Journal of Materials Chemistry C, 7*(42), 12991–13009.

Liu, S., Liu Y.,. Cebeci, H., de Villoriam R. G., Lin, J.-H., Wardle, B. L., & Zhang, Q. M. (2010). High electro-mechanical response of ionic polymer actuators with controlled-morphology aligned carbon nanotube/ Nafion nanocomposite electrodes. *Advanced Functional Materials, 20*(19), 3266–3271.

Liu, X., Zhang, X., Cong, J., Xu, J., & Chen K. (2003). Demonstration of etched cladding fiber Bragg grating-based sensors with hydrogel coating. *Sensors and Actuators B: Chemical, 96*(1–2), 468–472.

Liu, Y., Liu, L., Zhang, Z., & Leng, J. (2009). Dielectric elastomer film actuators: Characterization, experiment and analysis. *Smart Materials and Structures, 18*(9), 095024.

Liu, Z., & Calvert, P. (2000). Multilayer hydrogels as muscle-like actuators. *Advanced Materials, 12*(4), 288–291.

Lovinger, A. J. (1983). Ferroelectric polymers. *Science, 220*(4602), 1115–1121.

Lu, L., & Chen, W. (2010). Biocompatible composite actuator: A supramolecular structure consisting of the biopolymer chitosan, carbon nanotubes, and an ionic liquid. *Advanced Materials, 22*(33), 3745–3748.

Ma, D. (1995). *Autonomous Torque Sensor Calibration and Gravity Compensation for Robot Manipulators.* Montréal: McGill University.

Ma, S., Zhang, Y., Liang, Y., Ren, L., Tian, W., & Ren, L. (2020). High-performance ionic-polymer-metal composite: Toward large-deformation fast-response artificial muscles. *Advanced Functional Materials, 30*(7), 1908508.

Melling, D., Martinez, J. G., & Jager, E. W. H. (2019). Conjugated polymer actuators and devices: Progress and opportunities. *Advanced Materials, 31*(22), 1808210.

Mirfakhrai, T.,. Madden, J. D. W., & Baughman. R. H. (2007). Polymer artificial muscles. *Materials Today, 10*(4), 30–38.

Mirvakili, S. M, & Hunter, I. W. (2018). Artificial muscles: Mechanisms, applications, and challenges. *Advanced Materials, 30*(6), 1704407.

Nam, J., Hwang, T., Kim, K. J., & Lee, D.-C. (2017). A new high-performance ionic polymer-metal composite based on Nafion/polyimide blends. *Smart Materials and Structures, 26*(3), 035015.

Nemat-Nasser, S., Zamani, S., & Tor, Y. (2006). Effect of solvents on the chemical and physical properties of ionic polymer-metal composites. *Journal of Applied Physics, 99*(10), 104902.

Nickel, V. L., Savill, D. L., Karchak, A., & Allen, J. R. (1965). Synthetically powered orthotic systems. *The Journal of Bone and Joint Surgery British, 47*(3), 458–464.

Niu, X., Stoyanov, H., Hu, W., Leo, R., Brochu, P., & Pei. Q. (2013). Synthesizing a new dielectric elastomer exhibiting large actuation strain and suppressed electromechanical instability without prestretching. *Journal of Polymer Science Part B: Polymer Physics, 51*(3), 197–206.

Oguro, K. (1992). Bending of an ion-conducting polymer film-electrode composite by an electric stimulus at low voltage. *Journal of Micromachine Society, 5,*27–30.

Oktar, O., Caglar, P., & Seitz, W. R. (2005). Chemical modulation of thermosensitive poly (N-isopropylacrylamide) microsphere swelling: A new strategy for chemical sensing. *Sensors and Actuators B: Chemical, 104*(2), 179–185.

Osada, Y., Okuzaki, H., & Hori, H. (1992). A polymer gel with electrically driven motility. *Nature, 355*(6357), 242–244.

Otero, T. F., Angulo, E., Rodriguez, J., & Santamaria, C. (1992). Electrochemomechanical properties from a bilayer: Polypyrrole/non-conducting and flexible material-artificial muscle. *Journal of Electroanalytical Chemistry, 341*(1–2), 369–375.

Palza, H., Zapata, P. A., & Angulo-Pineda, C. (2019). Electroactive smart polymers for biomedical applications. *Materials, 12*(2), 277.

Pei, Q., & Inganläs, O. (1992). Conjugated polymers and the bending cantilever method: Electrical muscles and smart devices. *Advanced Materials, 4*(4), 277–278.

Pelrine, R., Kornbluh, R., Joseph, J., Heydt, R., Pei, Q., & Chiba, S. (2000). High-field deformation of elasto-meric dielectrics for actuators. *Materials Science and Engineering: C, 11*(2), 89–100.

Pelrine, R., Kornbluh, R., Pei, Q., & Joseph, J. (2000). High-speed electrically actuated elastomers with strain greater than 100%. *Science*, *287*(5454), 836–839.

Pelrine, R. E.,. Kornbluh, R. D., Pei, Q., & Joseph, J. P. (2003). *Electroactive polymer electrodes*. (U.S. Patent No. 6,583,533 B2).

Peng, Z., Shi, Y., Chen, N., Li, Y., & Pei, Q. (2021). Stable and high-strain dielectric elastomer actuators based on a carbon nanotube-polymer bilayer electrode. *Advanced Functional Materials*, *31*(9), 2008321.

Petit, L., Guiffard, B., Seveyrat, L., & Guyomar, D. (2008). Actuating abilities of electroactive carbon nanopowder/polyurethane composite films. *Sensors and Actuators A: Physical*, *148*(1),105–110.

Punning, A., Kim, K. J., Palmre, V., Vidal, F., Plesse, C., Festin, N. A.… Alici, G. (2014). Ionic electroactive polymer artificial muscles in space applications. *Scientific Reports*, *4*(1), 1–6.

Qiu, Y., Zhang, E., Plamthottam, R., & Pei, Q. (2019). Dielectric elastomer artificial muscle: Materials innovations and device explorations. *Accounts of Chemical Research*, *52*(2), 316–325.

Qin, X., & Shangyin, Y. (1996). Recent advances in research on electrostrictive materials. *Piezoelectrics & Acoustooptics*, *18*(2), 129–133.

Richter, A., Kuckling, D., Howitz, S., Gehring, T., & Arndt, K.-F. (2003). Electronically controllable microvalves based on smart hydrogels: Magnitudes and potential applications. *Journal of Microelectromechanical Systems*, *12*(5), 748–753.

Röntgen, W. C. (1880). About the Changes in shape and volume of dielectrics caused by electricity. *Annual Review of Physical Chemistry*, *11*, 771–786.

Rothemund, P., Kellaris, N., Mitchell, S. K., Acome, E., & Keplinger, C. (2021). HASEL artificial muscles for a new generation of lifelike robots—recent progress and future opportunities. *Advanced Materials*, 33(19), 2003375.

Shahinpoor, M. (1992). Conceptual design, kinematics and dynamics of swimming robotic structures using ionic polymeric gel muscles. *Smart Materials and Structures*, *1*(1), 91–94.

Shahinpoor, M., Bar-Cohen, Y., Simpson, J. O., & Smith, J. (1998). Ionic polymer-metal composites (IPMCs) as biomimetic sensors, actuators and artificial muscles: A review. *Smart Materials and Structures*, *7*(6), R15-R30.

Shankar, R., Ghosh, T. K., & Spontak, R. J. (2007a). Electroactive nanostructured polymers as tunable actuators. *Advanced Materials*, *19*(17), 2218–2223.

Shankar, R., Ghosh, T. K., & Spontak, R. J. (2007b). Electromechanical response of nanostructured polymer systems with no mechanical pre-strain. *Macromolecular Rapid Communications*, *28*(10), 1142–1147.

Shian, S., Bertoldi, K., & Clarke, D. R. (2015). Dielectric elastomer based "Grippers" for soft robotics. *Advanced Materials*, *27*, 6814–6819.

Shintake, J., Cacucciolo, V., Floreano, D., & Shea, H. (2018). Soft robotic grippers. *Advanced Materials*, *30*, 1707035.

Shirakawa, H., Louis, E. J., MacDiarmid, A. G., Chiang, C. K., & Heeger, A. J. (1977). Synthesis of electrically conducting organic polymers: Halogen derivatives of polyacetylene, (CH)$_x$. *Journal of the Chemical Society, Chemical Communications*, 16, 578–580.

Spinks, G. M. (2020). Advanced actuator materials powered by biomimetic helical fiber topologies. *Advanced Materials*, *32*(18), 1904093.

Spinks, G. M, Mottaghitalab, V., Bahrami-Samani, M., Whitten, P. G., &. Wallace, G. G. (2006). Carbon-nanotube-reinforced polyaniline fibers for high-strength artificial muscles. *Advanced Materials*, *18*(5), 637–640.

Sun, Z., Yang, L., Zhao, J., & Song, W. (2020). Natural cellulose-full-hydrogels bioinspired electroactive artificial muscles: Highly conductive ionic transportation channels and ultrafast electromechanical response. *Journal of the Electrochemical Society*, *167*(4), 047515.

Tian, S., & Wang, X. (2016). Establishing mechanical model of conducting polymer actuator. *Mechanical Science and Technology for Aerospace Engineering*, *35*(6), 870–876.

Urayama, K., Kondo, H. , O. Y., & Takigawa, T. (2005). Electrically driven deformations of nematic gels. *Physical Review E*, *71*(5), 051713.

Valasek, J. (1921). Piezoelectric and allied phenomena in Rochelle salt. *Physical Review*, *17*,475–481.

Van Loocke, M., Lyons, C. G., & Simms, C. (2004). The three-dimensional mechanical properties of skeletal muscle: Experiments and modelling. In P. J. Prendergast & P. E. McHugh (Eds.) *Topics in bio-mechanical engineering*. (pp. 216–234). Dublin: Trinity Centre for Bioengineering.

Wang, H.-M., & Qu, S.-X. (2016). Constitutive models of artificial muscles: A review. *Journal of Zhejiang University-SCIENCE A, 17*(1), 22–36.

Wang, H. S., Cho, J., Song, D. S., Jang, J. H., Jho, J. Y., & Park, J. H. (2017). High-performance electroactive polymer actuators based on ultrathick ionic polymer-metal composites with nano dispersed metal electrodes. *ACS Applied Materials & Interfaces, 9*(26), 21998–22005.

Wang, J., Gao, D., & Lee, P. S. (2021). Recent progress in artificial muscles for interactive soft robotics. *Advanced Materials, 33*(19), 2003088.

Wang, X.-X., Yu, G.-F., Zhang, J., Yu, M., Ramakrishna, S., & Long, Y.-Z. (2021). Conductive polymer ultrafine fibers via electrospinning: Preparation, physical properties and applications. *Progress in Materials Science, 115*, 100704.

Wongwirat, T., Wang, M. Huang, Y., Treufeld, I., Li, R., Laoratanakul, P., Manuspiya, H., & Zhu, L. (2019). Mesophase structure-enabled electrostrictive property in nylon-12-based poly (ether-block-amide) copolymers. *Macromolecular Materials and Engineering, 304*(9), 1900330.

Xie, X., Qu, L., Zhou, C., Li, Y., Zhu, J., Bai, H. …Dai, L. (2010). An asymmetrically surface-modified graphene film electrochemical actuator. *ACS Nano, 4*(10),6050–6054.

Yang, C., Liu, Z., Chen, C., Shi, K.. Zhang, L,. Ju, X. J. …and Chu, L. Y. (2017). Reduced graphene oxide-containing smart hydrogels with excellent electro-response and mechanical properties for soft actuators. *ACS Applied Materials & Interfaces, 9*,15758–15767.

Zhang, Q. M., Bharti, V., & Zhao, X. (1998). Giant electrostriction and relaxor ferroelectric behavior in electron-irradiated poly(vinylidene fluoride-trifluoroethylene) copolymer. *Science, 280*(5372), 2101–2104.

11 Electroactive Polymers for Electrochromic Applications

Mahmoud H. Abu Elella, Emad S. Goda, Heba M. Abdallah, Shaimaa Elyamny, Heba Gamal, Esraa Samy Abu Serea, and Ahmed Esmail Shalan

CONTENTS

11.1 INTRODUCTION

Recently electrochromic devices (ECD) have gained interest from academic and industrial researchers, and they have potential industrial applications in rearview mirrors, low power consumption displays, versatile color-changing textiles, smart windows, and glare reduction (1–6). It is well known that electrochromism is a physicochemical phenomenon where materials can controllably tune their color in absorbance, or transmittance, or both as a result of the application of an external electrical potential, which causes electrochemical oxidation or reduction (7, 8). Therefore, it is usually found when applying potential, and the materials can change their optical properties across the entire electromagnetic spectra, such as 400–800 nm, and 1,000–2,000 nm for the visible region and near-infrared (NIR) region, respectively. Similarly, the electrochromism phenomenon offers many outstanding properties unlike other display approaches, such as electroluminescence. Thanks to the breakthroughs in designing novel nanomaterials, a few ECDs have emerged in the market (9, 10).

In addition to the previous points, for the fabrication of ECDs, the system switches from the state of full bleaching to the complete coloring as a result of the generation of two redox couples in the reaction. Between the two established electrodes, one material is reduced, and the other is oxidized to form the redox couple to ensure that the produced device is working properly, with a prominent change in the color of the system. As shown in Figure 11.1, the working principle of a laminated

DOI: 10.1201/9781003173502-11

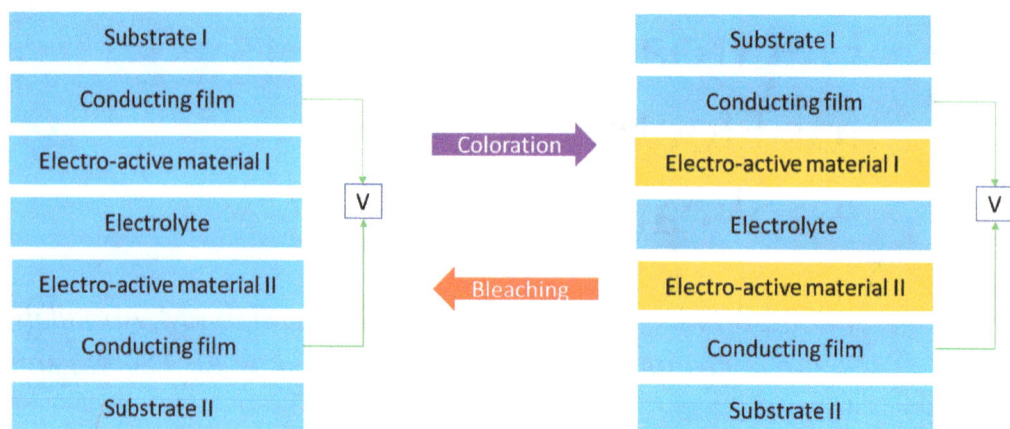

FIGURE 11.1 Configuration of laminated transmission ECD.

(From Thakur et al. 2012 (14). With permission.)

device is composed of two electroactive materials (EC1 and EC2) including one as electrochromic, two transparent conductive substrates, and an electrolyte. The device is wrapped with sealant to prevent electrolyte leakage and ease of handling. There were important factors when evaluating the electrochromic performance of the materials, such as optical density, contrast ratio (CR), coloration efficiency (CE (η)), cycle life, switching time, and stability. In particular, organic materials typically achieve high coloration efficiencies compared with inorganic ones due to their higher molar absorption. Therefore, good ECD performance should have higher electrochromic efficiency, optical contrast, stability, and a shorter response time (9–13).

Of note, the film made by EC materials is an important component of ECDs, a good ECD must have a short response time, high optical contrast (ΔT%), optical memory, high CE, and good stability (14). Various EC materials have been shown to undergo EC behavior and can be classified into two categories: inorganic EC metal oxides (13, 15, 16) and organic EC materials with conjugating polymers as the conducting polymers and small organic conjugated viologen (17, 18). Unlike its counterparts, EC polymer materials have been extensively used in ECDs thanks to their remarkable properties, such as their low cost and easy processing by lamination, blending and copolymerization techniques, high stability, and optical contrast, in addition to high flexibility that improved their application in foldable devices, and they can show more than two redox states and produce many colors. Polymers are increasingly used in display organic light-emitting diodes, cathode ray tubes, thin film transistors, optical shutters (e.g., mirrors, smart windows, and sunglasses), electric operation in the cell membrane, thermal exposure indicators for frozen foods, EC printing, and smart papers due to their outstanding advantages (19–21).

11.2 CLASSIFICATION OF ELECTROCHROMIC ORGANIC MATERIALS

11.2.1 CONJUGATED CONDUCTIVE POLYMERS

Recently, conjugated polymers (CPs) such as polyanilines (PANI), polythiophenes (PTh), and polypyrrole (PPy). In addition, polyamides (PA), and polycarbazoles (PCz) (Scheme 11.1), are prospects for practical applications because of their benefits, such as lightweight, flexibility, low cost, high optical contrast, and ease of deposition, fast switching speed, and processability compared with inorganic EC materials (22–24). Of interest, conductive polymer EC efficiency depends strongly on their chemical structure and aggregated state, which are sensitive to the transport of charges and

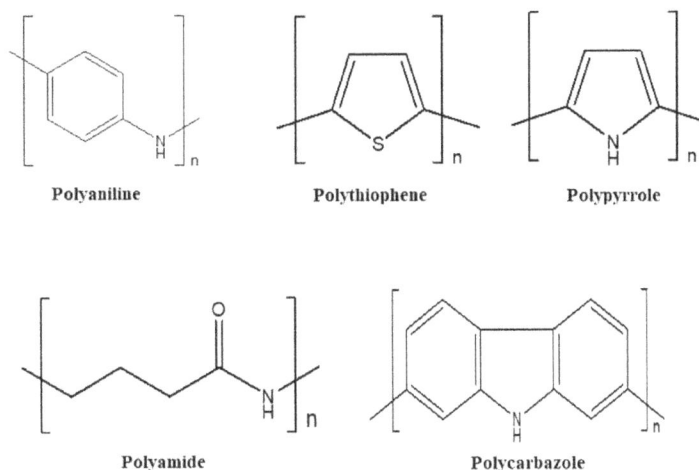

SCHEME 11.1 Chemical structures of some conducting polymers.

optical characteristics (25–27). In addition, two mechanisms are often used to produce multicolor chromophores to customize conducting polymers EC features. The first is polymer structure modification, which involves changing the bandgap of the CP along the mainstay or side chain operational amendment, the second is a copolymerization of various monomers (28).

Furthermore, CPs have conductivity due to π-π stacking interactions along the polymer backbone that allows the overlapping of π-bonded electrons along the polymeric chain (29). There are two essential parameters for conjugated macromolecules: the bandgap and the energy levels between the highest occupied molecular orbital (HOMO) and lowest unoccupied molecular orbital (LUMO). Therefore, the transition from HOMO to LUMO energy levels and the narrowing of the bandgap alter the key characteristics of EC polymers, such as ample optical contrast, the higher onset of absorbance, lower oxidation potential, and improved doped state stability (30).

11.2.1.1 Polyanilines

Because of their ease of synthesis, multiple colors, and high electroactivity, PANIs have been extensively studied as EC CP. PANIs molecular style and engineering have received a lot of attention to improve its EC properties (31, 32). Chemical and electrochemical polymerization is widely used to make copolymers. Chemical polymerization, which is known as step-growth or chain-growth polymerization, is a doping process for producing various functionalized polymers. In electrochemical copolymerization Scheme 11.2, the copolymer synthesized in the thin film is used in various optical and electronic devices. The benefits of using the electrochemical method are its good processibility at room temperatures and the ease of formation of a uniform film of desired thickness (33, 34).

Recently, a novel feature of PANI electrochemical deposition was studied by the potential dynamic cyclic voltammetry (CV) technique using UV-Vis spectroscopy and electrochemistry as correlative experiments (36). Three voltammetric redox pairs were noticed from the CV results that were consistent with aniline monomer oxidation, the oxidation or reduction of gold (Au) and PANI redox behavior. Spectroscopy provides insight into PANI's electronic structure compared with the electrochemical evidence obtained from CV, the existence of sub-bands was demonstrated by the absorbance bipolaron transition, π-polaron, and polaron-π*, which decreased the bandgap of PANI and produced a higher conductivity. Moreover, the electrochemical polymerization of aniline was carried out in 1 M HCl with 0.1 M aniline to avoid conjugated oligomers, obtain a polymer in a conducting form and enhance the monomer solubility. As shown in Figure 11.2(a), cleaning and initiating the electrode surface is the first step of polymerization on an inert screen-printed gold

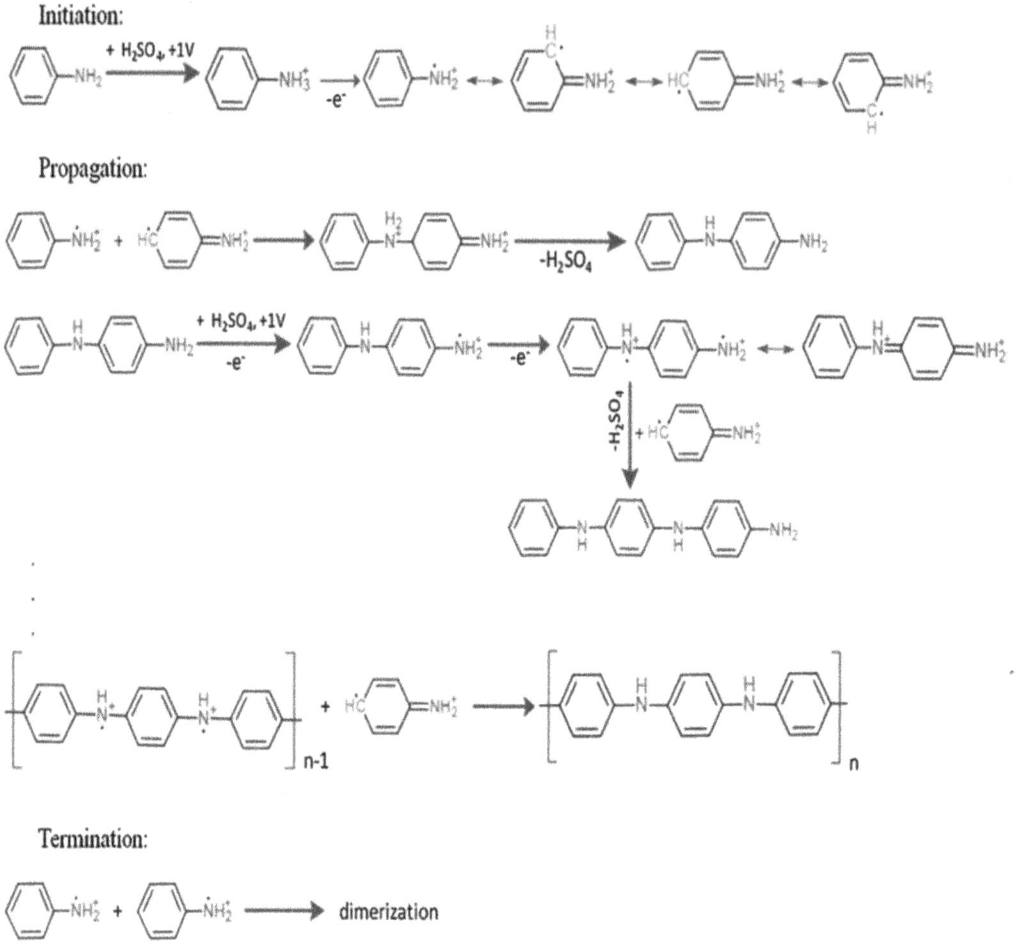

Initiation:

Propagation:

Termination:

SCHEME 11.2   Electrochemical polymerization mechanism of aniline to PANI.

(From Gvozdenovic et al. 2014 (35). With permission.)

electrodes (Au-SPE) in an acid-monomer suspension. Furthermore, two polymerization cycles were completed, both had identical anodic (E a_2=0.9 V, E a_3=1 V) and cathodic peaks (E c_1=0.51 V). The irreversible monomer oxidation that created the anodic peak (a_2) at 0.9 V led to the formation of cation radicals. Because it is unrelated to oxygen development, the anodic peak (a_1) almost reflected Au oxide synthesis. The oxygen appeared at potentials >1.2 V (e.g., versus silver/silver chloride (Ag/AgCl) and versus Ag), and the cathodic peak (c_1) showed Au oxide reduction. At the cycle's starting potential (−0.3 V) (Figure 11.2(b)), there are no absorbance peaks; however, as oxidation progresses, the amplitude of the absorbance peaks increases until the anodic scan ends at 1 V. Following the reduction step, the strength of the absorption peaks decreased during the cathodic scan between 500 and 545 nm, which suggested the existence of a small concentration of quinoid rings. Based on this, as the emeraldine to pernigraniline oxidation form progressed, the absorbances at approximately 500 nm started to increase. The absorbance peaks decreased to zero as emeraldine converted into leucoemeraldine at 0 V. The band at 680 nm was due to the exciton's absorption of the quinoid rings, which was due to interchain or intrachain charge transport in PANI (1.81 eV).

A novel synthesis method was proposed (37), which incorporated the electrochemistry of PANI and electrospinning to produce innovative multi-layered materials suitable for biomedical applications (Figure 11.3(a)). Biological tests showed that all PANI-coated samples were compatible

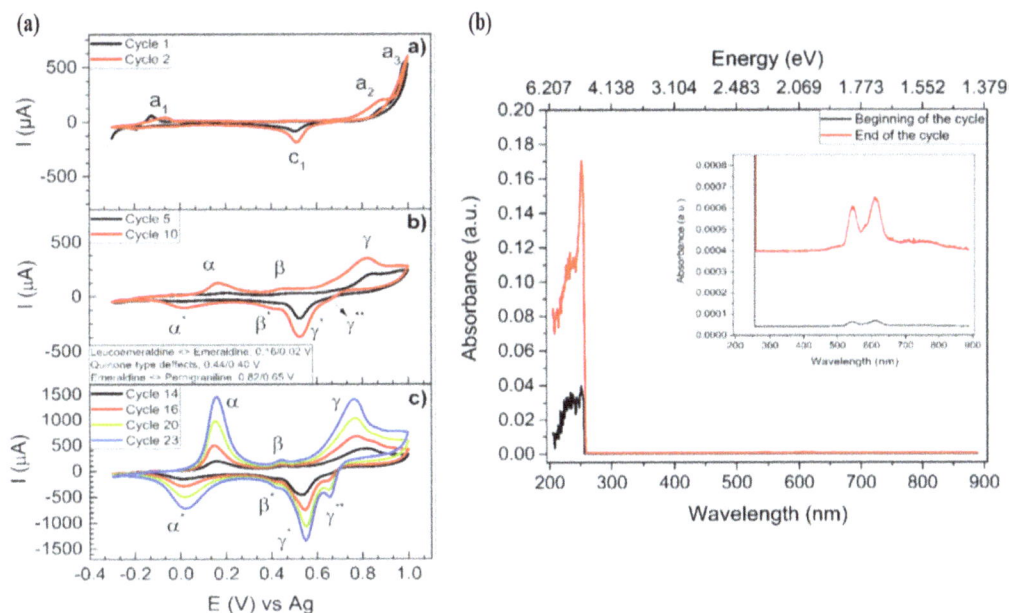

FIGURE 11.2 Showing: (a) electrochemical deposition of PANI on Au-SPE through CV starting cycles (b), first occurrence of PANI formation: (c) further growth of PANI on Au-SPE; and (d) absorbance spectra at the start and end of the first electrochemical polymerization cycle of PANI, Inset: Close up of absorbance spectra.

(From Korent et al. 2020 (36). With permission.)

with eukaryotic cells. The color and oxidation state of PANI can be altered by varying the voltage used on the resulting configurations in the presence of an electrolyte. A highly transparent metalized polymer fibered web served as the working electrode for PANI shell precipitation via electrochemical polymerization, with a platinum (Pt) plate serving as the counter electrode and a saturated calomel electrode serving as the reference electrode (Figure 11.3(b)). Furthermore, the processing parameters used in the electrospinning method affected the optical features of the fiber electrodes, the most important of which is the deposition period, which was calculated in increments of 100 s from 150 to 550 s. The electrospun Au cover samples were transparent from 325 to 700 nm with transmission of 65%–80%. Furthermore, samples with the highest fiber densities provided slightly lower transmission and those with lower fiber density had the highest light transmission values. Therefore, this system could allow to produce high-quality transparent electrodes with accurate light transmission. Throughout the EC layer deposition, potential–time and current–time curves were reported for 350 cycles (1 cycle/s) as shown in Figure 11.3(c). When the characteristics of the formed polymer film depended on the applied potential and oxidizing agent, the chronoamperogram was used. When the used potential was modified from 0 to 1 V, the monomer polymerization and polymer deposition began. In the deposition chronoamperogram, the peaks attributed to the reduction and oxidation states were visible. Therefore, the monomer and polymerization period influenced the color and morphology of PANI shells. In contrast, Figure 11.3(d) shows the EC properties of PANI-coated fiber webs, which demonstrated that the polymer layer exhibited reversible and repeatable color alteration during testing. The emeraldine salt was green at 0 V and the emeraldine base was blue at 1 V.

In addition to the applications mentioned previously, PANI can restrain light in the mid-to far-IR range (38). PANIs IR EC systems have a wide variety of possible applications in thermal control and IR camouflage, according to previous studies. A study (39) highlighted how a PANI porous film doped in perchloric acid ($HClO_4$) achieved greater IR control capacity about conversion efficiency

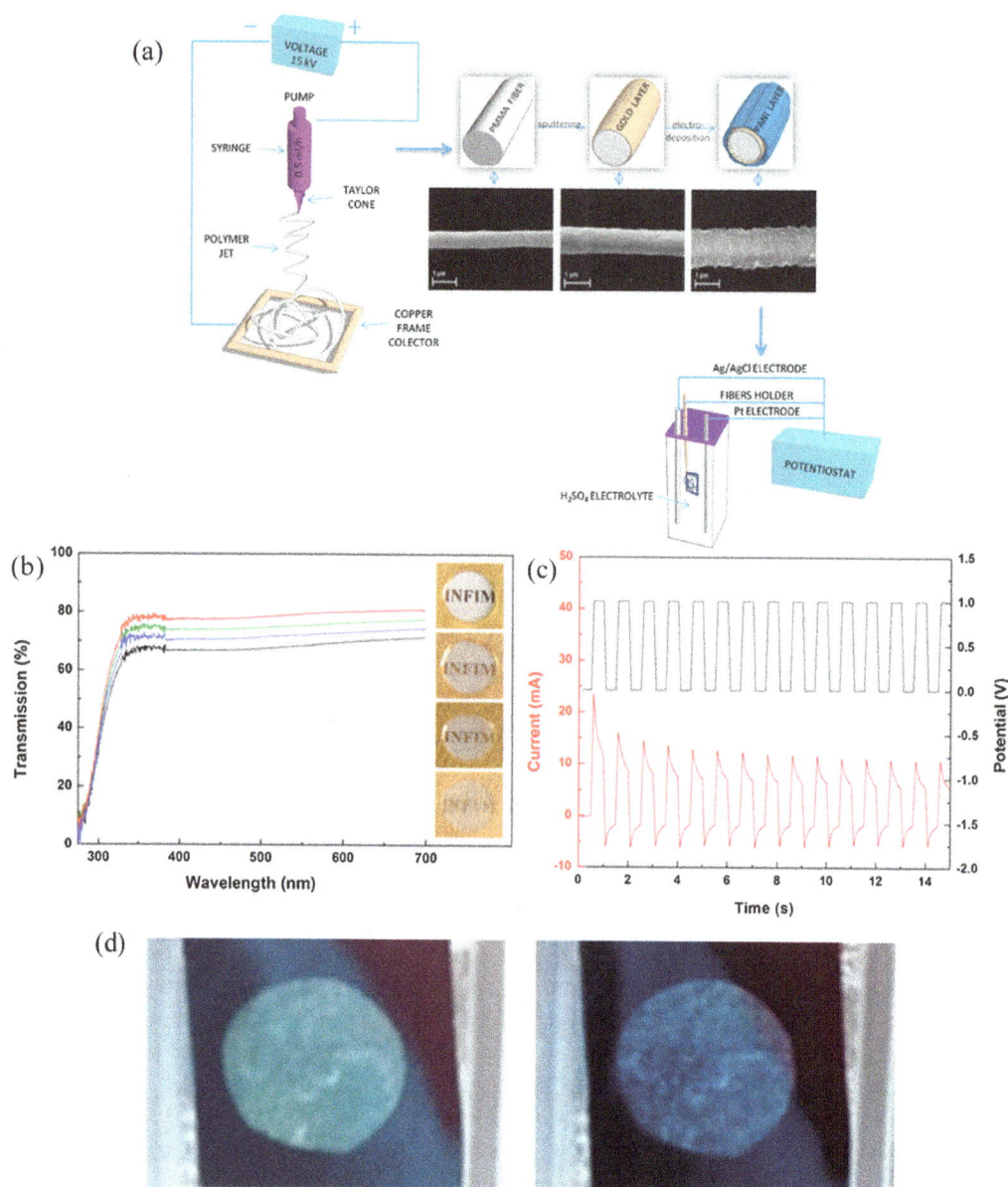

FIGURE 11.3 Showing: (a) fabrication scheme for the ECD; (b) digital images and transmittance spectra for four metalized fibers; (c) chronoamperegram deposition for PANI, and digital images of coated fiber with PANI in bleached and colored states.

(From Beregoi et al. 2016 (37). With permission.)

in different PANI states and what the key factor was that affected IR emission. In situ electrochemical deposition of aniline in 1M $HClO_4$ aqueous solution on an Au/porous substrate under various polymerization conditions, which used Ag/AgCl electrodes as a reference electrode and Pt foil as a counter electrode, formed PANI porous films doped in $HClO_4$ (Figure 11.4(a)). The CV curves of PANI porous films doped in $HClO_4$ under various polymerization conditions with a scan rate of 50 mV/s can be seen in Figure 11.4(b). The curves all have the same form, with two broad redox peaks associated with the transition between an emerald salt (ES) and a leucoemeraldine salt (LE)

(LE). Furthermore, Figure 11.4(c) depicts the surface morphologies of $HClO_4$-doped PANI porous films deposited with different polymerization charges. As the polymerization charge increased, PANI gradually fills the cavities of the porous film, resulting in an increased PANI loading on the porous films. Furthermore, as the polymerization charge increased, the accumulation of PANI on the porous film can be seen in three phases. The values of PANI porous films doped in $HClO_4$ at voltages between −0.25 and 0.5 V increased as the polymerization charge increased, displaying a ladder-type and S-type growth, respectively, as shown in Figure 11.4(d). The ideal polymerization charge in the grid region was 9.0–11.0 C with a value of 0.4, according to the findings. The IR ECDs in the LE and

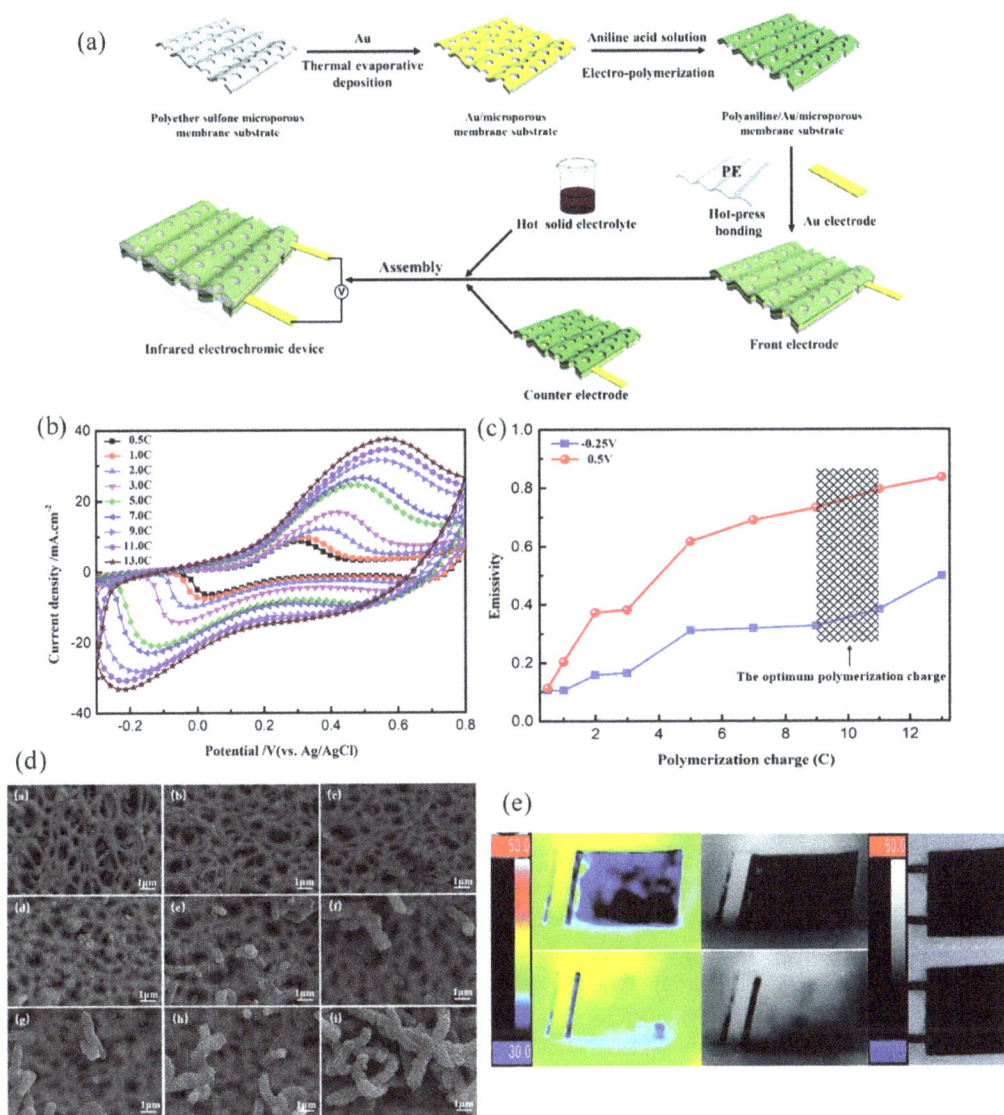

FIGURE 11.4 Showing: (a) in situ electrochemical fabrication process of EC device for deposition of aniline; (b) PANI porous films doped in $HClO_4$ CV curves with a scan rate of 50 mV/s; (c) SEM images of PANI porous films doped in $HClO_4$ with different polymerization charges: 0.5 C, 1.0 C, 2.0 C, 3.0 C, 5.0 C, 7.0 C, 9.0 C, 11.0 C, and 13.0 C; (d) ε data of PANI porous films doped in $HClO_4$ at −0.25 V and 0.45 V; and (e) IR thermal images and digital photographs and of the IR ECDs at LB and ES states.

(From Zhang et al. 2019 (39). With permission.)

ES states are shown in IR thermal images and digital photographs in Figure 11.4(e). The device's color ranges from golden yellow to dark green. It stands out as controllable IR camouflage when transitioning from LE to ES states.

11.2.1.2 Polythiophenes

PTh is a type of conducting polymer that has been used in coatings, batteries, non-linear optical instruments, smart windows, sensors, light-emitting diodes, transistors, electromagnetic shielding devices, artificial noses and muscles, solar cells, and microwave absorption materials among other applications (40). As shown in Figure 11.5(a), the incorporation of a charged functional group results in water-soluble PThs, these may be cationic, such as imidazolium and ammoniums or anionic as carboxylates and sulfonates or neutral hydrophilic functional groups, such as glycol (41, 42). Moreover, its color changes from red to blue due to EC changes during oxidation (43). As a result of their colorless to blue EC switches, derivatives of PTh, such as poly(3,4-ethylene dioxythiophene) (PEDOT), is widely used in ECDs due to their significantly lower bandgap, excellent conductivity, short redox potential, high transparency, and good environmental stability (44).

In a study (45), water-soluble PEDOT–methylamine chlorate was synthesized by electro-chemical polymerization of methylamine hydrochloride and PEDOT in the presence of 0.01 M monomer in 1 M $HClO_4$ aqueous solution (Figure 11.5(b)). By electrodepositing PEDOT–$CH_2NH_3^+$ A^- (ammonium perchlorate) (where $A^-=Cl^-$ and ClO_4^-) and PEDOT–$CH_2NH_3^+ClO_4^-$ onto an ITO electrode, polymer films of PEDOT–$CH_2NH_3^+A^-$ and PEDOT–$CH_2NH_3^+ClO_4^-$ were formed. The films of PEDOT–$CH_2NH_3^+ClO_4^-$ and PEDOT–$CH_2NH_3^+A^-$ when neutral is red and have absorption peaks at 555 and 603 nm, respectively (Figure 11.5(c)). As a result of stepwise oxidation ≤0.8 V at 0.1 V cycles, the absorption bands at 603 nm and 555 nm disappeared, resulting in increased absorption λ_{max}>730 nm and a color change to blue. By cycling from –0.5 V to 0.8 V with a 10 s switching time, the optical contrast, response times, and CE were all calculated (Figure 11.5(d)). With coloration efficiencies of 31 cm²/C and 8 cm²/C, PEDOT–$CH_2NH_3^+ClO_4^-$ had a 13.82% optical contrast at 555 nm and a 9.5% optical contrast at 1,000 nm. PEDOT–$CH_2NH_3^+A^-$ optical contrast was 26.38% at 603 nm and 17.47% at 1,100 nm with CEs of 156 cm²/C and 55 cm²/C, respectively. PEDOT–$CH_2NH_3^+A^-$ reaction times were 3.2 s for oxidation and 6.0 s for reduction at 1,100 nm, 1.4 s for oxidation and 9.6 s for reduction at 603 nm, and 4.4 s for oxidation and 7.8 s for reduction at 555 nm, and 8.0 s for oxidation and 7.2 s for reduction at 1,000 nm, respectively.

In addition, the synthesis of three-dimensional (3D) ordered microporous (3DOM)/PEDOT films have been used to improve the EC optical stability of PEDOT (46), which was electropolymerized using a template of colloidal crystal polystyrene in an ionic liquid, 1-butyl-3-methylimidazolium hexafluorophosphate ([Bmim]PF_6). The scanning electron microscopy (SEM) images of the 3DOM/PEDOT films with dissimilar pore sizes are shown in Figure 11.6(a), which shows that films possess a strongly well-arranged porous architecture over their entire volumes, in addition, to having a thickness of approximately 1 μm. Results from the aforementioned study suggest that the 3D ordered construction is valuable in ion diffusion (Figure 11.6(b)), because, during redox reactions, the effective diffusion coefficients of the 3DOM 400 film increased by ≤60 times compared with the dense film that resulted in the inherent structural benefits, such as larger surface areas and to 3D electron and ion transfer, the 3DOM PEDOT film with tiny pores displayed the maximum effective diffusion coefficients and, therefore, the fastest response time. Figure 11.6(c) shows the 3DOM PEDOT films and the dense film CV curves at different scan rates between 0.6 and –1 V, a significant reduction peak formed when the reduction peak at –0.6 V moved to a higher voltage and overlapped the peak at 0.05 V, which indicated an easy reduction mechanism. The subsequent film had a more porous morphology that enabled cations and anions to migrate more easily and resulted in higher peak current values. Digital images are shown in Figure 11.6(d) of PEDOT films in their colored and bleached states due to various nanostructures, these films show altered colors (e.g., dark blue, black, gray, and purple).

The CV method (47) was used to synthesize poly(3-methyl thiophene-co-3,4-ethylenedioxythiophene) P(3MT-co-EDOT) P(3MT-co-EDOT) by the electrochemical

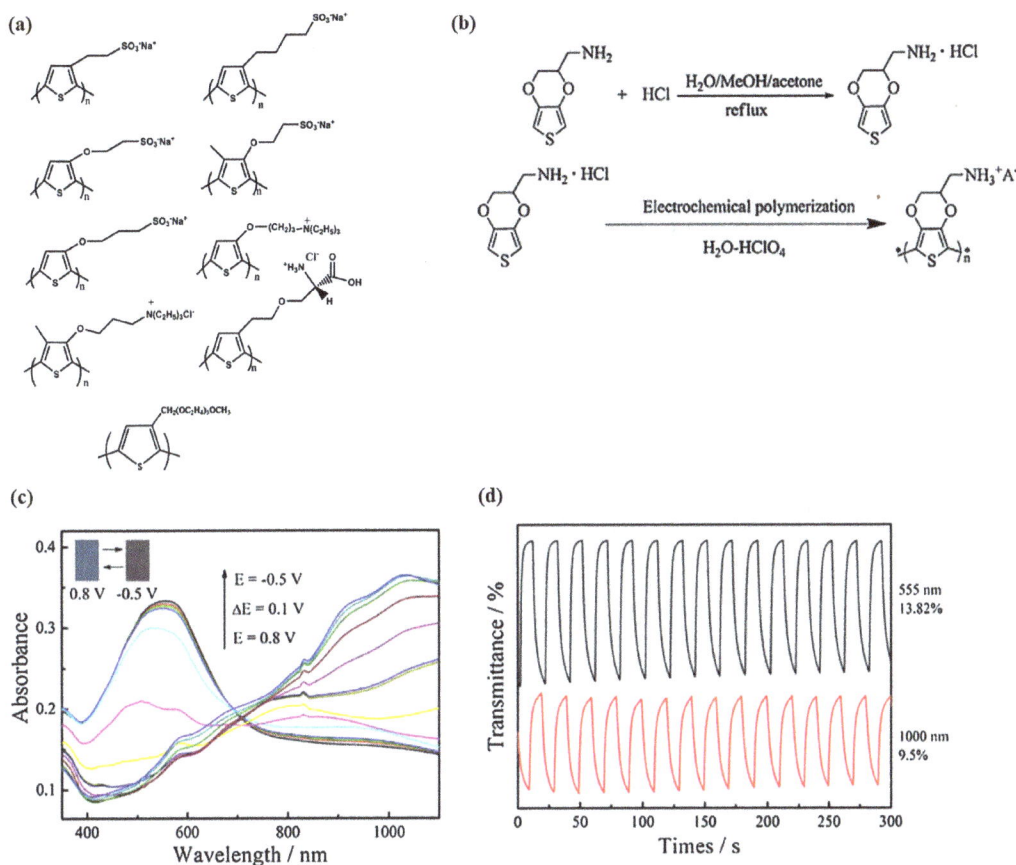

FIGURE 11.5 Showing: (a) different water-soluble PTh derivatives with anionic, and cationic, in addition, later neutral functional groups; (b) preparation of PEDOT–methylamine neutral films through electrochemical polymerization; (c) spectrophotometer of PEDOT–methylamine films on glass electrode (ITO) between –0.5V and 0.8V; and (d) switched time/transmittance for PEDOT films through ranged potential within 10 s interval.

(Panel (a) from Sung et al. 2003 (40); and Das et al. 2015 (41). With permission.)

(Panels (b–d) from Sun et al. 2016 (45). With permission.)

polymerization of 0.05 M3-methyl thiophene 3MT solution and 0.01M 3,4-ethylenedioxythiophene EDOT in 0.1 MTB (Figure 11.7). The UV-Vis spectrum of poly(3MT-co-EDOT) coated on ITO/glass shown in Figure 11.7(a) has peaks at λ_{max} at 286 and 368 nm due to π-π^* transition in the molecule, according to the findings. The polaron band transition was responsible for the wide absorption band at λ_{max} at 634 nm. The direction of the peak for the -* transition shifted dramatically. This showed that conjugation significantly increased during copolymerization. As shown in Figure 11.7(b), the polymer colors can be modified from oxford blue (–600–200 mV), teal (400–800 mV), and rosewood (400–800 mV). Furthermore, when the used potential is altered from reduction to oxidation, all the band absorption values increase.

11.2.1.3 Polypyrrole

PPy is a ubiquitous conducting polymer due to its low oxidation potential, good electrical conductivity, low cost of its monomer pyrrole water solubility, and high air stability (48). PPy derivatives exhibit features, such as good redox properties, compatibility in aqueous systems, and moderate

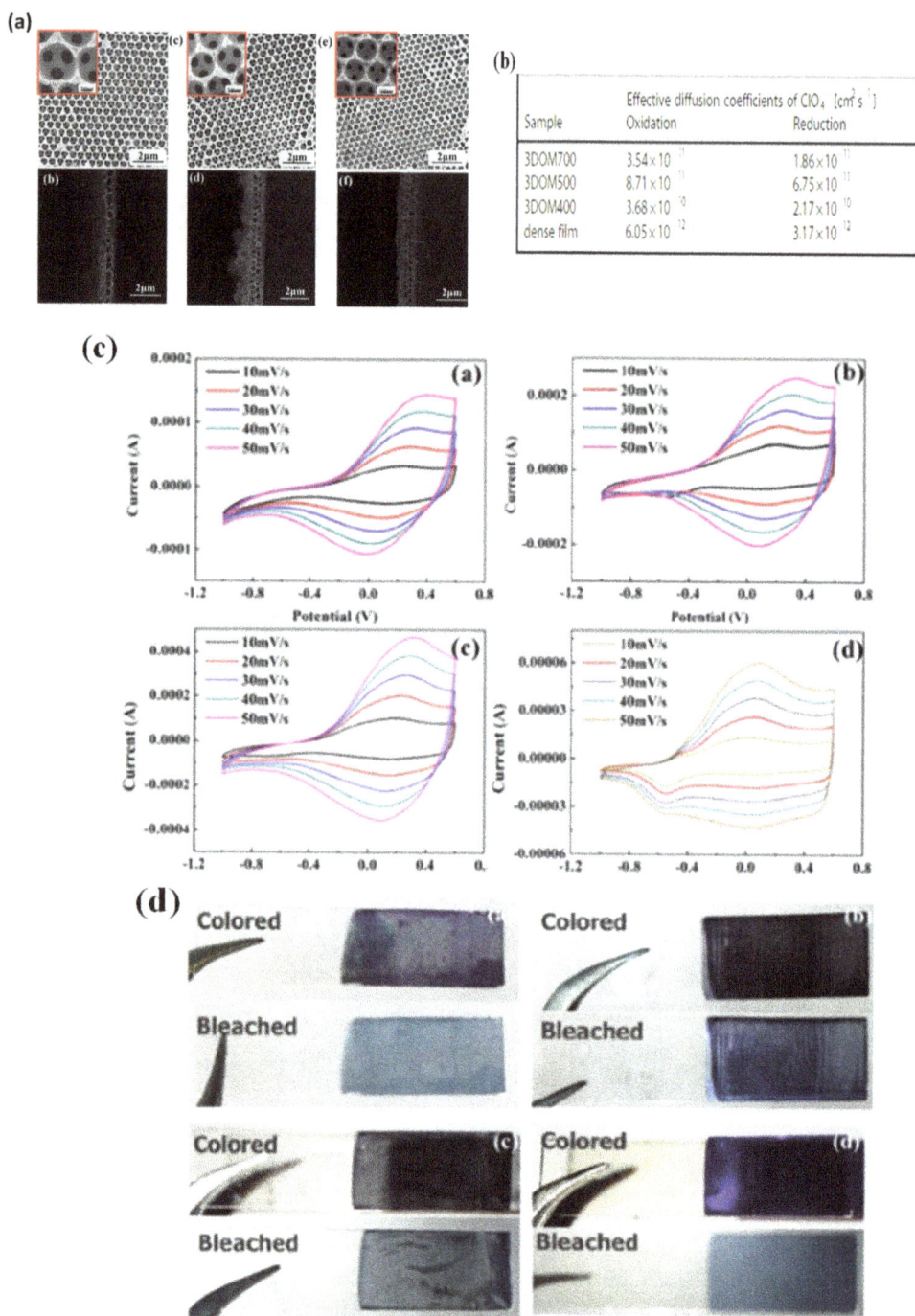

FIGURE 11.6 Showing: (a) SEM images of the 3DOM PEDOT films: 3DOM700 (a and b); 3DOM500 (c and d); and 3DOM400 (e, f), b) the effective diffusion coefficients for PEDOT films compared to their dense, c) CV cycles for PEDOT films through scan rates ranged in 10–50 mV/s for 3DOM700 (a), 3DOM500 (b), 3DOM400 (c), and dense film (d), and (e) photograph images of the PEDOT films in colored and bleached states: 3DOM700, 3DOM500, 3DOM400, and dense film.

(From Zhang et al. 2016 (46). With permission.)

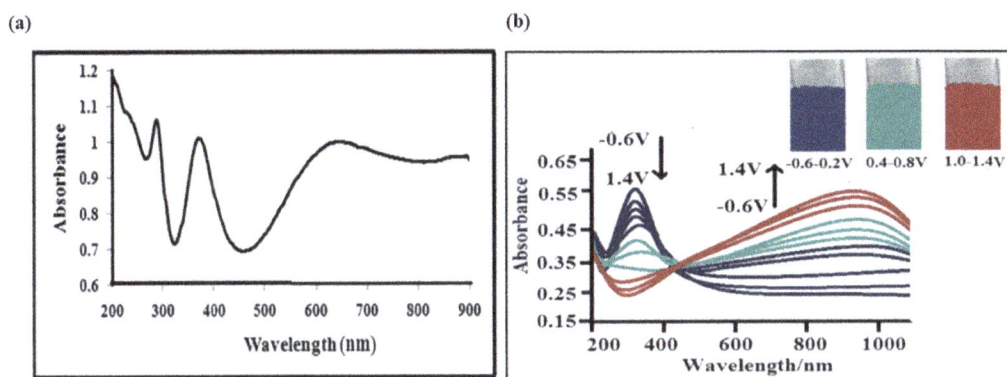

FIGURE 11.7 Showing: (a) UV-vis spectrum of poly(3MT-co-EDOT); and (b) spectrophotometer spectra of poly(3MT-co-EDOT) as applied potentials in range of 0.6–1.4 V in neutral pH.

(From Authidevi et al. 2020 (47). With permission.)

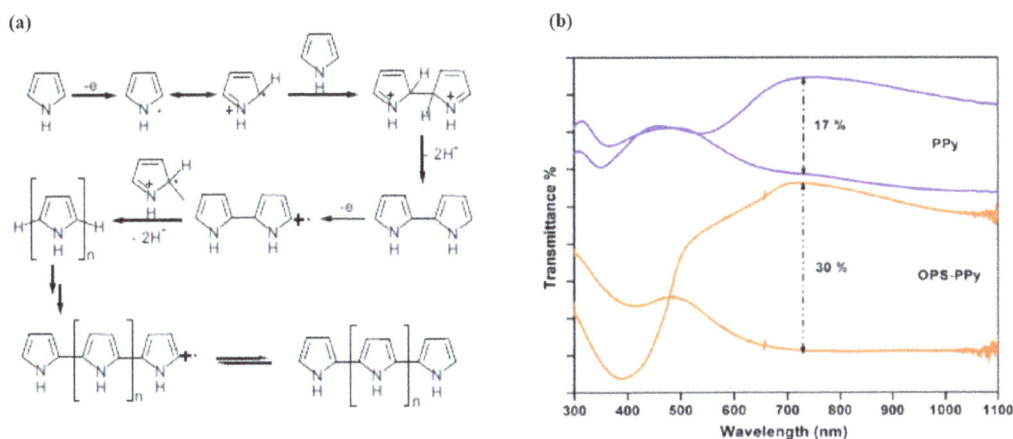

FIGURE 11.8 Showing: (a) preparation of PPy via electrochemical polymerization technique, and (b) transmittance spectra of OPS–PPy compared with PPy in neutral and oxidized states.

(Panel (a) from Gvozdenovic et al. 2014 (35). With permission.)

(Panel (b) from Ak et al. 2008 (55). With permission.)

environmental stability, that enable them to be used in a wide range of applications as sensors, supercapacitors, batteries, drug delivery systems, electromagnetic resistance insulation, mechanical actuators, and optoelectronic instruments (49).

In addition, PPy can be prepared chemically or electrochemically. Electrochemical polymerization is commonly used, because as it requires a small amount of monomer, and it offers an efficient mechanism for investigating the in situ development route of the polymer (Figure 11.8(a)) and additional study using spectroscopic and electrochemical analyses. It is a straightforward method that allows an inherently conducting polymer to be directly grafted onto the surface of an electrode without the need for additional doping (35, 50, 51). Furthermore, numerous methods have been established to enhance the electrical and mechanical features of PPy, one of which is the preparation of conducting star, graft, and block copolymers with anticipated end groups (52, 53).

Copolymerization, mixing, adjustment, and lamination are the procedures used to change the color of directing polymers by controlling the band distance (54).

Pyrrole copolymerization with octa(thiophenephenyl) silsesquioxane (OPS) was investigated electrochemically (55) to improve the electrochromic properties of PPy. The (OPS–PPy) copolymer was multichromic with color switches from yellow, red, green-gray, and blue. Compared with PPy (1.1 s), OPS–PPy (0.4 s) shifted faster and had an improved optical contrast (Figure 11.8(b)) from 17% to 30% at 730 nm. The introduction of OPS units loses bulk from the PPy chains, which allows for easier ion association throughout the redox transfer and active doping sites.

Derivatives of 2, 5-dithienylpyrrole (DTP) are well known for their excellent electrochemical film-forming performance, low oxidized potential, and ease of structure modification through the N-position of pyrrole units (56, 57). In addition, 2,2-bithiophene oxidation potentials are greater than DTP derivatives (approximately 0.7 V) because of the introduction of a pyrrole unit into two thiophenes, which improves the stability of the polymer (58).

Recently, (59) new EC materials, including new naphthalene modified with DTP monomer have been synthesized and polymerized onto the surface of an ITO-coated glass electrode (Figure 11.9(a)) have been developed. The potentiodynamic method was used to electrochemically polymerize DPN in an acetonitrile (ACN) solution that contained 0.1M lithium perchlorate (LiClO$_4$) with Ag wire and platin wire as reference and counter electrodes with the voltage from −0.5 V to +1.5 V, the PDPN film showed multicolor electrochromic, the color of the PDPN film improved from yellowish-green to blue as the applied potential was increased. This was due to the construction of lower energy charge carriers as polaron and bipolarons on the polymer backbone in the visible and near-IR regions. In addition, the reported results showed an increase in interchain interactions with a significant H-bonding, which affords structural inflexibility and forces it to planarity due to the existence of a bifunctional amide group in the polymer, as shown in Figure 11.9(b). The PDPN film showed high stability that was maintained at approximately 99.98% after 1,000 s and had a high optical contrast of 60% at 900 nm. In addition, the polymer film had a broad redox peak in a monomer-free electrolyte solution due to the presence of a polymer film that was electroactive in addition to reversible features at the electrode surface, and the electron transfer mechanism was not diffusion-controlled (Figure 11.9(c)). The PDPN showed an absorption peak that moved to a longer wavelength at 850 nm during in situ electrochemical polymerizations, as shown in Figure 11.9(d). The transitions from the thiophene-based valence band to its antibonding band, which were primarily absorbed in the UV region, were due to the absorption peak at approximately 430 nm in the electronic absorption spectra in the neutral form of the PDPN film, as shown in Figure 11.9(e). Because of the presence of two-sided polymerization, the chain structure had a longer π conjugation.

11.2.1.4   Polycarbazoles

PCz and its derivatives are good hole-transporting and light-emitting units (60). Due to their electroactive and photoactive properties, good hole-transporting and adequate high triplet energy the derivatives of carbazole have been applied as efficient host materials in phosphorescent light-emitting diodes (61). Carbazoles can be modified with a variety of aryl and alkyl chains on N-, 3,6- and 2,7 positions without changing the planar conformation of the resulting polymer (62). More recently, (63), a CP-based on N-substituted carbazole derivates by electrochemical polymerization of the N-positions occupied by cyanoethyl (carbazol-9-yl-cyanoethyl), the carboxylic acid (carbazol-9-yl- carboxylic acid), and methanol (carbazol-9-yl-methanol) was developed. As shown in Figure 11.10(a), the SEM images of PCz–OH showed some large clusters; however, PCz–COOH have an amorphous porous surface. Furthermore, the PCz–CN film had non-uniform particles with few clusters and the electrical conductivity of PCz at room temperature was quite low, between 10^{-4} and 5×10^{-7} S/cm. In addition, the conductivity of dried carbazole derivatives was detected at 25°C via four-probe conductivity techniques as $4.3 \times 10^{-2}, 7.6 \times 10^{-3}$, and 1.62×10^{-4} S/cm for PCz–COOH, PCz–OH, and PCz–CN, respectively. Figure 11.10(b and c) show that in the reduced form of the

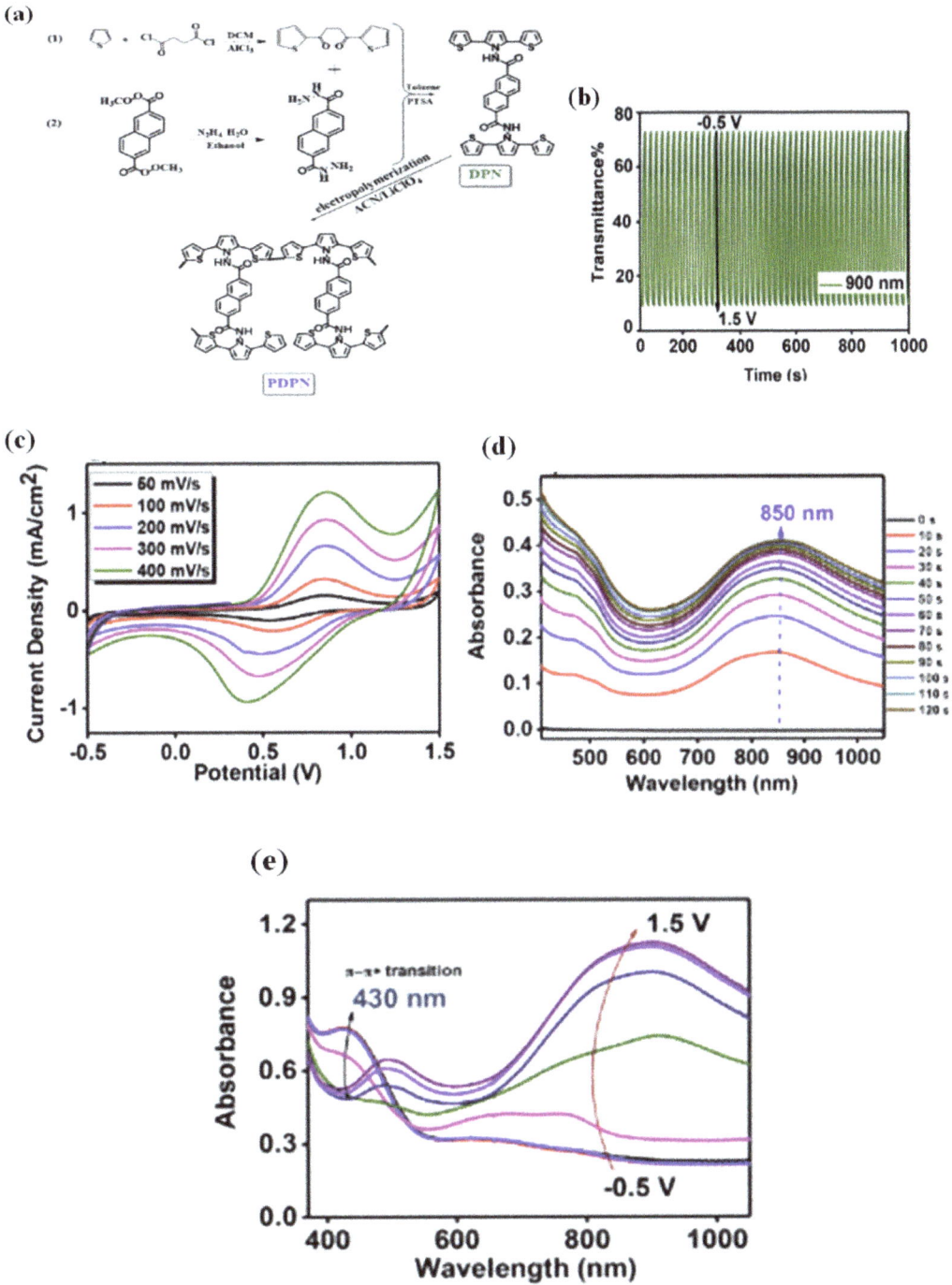

FIGURE 11.9 Showing: (a) electrochemical preparation of PDPN: (b) color change of PDPN films through applied voltage from –0.5V to 1.5V; (c) CV curves for PDPN films in LiClO₄/ACN within various scan rates; (d) in situ electrochemical polymerization of PDPN films; and (e) spectrophotometer spectra for electrochemical polymerization of PDPN on ITO electrode in LiClO₄/can.

(From Soyleyici 2019 (59). With permission.)

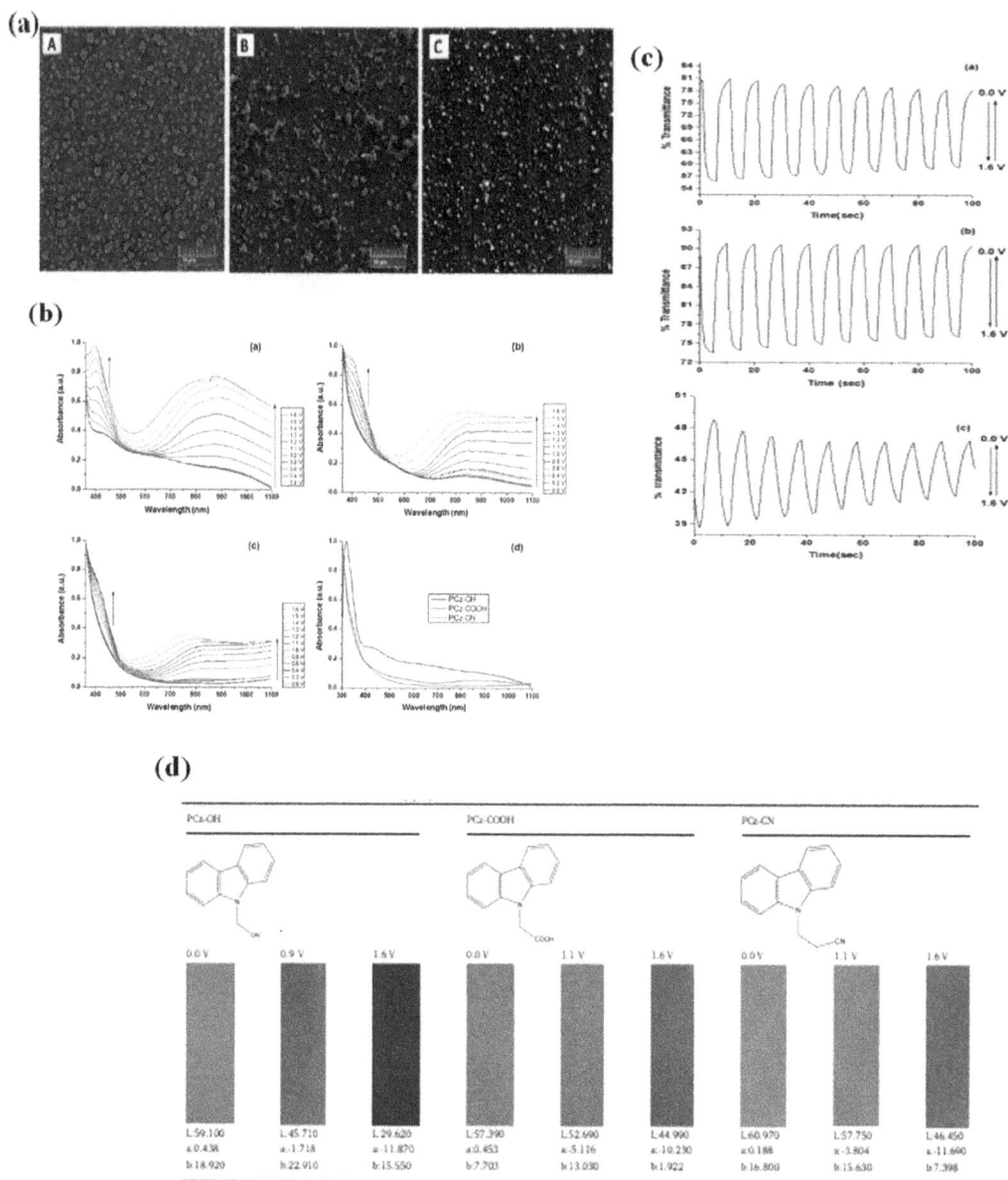

FIGURE 11.10 Showing: (a) SEM microimages for PCz films (PCz–COOH, PCz–OH and PCz–CN); (b) spectrophotometer behavior for electrochemical polymerization for PCz films (PCz–COOH, PCz–OH and PCz–CN) in NaClO4/LiClO4/ACN solution, in addition, to compare their λmax of π–π* transitions for the reduced state; (c) electrochromic switching responses: optical absorbance alteration observed at (a) 875 nm for PCz–OH; (b) 845 nm for PCz–COOH; and (c) 770 nm for PCz–CN; and (d) color changes and their coordinates for PCz–OH, PCz–COOH, and PCz–CN according to CIE standards.

(From Elkhidr et al. 2020 (63). With permission.)

polymers there is no strong absorption peak in the visible region and the polymer films are nearly transparent. Therefore, an absorption peak attributed to the π–π* was observed in the UV region at 300 nm. The polymer films displayed the maximum absorption band of 305 nm for PCz–COOH and 303 nm for PCz–CN. However, –OH substituted PCz–OH had longer absorption maxima at 318 nm. Finally, in situ spectroelectrochemical studies demonstrated that PCz–COOH, PCz–OH, and PCz–CN films had multichromic behavior by applying various potentials (Figure 11.10(d)).

11.2.1.5 Polyamides

PAs are high-performance polymers due to their EC properties, fluorescence in solvents (PL), and internal charge transfer. In addition, PAs are used in plastic engineering due to their thermal stability, plasticity, corrosion resistance, flame retardment, and improved mechanical strength (64, 65). The incorporation of bulky groups inside the leading chain or as the suspended groups of triphenylamine (TPA) units into polyamides is a method to solve these problems that preserve the glass transition temperature and increase the solubility of the polyacrylamides (66, 67). MeOTPA–(NPC)$_2$ and MeOTPA–(TPA)$_2$ were synthesized (68) by combining N-(4-carboxyphenyl) carbazole (NPC–COOH) and 4-carboxytriphenylamine (TPA–COOH) with 4,4´-diamino4´-methoxytriphenylamine (MeOTPA–NH$_2$)$_2$. The behavior of electrosynthesized polyamide films (P1 and P2) was investigated using the potential dynamic CV technique on an Ag/AgCl reference electrode and an ITO-glass substrate as the working electrode in a 0.1 M Bu$_4$NClO$_4$/CH$_2$Cl$_2$ solution. There are three reversible reduction–oxidation processes in the polymer films as a result of arylamine oxidation, with potentials ($E_{1/2}$) of 1.04, 0.93, and 0.61 V for P1 and 1.07, 0.80, and 0.62 V for P2. P1 and P2s oxidation onset potentials (E_{onset}) were measured at 0.47 and 0.49 V, respectively. The E_{onset} values and UV-Vis absorption edge were used to determine the energy levels of the HOMO and LUMO of the related polymers. The absorption peak at 362 nm was due to π–π* transitions for the neutral form of the P1 film, and the electronic absorption profile exhibited a maximum absorption peak at 362 nm, as shown in Figure 11.11(a). By steadily increasing the applied potential to 0.8 V, the peak at 362 nm decreased marginally and a new band emerged at 775 nm, and the color shifted to pale green. Furthermore, a new absorption peak at 502 nm was observed due to the formation of a MeOTPA radical cation, and the film color changed to orange as a result of further oxidation at 1.1 V. When the voltage was increased to 1.3 V, a new absorption band at 716 nm emerged, the absorption peak at 502 nm decreased, and the color shifted to blue. The formation of the tetraphenyl benzidine radical cation (TPB$^+$) moieties is linked to spectral changes. The same spectral shift was detected for the

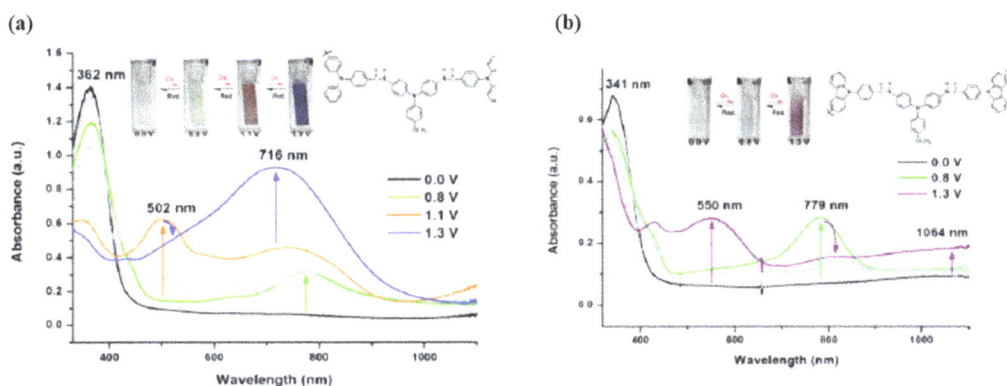

FIGURE 11.11 Spectroelectrochemical polymerization of: (a) P1 films and b) P2 films in Bu$_4$NClO$_4$/CH$_2$Cl$_2$ on the ITO-coated glass substrate at different applied voltages.

(From Hsiao & Lu 2017 (68). With permission.)

P2 polymer film at the first stage of oxidation, as shown in Figure 11.11(b). The electrochemically synthesized polymers had good cycling stability and coloration performance, which suggested that they could be good candidates for EC materials.

11.2.2 Viologen-Based Electrochromes

Viologens are conducting organic materials. In general, a sharp color is obtained by reducing the viologen dication, which led to intense absorption at the radical in its cation state. The color was mainly altered by the replacement groups presented on the nitrogen atoms of the bipyridinium salt. As shown in Figures 11.12(a and b), the absorption band for the radical cationic of ethyl viologen (V2) had a sharp peak detected at approximately 608 nm and a weak absorption band located in the near-UV irradiation region of 400–450 nm leading to a blue color after reduction (69). Other viologens with the substitution of alkyl chains behave optically the same as V2, such as like heptyl viologen (V3), methyl viologen (V1), vinyl benzyl viologen (V6), and benzyl viologen where the color of the radical cation at 600 nm might be a typical blue or violet-blue (70–72). However, when the aryl groups were used to substitute nitrogen contained bipyridinium as in cyanophenyl viologen (V5), phenyl viologen (V4) as shown in Figure 11.12, the viologens produced were colored green after reduction.

For the V5, two strong absorption peaks were identified at 420 and 600 nm as shown in Figures 11.12(c and d). Viologens are supposed to perform two alterable one-electron reduction steps to obtain neutral or radical cation species of various colors. For instance, in V5, the radical cation could reduce to give a neutral state with a red color and one broad absorption peak at approximately 500 nm (73).

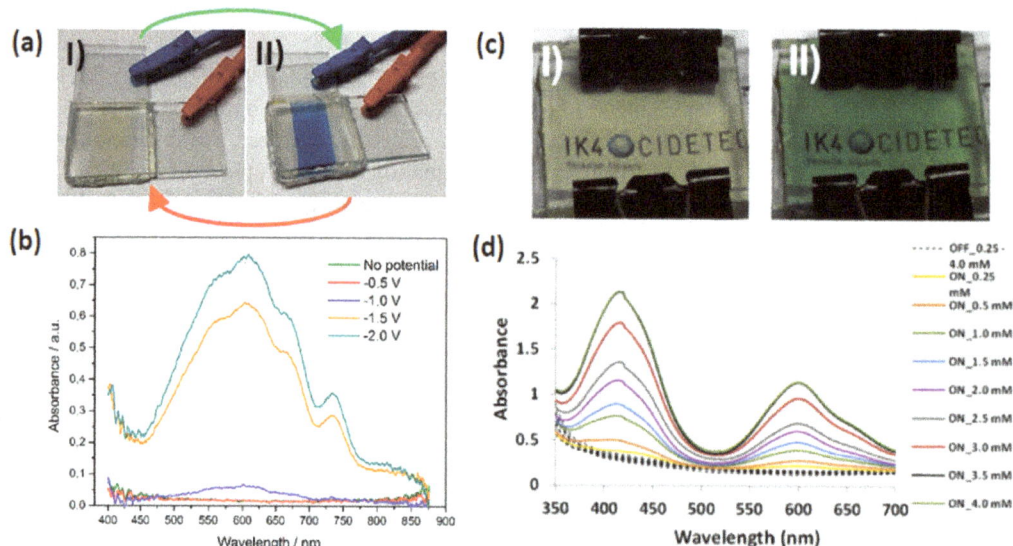

FIGURE 11.12 Showing: (a) color contrast test for ECD containing ethyl viologen in bleached (I) and colored (II) states; (b) UV-vis spectra for ECD containing ethyl viologen-based ECD obtained at different potentials; (c) Color contrast test for ECD containing V5 under bleached (I) and colored (II) conditions; and (d) UV-vis spectra of ECDs from V5 at different concentrations in their radical states.

(Panels (a)–(b) from Gelinas et al. 2017 (69). With permission.)

(Panels (c)–(d) from Dmitrieva et al. 2018 (73). With permission.)

Furthermore, the electrolytes used could alter the coloring and bleaching processes of the viologens (71). The neutral alkyl viologen (V^0) in aqueous media can interact with V^{2+} to produce undesirable radical cation of the dimerized viologen ($2V^{+\bullet}$) with crimson color and low ability to write–erase due to the unwanted irreversible comproportionation (71). In contrast, in most organic solvents, $2V^{+\bullet}$ could spontaneously dissociate, because the solvation energy was weak. The presence of ionic liquid-based electrolytes could control the formation of $2V^{+\bullet}$. In addition, the bulky substituents on the chain of a viologen are enough to prevent the molecules from moving resulting in decreased chances for dimerization or aggregation, and therefore, increasing the cyclic stability of the final device (74, 75).

Recently, a novel viologen electrochrome was introduced (71) based on 4,4′-bipyridinium diperchlorate (V7), which led to the formation of a 4,4′-bipyridyl as a core that was inserted between two indole moieties. Of note, bipyridyl radical cation systems possibly form radical dimers on reduction because of the slow reoxidation process and aging. However, in this work, the 4,4′-bipyridyl system attached with bulky indole groups made the radical cation dimerization hard to happen due to steric hindrance, in addition to the π-electron density that was added from the indole rings to the bipyridyl system and assisted in a rapid color bleaching the electron-rich viologen. Therefore, the V7 ECD system could achieve coloration and bleaching times of approximately 2 s, at 0.33 Hz and the write–erase efficiency was excellent at 95%, 86%, and 79% when the switching time was 3, 5, and 10 s, respectively.

In addition, the hydroxyl (OH) group is recognized as an auxochrome. Other research attempted to design a viologen molecule containing substituents from OH groups as shown in the structure of V8 (Figure 11.13) for fabricating a single-layer all-in-one ECD using an electrolyte from a polymer gel. The resultant ECD V8 operated at a low driving voltage of 0.9 V with a higher optical contrast ≤82% and an outstanding pattern efficiency of >240 cm²/C (76).

11.3 CONDUCTIVE COMPOSITE FILMS

The EC features for these materials could be manipulated by designing new nanocomposites. By preparing the nanostructured composites, the interfacial bonding between the organic polymer and nanofiller is related to the formation of stable structures that improves the ion transport and electron conduction and leads to enhancing the final EC properties (77–79). Recently, interest has

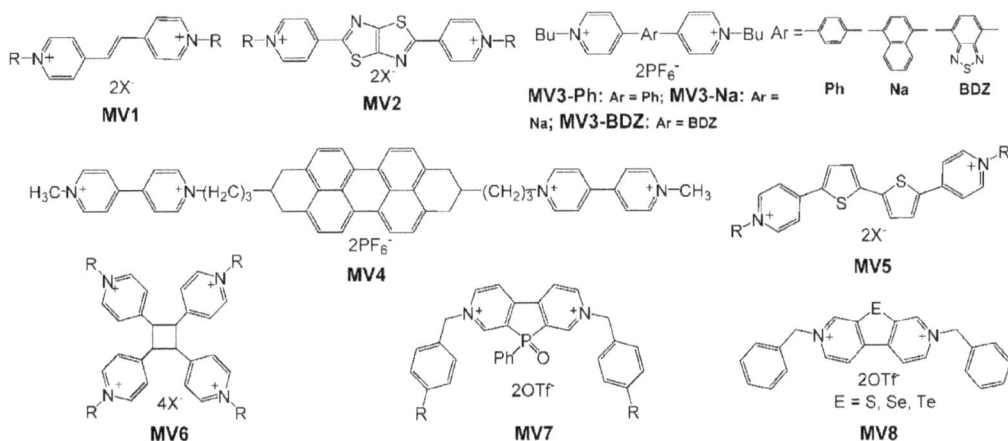

FIGURE 11.13 Chemical structure for different derivatives of viologens with a spacer between two pyridyl groups and a bridging group.

been directed to conducting polymers for developing EC devices. In general, the mechanism of the conduction action for these polymers was via two-step processes. Formation of a cation or anion that is defined as a polaron, where the transfer of a second electron causes the production of a dication (dianion) known as a bipolaron. In addition, after the typical first redox step, complexes of charge allocation can be formed between neutral and charged segments of the polymer. During the color-changing process, conducting polymers undergo p-doping/de-doping steps. The p-doped state occurs during oxidation of the polymer chain and the counter anions help in balancing the charge that is utilized by lower energy intraband transitions to form delocalized π-electron band structures. When the reduction of the p-doped structure occurs, the presence of counter anions or cations from the electrolytes could restore the novel neutral state of the structure (9, 10, 77, 80). Due to the formation of extensive delocalized π transitions that is a characteristic of most conducting polymers, they have received attention from many researchers for use in batteries, supercapacitors, and EC devices. These polymers obtain a higher coloration performance, a lower potential of redox switching, and faster response speed compared with metal oxides. Since polymer nanocomposites were discovered, hybrid organic–inorganic materials have attracted attention due to the combination of functionality and elasticity of the organic polymers with chemical stability and the high thermal state of the inorganics to obtain EC conductive polymer nanocomposites with outstanding features (81–85).

11.3.1 Metal Coordination Complex-Based Composite Films

Metal complexes as EC materials often experience multicolor switching due to the electro–redox of the central metal ions and the use of various ligands. Polypyridine metal complexes include Prussian blue (PB) and its analogues (PBAs), metal phthalocyanine (MPc), and other compounds (86).In particular, PB gives an EC performance, because at a low potential a color change is obtained from colorless and blue even at 5×10^6 cycles (87). The anodic function of PB leads to a mixed with cathodic material in an EC device with outstanding efficiency. Recently, many trials have been carried out to fabricate PB films of superior contrast and switching speed (78, 88, 89). The viologen can be composited with PB/antimony tin oxide with a sandwiched polymer gel electrolyte to obtain a dark-colored EC device (90). The device could offer 64.8% contrast at 600 nm with a rapid switching time of 600 and 720 ms for the coloration and bleaching, respectively, minimum transmittance operated at the colored stage (0.1% at 600 nm). In other research, (91) a solid-state electrochromic device was prepared using a water-dispersible PB thin film as the anode to color the electrode and a poly(butyl viologen) (PBV) film as the cathode. As a solid electrolyte, succinonitrile, silicon dioxide (SiO_2), and potassium bis(trifluoromethanesulfonyl)imide at 0.1 M were used. The system could reversibly switch between blue and violet colors when remaining transparent at potentials of 1.7 V and 1.0 V, with an initial transmittance shift of 62.5% and a CE of 157 cm^2/C at 545 nm.

Altering the metal ion or ligand structure will enhance transition metal (TM) complexes and because of the presence of a large number of electronic transitions, results in intense coloration. Coloration stability and efficiency, colorimetric analysis, EC contrast, switching rate, and optical memory are all parameters that can be used to characterize EC materials (92, 93). The d-d transitions, ligand-to-metal charge transfers, and metal-to-ligand charge transfers (MLCT) are all types of electronic transitions that occur in metal complexes (94).

In polynuclear mixed-valence complexes, a charge transfer band from metal to metal (MMCT) can be detected (95). Furthermore, in the UV-Vis-NIR spectra, the existence of a ligand-to-ligand charge transfer band is possible for complexes that enclose >1 ligand molecule. Low energy MLCT, intra ligand excitation, intervalence CT, in addition to related electronic transitions to the visible region result in chromophoric properties. Because the valence electrons are involved in these transitions, chromophoric physical characteristics are changed or ignored after the complex is oxidized or reduced. In the interior of the potential window of a given electrolyte solution, any redox-switchable colored TM coordination complex will undergo a color change, making it an EC.

Many types of EC materials with unique construction and features have been established and are being used in prototype ECDs (96).

Many factors, including the TM ion type, the ligand molecule structure, and the nature of the anions, have been shown to influence the EC properties of TM complexes. Color mixing theory states that tuning the EC metal complex properties can be accomplished by incorporating an electrically activated anion or ligand. The nanocomposites preparation with different inorganic resources, such as graphene or titanium dioxide (TiO_2), can be used to modify the physicochemical properties, such as optical modulation and electrochemical stability to improve the TM complexes. Many researchers have written about different types of TM complexes that can be used in ECDs as active materials. With the ability to fine-tune colors by manipulating the energy band configuration of TM complexes, the potential of synthesizing stable EC materials with a variety of colors in addition to functioning >10,000 switching cycles has emerged (9, 92).

TM coordination complexes have sufficient spectroscopic and redox properties for direct use in solution–phase ECDs. Polymeric systems that depend on the coordination of complex monomer units are frequently studied because they have the potential to be used in all-solid-state systems. The synthesis of various types of TM complexes for use in material chemistry (97), catalysis (98, 99), and medicine (100) are possible by the coordinated self-assembly of inorganic and organic materials. On electrode surfaces, electrochemical polymerization is commonly applied to arrange redox-active polymer films (101). The advantage of this method is that it is possible to use a wide range of conducting substrates and the electrochemical potential range, potential scan rate, polymerization time, monomer concentration, and electrolysis can be used to control the film thickness (4). Figure 11.14 shows the reductive electropolymerization of vinyl derivatives as an example. Direct electron transfer between EC units and electrodes, with or without operative ionic doping, is an organic EC mechanism. The disadvantage of polymers with metal complexes in the leading sequence is that they frequently have low solubility, which makes processing difficult. This can be overcome by polymerizing solution-processable monomers in a straight line on an electrode surface to form an insoluble thin film, which is called electropolymerization (92). On the electrode surface, a polymeric film is electrodeposited using a potential to initiate the polymerization of substrates in this surface modification procedure. This requires a good monomer solubility in the electrolyte solution (102). Reductive and oxidative electropolymerization is possible with applicable electropolymerizable groups, such as triphenylamine, pyrrole, thiophene, and vinyl functional groups (103–107, 92).

In addition, the electrochemical and the optical features of the reduced and oxidized electropolymerized thin films are listed in Table 11.1.

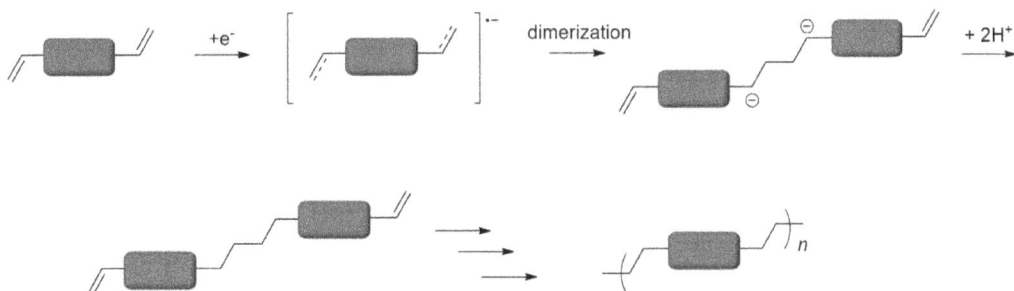

FIGURE 11.14 Reductive electropolymerization of vinyl derivatives mechanism.

(From Banasz & Walesa-Chorab 2019 (92). With permission.)

TABLE 11.1
Electrochromic features of redox electropolymerized films

Material	Structure	E_{ox}. (V)	Color change	ΔT % λ (nm)	Process	Reference
p-Benz-3TPA		0.95 0.83	Colorless, brown, blue	71% (790)	Oxidation	(108)
p-Benz-3CNTPA		0.86 0.97	Yellow, orange, brown, blue	65% (850)	Oxidation	(108)
[Ru2(dpb)(vbpy)4](PF$_6$)$_2$		+0.59	Not given	Not given	Reduction	(109)
phen-1,4-diyl-bridged tris-bidentate diruthenium complex 3(PF$_6$)$_2$		+0.16 +0.60	Wine, orange, dark cyan	41% (1300)	Reduction	(109)

Compound	Structure	Potential	Color change	% (nm)	Process
6(PF$_6$)$_2$ ([(dvtpy)Ru(tpy)](PF$_6$)$_2$)		+1.29	Orange-red, bleached	40% (490)	Reduction (110)
4(PF$_6$) ([(dvtpy)Ru(dpb)](PF$_6$))		+0.59	Purple-red, bleached	33% (520)	Reduction (110)
5(PF$_6$)$_2$ ([(dvtpy)Ru(Mebip)](PF$_6$)$_2$)		+1.05	Orange-red, bleached	27% (500)	Reduction (110)
3(PF$_6$)$_2$ ([(dvtpy)Ru(Mebip)](PF$_6$)$_2$)		+0.51 +1.05	Purple-red, bleached	12% (516)	Reduction (110)

(continued)

TABLE 11.1 (Continued)
Electrochromic features of redox electropolymerized films

Material	Structure	$E_{ox.}$ (V)	Color change	ΔT % λ (nm)	Process	Reference
A vinyl substituted cyclometalated ruthenium–amine hybridized		+0.68 +0.54	Purple, brown, sky-blue	52% (1070)	Reduction	(111)
biscyclometalated ruthenium complex [(vtpy)Ru(tpb)Ru-(vtpy)]2+		+0.14	Deep-blue, pink, green	40% (1165)	Reduction	(112)
biscyclometalated ruthenium complex 12+ bridged by tppyr		+0.47 +0.67	Blue, brown, orange	35% (2050)	Reduction	(113)
diruthenium complex with a redox-active amine bridge		+0.21 +0.44	Magenta, chocolate, violet-red, light blue	42% (1680)	Reduction	(114)

cyclometalated triruthenium complex 2(PF$_6$)n with a triarylamine core		+0.082 +0.31	Purple, brown, chocolate, olive, blue	63% (1550)	Reduction	(114)
[(C^N)M(O^N)], M= Pd(II) H(C^N) = 2-phenylpyridine H(O^N) = a triphenylamine functionalized Schiff	Monomer	+0.36 +0.57	Bleached, dark	24% (886)	oxidation	(115)
[(C^N)M(O^N)], M=Pt(II), H(C^N) = 2- phenylpyridine H(O^N) = a triphenylamine functionalized Schiff	Monomer	+0.38 +0.48	Bleached, dark	65% (864)	oxidation	(115)
[(C^N)M(O^N)], M= Pd(II) H(C^N) =, 2-thienylpyridine H(O^N) = a triphenylamine functionalized Schiff	Monomer	+0.34 +0.52	Bleached, dark	24% (860)	oxidation	(115, 116)

(continued)

TABLE 11.1 (Continued)
Electrochromic features of redox electropolymerized films

Material	Structure	E_{ox} (V)	Color change	ΔT % λ (nm)	Process	Reference
[(C^N)M(O^N)], M= Pt(II), H(C^N) = 2-thienylpyridine H(O^N) = a triphenylamine functionalized Schiff	Monomer	+0.35 +0.51	Bleached, dark	28% (834)	oxidation	(115, 116)
[(L1)PtCl] HL1 = 4-[p-(N-octyl-N-phenyl)amino]-phenyl-6-phenyl-2,2'-bipyridine	Monomer	+0.83 +1.14	Orange, bluish-black	70.5% (664)	oxidation	(117)
[Fe(L)2](PF6)2 {L = 4'-{[p-(N-butyl-N-phenyl)amino]-phenyl}-2,2':6',2''-terpyridine}	Monomer	+0.76 +0.83	Purple, pale yellow	20.3% (597)	oxidation	(118)
bithiophene–pyridine complex of ruthenium(III)–porphycene [Ru(TPrPc)(btp)2]PF6 (1)	Monomer	0.00	Blue, olive-green	Not given	oxidation	(105)

Name	Structure	Potentials	Colors	Yield		Ref.
[Ru(terpy)2 selenophene]	Monomer	+0.95 +1.50	Red, yellow	Not given	oxidation	(106)
P(ZnPc2-co-HKCN)	Monomer	+0.27	Greenish yellow, dark blue	44% (471)	oxidation	(119)
Cobalt phthalocyanine (CoPc)	Monomer	+0.87	Cyan, purple	59% (560)	oxidation	(120)
Metallophthalocyanines (MPcs) (M = Zn²⁺) bearing diethylaminophenoxy and chloro substituents at the peripheral positions (MPc-tdea-tCl)	Monomer	+1.10 +1.46	Green, orange	36% (525)	oxidation	(121)

(*continued*)

TABLE 11.1 (Continued)
Electrochromic features of redox electropolymerized films

Material	Structure	E_{ox} (V)	Color change	ΔT % λ (nm)	Process	Reference
Metallophthalocyanines (MPcs) ($M = Cu^{2+}$) bearing diethylaminophenoxy and chloro substituents at the peripheral positions (MPc-tdea-tCl)	Monomer	+1.03 +1.33	Greenish blue, light orange	20% (680)	oxidation	(121)
Metallophthalocyanines (MPcs) ($M = Co^{2+}$) bearing diethylaminophenoxy and chloro substituents at the peripheral positions (MPc-tdea-tCl)	Monomer	+0.75 +1.00 +1.25 +1.35	Blue, red	55% (673)	oxidation	(121)

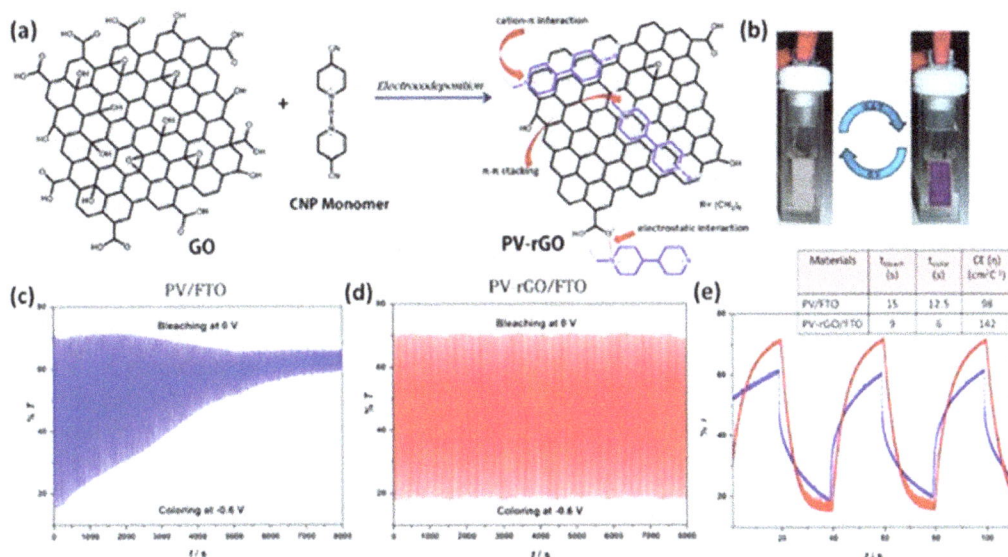

FIGURE 11.15 Showing: (a) electrodeposition of GO with CNP monomer to produce the PV–rGO composite, and the non-covalent binding that might occur in the PV–rGO structure; (b) transmittance changes regulated by voltage at bleached (0 V) and colored (0.6 V) states for PV/FTO; (c) PV–rGO/FTO; (d) films at 525 nm in 0.1 M KCl aqueous solution; and (e) transmittance response of PV/FTO (blue line) and PV–rGO/FTO (red line) films during potential switching between 0 (bleach) and 0.6 V (color) for 20 s. The times of color switching and CE values for both composite films are shown in the inset table.

(From Gadgil et al. 2015 (123). With permission.)

11.3.2 COMPOSITES WITH CARBON NANOMATERIALS

The nanocomposites prepared from mixing inorganic carbon materials and CPs had enhanced EC performance, for example, faster response speed, and higher color efficiency compared with pure CPs. Of interest, the existence of carbon nanotubes (CNTs) and graphene in the crystalline chains of CPs could have an important role in enhancing the CE, because these conductive nanomaterials could decrease the resistance that opposes the charge transfer, and this factor is known as the bridging effect (122). Graphene is a promising two-dimensional nanomaterial for applications in electronics and optics due to its exceptional features, for example, high transparency, surface area, flexibility, electrical conductivity, and thermal stability. However, the response time and cyclic reversibility of EC devices that contain a viologen polymer need to be enhanced for practical aspects. Of interest, graphene nanosheets can be mixed with polyviologen (PV) to generate EC composite materials of improved cyclic stability and low response times (123–134). The preparation of nanocomposite thin films from PV and reduced graphene oxide (rGO) using an electrochemical deposition approach (123) was studied, and the water-dispersible GO solution was electrochemically reduced via the in situ electropolymerizations of a cyanopyridinium as shown in Figure 11.15(a). The PV–rGO nanocomposite was highly stable due to the formed π–π stacking, and electrostatic binding between the PV and graphene sheet structures as the viologen unit had a cation–π electron and the reduced graphene sheets had a negative π- π system. When the nanocomposite was deposited on the fluorine-doped tin oxide (FTO) substrates, the EC device obtained a higher contrast color perform-ance from colorless to purple at a low operating voltage of 0.6 V (Figure 11.15(b)). Furthermore, when associated with the PV/FTO film, the PV–rGO/FTO substrate achieved significantly higher switching stability. The kinetic transmittance for PV–rGO/FTO films was steady at the switching step of 20 s, as shown in Figures 11.15(c and d), with no detectable shift over 8,000 s as the scanning time. However, the transmittance of PV/FTO films was reduced significantly to 43% in the colored state compared with 3.5% in the bleached state, which indicated that the cation radical rapidly degraded in pure PV films. Furthermore, the heavy electrostatic binding between rGO and PV is responsible for the long-term color stability of the viologen cation radical species. After performing the kinetic switching test, the response time for both films are shown in Figure 11.15(e). Compared with the blank PV/FTO films (17 and 12.5 s, respectively), the PV–rGO/FTO films have a superior CE value of 142 cm^2/C.

Another report offered a new flexible EC device without using an electrolyte where the EC material was composited with a conductive nanomaterial stabilized via various intermolecular forces, such as electrostatic, π–π stacking and cation-π electron interactions. The device had excellent stability when performing the switching by changing voltage between a colored state and a bleached state. The device was composed of methyl viologen (MV$^+$) as the cation and graphene quantum dots (GQDs) as the conductive anion that led to the formation of strong electrostatic and π–π binding and they bonded well together (135). Scheme 1 in Figure 11.16(a) shows the previously mentioned communication between MV2$^+$ and GQDs where both benzene rings for the viologen derivative (π electron-deficient bipyridinium units) can attach the GQDs (π electron-rich) by face-to-face stacking at that leads to accepting the electrons from the structure of the GQDs. Under a harsh environment, the stability of the MV2$^+$–GQD EC device was examined against heat and mechanical stress tests and the data suggested a highly operationally stable device. In addition, CV using a three-type electrode was carried out to check the electrochemical steps of electron transfer for MV^{2+} in a solution from GQD without an electrolyte and with a potassium chloride (KCl) electrolyte. The potential was between 0 and −1.8 V, which revealed two redox reactions related to MV^{2+}/MV$^{+\bullet}$ and MV$^{+\bullet}$/MV^0 (blue line, Figure 11.16(b)) and as a result, a color change occurred from the colorless (MV^{2+}) species to the purple (MV$^{+\bullet}$) one. The redox peaks agreed well with the obtained peaks when using KCl electrolyte (red line, Figure 11. 16(b)), which confirmed the electrochemical nature of V1 in the electrolyte solutions as indicated from the UV-Vis spectra of the devices in Figure 11.16(c). The GQDs could act as an electron transfer medium for facilitating the redox peaks of organic species and performing as a stable electrolyte for the transport of charge in a solution.

CNTs are interesting one-dimensional carbon materials that can be exploited as conductive substrates for improving the response times of EC polymers. The main drawback of CNTs processing is the π–π assembling and Van der Waals forces can increase the agglomeration of the individual particles. Recently(136), prepared a new composite as an EC material from multiwalled carbon nanotubes (MWCNTs) and poly(4,4′-(1,4-phenylene)bis (2,6-diphenylpyridinium) triflate) (PPDP) through simple physical mixing. The benzene ring of the PPDP chain helps to produce an effective π–π stacking interaction with MWCNTs that results in the stable adsorption of PPDP molecules on MWCNTs surface.

11.3.3 Metal Oxide Composite Films

Because different EC materials have different advantages and disadvantages, the advantages of organic–inorganic metal oxide EC hybrids can be combined, which results in enhanced EC properties. TM oxides, which exhibit a variety of valence states on reduction, are the most common inorganic EC materials (137). Inorganic materials with different bandgaps have different light absorption and color changes due to electron delocalization between mixed-valence states (138). Metal oxides have a variety of properties, such as high stability and CE, but due to their limited solubility, only the vapor deposition technique is possible to form thin films, which can be problematic. Furthermore, only the formation of nanocomposites can adjust the EC properties of metal oxides (79). Conducting viologen molecules have good structural and color tunability, and a high color purity and CR (12). However, conducive CPs, demonstrate film-forming properties, a wide variety of species, processing

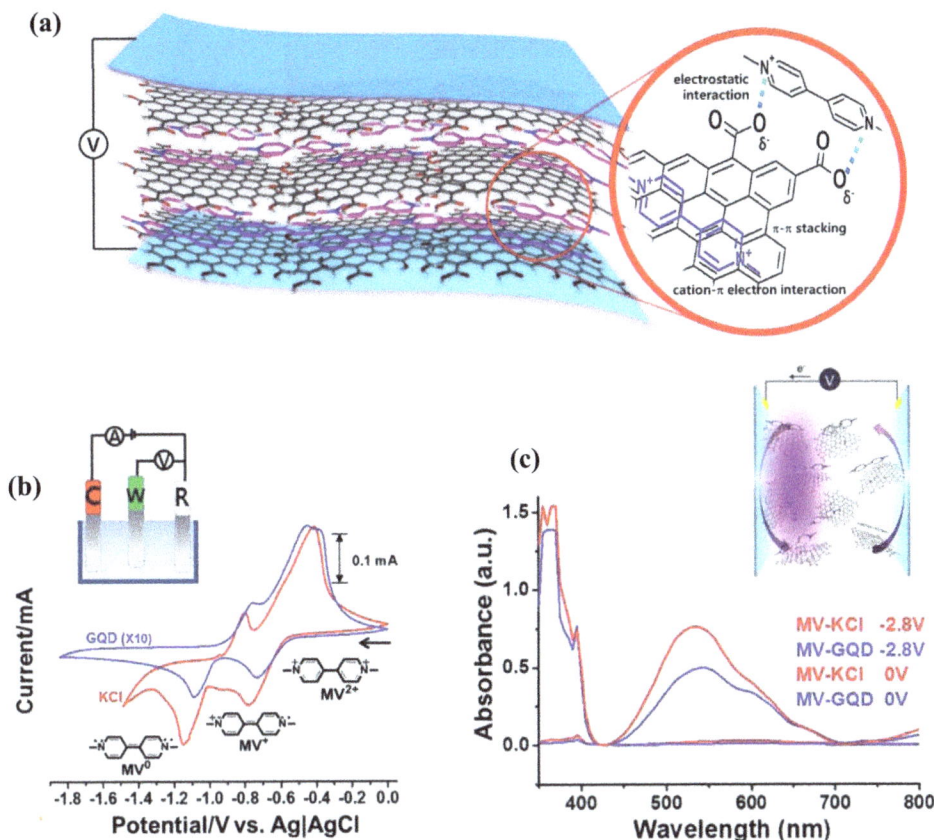

FIGURE 11.16 Showing: (a) CVs of 5 mM MV^{2+} at an ITO electrode in an aqueous solution containing 8 mg/mL GQD (blue line) and 0.1 M KCl (red line) at a scan rate of 100 mV/s in an aqueous solution containing 8 mg/mL GQD (blue line) and 0.1 M KCl (red line) at a scan rate of 100 mV/s. A three-electrode cell with a functioning electrode (W), a counter electrode (C), and a reference electrode (R) is depicted in the inset; (b) changes in the UV-Vis absorption spectra of 50 mM MV^{2+} with 0.1 M KCl in water (red line) and 8 mg/mL GQD in water from initial states (black line) to colored states (blue line); and (c) ECDs is exposed to a voltage of 2.8 V. ITO/MV2+GQD@PVA or MV2+KCl@PVA/ITO is used to make the ECDs. A two-electrode cell is depicted in the inset.

(From Sesuraj 2014 (135). With permission.)

simplicity, coloration fine-tuning, high coloration performance, quick switching speeds, high contrast capacity, and good processability (20, 139–141). In addition, they have a low operating voltage and a varied assortment of colors, from UV to NIR, in neutral and charged states (139).

Previously, TM oxides dominated EC materials and the expensive physical vapor deposition methods were used in the thin film preparation (142). Later, hybrid materials that contained two TM oxides, a TM oxide and organic molecules, or conducting polymers were developed, often exhibiting multi-electrochromism, and the fabrication methods were diverse and less expensive methods were discovered (143). Under a realistic voltage, the EC characteristics of TM oxides were determined by the electron–ion double injection or extraction reactions of TM ions, which are reversible redox reactions. The EC performances of inorganic materials are primarily governed by redox reaction characteristics. Figure 11.17 shows the transition metals whose oxides have EC features. In an EC TM oxide, there are two types of coloration mechanisms: cathodic EC materials that are colored by

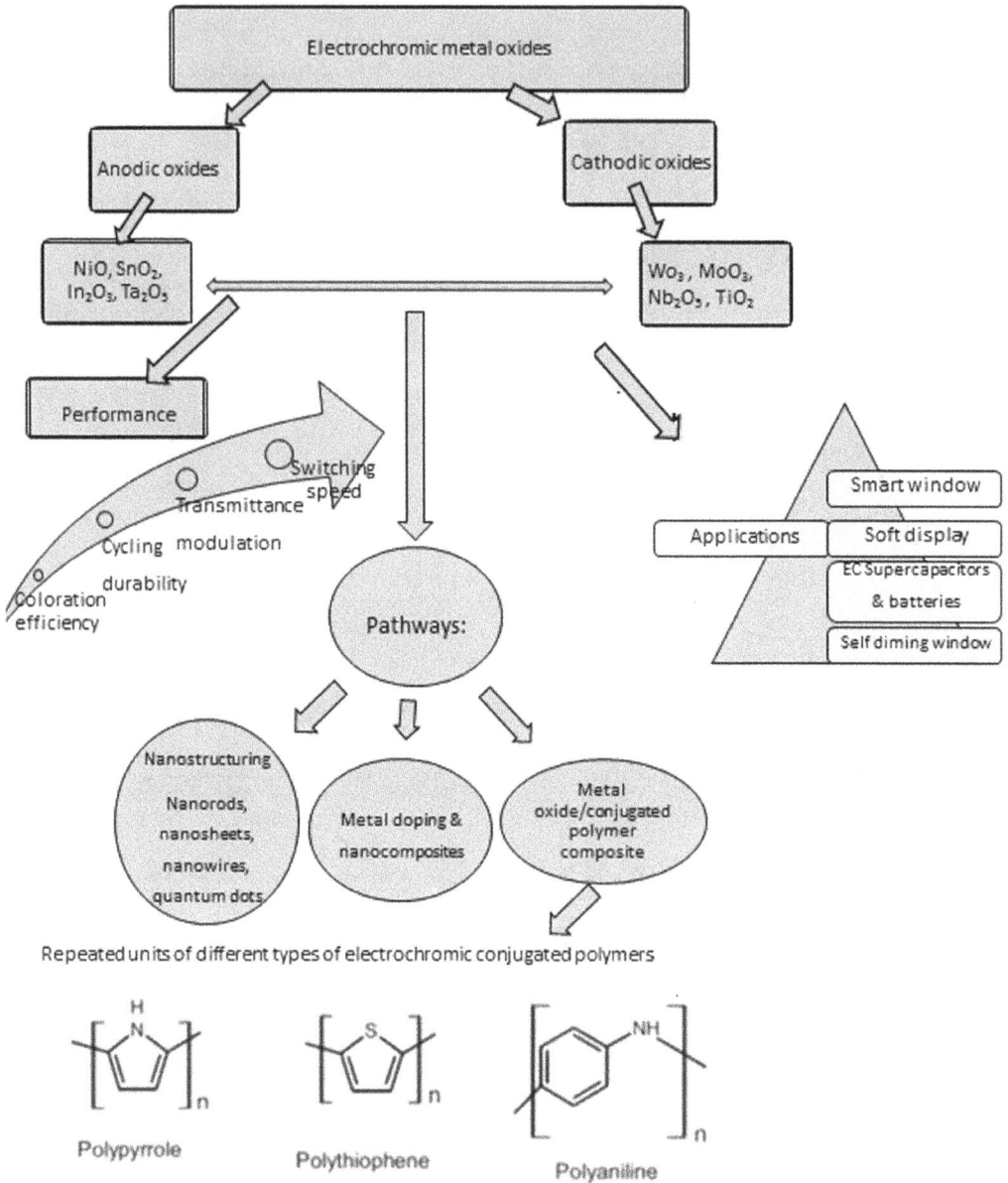

FIGURE 11.17 EC features of TM oxides composite.

guest ion injection (i.e., reduction process) and anodic EC materials colored by guest cation extraction (i.e., oxidation process) (144).

The inorganic materials, such as titanium (Ti), niobium (Nb), vanadium (V) and molybdenum (Mo) oxides and PB (i.e., cathodically coloring) and iridium (Ir), nickel (Ni) and cobalt (Co) oxides (i.e., anodically coloring) have been studied for their EC properties. The electronic arrangements of TM oxides are similar, with unfilled d bands that fill when cathodic charge injection occurs, and the color shift is caused by interband transitions (143). TM oxides in nanomaterial form have been shown to have shorter response times and, in some cases, improved CE (143). Some researchers have argued that nanostructuring does not have novel functionalities compared with its bulk counterparts (145). According to some researchers, EC switching is expected to be fast and stable in crystalline mesoporous structures, ultrathin crystalline nanowires, nanorods or nanotubes, and other large specific surface areas nanostructures. To display multicolor and improve the stability of devices and the CE, different types of materials must be combined (143). Specifically, CPs can be filled with metal oxides nanomaterials, such as NiO, TiO_2, IrO_2, WO_3, in addition to Au and Ag nanoparticles to obtain EC conductive polymer nanocomposites of outstanding features (81–85). Meanwhile, the fast switching speed can be conducted, because the diffusion length is supposed to shorten in a highly porous surface that is found in inorganic nanofillers, for example, metal oxides and metal nanoparticles (122).

In addition, an organic macromolecule with an alternating single and double bond backbone is known as a CP. A delocalized band of π-electrons is formed by the combination of overlapping p-orbitals with each other. Doping can improve the conductivity of CPs, which can be either conductors or semiconductors. The majority of CP research focused on those with aromatic and heterocyclic aromatic structures (146). CPs have been studied extensively for EC applications, for their redox reactions, which are a p-doping/de-doping route, which causes the electrochromism of CPs (14).

The CP is doped (p-doping) with counter anions in the oxidized state and has a delocalized π-electron band. The band structure of the CP is recovered after reduction by the incorporation of cations (de-doping) or exit of counter anions. Due to the redox reactions of CPs, a new optical absorption band is created and removed, which results in color changes. Fast switching speed, high contrast, good processability, and flexibility are all advantages of EC CPs (122, 147, 148). Because composite/hybrid materials have benefits from both components, CP-based composites have been extensively considered for EC applications. Numerous reports of CPs composites with TM oxides such as, WO_3 (149), MoO_3 (150), NiO (151), TiO_2 (152), and Fe_2O_3 (153), have been described. The interfacial connections between the organic and the inorganic phases in organic–inorganic nanocomposites have a significant impact on the nanocomposites' properties (154). CE and switching speed are generally higher in CP/inorganic material composites. In addition, the reduced charge transfer resistance is due to the bridging effect of conductive particles, which results in a higher CE. The porous construction produced via the inorganic materials causes the diffusion length to be shorter and increased the switching speed (147).

An electrochromic film was prepared from a WO_3/poly(p-phenylenebenzobisthiazole) nanocomposite and WO_3 loading was investigated in detail (155). In addition, during the formulation of the thin film, chlorosulfuric acid (CSA) was used as a solvent to dissolve the p-phenylenebenzobisthiazole polymer and as a dopant to expand the electrical conductivity of the resulting polymer composite. In addition, cadmium selenide (CdSe) QDs were prepared via a chemical capping technique (156), and a stable gold nanparticles (Au NPs) solution in toluene was achieved using a previous method (157). Organic acids were used as compatibilizers to advance the spreading and interfacial interactions of the AuNPs/CdSe QDs with CPs after they were prepared. Compared with the pure PEDOT film, which had a coloring efficiency of 191 cm^2/C and the kinetic response of 4.5 s to color and 1.5 s to bleach, the optimum added Au–CdSe film had a superior coloring efficiency of 300 cm^2/C and quick kinetic response of 4.5 s to color and 1.5 s to bleach (colors in 5.2 s and bleaches in 3.1 s).

11.4 CONCLUSIONS AND OUTLOOK

Recently, significant advances in the use of EC materials in the application of CP-based ECDs (e.g., PANI, PPy, PTh, PA, PCz, and PVs) due to their properties, such as their flexibility, low costed processes, easily abundant and available, high thermal, mechanical and electrochemical stability. Due to altering their molecular structure and inserting electroactive functional groups into the polymer backbone or side chain groups led to a change in EC performance, in addition to the properties of EC polymers. Furthermore, electroactive functional groups have high stability; therefore, EC polymers have excellent thermal and mechanical stability. They need to be optimized in ECDs through different areas including processing techniques, architecture, and device components. In addition, EC polymers have increased the thermal and UV stability of ECDs.

However, there are still some issues, such as choice of charge storage layers and their compatibility with each ECD component; the transfer from laboratory conditions to industrial practice; and eco-friendly processes in preparation and manufacturing. These need to be discussed for the future of EC polymers. Therefore, there is increased research into the commercialization of EC polymer-based ECDs.

ACKNOWLEDGMENTS

The authors will consider the support of BC Materials in Spain and Cairo University in Egypt to follow-up this study using their facilities. In addition, AES thanks the National Research grants from MINECO Juan de la Cierva [FJCI-2018-037717] and they are currently on leave from CMRDI.

REFERENCES

1. Shah KW, Wang S-X, Soo DXY, Xu J. Viologen-based electrochromic materials: from small molecules, polymers and composites to their applications, Polymers. 2019; 11(11):1839.
2. Mortimer RJ. Switching colors with electricity: Electrochromic materials can be used in glare reduction, energy conservation and chameleonic fabric. Am Sci. 2013; 101(1): 38–46.
3. Granqvist CG. Electrochromics for smart windows: Oxide-based thin films and devices. Thin solid films. 2014; 564: 1–38.
4. Mortimer RJ, Dyer AL, Reynolds JR. Electrochromic organic and polymeric materials for display applications. Displays. 2006; 27(1):2–18.
5. Ding Y, Invernale MA, Sotzing GA. Conductivity trends of PEDOT-PSS impregnated fabric and the effect of conductivity on electrochromic textile, ACS Appl Mat Interfaces 2(6) (2010) 1588–93.
6. Beaupré S, Dumas J, Leclerc M. Toward the development of new textile/plastic electrochromic cells using triphenylamine-based copolymers. Chem Mat. 2006; 18(17):4011–18.
7. Lahav M, van der Boom ME. Polypyridyl Metallo-Organic Assemblies for Electrochromic Applications, Adv Mater. 2018; 30(41):1706641.
8. Zhan Y, Tan MRJ, Cheng X, Tan WMA, Cai GF, Chen JW. et al. Ti-Doped WO 3 synthesized by a facile wet bath method for improved electrochromism. J Mater Chem C. 2017; 5(38):9995–10000.
9. Amb CM, Dyer AL, Reynolds JR. Navigating the color palette of solution-processable electrochromic polymers. Chem Mater. 2011; 23(3):397–415.
10. Balan A, Baran D, Toppare L. Benzotriazole containing conjugated polymers for multipurpose organic electronic applications. Polym Chem. 2011; 2(5):1029–43.
11. Zhang W, Li H, William WY, Elezzabi AY. Transparent inorganic multicolour displays enabled by zinc-based electrochromic devices. Light: Sci Appl. 2020; 9(1):1–11.
12. Yang G, Zhang Y-M, Cai Y, Yang B, Gu C, Zhang SX-A. Advances in nanomaterials for electrochromic devices. Chem Soc Rev. 2020; 49: 8687–720.
13. Wang Z, Wang X, Cong S, Chen J, Sun H, Chen Z. et al. Towards full-colour tunability of inorganic electrochromic devices using ultracompact fabry-perot nanocavities, Nature Comm. 2020; 11(1):1–9.
14. Thakur VK, Ding G, Ma J, Lee PS, Lu X. Hybrid materials and polymer electrolytes for electrochromic device applications. Adv Mater. 2012; 24(30):4071–96.
15. Lamsal C, Ravindra N. Optical properties of vanadium oxides-an analysis. J Mater Sci. 2013; 48(18):6341–51.
16. Wang H, Tang C, Shi Q, Wei M, Su Y, Lin S. et al. Influence of Ag incorporation on the structural, optical and electrical properties of ITO/Ag/ITO multilayers for inorganic all-solid-state electrochromic devices, Ceram Inter. 2021; 47(6):7666–73.
17. Li M, Wei Y, Zheng J, Zhu D, Xu C. Highly contrasted and stable electrochromic device based on well-matched viologen and triphenylamine, Org Electron. 2014; 15(2):428–34.

18. Ghoorchian A, Tavoli F, Alizadeh N. Long-term stability of nanostructured polypyrrole electrochromic devices by using deep eutectic solvents. J Electroanal Chem. 2017; 807: 70–5.

19. Wang H, Barrett M, Duane B, Gu J, Zenhausern F. Materials and processing of polymer-based electrochromic devices, Mater Sci Eng: B. 2018; 22: 167–74.

20. Jensen J, Hose M l, Dyer AL, Krebs FC. Development and manufacture of polymer-based electrochromic devices. Adv Funct Mater. 2015; 25(14):2073–90.

21. Abidin T, Zhang Q, Wang K-L, Liaw D-J. Recent advances in electrochromic polymers, Polymer. 2014; 55(21):5293–4.

22. Unver EK, Tarkuc S, Udum YA, Tanyeli C, Toppare L. The effect of the donor unit on the optical properties of polymers. Org Electron. 2011; 12(10):1625–31.

23. Neo WT, Ye Q, Chua S-J, Xu J. Conjugated polymer-based electrochromics: materials, device fabrication and application prospects. J Mater Chem C. 2016; 4(31):7364–76.

24. Lv X, Li W, Ouyang M, Zhang Y, Wright DS, Zhang C, Polymeric electrochromic materials with donor–acceptor structures. J Mater Chem C. 2017; 5(1):12–28.

25. Xu Z, Kong L, Wang Y, Wang B, Zhao J. Tuning band gap, color switching, optical contrast, and redox stability in solution-processable BDT-based electrochromic materials, Org Electron. 2018; 54: 94–103.

26. Xiong S, Li S, Zhang X, Wang R, Zhang R, Wang X. *et al.* Synthesis and performance of highly stable star-shaped polyaniline electrochromic materials with triphenylamine core, J Electron Mater. 2018; 47(2):1167–75.

27. Lee JY, Han S-Y, Cho I, Lim B, Nah Y-C, Electrochemical and electrochromic properties of diketopyrrolopyrrole-based conjugated polymer, Electrochem Comm. 2017; 83: 102–5.

28. Argun AA, Aubert P-H, Thompson BC, Schwendeman I, Gaupp CL, Hwang J. *et al.* Multicolored electrochromism in polymers: structures and devices. Chem Mater. 2004; 16(23):4401–12.

29. Perera K, Yi Z, You L, Ke Z, Mei J. Conjugated electrochromic polymers with amide-containing side chains enabling aqueous electrolyte compatibility. Polymer Chem. 2020; 11(2):508–16.

30. Zhang Y, Kong L, Zhang Y, Du H, Zhao J, Chen S. *et al.* Ultra-low-band gap thienoisoindigo-based ambipolar type neutral green copolymers with ProDOT and thiophene units as NIR electrochromic materials. Org Electron. 2020; 81: 105685.

31. Otero TF. Biomimetic conducting polymers: synthesis, materials, properties, functions, and devices. Polymer Rev. 2013; 53(3):311–51.

32. X. Fu, C. Jia, Z. Wan, X. Weng, J. Xie, L. Deng, Hybrid electrochromic film based on polyaniline and TiO2 nanorods array, Organic Electronics 15(11) (2014) 2702–2709.

33. Tang Y, Pan K, Wang X, Liu C, Luo S. Electrochemical synthesis of polyaniline in surface-attached poly (acrylic acid) network, and its application to the electrocatalytic oxidation of ascorbic acid. Microchim Acta. 2010; 168(3):231–7.

34. Mascaro LH, Berton AN, Micaroni L. Electrochemical synthesis of polyaniline/poly-o-aminophenol copolymers in chloride medium. Int J Electrochem. 2011; Article ID 292581.

35. Gvozdenovic MM, Jugovic B , Stevanovic JS, Grgur B. Electrochemical synthesis of electroconducting polymers. Hem Ind. 2014; 68(6):673–84.

36. Korent A, Soderznik KZ, Sturm S, Rozman KZ. A correlative study of polyaniline electropolymerization and its electrochromic behavior. J Electrochem Soc. 2020; 167(10):106504.

37. Beregoi M, Busuioc, Evanghelidis A, Matei E, Iordache F, Radu M. Electrochromic properties of polyaniline-coated fiber webs for tissue engineering applications. Int J Pharmaceutics. 2016; 510(2):465–73.

38. Chandrasekhar P, Dooley T. Far-IR transparency and dynamic infrared signature control with novel conducting-polymer systems, Optical and photonic applications of electroactive and conducting polymers. Int Soc Optic Photon. 1995:169–80.

39. Zhang L, Wang B, Li X, Xu G, Dou S, Zhang X. *et al.* Further understanding of the mechanisms of electrochromic devices with variable infrared emissivity based on polyaniline conducting polymers. J Mater Chem C. 2019; 7(32):9878–91.

40. Sung J-H, Kim S-J, Lee K-H. Fabrication of microcapacitors using conducting polymer microelectrodes. J Power Source. 2003; 124(1):343–50.

41. Das S, Chatterjee DP, Ghosh R, Nandi AK. Water soluble polythiophenes: preparation and applications. RSC Adv. 2015; 5(26):20160–77.

42. Shao M, He Y, Hong K, Rouleau CM, Geohegan DB, Xiao K. A water-soluble polythiophene for organic field-effect transistors. Polymer Chem. 2013; 4(20):5270–74.

43. Moss KC, Bourdakos KN, Bhalla V, Kamtekar KT, Bryce MR, Fox MA. *et al.* Tuning the intramolecular charge transfer emission from deep blue to green in ambipolar systems based on dibenzothiophene S, S-dioxide by manipulation of conjugation and strength of the electron donor units. J Org Chem. 2010; 75(20):6771–81.

44. Groenendaal L, Jonas F, Freitag D, Pielartzik H, Reynolds JR. Poly (3, 4-ethylenedioxythiophene) and its derivatives: past, present, and future. Adv Mater. 2000; 12(7): 481–94.

45. Sun H, Zhang L, Dong L, Zhu X, Ming S, Zhang Y. *et al.* Aqueous electrosynthesis of an electrochromic material based water-soluble EDOT-MeNH2 hydrochloride. Synth Met. 2016; 211: 147–54.

46. Zhang H, Qu H, Lv H, Hou S, Zhang K, Zhao J. Improved electrochromic performance of poly(3, 4-ethylenedioxythiophene) by incorporating a three-dimensionally ordered macroporous structure. Chem Asian J. 2016; 11(20):2882–8.

47. Authidevi P, Kavitha G, Kanagavel D, Vedhi C. Studies of electrochromic behavior of conducting copolymer of 3-methylthiophene with 3, 4-ethylenedioxythiophene. Mater Today. 2020. doi.org/10.1016/j.matpr.2020.07.112

48. Wang L-X, Li X-G, Yang Y-L. Preparation, properties and applications of polypyrroles. React Funct Polym. 2001; 47(2):125–39.

49. Tarkuc S, Sahin E, Toppare L, Colak D, Cianga I, Yagci Y. Synthesis, characterization and electrochromic properties of a conducting copolymer of pyrrole functionalized polystyrene with pyrrole. Polymer. 2006; 47(6):2001–9.

50. Zotti G, Zecchin S, Schiavon G, Vercelli B, Berlin A, Dalcanale E, Groenendaal LB. Potential-driven conductivity of polypyrroles, poly-n-alkylpyrroles, and polythiophenes: role of the pyrrole nh moiety in the doping-charge dependence of conductivity. Chem Mater. 2003; 15(24):4642–50.

51. Bazzaoui M, Martins J, Bazzaoui E, Martins L, Machnikova E. Sweet aqueous solution for electrochemical synthesis of polypyrrole part 1B: On copper and its alloys. Electrochim Acta. 2007; 52(11):3568–81.

52. Alkan S, Toppare L, Hepuzer Y, Yagci Y. Block copolymers of thiophene-capped poly (methyl methacrylate) with pyrrole, J Polym Sci A: Polymer Chem. 1999; 37(22): 4218–25.

53. Bengu B , Toppare L, Kalaycioglu E. Synthesis of conducting graft copolymers of 2-(N-pyrrolyl) ethylvinyl ether with pyrrole, Des Monomers Polym. 2001; 4(1): 53–65.

54. Brotherston ID, Mudigonda DS, Osborn JM, Belk J, Chen J, DC Loveday. *et al.* Tailoring the electrochromic properties of devices via polymer blends, copolymers, laminates and patterns., Electrochim Acta. 1999; 44(18):2993–3004.

55. Ak M, Gacal B, Kiskan B, Yagci Y, Toppare L. Enhancing electrochromic properties of polypyrrole by silsesquioxane nanocages. Polymer. 2008; 49(9):2202–10.

56. Soganci T, Soyleyici S, Soyleyici HC, Ak M. High contrast electrochromic polymer and copolymer materials based on amide-substituted poly (dithienyl pyrrole). J Electrochem Soc. 2016; 164(2):H11.

57. Wu T-Y, Tung Y-H. Phenylthiophene-containing poly(2, 5-dithienylpyrrole)s as potential anodic layers for high-contrast electrochromic devices. J Electrochem Soc. 2018; 165(5):H183.

58. Cihaner A, Algi F. A processable rainbow mimic fluorescent polymer and its unprecedented coloration efficiency in electrochromic device. Electrochim Acta. 2008; 53(5):2574–8.

59. Soyleyici HC. Electrochromic properties of multifunctional conductive polymer based on naphthalene. Opt Mater. 2019; 90: 208–14.

60. Tao Y, Yang C, Qin J. Organic host materials for phosphorescent organic light-emitting diodes. Chem Soc Rev. 2011; 40(5): 2943–70.

61. Jiang W, Duan L, Qiao J, Dong G, Zhang D, Wang L. *et al.* High-triplet-energy tri-carbazole derivatives as host materials for efficient solution-processed blue phosphorescent devices. J Mater Chem. 2011; 21(13):4918–26.

62. Witker D, Reynolds, JR. Soluble variable color carbazole-containing electrochromic polymers. Macromolecules. 2005; 38(18): 7636–44.

63. Elkhidr HE, Ertekin Z, Udum YA, Pekmez K. Electrosynthesis and characterizations of electrochromic and soluble polymer films based on N-substituted carbazole derivates. Synth Met. 2020; 260: 116253.

64. Li Z, Cheng X, He S, Shi X, Gong L, Zhang H. Aramid fibers reinforced silica aerogel composites with low thermal conductivity and improved mechanical performance. Compos A. Appl Sci Manuf. 2016; 84: 316–25.

65. Patterson BA, Sodano HA. Enhanced interfacial strength and UV shielding of aramid fiber composites through ZnO nanoparticle sizing. ACS App Mater Interfaces. 2016; 8(49):33963–71.

66. Lu Z, Si L, Dang W, Zhao Y. Transparent and mechanically robust poly (para-phenylene terephthamide) PPTA nanopaper toward electrical insulation based on nanoscale fibrillated aramid-fibers. Compos - Part A: Appl Sci Manuf. 2018; 115: 321–30.

67. Wang HM, Hsiao SH. Enhancement of redox stability and electrochromic performance of aromatic polyamides by incorporation of (3, 6-dimethoxycarbazol-9-yl)-triphenylamine units. J Polym Sci A: Polym Chem. 2014; 52(2):272–86.

68. Hsiao S-H, Lu H-Y. Electrosynthesis of aromatic poly (amide-amine) films from triphenylamine-based electroactive compounds for electrochromic applications. Polymers. 2017; 9(12):708.

69. Gelinas B, Das D, Rochefort D. Air-Stable, Self-Bleaching Electrochromic Device Based on Viologen- and Ferrocene-Containing Triflimide Redox Ionic Liquids. ACS Appl Mater Interf. 2017; 9(34):28726–36.

70. Watanabe T, Honda K. Measurement of the extinction coefficient of the methyl viologen cation radical and the efficiency of its formation by semiconductor photocatalysis. J Phys Chem. 1982; 86(14):2617–19.

71. Monk PMS. The effect of ferrocyanide on the performance of heptyl viologen-based electrochromic display devices. J Electroanal Chem. 1997; 432(1):175–9.

72. Wardman P. The Reduction Potential of Benzyl Viologen: An Important Reference Compound for Oxidant/Radical Redox Couples. Free Radic Res Commun. 1991; 14(1):57–67.

73. Dmitrieva E, Rosenkranz M, Alesanco Y, Vinuales A. The reduction mechanism of p-cyanophenylviologen in PVA-borax gel polyelectrolyte-based bicolor electrochromic devices. Electrochim Acta. 2018; 292: 81–7.

74. Monk PMS. Comment on: Dimer formation of viologen derivatives and their electrochromic properties. Dyes Pigm. 1998; 39(2):125–8.

75. Moon HC, Kim C-H, Lodge TP, Frisbi CD. Multicolored, Low-Power, Flexible Electrochromic Devices Based on Ion Gels, ACS Appl Mater Interf. 2016; 8(9):6252–60.

76. Zhao S, Huang W, Guan Z, Jin B, Xiao D. A novel bis(dihydroxypropyl) viologen-based all-in-one electrochromic device with high cycling stability and coloration efficiency. Electrochim Acta. 2019; 298: 533–40.

77. Mortimer RJ. Organic electrochromic materials. Electrochim Acta. 1999; 44(18):2971–81.

78. Cheng K-C, Chen F-R, Kai J-J. Electrochromic property of nano-composite Prussian Blue based thin film. Electrochim Acta. 2007; 52(9):3330–5.

79. Wu W, Wang M, Ma J, Cao Y, Deng Y. Electrochromic metal oxides: Recent progress and prospect. Adv Electron Mater. 2018; 4(8):1800185.

80. Beaujuge PM, Amb CM, Reynolds JR. Spectral Engineering in π-Conjugated Polymers with Intramolecular Donor–Acceptor Interactions. Acc Chem Res. 2010; 43(11):1396–407.

81. Park JH, Ko IJ, Kim GW, Lee H, Jeong SH, Lee JY. et al. High transmittance and deep RGB primary electrochromic color filter for high light out-coupling electro-optical devices. Opt Express. 2019; 27(18):25531–43.

82. Ouyang M, Yang Y, Lv X, Han Y, Huang S, Dai Y. et al. Enhanced electrochromic switching speed and electrochemical stability of conducting polymer film on an ionic liquid functionalized ITO electrode. New J Chem. 2015; 39(7):5329–35.

83. Morita M. Multicolor electrochromic behavior of polyaniline composite films combined with tungsten trioxide, Macromol Chem Phys. 1994; 195(2):609–20.

84. Zhu J, Wei S, Zhang L, Mao Y, Ryu J, Karki AB. et al. Polyaniline-tungsten oxide metacomposites with tunable electronic properties. J Mater Chem. 2011; 21(2):342–8.

85. Elzanowska H, Miasek E, Birss VI. Electrochemical formation of Ir oxide/polyaniline composite films. Electrochim Acta. 2008; 53(6):2706–15.

86. Tieke B. Coordinative supramolecular assembly of electrochromic thin films. Curr Opin Colloid Interface Sci. 2011; 16(6): 499–507.

87. Itaya K, Ataka T, Toshima S. Spectroelectrochemistry and electrochemical preparation method of Prussian blue modified electrodes. J Am Chem Soc. 1982; 104(18):4767–72.

88. Tung T-S, Ho, K-C. Cycling and at-rest stabilities of a complementary electrochromic device containing poly(3,4-ethylenedioxythiophene) and Prussian blue. Sol Energy Mater Sol Cell. 2006; 90(4):521–37.

89. DeLongchamp DM, Hammond PT. Multiple-Color Electrochromism from Layer-by-Layer-Assembled Polyaniline/Prussian Blue Nanocomposite Thin Films. Chem Mater. 2004; 16(23):4799–805.

90. Rong Y, Kim S Su, F, Myers D, Taya M. New effective process to fabricate fast switching and high contrast electrochromic device based on viologen and Prussian blue/antimony tin oxide nanocomposites with dark colored state. Electrochim Acta. 2011; 56(17):6230–6.

91. Fan M-S, Kao S-Y, Chang T-H, Vittal R, Ho K-C. A high contrast solid-state electrochromic device based on nano-structural Prussian blue and poly(butyl viologen) thin films. Sol Energy Mater Sol Cells. 2016; 145: 35–41.

92. Banasz R, Walesa-Chorab M. Polymeric complexes of transition metal ions as electrochromic materials: synthesis and properties. Coord Chem Rev. 2019; 389: 1–18.

93. Camurlu P. Polypyrrole derivatives for electrochromic applications. RSC Advances. 2014; 4(99):55832–45.

94. Kettle S. Electronic spectra of transition metal complexes. Physical Inorg Chem. 1996; 156–84.

95. Domingo A, Carvajal MA, de Graaf C, Sivalingam K, Neese F, Angeli, C. Metal-to-metal charge-transfer transitions: reliable excitation energies from ab initio calculations. Theor Chem Acc. 2012; 131(9):1–13.

96. Acosta A, Zink JI, Cheon J. Ligand to ligand charge transfer in (hydrotris (pyrazolyl) borato) (triphenylarsine) copper (I), Inorg Chem. 2000; 39(3):427–32.

97. Roder R, Preib T, Hirschle P, Steinborn B, Zimpel A, Hohn M. Multifunctional nanoparticles by coordinative self-assembly of His-tagged units with metal–organic frameworks. J Am Chem Soc. 2017; 139(6):2359–68.

98. Zaranek M, Witomska S, Patroniak V, Pawluc P. Unexpected catalytic activity of simple triethylborohydrides in the hydrosilylation of alkenes. Chem Comm. 2017; 53(39):5404–7.

99. Huang J-P, Zhang P, Song J-G, Zhao J. Two Cu (II)-based coordination polymers: Photocatalytic dye degradation and treatment activity combined with BDNF modified bone marrow mesenchymal stem cells on craniocerebral trauma via increasing complement C3 expression Arabian J Chem. 2020; 13(9):7045–54.

100. Yue Z, Wang H, Li Y, Qin Y, Xu L, Bowers DJ. *et al.* Coordination-driven self-assembly of a Pt (IV) prodrug-conjugated supramolecular hexagon. Chem Comm. 2018; 54(7):731–4.

101. Compton R, Hancock G. Dynamic processes in polymer modified electrodes. Res Chem Kinet. 2012; 261.

102. Cosnier S, Karyakin A. *Electropolymerization: Concepts, Materials, Applications.* 2010; Wiley-VCHA.

103. Walesa-Chorab M, Banasz R, Kubicki M, Patroniak V. Dipyrromethane functionalized monomers as precursors of electrochromic polymers. Electrochim Acta. 2017; 258: 571–81.

104. Tang J-H, He Y-Q, Shao J-Y, Gong Z-L, Zhong Y-W. Multistate redox switching and near-infrared electrochromism based on a star-shaped triruthenium complex with a triarylamine core. Sci Rep. 2016; 6(1):1–9.

105. Abe M, Futagawa H, Ono T, Yamada T, Kimizuka N, Hisaeda Y. An electropolymerized crystalline film incorporating axially-bound metalloporphycenes: remarkable reversibility, reproducibility, and coloration efficiency of ruthenium (II/III)-based electrochromism, Inorg Chem. 2015; 54(23):11061–3.

106. Taouil AE, Husson J, Guyard L. Synthesis and characterization of electrochromic [Ru (terpy) 2 selenophene]-based polymer film. J Electroanal Chem. 2014; 728: 81–5.

107. Icli M, Pamuk M, Algi F, Onal AM, Cihaner A. A new soluble neutral state black electrochromic copolymer via a donor–acceptor approach. Org Electron. 2010; 11(7):1255–60.

108. Santra DC, Nad S, Malik S. Electrochemical polymerization of triphenylamine end-capped dendron: Electrochromic and electrofluorochromic switching behaviors. J Electroanal Chem. 2018; 823: 203–12.

109. Nie H-J, Zhong Y-W. Near-infrared electrochromism in electropolymerized metallopolymeric films of a phen-1, 4-diyl-bridged diruthenium complex. Inorg Chem. 2014; 53(20):11316–22.

110. Cui B-B, Nie H-J, Yao C-J, Shao J-Y, Wu S-H, Zhong Y-W. Reductive electropolymerization of bistridentate ruthenium complexes with 5, 5″-divinyl-4′-tolyl-2, 2′: 6′, 2″-terpyridine. Dalton Trans. 2013; 42(39):14125–33.

111. Cui B-B, Yao C-J, Yao J, Zhong Y-W. Electropolymerized films as a molecular platform for volatile memory devices with two near-infrared outputs and long retention time. Chem Sci. 2014; 5(3):932–41.

112. Yao C-J, Zhong Y-W, Nie H-J, Abruna HCD, Yao J. Near-IR electrochromism in electropolymerized films of a biscyclometalated ruthenium complex bridged by 1, 2, 4, 5-tetra (2-pyridyl) benzene. J Am Chem Soc. 2011; 133(51):20720–23.

113. Yao C-J, Yao J, Zhong Y-W. Metallopolymeric films based on a biscyclometalated ruthenium complex bridged by 1, 3, 6, 8-tetra (2-pyridyl) pyrene: applications in near-infrared electrochromic windows. Inorg Chem. 2012; 51(11):6259–63.

114. Cui BB, Tang JH, Yao J, Zhong YW. A Molecular Platform for Multistate Near-Infrared Electrochromism and Flip-Flop, Flip-Flap-Flop, and Ternary Memory. Angew Chem Int Ed. 2015; 54(32):9192–7.

115. Zhao M, Wu W, Su B. pH-controlled drug release by diffusion through silica nanochannel membranes. ACS Appl Mater Interf. 2018; 10(40):33986–92.

116. Ionescu A, Lento R, Mastropietro TF, Aiello I, Termine R, Golemme A. *et al.*, Electropolymerized highly photoconductive thin films of cyclopalladated and cycloplatinated complexes. ACS Appl Mater Interf. 2015; 7(7):4019–28.

117. Qiu D, Bao X, Zhao Q, Feng Y, Wang H, Liu K. Electrochromic and proton-induced phosphorescence properties of Pt (II) chlorides with arylamine functionalized cyclometalating ligands. J Mater Chem C. 2013; 1(4):695–704.

118. Bao X, Zhao Q, Wang H, Liu K, Qiu D. Metallopolymer electrochromic film prepared by oxidative electropolymerization of a Fe (II) complex with arylamine functionalized terpyridine ligand. Inorg Chem Comm. 2013; 38: 88–91.

119. Goktug O, Soganci T, Ak M, Sener MK. Efficient synthesis of EDOT modified ABBB-type unsymmetrical zinc phthalocyanine: optoelectrochromic and glucose sensing properties of its copolymerized film. New J Chem. 2017; 41(23): 14080–7.

120. Arican D, Aktas A, Kantekin H, Koca A. Electrochromism of electropolymerized cobaltphthalocyanine–quinoline hybrid, Sol Energy Mater Sol Cells. 2015; 132: 289–95.

121. Kobak RZU, Akyuz D, Koca A. Substituent effects to the electrochromic behaviors of electropolymerized metallophthalocyanine thin films. J Solid State Electrochem. 2016; 20(5):1311–21.

122. Zhan C, Yu G Lu, Y, Wang L, Wujcik E, Wei S. Conductive polymer nanocomposites: a critical review of modern advanced devices. J Mater Chem C. 2017; 5(7):1569–85.

123. Gadgil B, Damlin P, Heinonen M, Kvarnstrom C. A facile one step electrostatically driven electrocodeposition of polyviologen–reduced graphene oxide nanocomposite films for enhanced electrochromic performance. Carbon. 2015; 89: 53–62.

124. Wang N, Lukaacs Z, Gadgil B, Damlin P, Janaky C, Kvarnstrom C. Electrochemical deposition of polyviologen-reduced graphene oxide nanocomposite thin films. Electrochim Acta. 2017; 231: 279–86.

125. Lee J, Goda ES, Choi J, Park J, Lee S. Synthesis and characterization of elution behavior of nonspherical gold nanoparticles in asymmetrical flow field-flow fractionation (AsFlFFF). J Nanopart Res. 2020; 22(9):256.

126. Abu Elella MH, ElHafeez EA Goda ES, Lee S, Yoon KR. Smart bactericidal filter containing biodegradable polymers for crystal violet dye adsorption. Cellulose. 2019; 26(17):9179–206.

127. Goda ES, Lee S, Sohail M, Yoon KR. Prussian blue and its analogues as advanced supercapacitor electrodes. J Energy Chem. 2020; 50: 206–29.

128. Abu Elella MH, Goda ES, Yoon KR, Hong SE, Morsy MS, Sadak RA. *et al.* Novel vapor polymerization for integrating flame retardant textile with multifunctional properties Compos Commun. 2021; 24, 100614a.

129. Abu Elella MH, Goda ES, Abdallah HM, Shalan AE, Gamal H, Yoon KR. Innovative bactericidal adsorbents containing modified xanthan gum/montmorillonite nanocomposites for wastewater treatment. Int J Biologic Macromol. 2021; 167: 1113–25.

130. Goda ES, Gab-Allah MA, Singu BS, Yoon KR. Halloysite nanotubes based electrochemical sensors: A review, Microchem J. 2019; 147: 1083–96.

131. Goda ES, Yoon KR, El-sayed SH, Hong SE. Halloysite nanotubes as smart flame retardant and economic reinforcing materials: A review. Thermochim Acta. 2018; 669: 173–84.

132. Goda ES, Singu BS, Hong SE, Yoon KR. Good dispersion of poly(δ-gluconolactone)-grafted graphene in poly(vinyl alcohol) for significantly enhanced mechanical strength. Mater Chem Phys. 2020; 254: 123465.

133. Goda ES, Hong SE, Yoon KR. Facile synthesis of Cu-PBA nanocubes/graphene oxide composite as binder-free electrodes for supercapacitor. J Alloys and Compd. 2020; 859, 157868.

134. Singu BS, Goda ES, Yoon KR. Carbon Nanotube–Manganese oxide nanorods hybrid composites for high-performance supercapacitor materials. J Ind Eng Chem. 2021; 97, 239–49.

135. Sesuraj RS. Nano Focus: Electrolyte-free electrochromic device fabricated using graphene quantum dot-viologen nanocomposites. MRS Bull. 2014; 39(10):839.

136. Pichugov RD, Makhaeva EE, Keshtov ML. Fast switching electrochromic nanocomposite based on Poly(pyridinium salt) and multiwalled carbon nanotubes, Electrochim Acta. 2018; 260: 139–49.

137. Zhang W, Li H, Hopmann E, Elezzabi AY. Nanostructured inorganic electrochromic materials for light applications. Nanophotonics. 2021; 10(2):825–50.

138. Patel K, Bhatt G, Ray J, Suryavanshi P, Panchal C. All-inorganic solid-state electrochromic devices: a review, J Solid State Electrochem. 2017; 21(2):337–47.

139. Beaujuge PM, Reynolds JR. Color control in π-conjugated organic polymers for use in electrochromic devices. Chem Rev. 2010; 110(1):268–320.

140. Dyer AL, Thompson EJ, Reynolds JR. Completing the color palette with spray-processable polymer electrochromics. ACS Appl Mater Interf. 2011; 3(6):1787–95.

141. Gorkem G, Levent T. Electrochromice Conjugated Polyheterocycles and Derivatives-Highlights from The Last Decade towards Realization of Long Lived Aspirations. Chem Commun. 2012; 48: 1083.

142. Yin Y, Lan C, Guo H, Li C. Reactive sputter deposition of WO3/Ag/WO3 film for indium tin oxide (ITO)-free electrochromic devices. ACS Appl Mater Interf. 2016; 8(6):3861–7.

143. Li X-K, Ji W-J, Zhao J, Wang S-J, Au C-T. Ammonia decomposition over Ru and Ni catalysts supported on fumed SiO2, MCM-41, and SBA-15. J Catal. 2005; 236(2):181–9.

144. Yang P, Sun P, Mai W. Electrochromic energy storage devices. Mater Today. 2016; 19(7):394–402.

145. Wang JM, Sun XW, Jiao Z. Application of nanostructures in electrochromic materials and devices: recent progress. Mater. 2010; 3(12):5029–53.

146. Monk PM, Mortimer RJ, Rosseinsky DR. *Electrochromism and electrochromic devices.* 2007; Cambridge: Cambridge University Press.

147. Yen H-J, Liou G-S. Solution-processable triarylamine-based electroactive high performance polymers for anodically electrochromic applications. Polym Chem. 2012; 3(2):255–64.

148. Osterholm AM, Shen DE, Kerszulis JA, Bulloch RH, Kuepfert M, Dyer AL. *et al.* Four shades of brown: tuning of electrochromic polymer blends toward high-contrast eyewear. ACS Appl Mater Interf. 2015; 7(3):1413–21.

149. Lyu H. Triple layer tungsten trioxide, graphene, and polyaniline composite films for combined energy storage and electrochromic applications. Polymers. 2020; 12(1):49.

150. Zhang K, Wang Y, Ma X, Zhang H, Hou S, Zhao J. *et al.* Three dimensional molybdenum oxide/ polyaniline hybrid nanosheet networks with outstanding optical and electrochemical properties. New J Chem. 2017; 41(19): 10872–9.

151. Da Rocha M, He Y, Diao X, Rougier A. Influence of cycling temperature on the electrochromic properties of WO3//NiO devices built with various thicknesses. Sol Energy Mater Sol Cells. 2018; 177: 57–65.

152. Danine A, Manceriu L, Fargues A, Rougier A. Eco-friendly redox mediator gelatin-electrolyte for simplified TiO2-viologen based electrochromic devices. Electrochim Acta. 2017; 258: 200–7.

153. Levasseur D, Mjejri I, Rolland T, Rougier A. Color tuning by oxide addition in PEDOT: PSS-based electrochromic devices. Polymers. 2019; 11(1):179.

154. Atak G, Pehlivan IB, Montero J, Granqvist CG, Niklasson GA. Electrochromic tungsten oxide films prepared by sputtering: Optimizing cycling durability by judicious choice of deposition parameters. Electrochim Acta. 2021; 367: 137233.

155. Zhu J, Wei S, Alexander MJ, Dang TD, Ho TC, Guo Z. Enhanced electrical switching and electro-chromic properties of poly(p-phenylenebenzobisthiazole) thin films embedded with nano-WO3. Adv Funct Mater 2010; 20(18):3076–84.

156. Sharma SN, Pillai ZS, Kamat PV. Photoinduced charge transfer between CdSe quantum dots and p-phenylenediamine. J Phys Chem B. 2003; 107(37):10088–93.

157. Brust M, Walker M, Bethell D, Schiffrin DJ, Whyman R. Synthesis of thiol-derivatised gold nanoparticles in a two-phase Liquid–Liquid system. J Chem Soc Chem Comm. 1994; (7):801–2.

12 Electroactive Polymers for Batteries

Roger Gonçalves, Kaique Afonso Tozzi, and Ernesto Chaves Pereira

CONTENTS

12.1 INTRODUCTION

Portable electrical energy storage is a method that has increasing importance in different technological applications. Portable electronics are the most common; however, demands that require greater robustness, such as electric vehicles, are increasingly popular. Such broad applications have led to the need for a range of storage systems with different capacities for energy storage and delivery. Thus, batteries and supercapacitors have complementary properties.[1] In general, in energy-related electrochemical devices, for storage or conversion, two characteristics are of significant importance: the energy stored, which is given in watt-hour (Wh), and the maximum power, which is given in watts (W). The first translates how much energy the device can supply or transform. The second gives the amount of work performed in a given time period, which relates to mechanical power.[2] These properties are the main difference between supercapacitors and batteries, which are the most common electrochemical devices for energy storage.[3] Supercapacitors are capable of small charge storing and they can deliver it quickly (peak of power).[4] In addition, batteries can retain a large amount of charge, and release it gradually (long term) and at low power.[5] However, electronic equipment requires both types of charge storage devices to work together, obtaining a peak of power from a supercapacitor that is regularly recharged from a primary energy storage unit, for instance, a battery.

12.2 HISTORY

12.2.1 Batteries

Electricity has been known of since ancient times; although in 1800 equipment capable of creating it was developed (Figure 12.1). The first recorded device that resembled a battery was built approximately 2,000 years ago and was found near Baghdad in the 1930s. Known as the Baghdad Battery, it consists of

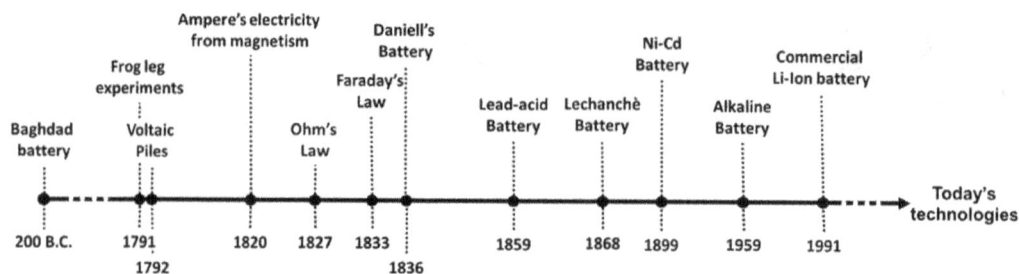

FIGURE 12.1 Timeline of major events in the history of battery development.

a clay jug with an iron rod inside a copper cylinder.[6] Observing how the metals were worn, the jug was probably filled with some standard acid solution, such as vinegar or lemon juice. However, there is no record of these devices actual use, whether they were batteries or who assembled them.[7]

During the 1770s, biologist Luigi Galvani conducted a series of experiments that were called "bioelectrogenesis".[8] The electric current caused the contraction of the leg muscles of a frog and many other animals, either by applying the charge on the muscle or the nerve. To theorize the origin of Galvani's phenomena in the experiment with frog legs, Alessandro Volta started a series of experiments that led to the invention of the first battery in approximately 1800.[9,10] The device consisted of a series of metal disks of two or more types, which were separated by cardboard disks soaked in acidic or saline solutions.

In 1836, French engineer Georges Leclanché invented the precursor of the battery we know today.[11] This was formed by a metallic zinc cylinder (anode) and a graphite cylinder covered by a layer of manganese dioxide and powdered coal (cathode). An ammonium chloride solution was used as electrolyte. Since the reactions involved were not reversible, the Leclanché battery is not rechargeable; therefore, the battery ceased to function when no more manganese dioxide was available consumed.[12] The first rechargeable battery created, which is still widely used today, was invented in 1859 by French physicist Gaston Planté.[13] The lead acid battery has an extremely low energy-to-weight, and energy-to-volume ratio, which does not make it attractive for portable electronics and its most popular use is in automobiles.[14] However, it has an enormous capacity to supply high surge currents; therefore, it has an excellent power-to-weight ratio, which explains its primary use in motor vehicle to supply a large amount of current to start engines.

Table 12.1 summarizes the properties of commercial batteries today. Of note, the properties change in a wide range of applications. Lead acid batteries are the best choice for automotive vehicles, and lithium batteries are common in portable rechargeable electronics. Finally, alkaline batteries are the cheapest to be used as single-use disposable batteries.

Today, the trend in battery development is the balance between power capacity, low cost, chemical and mechanical stability, and low weight. Furthermore, the use of environmentally friendly materials is a vital characteristic of these devices.[15] To develop innovative technologies, such as wearable devices, auto robots, and Internet of Things artifacts with all the previously mentioned characteristics, the research into organic polymers for batteries started to increase as these materials might address these needs, since they could have all of them. Although the most challenging task is to modulate the ionic conductivities at a value high enough to enable polymer batteries' usage at room temperature, there are many physical and chemical modifications of these materials structure that allows their application.[3]

12.2.2 POLYMERS

The history of polymers is quite extensive, it will be briefly summarized, along with the techniques that allowed the discovery of conducting polymers. Polymeric materials have many significant advantages compared with different ones, and they have been known since 1830.[16] The polymer concept was

TABLE 12.1
Energy storage properties of popular commercial batteries[14]

Model	Cell voltage (V)	Capacity (mAh)	Specific energy (Wh/Kg)	Specific power (W/Kg)
Leclanché	1.5	400–1,700	55–75	
Acid-lead	2.1	*	35–40	180
Alkaline	1.5	1,800–2,850	*	*
Ni-Cd	1.2	600–1,000	40–60	150
NiMH	1.2	600–2,750	60–120	250–1,000
Li-ion	3.6	600–840	100–265	250–340

* Depends on format, model, or size, or both

established mainly in 1930 with Staudinger, Kuhn, and Carothers's research.[17] The main advantages of polymers are low density, processability, chemical stability (compared with metals), and various applications.[18] For electrical properties, polymers are insulating. However, after the discovery of conductive polymers (CPs) by Alan MacDiarmid, Hideki Shirakawa, and Alan Heeger,[19] this has changed through the years. Electrically CPs present mechanical properties of conventional polymers and electrical conductivity compared with metals.[20] Under specific synthesis conditions, these polymers can exhibit approximately 105 S/m in conductivity, and copper has 107 S/m.[21] CPs have been extensively researched because, in addition to exhibiting high conductive values (and they are still considered semiconductors) they maintain lightness, processability, and mechanical properties of conventional polymers.[22] The discoverers of CPs won the Nobel Prize for Chemistry in 2000.[19,23]

Another name for a CP is conjugated polymers, because they have unsaturated bonds interspersed with simple bonds (C=C–C), this structural property is responsible for the electrical conductivity,[24] when electronic defects are introduced to the structure. In polymers with a saturated structure, the four valence electrons of the carbon atom undergo hybridization, forming sp^3-type orbitals, which allows carbon to make four σ-type covalent bonds.[25] However, in conjugated polymers, hybridization leads to the formation of sp^2-type orbitals, which allows carbon to bond to three other atoms.[25] Therefore, three of the four valence electrons are located in σ-type orbitals. In contrast, the remaining electron remains in a p_z-type orbital, making it possible to form a double bond with another atom (type π).[24] Since the π bond characterizes a state of greater electronic relocation, added to the macromolecule's long structure, there is the formation of π bands that in turn give a metallic or semiconductor characteristics to the conjugated polymer, which depends on their level of occupation.[26] Therefore, unlike conventional polymers with conductive particles, CPs present their conduction as an intrinsic property.[24] Today, polymers are associated with electronic and technology devices, because these materials are not used to attain the mechanical demands.[27]

12.3 SYNTHESIZING POLYMERIC FILMS

In this section, some of the most used techniques to manufacture polymeric films will be discussed. Obtaining CPs is simple, fast, and cheap. It can be achieved electrochemically or chemically. Each method has a set of parameters that can be optimized to improve specific properties because the conducting polymers is quite sensitive to the method and variable values used in the synthesis.[28–33] These changes occur due to variations in the density of the electronic defects, in addition to alterations in the chain size, and in the folding of the whole material. Therefore, films can be obtained for specific applications by controlling the synthesis variables and modulating the polymer properties.[34] As noted by Maia et al.,[35] the monomer concentration, temperature, and nature of the electrolytic medium are some of the parameters that can be controlled to obtain polymers with suitable properties for a given application. Besides, with the adjustment of some simple parameters, such as temperature, electrical potential, or current values applied, or both during electrochemical formation, the monomer, and electrolyte concentration, it is possible to control the morphology, polymerization kinetics, and therefore, adjust properties as desired.[30,36,37]

Electrochemical synthesis is interesting because it allows thin films to be produced on an appropriate substrate in a short time.[38] In addition, homogeneous films can be obtained with high chemical stability in most solvents.[39,40] This approach is environmentally friendly, because it produces a smaller amount of waste.[41,42]

Polymer casting from the solution is a manufacturing process that is used to make flexible polymeric components.[43] It is very cheap, and another advantage of this method is the potential for large production yields. However, some disadvantages, such as the polymer needs to be easily solubilized, which is not the case for most CPs, such as polythiophenes without a side group, and polypyrrole (PPy). Although modifications to the polymer can improve its solubility, changes in its properties and manufacturing costs are inevitable.

The layer-by-layer[44] (LBL) deposition technique can be used for the manufacture of thin films. The films are formed by depositing alternating layers of materials with opposite charges, with washing steps between them. Oppositely charged materials adhere to each other due to electrostatic interactions. The washing step is used to remove material that has not electrostatically adhered to the previous layer and allows a thin layer of material to be deposited in a uniform and structured manner. The LBL can be carried out using various techniques, such as immersion, rotation, spray, or any other allied technique that allows the deposition of small amounts of material alternately.[45] This technique is fascinating, because it allows the film to be made in a very orderly manner, which leads to the enhancement of properties. However, due to the small amount of deposited material on each layer, it is usually necessary to repeat the deposition steps several times, then the film formation velocity.

12.4   POLYMERS AS REDOX MATERIALS

Polymeric chains can present structural defects either due to the way they are synthesized or by the processes that lead to premature aging, chain twists, and unconjugated segments.[46] The most common types of defects are the cis defect (i.e., chain twist), the sat defect (e.g., saturation of the double bond and breakdown of conjugation), and superoxidation (i.e., oxidation of the double bond to carbonyl). However, the cis defects are the most worrying, as sat and oxidation defects can be eliminated during synthesis.

These defects, or torsions, or both hinder (or even prevent) the superposition of the molecular orbitals to extend along the entire chain. Therefore, the chain is composed of several segments, and depending on the conjugated segment length there are wells of different lengths, which directly influence the band gap value.[47,48]

The charge carriers are confined to different energy levels leading to a heterogeneous electronic structure along the polymeric chain. This makes the absorption or luminescence spectra comprehensive.[24] It causes a structural rearrangement of the polymer chain and an electronic rearrangement during the redox process. Consequently, it has a significant impact on the properties of the polymer immediately and longterm.[26]

12.5   ELECTROCHEMICAL AGING OF CONDUCTING POLYMERS

The natural aging of CPs that occurs due to several oxidation or reduction cycles must be discussed. The term aging refers to the degradation in the conductivity and electroactivity of a CP in the oxidized state due to the material reaching very positive potentials, which is a state called "superoxidation" of the polymer.[49,50] In this state, polymer degradation occurs due to the breakdown of chemical bonds, or reaction or both with nucleophiles present in the solution.

One of the first published works to study the variables that affected the polymer degradation was described by Tourillon,[51] who subjected thiophene films to an electrochemical aging process under different experimental conditions, such as the absence and presence of water and different pH values. In 1991, Harada[52] demonstrated through several redox cycles in the superoxidation region, that it could no longer be recovered once the films degrade. There is a gradual decrease in the current as the voltametric cycles are repeated. Wang[53] related the process of deactivating the poly(3-methyl thiophene) with the solvent's pKa value. Solvents with higher pKa values accelerated the deactivation process and the presence of water in the medium, which acted with a strong nucleophile.

Whenever a CP is used as a redox material, electrochemical aging must be considered, either by superoxidation or by the extensive use of the redox reaction. Therefore, studies in this area are required.[54] This section briefly discusses studies into the use of CPs as active materials in electrochemical devices.

The use of electrochemical impedance spectroscopy (EIS) in this type of investigation was described in 1999.[55] The work involved evaluating the impedance parameters of thiophene films in lithium trifluoromethanesulfonate, treating the EIS data using the equivalent circuit model, and considering the polymer morphology, electrode geometry, and ion intercalation process. The electrochemical aging of poly(3-hexylthiophene) using EIS data was studied.[56,57] In this work, the authors compared the effect of aging films that were obtained using different parameters of electrosynthesis. The initial properties interfered with the shape and speed as the films aged, and therefore, they did not depend on the morphology, but on the thickness, since films of less thickness ended up losing the essential properties quickly. Similarly, films with better starting properties lose properties faster than those with lower properties.

In this section, some works related to electrochemical aging were discussed. Their importance was considered when analyzing new materials for energy storage since the main property required for this purpose is the redox properties of CPs. Understanding the natural aging of polymers is a crucial step in the development of new materials. Therefore, this aspect of aging will be addressed in this chapter.

12.6 IMPEDANCE SPECTROSCOPY AS A CHARACTERIZATION METHOD

In this section, a brief history of the development of models for the treatment of EIS data describes the unique characteristics of electrical and ionic conductance of CPs.

EIS is a characterization technique that uses Ohm's Law[58] for real physical systems disturbed by AC to characterize various processes that occur in the mediation and inside a material deposited under a current collector (modified electrode).[59] EIS can provide information on the kinetics of processes on the electrode surface and the double electrical layer structure; therefore, being quite versatile in its application.[60]

The impedance in electrochemical systems is usually measured by applying a sinusoidal waveform potential to an electrochemical cell, and an alternating current signal is obtained in response.[60] This applied disturbance is generally from 5 to 10 mV in amplitude; therefore, ensuring that the cell response was linear.[58] The current response for a sinusoidal potential will be a sinusoid for the same frequency altered phase in a linear system.

In addition, the use of equivalent circuits is most used to describe the impedance spectra obtained. In this model, circuit elements are used to describe the processes that occur on an electrode's surface, which models well flat or well-polished electrodes. However, it is more usual to use than an ideal parallel plate capacitor, a constant phase element (CPE). A CPE is used to describe a non-ideal capacitor.[61] As the name suggests, it always has a constant phase, in which $n=1$ represents an ideal capacitor and $n=0$ represents a pure resistor.

This representation is straightforward and often does not correctly translate what is actually occurring around the electrode, for example, a diffusion-controlled processes, which was explained in a circuit proposed by Randles, in which an element that does not have an analog in an electronic systems called the Warburg Element (Z_w),[62] is a CPE ($\varphi=45°$) and with a magnitude inversely proportional to the frequency value; therefore, at low frequencies, where $\omega\rightarrow 0$, the transport processes of the electroactive species contributes significantly.

Numerous circuit elements can make up an equivalent circuit to become complete and closer to the phenomena that occurs in some systems. However, they are complex and work very well to describe polished or rough electrodes. In porous electrodes, which are types in which the films of CPs fit together, another model is required to describe the porosity of the system. Therefore, the transmission lines stand out among the possible forms of treatment for impedance spectra.[63,64]

To understand the reasons that lead to one form of data processing in relation to another, it is necessary to differentiate rough electrodes from porous electrodes. For a rough surface,[65,66] (Figure 12.2(a)), at the interface there is a CPE that can represent the roughness of the material or some other factor that can lead to the existence of interfacial impedance.[67–69]

Considering a generic model, the electrolyte impedance in which the electrode is located is χ_1, which is represented by the resistance of the solution between the surface of the material and the reference electrode, and χ_2 is the impedance of the compact rough material deposited on the substrate. In ζ, the existing processes on the electrode surface and the CPE are usually represented, there are resistances associated with electronic load transfer processes. This type of approach works very well for rough electrodes, because it is possible to approximate that the surface processes occur identically throughout the material, obtaining simple equivalent circuits.[62]

The same analogy can be made for porous electrodes (Figure 12.2(b)) since the processes are calculated locally and the interface between the rough material and the electrolyte. As the flow of electrons is the same across the surface and the potential difference is found in it, this region's impedance can be described by a combination of resistances and capacitors in series, or in parallel, or both, which describes the electrochemical processes at this interface. The behavior of electrochemical phenomena in porous electrodes can be explained using equivalent circuits; however, based on a different approach. This is true for pore electrodes, because it is necessary to take geometry into account when analyzing impedance spectra.[70] This is possible with a model in which there is a coherent physical description for this type of response's geometric effect. Unlike in rough electrodes, the holes are of such magnitude that they can be considered as channels that are scattered throughout the material. Figure 12.2(b) shows a porous electrode on a microscopic scale, in which two phases are considered, the electrolyte and porous material.

Of note, elements of equivalent circuits are used generically to represent the electrical transport characteristics between both phases, for example, χ_1 and χ_2 describe the ohmic drop at each point along the channels. The ζ element describes the charge transfer at the pore interface and is related to faradaic and polarization currents. Because the branching of the elements of equivalent circuits is continuous, the calculation occurs through differential equations, which makes the system's total impedance physically analogous to a transmission line.

The current resulting from the AC disturbance in the two interfaces can flow in either direction along both methods considered. Therefore, the system's total electrical current is the sum of all currents that pass through the electrode. In a particular case where ionic and electronic conductivities are comparable, for example, χ_1 and χ_2 have comparable impedances, and the system can be represented by two-channel TL, which describes the transport in the solution and electrode, and transport through the interface. Therefore, χ_1, χ_2, and ζ must be considered; therefore, the adjustment of impedance spectra can physically and coherently describe electrochemical phenomena.[71,72] This can be achieved using equivalent circuit elements analogous to the processes that might be occurring.

This type of approach to CPs has been described several times, since the redox process in these materials occurs in conjunction with the polymer's ionic deintercalation process.[73] In this case, mass transport does not depend solely on diffusion processes, and anomalous responses are seen during EIS measurement.

Using the transmission line approach, [74] ionic and electronic transport in poly-[1-methyl-3-(pyrrole-l-ylmethyl)pyridinium]$^+$ electrodes by modifying the conductivity of the polymer based on the thickness of the deposited film were correlated. In addition, the film's behavior was verified in two different non-aqueous media, acetonitrile, propylene carbonate, and water. Analyzing the behavior in the different media, in the reduced and oxidized states of the polymer, when a polymer is in a reduced state and is insulating, and the ionic transport is maximum. This fact was associated ionic deintercalation, which is hindered by the transport of polarons and bipolarons in polymer chains. This explanation makes sense because ions must be intercalated to ensure neutrality in the polymer matrix that is disturbed when creating these species due to polymer oxidation.

The cationic and anionic transport in PPy/poly(styrene sulphonate) composite films was studied.[75,76] Such studies are crucial for developing innovative technologies for polymeric batteries since the amount of charge stored by these devices is related to the volume of intercalated ions, as will be discussed in the following sections.

Dealing with models of transmission lines applied to porous materials, the quality of the adjustment of experimental data for PPy and poly(3-methylthiophene) using Randles model and three

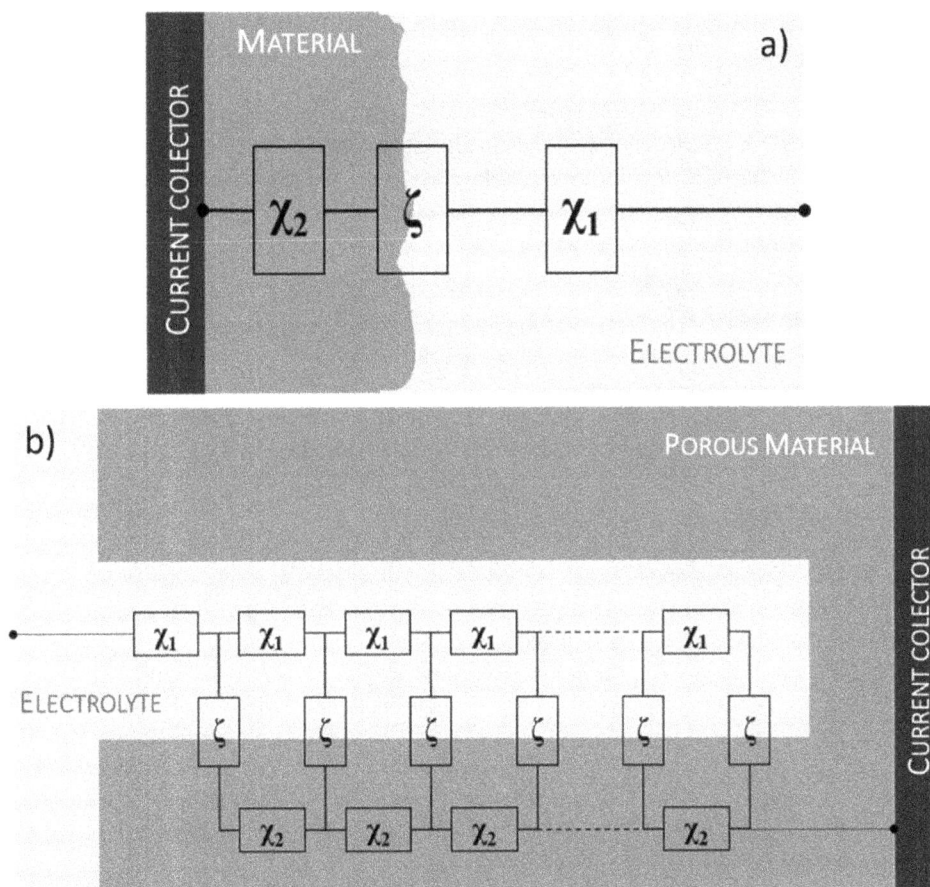

FIGURE 12.2   Generic model for: (a) rough electrode; and (b) porous electrode.

transmission line models for linear diffusion and linear diffusion–migration and an adaptable system were compared. Comparing the results, the biggest differences in the quality of the adjustment were in the low-frequency region, which is the region that describes ionic transport.

In 1998,[77] the anomalous behavior of diffusion in electrodes with a blocking interface with intercalation of charged species were studied. In 1999,[78,79] together with other researchers, a model of impedance treatment with transmission lines for metallic electrodes covered with a polymer film that presented finite diffusion was proposed. In this model, low-frequency dispersions could be explained due to mass transfer and storage. This research dealt with CPs that could be explained very well in the model mentioned under conditions in which dispersion occurs.

Then, models were developed based on the characterization of CP films using EIS. One of the motivations for this model to be proposed was, although several works in the literature used different models, all treated polymers in two dimensions and considered a single interface (i.e., polymer/ electrolyte). However, although this model was more complete and more valuable for adjusting experimental data, as tested in the research itself with a poly(thiophene acetic acid) film, it does not allow for a good fit in a mixed conduction condition, which is when the polymer is in transition from insulator to conductor. However, in other research, this transition region was studied to improve the model previously proposed. In an independent study,[80] a complete model for the description of homogeneous and a rough two-phase system for CPs was proposed.

In 2011,[81,82] an additional pseudo capacitance was introduced to describe mass transport processes through the pore and polymer interface. The model has a resistor and CPE in series, in parallel with the double layer's capacitance. This resistor represents the charge transfer, which in the polymer is translated as resistance to the ionic charge transfer of the counter ions through the solution and polymer interface to neutralize the electrical charges generated by oxidation. In addition, the additional capacitance is related to the slowness of the ion transport process through the polymer and the time needed to find a suitable place for the interaction. This description is essential, because the transfer of charge in interleaving materials involves the transfer of mass, a process that is much slower than the transfer of a carrier of negligible mass (e.g., electron and hole). In addition, from the insertion of this other circuit element, other behaviors could be explained and related to other phenomena.

12.7 STATE OF THE ART

12.7.1 Pristine Polymers

Through the study of new composites, the efficiency of devices can be improved, and the study of materials in their pristine form is of paramount importance to understand their properties and the limits of development. In addition, the study of a pure material allows for the development of new manufacturing techniques and studies the effects of obtaining new morphologies. Therefore, this section presents some relevant works, and some PANi-based devices are summarized in Table 12.2, it is possible to see the influence of the morphology in the performance of the material.

Counterbalancing low ionic conductivity below $80°C$ without losing mechanical stability is something that many researchers have investigated. Sidechain modifications with crosslinking of chemical groups that can be ionic plasticizers, such as allyl-ethers in polytrimethylene carbonates could raise the ionic conductivity by one order of magnitude,[85] since the plasticizing effect decrease the T_g and can improve the ion-pair dissociation, which is related to ionic conductivity. This method increases the free volume; however, it is associated with loss of mechanical stability. Therefore, tailoring the backbone is a trend because it can enhance the ionic conductivity by two orders of magnitude[86,87]; however, the challenge when using this approach is to maintain the polymer with an environment-friendly label.

Nanotechnology is one of the most exciting polymer science subjects, and it is common to see nano applications in the battery field. Silica nanoparticles are used to design a crosslinked structure

TABLE 12.2

Performance of some pristine PAni-based materials for energy storage

Material	Preparation method	Capacitance (F/ g)	Capacitance Retention	CnD cycles	Ref.
PAni nanowires	electrodeposition and freeze-dry treatment	950	84%	100	[83]
PAni nanowires (in ionic liquid)	electrodeposition and freeze-dry treatment	440	88%	500	[83]
PAni nanobelts	12 h of electrodeposition at 70°C	873	93%	1,000	[84]

where the polymer provides the ionic conduction pathways. The nano silica enhances the mechanical properties and oxidative stability, which is known as soft colloidal glass.[88] Similarly, titanium and aluminum oxide additions in polymer-salts could make the ionic conductivity achieve values of 10^{-4} S/cm at 50°C.[89,90]

On the chemical stability demand, polyoxyethylene (POE) is a suitable polymer since it has some stability at 3 V, and it is difficult to surpass its conductivity. Although promising, multi-phase polymers need to be better understood for their conductivity mechanisms through the different phases and how each phase could be tailored to design the electrochemical properties.[91]

Another approach that can be used to tune up the electrical properties of polymers is ion selection. Since lithium (Li^+) and sodium (Na^+) have similar physicochemical properties, these ions can be used. However, several studies showed that although the smaller radius and higher chemical affinity with sulfonic and oxygen groups, the lower Lewis acidity of Na^+ can facilitate mechanisms, such as desolvation through ionic microchannels.[92] This is a trend seen in inorganic batteries, such as NASICON.[93] Ionic liquids can be used in this approach, but for other purposes, such as providing electrochemical stability or as electrolytes for a broad electrochemical window. The mixing of electrolytes could be of interest in the future, but theoretical studies need to be combined with experimental data; therefore, it could be possible to model these devices with the electrolytes of interest.

Although the polymers can be used in all the ways discussed previously, the problems associated with their sole use cannot be solved without using other materials to enhance their properties. Therefore, hybrid materials will be discussed in the following section, which are composed of polymer and nanomaterials, or oxides, or both and could be strategy for future batteries.

12.7.2 Composite Materials

Several studies investigated property and lifetime improvement. Although hybrid materials have good durability, it is essential to note that organic materials have advantages because of their low density and low cost and are sustainable materials.[94] Therefore, the combined use of CPs and advanced carbon compound composites, such as nanotubes or graphene, is a promising combination to build electronic and electrochemical devices.[95] Despite the poor adhesion of the carbon films, CPs improve the conductivity and mechanical strength and promote an increase in surface area. The in situ polymerization represents a promising approach from an environmentally friendly point of view. The monomer is polymerized in the other components' presence and avoids additional experimental steps, which leads to a homogeneous composite.[96]

Electrochemical polymerization has advantages over chemical synthesis for the preparation of homogeneous films with controlled morphology and high chemical stability in several solvents.[39,40] In addition, this approach is considered to be environmentally friendly, because it markedly reduces

TABLE 12.3

Some materials of interest found in the literature and their properties

Material	Preparation method	Capacitance (F/ g)	Capacitance Retention	CnD cycles	Ref.
Ppy/MWCNTs on ceramic fabrics	chemical polymerization	282	85%	5 000	95
Ppy/ZnO/MWCNTs	electro-co-deposition	279	99%	1 000	99
Ppy/V$_2$O$_5$	electro-co-deposition	412	80%	5 000	100
Ppy/rGO	chemical polymerization	375	87%	4 000	101
Ppy/rGO	electro-co-deposition	152	88%	10 000	102
Ppy/GO:CNT-COOH	electro-co-deposition	142	97%	5 000	96
PEDOT/GO:CNT-COOH	electro-co-deposition	99	99%	5 000	96
POMA/PTAA	LBL deposition	140	99%	3 000	103
PAni nanorods/rGO	Electrodeposition on rGOmicroelectrodes	970	90%	1 700	104

the amount of waste generated in the process.[97,98] Table 12.3 summarizes some organic based polymers deposited in different substrates reported in the literature. Furthermore, the capacitance retention and charge–discharge cycles are presented for comparison.

To build supercapacitors using CP, PANi, and its derivatives are the most studied. PANi-based organic supercapacitors exhibit exceptional capacitance values. A nanostructured PANi material, using a commercial carbon cloth as substrate was developed.[105,106] Considering that PANi is the active material, the normalized values using PANi mass, which was a specific capacitance of 1,079 F/g with a retention of 86% after 2,100 CnD cycles. A macroporous carbon (MC),[107] was prepared to act as a substrate for the electrodeposited PANi. This composite showed 1,490 F/g of capacitance and retention of 89% after 1,000 cycles normalizing the data by PANi mass. If all the composite mass was considered, the capacitance decreased to 480 F/g.

Similarly, PANi deposited on hierarchically porous carbon monolith (HPCM),[108] was studied and considering the PANi mass, a capacitance of 2,200 F/g was achieved. However, normalizing the whole composite mass, this value was 490 F/g, with retention of 93% after 500 CnD cycles. These studies were groundbreaking considering the high values of capacitance obtained and represent a step forward for new and sustainable supercapacitor devices; however, it is essential to note that the work did not provide a full explanation of the capacitance values.

Apart from the techniques and polymers already discussed, POE is the most promising. Its low cost and environmentally friendly characteristics address the need for versatile materials, and combine the easiness of production with interesting electrochemical properties that can be applied in solid state batteries, which is an example of its high ionic conductivity.[109,110]

Besides its high charge energy, batteries made from this material must have electrodes with high capacity. Many researchers attempted to develop nickel-manganese-cobalt oxide (NMC) as an electrode due to its high charge capacity and developed batteries made from POE/NMC.[111–116] However, problems such as poor performance from an applied electrical potential of 4 V limited their use.[117,118] Furthermore, at 4.3 V, the polymer started to decompose, and the instabilities of the electrode on cycling that arose from nickel dissolution and phase transformation have reduced the applications. Most of the studies tried to apply coatings or doping elements to increase this hybrid material's stability.[119–124] Based on this, the atomic layer deposition (ALD) of lithium niobium oxide in the interface between an NMC electrode and solid state POE was studied. The electrode was assembled using a doctor blade and an ALC reactor, coating the Li particles directly.[125] This layer was chosen because of its stability and high ionic conductivity, and the technique provided a uniform layer in a low-temperature process. Through electrochemical measurements, such as EIS and CV, the stability of the ALD-NMC-POE was compared with the bare NMC-POE with 200 cycles of charging and discharging, where the stability was increased, and the midpoint voltage after all cycling was 20% higher than the raw material.

Furthermore, due to higher capacity retention, ALD-NMC-POE had a Coulombic efficiency of 99.2% after 200 cycles, and the raw material had 92.3%, which indicated enhancement of cathode stability. The EIS results corroborated these characteristics, since the cell resistance of the raw material increased from 4 kΩ after 30 cycles (e.g., charging and discharging) to 10 kΩ after 200 cycles, which was related to the decomposition of the NMC electrode and polymer, and the ALD material was approximately 1.5 kΩ and 2.8 kΩ after the same amount of cycling. The main conclusion was that lithium niobium oxide has good ionic conductivity and poor electronic conductivity, and the oxidation window is broad. Therefore, the electrochemical properties could stabilize the interface, for instance, the oxygen release was quite reduced. In addition, the low electronic conductivity might reduce the electrochemical decomposition of POE at voltages >4 V. In conclusion, an ALD with lithium niobium oxide appears to be an excellent strategy to attain electrochemical stability and to enhance hybrid material battery efficiency.

As crucial as they search for new composites is developing more advanced techniques, it is also accessible for obtaining increasingly functional composites. Therefore, for a long time the LBL assembly technique was a method to produce self-assembled films with unique characteristics.

In 2004,[126] the effects of self-doping of self-assembled films using the LBL technique of poly(o-methoxyaniline)/poly(thiophene acetic acid) composite (POMA/PTAA) was studied. In this work, a film with 80 layers interleaved from both polymers (40 bilayers) were described and monitored the increase in the number of bilayers by UV-Vis spectroscopy. The amount of mass transport of the films obtained versus PANi films was evaluated. For comparison, some precautions were taken: (1) in the studied potential window, the PTAA had no electrochemical activity; therefore, it did not undergo oxidation or reduction and, therefore, did not intercalate ions; and (2) the PANi mass in both samples was maintained the same. Therefore, using quartz crystal electrochemical microbalance (EQCM) data, the extent of ion intercalation during the redox process of both samples was evaluated. In the composite film, the injection of ionic species was different from a pure POMA film. For the LBL film, oxidation caused the positive charges created in the POMA to be readily compensated for by the sulfonic groups ($-SO_3^-$) present in the PTAA, which caused positive species to leave, which had previously neutralized the $-SO_3^-$ groups alongside the entry of negative species; however, this intercalation or deintercalation occurred in the layers of PTAA.

During oxidation, the bare POMA film requires anions to be injected into the film to neutralize the charges created. During reduction, in both samples, the inverse intercalation or deintercalation process was observed. The simultaneous intercalation of positive and negative species in self-doped films has been reported for poly(sulfonated aniline) films. However, this differentiated ion has important implications for the film's general properties, such as increased electrochemical response and electrical conductivity.

Using the same film-making technique, obtained a new capacitor material was obtained.[127] The performance of composites was evaluated using EIS and charge/discharge curves. An increase in the electrodes' capacitance as a function of the number of deposited bilayers was observed and compared with pure POMA films produced via casting with a similar electroactive mass. As the specific capacitance increased linearly with the number of layers, the same was not observed for films produced via casting, which lost capacitance as the amount of mass on the electrode increased. In addition, with the same amount and electroactive mass, the self-assembled films' capacitance was far superior to that of the POMA film. A explanation was found using EIS data, the composite film had so little resistivity compared with the electrolyte branch, which was considered to be a short circuit, and the POMA film did not have this characteristic. As the amount of polymer increased, the resistivity of the film also increased, which directly influenced the capacitance of the material.

In the literature, devices based on polymers have low durability due to their inherent electrochemical aging process. Therefore, the retention of the capacitive property of the self-assembled films already described was evaluated.[103] The forced aging test was carried out via 3,000 cycles of charge and discharge at 1.5 A/g, the aging was monitored in addition to the capacitance, the resistance of the electrode, the change in the impedance spectra, the queries was treated with a model of transmission line previously proposed. The composite film maintained capacitance properties during 3,000 charge/discharge cycles, with episodes in which there was a loss. However, after some time, this value was recovered, which indicated that the film had auto property and excellent stability. The film recovery is a significant property for industrial applications, because the device is expected to last several operating cycles without significant loss of performance. The POMA film did not withstand the first hundred of cycles, losing >50% of its properties due, as shown by the EIS data, to the significant increase in the polymer's resistance and the transfer to charge resistance.

Due to the peculiar and exciting behaviors observed for self-assembled films with self-doping effect, the material could be interesting for other applications, such as electrochromic devices, which demonstrates the enormous versatility of polymer conductors. Therefore, how the polymeric chains behaved during the redox cycles in these composites from the molecular point of view through DFT and molecular dynamics simulations was studied.[128] The theoretical data was supported by EIS data in the different redox states of the samples. Of note, during oxidation the polymeric chains of POMA twist and lengthen due to defects produced in the chains by the creation of the polarons/bipolarons. In addition, in the opposite direction, during reduction, the chains return to their initial conformation. This structural and conformational modification process caused the chains to suffer mechanical stresses that became irreversible after a certain point and caused the polymer's electrochemical properties to decrease drastically. However, although undergoing such structural alteration, self-assembled films undergo it much less extensively, which contributes to the significant increase in the number of redox cycles. This type of study is significant to outline strategies to minimize the natural electrochemical aging in CPs.

Similarly, using theoretical data combined with experimental data,[99] significant results were found for supercapacitors based on PPy containing zirconium nanoparticles and carbon nanotubes. This ternary compound showed a specific capacitance value of 279.2 F/g with the durability of 1,000 charge/discharge cycles, and pure PPy lost approximately 77% of its capacitance after 100 cycles. In this work, the theoretical data were used to explain the mechanical strength of the reduced graphene oxide sheets of the composite and relate it to its durability in over the number of cycles. Through the simulations, mechanical deformations in the particles reduced due to other particles' insertion with better mechanical resistance to deformation. Therefore, this type of approach could be interesting for polymers.

Material was developed for supercapacitors using two conductive polymers with different aspect ratios, PPy, traditionally a one-dimensional material, and carbon nitride considered a two-dimensional CP. Both were codeposited by electrosynthesis of PPy in the presence of exfoliated carbon nitride in the synthesis solution. As previously discussed, codeposition and synthesis in one step are attractive due to their simplicity and high reproducibility. Excellent results were obtained with purely organic material, with high cyclability, 92% retention of its initial value after 6,000 charges/discharge cycles. The obtained capacitance value was 810 F/g. Of note, both materials' union resulted in a new three-dimensional material, which was highly porous and with an ion intercalation capacity 30% greater than the pure polymer. In addition, to explain these results, quantum mechanical theoretical methods were used, which showed an important PPy planning effect, leading to a significant increase in the composite's electronic conductivity. These data were corroborated by electrochemical impedance and EQCM data.

The use of mechanical–quantum or even classical computational methods combined with experimental results is vital for developing new materials that are robust explanation and to predict more innovative materials. In this way, it is salutary that this approach becomes more common. There could be more empirical–theoretical works, because this work is important and takes research and development to a new level.

12.8 THE NEXT CHALLENGES

To build and optimize polymeric materials for charge storage, the main challenge is to control the film aging process. This implies that there is a method to investigate this process. In addition, there must be a unified protocol for the characterization of new materials and devices aimed at energy. Today, it is difficult to compare them, because they use different standards, for example, area, mass, or different ranges of charge density in the charge/discharge experiment. Therefore, in this section, an analysis is made of the important parameters when choosing the best unified protocol to characterize electrical energy storage devices. Finally, new materials, which use composites or even low cost high building rates must be explored to optimize film properties.

In addition, there are many studies in the literature that have attempted to improve the properties; however, there is a lack of more basic and fundamental studies, which increase the lifetime operation of the devices based on CPs; therefore, to be commercially viable, they must have operating times >10,000 cycles.

REFERENCES

1. Nguyen TH, Fraiwan A, Choi S. Biosens Bioelectron. 2014; 54:640–9.
2. Pinnangudi B, Kuykendal M, Bhadra S. Power Grid. 2017; 93–135.
3. Arico AS, Bruce P, Scrosati B, Tarascon J-M, van Schalkwijk W. Nat Mater. 2005; 4: 66–77.
4. Wang G, Zhang L, Zhang J. Chem Soc Rev. 2012; 41:797–828.
5. Grande L, Chundi VT, Wei D, Bower C, Andrew P, Ryhanen T. Particuology. 2012; 10: 1–8.
6. Konig W. Forsch und Fortschritte. 1938; 1:8–9.
7. Keyser PT. J Near East Stud. 1993; *52:* 81–98.
8. Schoffeniels E, Margineanu D. Mol Basis Thermodyn. Bioelectrogenes. 1990; 5: 125–42.
9. Pancaldi G. in *Volta*. Princeton University Press, 2018; 73–109.
10. Elliott P. Notes Rec. R. Soc. Lond. 1999; *53,* 59–78.
11. Kordesch K, Taucher-Mautner W. in Encycl Electrochem Power Sources. Elsevier, 2009; 43–54.
12. Kordesch K, Taucher-Mautner W. in Encyc Electrochem Power Sources. Elsevier. 2009; 784–795.
13. Pavlov D. in Lead-Acid Batter. Sci. Technol. Elsevier. 2017; 3–32.
14. Enos DG. in Adv Batter Mediu Large-Scale Energy Storage. Elsevier. 2015; 57–71.
15. Zhang Y, Liu J, Wu G, Chen W. Nanoscale. 2012; 4: 5300.
16. Feldman D. Des Monomers Polym. 2008; 11:1–15.
17. Friedel R. Polymer Pioneers: A Popular History of the Science and Technology of Large Molecules. *Peter J.T. Morris*, 1987.
18. Utracki LA. Polym Eng Sci. 1995; 35: 2–17.
19. Shirakawa H. Rev Mod Phys. 2001; 73:713–18.
20. Shirakawa H, Chiang C, Park Y. Phys Rev Lett. 1977; 39: 1098–101.
21. Warren MR, Madden JD. J Electroanal Chem. 2006; 590: 76–81.
22. Balint R, Cassidy NJ, Cartmell SH. Acta Biomater. 2014; 10; 2341–53.
23. Shirakawa H. Synth Met. 2001; 125:3–10.
24. Faez R, Reis C, Freitas P, Kosima O. Química Nov Na Esc. 2000; 13–18.
25. Lee JD. Química Inorgânica Não Tão Concisa .1996.
26. Medeiros ES, Oliveira JE, Paterno LG, Mattoso LHC, Consolin-Filho N, Paterno LG. *et al.* Rev. Eletrônica Mater e Process. 2012; 2: 62–77.
27. Li C, Bai H, Shi G. Chem Soc Rev. 2009; 38: 2397.
28. McCullough RD, Lowe RD, Jayaraman M, Anderson DL. J Org Chem. 1993; 58:904–12.

29. Amou S, Haba O, Shirato K, Hayakawa T, Ueda M, Takeuchi K. *et al.* J Polym Sci Part A Polym Chem. 1999; 37: 1943–8.
30. Feast WJ, Tsibouklis J, Pouwer KL, Groenendaal L, Meijer EW. Polymer (Guildf). 1996; 37:5017–47.
31. Otero TF. Polym Rev. 2013; 53:311–51.
32. Gaudoin R, Burke K. Phys Rev Lett. 2004; 93:173001.
33. Kabasakaloglu M, Kiyak T, Toprak H, Aksu ML. Appl Surf Sci. 1999; 152:115–25.
34. de Alves MR, Calado HDR, Matencio T, Donnici CL. Quim Nova. 2010; 33:2165–75.
35. Maia DJ, De Paoli M-A, Alves OL, Zarbin AJG, das Neves S. Quim Nova. 2000; 23: 204–15.
36. Yao Y, Hou J, Xu Z, Li G, Yang Y. Adv Funct Mater. 2008; 18: 1783–9.
37. Kaeriyama K. In: Nalwa HS, editor. Handb Org Conduct Mol Polym. 2 Wiley; 1997, p. 888.
38. Stenger-Smith JD. Prog Polym Sci. 1998; 23: 57–79.
39. Berkes BB, Bandarenka AS, Inzelt G. J Phys Chem C. 2015; 119: 1996–2003.
40. Waltman RJ, Bargon J, Chemie P, Bonn U, Germany W, Waltman RJ. *et al.* Can J Chem. 1986; 64:76–95.
41. Asavapiriyanont S, Chandler GK, Gunawardena GA, Pletcher D. J Electroanal Chem. 1984; 177:229–44.
42. Chao F, Costa M, Tian C. Synth Met. 1993; 53:127–47.
43. Siemann U. In Scatt Methods Prop Polym Mater. Berlin:Springer; 2005. 1–14.
44. de Avila ED, Castro AGB, Tagit O, Krom BP, Lowik D, van Well AA. *et al.* App Surf Sci. 2019; 488:194–204.
45. Gormley AJ, Chandrawati R, Christofferson AJ, Loynachan C, Jumeaux C, Artzy-Schnirman A. *et al.* Chem Mater. 2015; 27:5820–4.
46. Chen Y-S, Meng H-F. Phys Rev B. 2002; 66:035202.
47. Baughman RH, Shacklette LW. J Chem Phys. 1989; 90:7492–504.
48. Roncali J. Chem Rev. 1992; 92:711–38.
49. Lewis TW, Wallace GG, Kim CY, Kim DY. Synth Met. 1997; 84:403–4.
50. Debiemme-Chouvy C, Tran TTM. Electrochem Commun. 2008; 10:947–50.
51. Tourillon G. J Electrochem Soc. 1983; 130: 2042.
52. Harada H, Fuchigami T, Nonaka T. J Electroanal Chem Interfacial Electrochem. 1991; 303: 139–50.
53. Wang J. Electrochim Acta. 1997; 42:2545–54.
54. Marmisolle WA, Florit MI, Posadas D. J Electroanal Chem. 2012; 669:42–9.
55. Refaey SAM, Schwitzgebel G, Schneider O. Synth Met. 1999; 98: 183–92.
56. Goncalves R, Correa AA, Pereira R, Pereira EC. Electrochim Acta. 2016; 190: 329–36.
57. Goncalves R, Pereira EC, Marchesi LF. Int J Electrochem Sci. 2017; 1983–91.
58. Oldham KB, Myland JC. In: Fundam Electrochem Sci. Oxford: Wiley; 1994. pp. 357–93.
59. MacDonald JR. Ann Biomed Eng. 1987; 20: 1–364.
60. Aloui W, Ltaief A, Bouazizi A. Superlattices Microstruct. 2014; 75:416–23.
61. Shoar Abouzari MR, Berkemeier F, Schmitz G, Wilmer D. Solid State Ionics. 2009; 180:922–7.
62. Buck RP, Mundt C. Electrochim Acta. 1999; 44: 1999–2018.
63. Meyers JP, Doyle M, Darling RM, Newman J. J Electrochem Soc. 2000; 147: 2930.
64. Schlesinger M, editor. Modern Aspects of Electrochemistry, Number 43, New York: Springer New York; 2009.
65. De Levie R. Electrochim Acta. 1965; 10:113–30.
66. De Levie R. J Electroanal Chem Interfacial Electrochem. 1990; 281:1–21.
67. Lang G, Heusler K, Sadkowski A. J Electroanal Chem. 2000; 481: 227–9.
68. Zoltowski P. J Electroanal Chem. 2000; 481: 230–1.
69. Sadkowski A. J Electroana. Chem. 2000; 481: 232–6.
70. Paasch G, Micka K, Gersdorf P. Electrochim Acta. 1993; 38: 53–2662.
71. Fletcher S. J Electroanal Chem. 1992; 337:127–45.
72. Fletcher S. J Chem Soc Faraday Trans. 1993; 89: 311.
73. Goncalves R, Paiva RS, Pereira EC. J Solid State Electrochem. 2021; doi:10.1007/s10008-021-04938-6.
74. Ren X, Pickup PG. J Chem Soc Faraday Trans. 1993; 89:321–6.
75. Ren X, Pickup PG. Electrochim Acta. 1996; 41:1877–82.
76. Ren X, Pickup PG. J Electroanal Chem. 1997; 420:251–7.
77. Bisquert J, Garcia-Belmonte G, Bueno P, Longo E, Bulhoes LO. J Electroanal Chem. 1998; 452:229–34.

78. Bisquert J, Garcia-Belmonte G, Fabregat-Santiago F, Bueno PR. J Electroanal Chem. 1999; 475:152–63.
79. Garcia-Belmonte G, Fabregat-Santiago F, Bisquert J, Yamashita M, Pereira EC, Castro-Garcia S. Ionics (Kiel). 1999; 5: 44–51.
80. Paasch G. Synth Met. 2001; 119:233–4.
81. Fabregat-Santiago F, Garcia-Belmonte G, Mora-Sero I, Bisquert J. Phys Chem Chem Phys. 2011; 13:9083–118.
82. Simoes FR, Pocrifka LA, Marchesi LFQP, Pereira EC. J Phys Chem B. 2011; 115:1092–7.
83. Wang K, Huang J, Wei Z. J Phy Chem C. 2010; 114:8062–7.
84. Li G-R, Feng Z-P, Zhong J-H, Wang Z-L, Tong Y-X. Macromolecules. 2010; 43:2178–83.
85. Mindemark J, Imholt L, Montero J, Brandell D. J Polym Sc Part A Polym Chem. 2016; 54:2128–35.
86. Guzman G, Nava DP, Vazquez-Arenas J, Cardoso J. Macromol Symp. 2017; 374:1600136.
87. Fan L-Z, Hu Y-S, Bhattacharyya AJ, Maier J. Adv Funct Mater. 2007; 17: 2800–7.
88. Choudhury S, Stalin S, Deng Y, Archer LA. Chem Mater. 2018; 30:5996–6004.
89. Croce F, Appetecchi GB, Persi L, Scrosati B. Nature. 1998; 394:456–8.
90. Lin Y, Wang X, Liu J, Miller JD. Nano Energy. 2017; 31:478–85.
91. Zhou W, Wang Z, Pu Y, Li Y, Xin S, Li X. et al. Adv Mater. 2019; 31: 1805574.
92. Zhao Q, Stalin S, Zhao C-Z, Archer LA. Nat Rev Mater. 2020; 5:229–52.
93. Hong H-P. Mater Res Bull. 1978; 13: 117–24.
94. In Org Electron. 2009; 263.
95. Lee H, Kim H, Cho MS, Choi J, Lee Y. Electrochim Acta. 2011; 56: 7460–6.
96. Zhou H, Han G. Electrochim Acta. 2016; 192:448–55.
97. Sarac AS, Sezai Sarac A. in Encycl Polym Sci Technol. 2004.
98. Vorotyntsev MA, Zinovyeva VA, Konev DV. Electropolymerization Concepts Mater Appl. 2010; 27–50.
99. Alves APP, Koizumi R, Samanta A, Machado LD, Singh AK, Galvao DS. et al. Nano Energy. 2017; 31:225–32.
100. Bai M-HH, Bian L-JJ, Song Y, Liu X-XX. ACS Appl Mater Interfaces. 2014; 6:12656–64.
101. Zhang F, Xiao F, Dong ZH, Shi W. Electrochim Acta. 2013; 114:125–32.
102. Zhou H, Han G, Xiao Y, Chang Y, Zhai H-JJ. J Power Sources. 2014; 263:259–67.
103. Christinelli WA, Goncalves R,Pereira EC. Electrochim Acta. 2016; 196:741–8.
104. Xue M, Li F, Zhu J, Song H, Zhang M, Cao T. Adv Funct Mater. 2012; 22:1284–90.
105. Horng Y-Y, Hsu Y-K, Ganguly A, Chen C-C, Chen L-C, Chen K-H. Electrochem Commun. 2009; 11:850–3.
106. Horng Y-Y, Hsu Y-K, Ganguly A, Chen C-C, Chen L-C, Chen K-H. J Power Sources. 2010; 195: 4418–22.
107. Zhang LL, Li S, Zhang J, Guo P, Zheng J, Zhao XS. Chem Mater. 2010; 22:1195–202.
108. Fan L-Z, Hu Y-S, Maier J, Adelhelm P, Smarsly B, Antonietti M. Adv Funct Mater. 2007; 17:3083–7.
109. Tan S-J, Zeng X-X, Ma Q, Wu X-W, Guo Y-G. Electrochem Energy Rev. 2018; 1:113–38.
110. Zhang H, Zhang J, Ma J, Xu G, Dong T, Cui G. Electrochem Energy Rev. 2019; 2:128–48.
111. Fergus JW. J Power Sources. 2010; 195:939–54.
112. Yoshizawa H, Ohzuku T. J Power Sources. 2007; 174:813–17.
113. Xiao B, Liu H, Liu J, Sun Q, Wang B, Kaliyappan K. et al. Adv Mater. 2017; 29:1703764.
114. Nitta N, Wu F, Lee JT, Yushin G. Mater Today. 2015; 18:252–64.
115. Myung S-T, Maglia F, Park K-J, Yoon CS, Lamp P, Kim S-J. et al. ACS Energy Lett. 2017; 2:196–223.
116. Lin F, Markus IM, Nordlund D, Weng T-C, Asta MD, Xin HL. et al. Nat Commun. 2014; 5:3529.
117. Rozier P, Tarascon JM. J Electrochem Soc. 2015; 162:A2490–A2499.
118. Yan P, Zheng J, Liu J, Wang B, Cheng X, Zhang Y. et al. Nat Energy. 2018; 3:600–5.
119. Xie J, Sendek AD, Cubuk ED, Zhang X, Lu Z, Gong Y. et al. ACS Nano. 2017; 11:7019–27.
120. M Chen, E Zhao, D Chen, M Wu, S Han, Q Huang, et al. Inorg Chem. 2017; 56:8355–62.
121. Cho W, Lim YJ, Lee S-M, Kim JH, Song J-H, Yu J-S. et al. ACS Appl Mater Interfaces. 2018; 10:38915–921.
122. Sun HH, Ryu H-H, Kim U-H, Weeks JA, Heller A, Sun Y-K. et al. ACS Energy Lett. 2020; 5:1136–46.
123. Ryu H-H, Park N-Y, Seo JH, Yu Y-S, Sharma M, Mucke R, et al. Mater Today. 2020; 36:73–82.

124. Park G-T, Ryu H-H, Park N-Y, Yoon CS, Sun Y-K. J Power Sources. 2019; 442:227242.
125. Liang J, Hwang S, Li S, Luo J, Sun Y, Zhao Y. *et al.* Nano Energy. 2020; 78:105107.
126. Trivinho-Strixino F, Pereira EC, Mello SV, Oliveira ON. Langmuir. 2004; 20: 3740–5.
127. Christinelli WA, Goncalves R, Pereira EC. J Power Sources. 2016; 303: 73–80.
128. Goncalves R, Christinelli WAA, Trench ABB, Cuesta A, Pereira ECC. Electrochim Acta. 2017; 228:57–65.

13 Electroactive Polymeric Membranes

Maria Wasim, Aneela Sabir, Muhammad Shafiq, and Rafi Ullah Khan

CONTENTS

13.1   INTRODUCTION

Due to technological progress, the use of traditional materials, such as metals or alloys has been changed to polymers in fields including aerospace industries, automobiles, electronic goods, and household materials(Wasim, Sagar *et al.*, 2017; Asim *et al.*, 2018). Due to the developments in the polymer field, numerous fabrication methods have been developed that allow for the processing of polymers that have tunable characteristics, such as electrical characteristics and mechanical strength, which allows for the development of cost-effective and light weight and small sized materials (Dutta, Bhattacharjee, and De, 2020; Wasim, Sabir *et al.*, 2019; Wasim *et al.*, 2017).

In contrast to inorganic materials, polymers have a wide range of characteristics. These include being inexpensive, light weight, easy to process, flexibile, and versatile (Wasim *et al.*, 2017). They can be developed in complex conformations and have tailormade properties according to their applications (Vatanpour and Khorshidi, 2020). The rapid development of these materials and their applications in applied science, means that materials at a molecular scale are rapidly advancing

(Wasim *et al.*, 2019). These materials can detect minor changes that occur in their surroundings, respond to stimuli or data and act according to that. That might include piezoelectric materials and shape-memory alloys. (Ye *et al.*, 2020). For several decades, studies have been conducted to explain their behavior, for example, the change in shape or dimensions in response to stimuli received. These can be changes in pH, temperature, electrical field, or light (Cao *et al.*, 2020). These intelligent polymers are termed as active polymers. The application of these active polymers started biomimetics, which uses impressions from nature and employs them in engineering and design. A number of designs has been developed that imitate insects, birds, and fish (Cao *et al.*, 2020).

Based on a variety of stimuli, a range of active polymers can be fabricated that have controllable characteristics. They can create reversible or irreversible reactions, can be active or passive by implementing a polymer network, and develop smart configurations. The durability and resilience of the host material can be beneficial for the development of the smart assembled structures with controlled conformations and self-sensing properties (Zhu *et al.*, 2020).

13.2   CLASSIFICATION

Dependent on the type of actuation, the material can be generally categorized as:

1. Non-electrically deformable polymers
2. Electroactive polymers (EAPs)

The first are actuated by non-electrical stimuli, such as changes in light, temperature, and pH and the second are actuated by electrical input or signals. Normally, the non-electrically deformable polymers are a chemically active, inflatable structures, light, thermally sensitive, or shape-memory polymers (Vatanpour and Khorshidi, 2020). The EAPs change size or dimension in response to an electric stimulus. They are further categorized depending on the mechanism that is responsible for the change as:

1. Electronic EAPs
2. Ionic EAPs

In the first, the driving stimuli is either an electrical field or Coulomb force, and in the second the change in dimension is caused by the mobility of ions, or conjugated substances, or their diffusion. The most prominent EAPs are listed in Table 13.1.

The electronic EAPs include piezoelectric, ferroelectric, electrostatic, and electrostrictive that usually entail a high activation field >150V/μm, that is close to the breakdown level of the material. The characteristic of these materials is to hold the induced dislocation when the DC voltage is applied. This forms them into suitable constituents for robotics that can perform in air without

TABLE 13.1
List of leading EAPs

Electronic EAPs	Ionic EAPs
Ferroelectric polymers	Conducting polymers
Electrostrictive paper	Ionic polymer–metal composite
Electroviscoelastic elastomers	Ionic polymer gels
Electrostrictive graft elastomers	Carbon nanotubes
Dielectric electroactive polymers	
LCEs	

any limitations. These materials have elevated energy density along with a prompt response (in milliseconds). Overall, electronic EAPs have a glass transition temperature (Tg) insufficient for low temperature range stimuli.

In contrast, ionic EAPs, such as ionic polymer–metal composites (IPMCs), gels, carbon nanotubes (CNTs), or conducting polymers necessitate a low driving voltage that is approximately 1–5 V. One of the drawback of these materials is that can perform under wet conditions or in solid state electrolytes. Ionic EAPs mainly yield a bending response that caused minor actuation forces in contrast to electronic EAPs. The process in wet conditions cause the hydrolysis of water. In addition, ionic EAPs have a slow response time but have large deformations and are more similar to biological muscle compared with electronic EAPs. The induced deformation in both types can be manipulated to contract, stretch, or bend (Saja *et al.*, 2020)

13.3 ELECTROACTIVE POLYMER MEMBRANES

For the last few decades, research has been carried out to develop and understand EAP membranes. The capacity of contactless stimuli along with a superior controller system that employs electrical energy shows promise for the development of variety of active polymers. These polymers are termed APs.

13.3.1 ELECTROACTIVE POLYMER MEMBRANES FOR SENSING

The points that must be considered when designing novel sensors are the selectivity of the sensitive layer and the transduction phenomena by which the chemical interaction is converted into a physical measurable signal.

The electromechanical coupling in an electroactive polymer has been widely conducted for actuation. In addition, it has been studied for sensing the chemical and mechanical response to stimuli. EAPs are materials that have a small modulus, extraordinary strain ability, and they can conform or transform to a substrate with various morphologies. These distinctive features make them favorable for applications, such as sensors that can be worn and have the interface connected to soft tissues. The major categories of EAPs depend on the type of carrier, which are ionic (e.g., conducting polymer, ionic metal/polymer composites) and electronic (e.g., di-electric elastomers, piezoelectric polymers).

13.3.1.1 Ionic Electroactive Polymers for Sensing

Ionic EAPs normally work at low voltages, which are approximately <5 V for the actuation (Jean-Mistral, Basrour, and Chaillout, 2010). Moreover, the materials employed for the fabrication of ionic EAPs include polypyrrole (PPy) and poly(3,4-ethylenedioythiophene) (PEDOT), which are biologically compatible. Therefore, the EAPs are appropriate for applications in biological systems (Carpi and Smela, 2009). This type of EAP need an ion pool to perform conventionally, which allows the migration of ionic or molecular species. The presence of ionic species in ionic EAPs, means that they share common features with responsive structures: (1) the strain/stress implemented to the structures causes the relocation of ions and a ruffle dispersal of ions; and (2) the characteristics of ionic transport into the electrodes or via the electronic separator affect the larger results in applications, such as actuation and sensing (Baughman *et al.*, 1999).

13.3.1.2 Conducting Polymer-Based Sensors

Conducting polymers are organic and can conduct electronically with moderate and reversible ion storage capability. The mechanism involved in actuation and mechanical sensing are similar and depend on the addition and rejection of ions from the polymer membrane, the membrane should be electrically conductive (Zhang *et al.*, 2009). This type of material can be employed as the sensor or the actuator, and depend on the output produced and the external stimuli. There are two types of configuration for a conducting polymer based sensing device: (1) the standing membrane of the conducting polymer that functions in an electrolyte solution; and (2) the trilayer type of structure that has a conducting polymer layer on the upper and lower sides, between the nonconductive separator layer. The separating layer is composed of an ionically conductive material that performs as an ion pool and an insulator. The trilayer type of sensors, along with an electrolyte that is present in a separate section, might perform in an ambient atmosphere and does not require an external electrolyte.

13.3.1.3 Conducting Polymer-Based Free-Standing Membrane

The mechano–chemo–electric effect (in which mechanical deformation causes the electronic and ionic currents) is shown in Figure 13.1 and is an isolated standing membrane. The membrane is composed of a conductive polymer fabricated in early 1990s (Takashima *et al.*, 1997). During this procedure, charge is induced mechanically and is transformed into electricity. This type of energy is directly proportional to the dimensional changes that occur in the membrane and on the magnitude of the load applied. The efficiency was approximately <0.01% for this transformation and the voltage used was a few microvolts (Takashima *et al.*, 1997). The mechano–chemo–electric effect is reversible and the charge induced for all stages was practically the same. The ability of a conducting polymer to hold large strains and low mechanical impedance makes them a better sensor for various

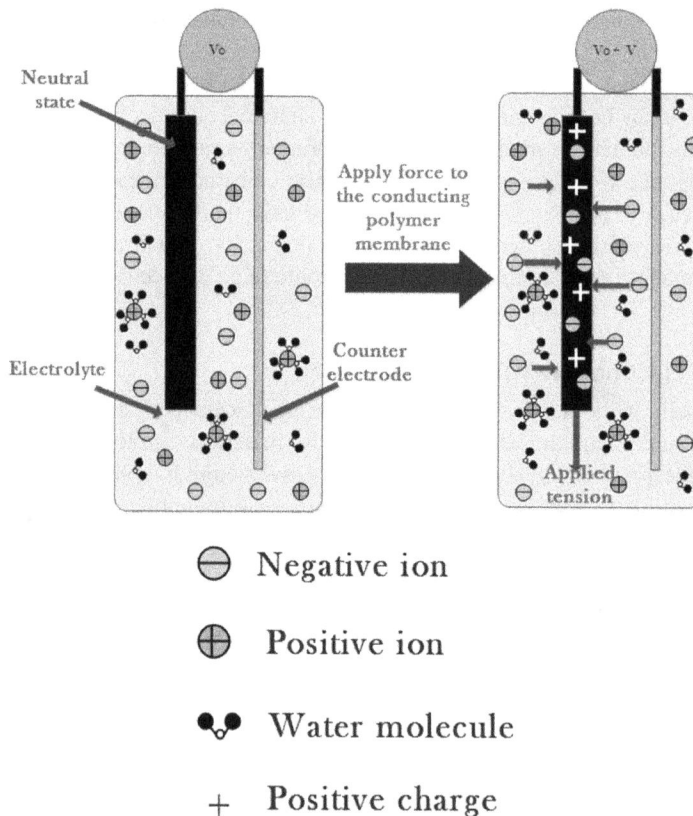

FIGURE 13.1 Free-standing membrane of a conducting polymer as the linear force in the sensor before and after applying tension.

applications that might include instrumentation to measure strain and force (Shoa *et al.*, 2010). In addition, the tensile strain can be studied via the free-standing membrane of the conducting polymer.

There are twin hypotheses related to the mechanism involved during the generation of voltage. In one of the hypotheses, it is stated that the if the conducting polymer membrane or the coating is compacted, then mechanical change trails the increment in the amount of movable ionic species compared with the ionic species present in the external electrolyte. Conventionally, the movable ionic species are either anionic or cationic, in which the charge is equalized, the electronic charge of the polymer or the bulk matrix or immobile counter ions. The difference in concentration and the mechanical deformation of the mobile ions results in the total ejection of the movable ions from the polymer membrane. This generates a voltage difference that can be detected using open circuit measurements (Wu et al. 2007). Therefore, the increase in mechanical deformation in the polymer assembly is directly related to the volume increase on the polymer followed by a reduction in ion content. This results in an influx of movable ions from one category into the polymer membrane (i.e., anions as shown in Figure 13.1) followed by difference in voltage. An another model suggested that the ionic species are added or ejected directly as a result of the mechanical stress and not because of the concentration change (Shoa *et al.*, 2010). In this case, the mechanical stress applied can be connected to the voltage produced via the conducting polymer sensors as an output. Other research exhibited the linear relationship between the stress implemented and the voltage produced for the PPy free-standing membrane (Shoa et al. 2010)

13.3.1.4 Conducting Polymer-Based Trilayer Structure

A trilayer structure is shown in Figure 3.2, it shows the bending trilayer sensor depicting the stretched layer being infiltrated by the movable ionic species. In addition, the mobile ionic species are ejected from the contracted layer on the oppose side mainly because of the mechanical deformation and increase in amount of ions. Deformation or the stress implemented can influence the ions present in the middle separator section. The difference in potential generated between both sheets could be measured using the open circuit method or by measuring the short circuit current. Variability in the volts is provided by the charge to strain ratio that is multiplied via the pressure implemented (Shoa *et al.*, 2010). The bending type of mechanical trilayer sensor used PPy as the conducting polymer was studied by Wu *et al.* (2007). They studied a linear relationship between the charge density and the induced strain through the bending deformation of the trilayer sensor.

13.3.2 Ionic Polymer–Metal Composite-Based Sensors

Analogues to conducting polymers, IPMC sensors and actuators mainly perform their function in response to the migration of ions; however, based on different approach and structures, IPMC membranes are composed of a trilayer structure that is composed of metal electrodes and an ionic polymer (Biddiss and Chau, 2006). To be employed as a sensor, the property that makes them

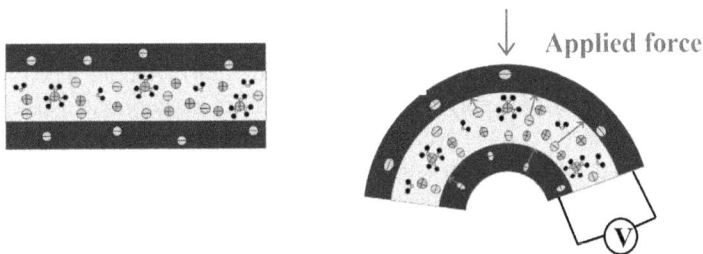

FIGURE 13.2 Bending trilayer sensor.

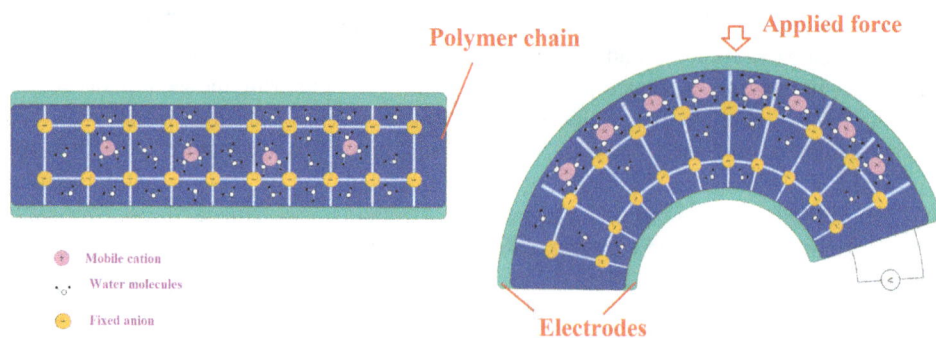

Polymer chain

Applied force

Mobile cation

Water molecules

Fixed anion

Electrodes

FIGURE 13.3 IPMC sensor. The application of force causes the bending of the sensor. The expansion on the top and contraction at the bottom led to a concentration gradient. Followed by the migration of cations toward the top. Due to differences in charge concentration carried by cations, which caused the potential difference between electrodes. The difference can be detected by open circuit voltage.

favorable are the are their outstanding sensitivity to stimuli (i.e., physical), such as mechanical force that can induce the bending phenomena on the IPMC (Figure 13.3) (Shahinpoor and Kim, 2001). In contrast to a sensor composed from a conducting polymer, IPMCs are mostly employed as mechanical sensors.

In the 1970s, Grodzinsky (2005) validated the electro–mechanical transduction in the membranes of collagen. They demonstrated experimentally that the electrical–mechanical and mechanical–electrical transductions were possible in a polyelectrolyte membrane. IPMC membrane based vibrational sensors are described in Sadeghipour, Salomon, and Neogi (1992).

An IPMC membrane is conventionally in a trilayer structure that has a middle section composed of an ionic polymer layer and the upper and bottom layers are composed of the metallic electrodes (Tiwari and Garcia, 2011). Ionic polymers are termed as polyelectrolytes that are synthesized via the fixed, covalently bonded immobile ions that are repeated and are hydrophilic. These repeating units are sometimes attached to the main chain of non-ionic polymer that are hydrophobic. The presence of a hydrophilic, microstructure network allows the formation of porosity; therefore, increasing the transport of oppositely charged (i.e., counter ions), which are mobile especially in a diluent that causes the swelling phenomena as shown in Figure 13.3 (Bahramzadeh and Shahinpoor, 2014). Ion exchange capacity or the acidity of the ionic polymer shows the capacity of oppositely charged species, capacity storage, and the ionic conductivity of the ionic based polymer exhibits the mobility of ions through the membrane (Bahramzadeh and Shahinpoor, 2014). For the ionic polymer to be used as an actuator and sensor, ionic conductivity and ion exchange capacity are the two main properties. Because both are dependent on to the structure of the membrane. In addition, the conductivity of ions is connected to the dimensions and the charge present on to the opposite ions along with the type of electrolyte and their uptake (Bennett *et al.*, 2006). However, the tensile modulus normally decreases when conductivity of the ions is enhanced by the higher uptake of the diluent. The main focus in this field is to fabricate ionic based polymers with enhanced conductivity of ions that can work in dry and wet environments, with improved chemical and thermal stability and increased mechanical characteristics (i.e., strength) (Bahramzadeh and Shahinpoor, 2014). A conventional type of ionic polymer is composed from perfluorinated alkene that has short side chains that are eliminated via ionic groups (that might include carboxylic acid or sulphonic acid groups for the exchange of cations or for anion exchange, ammonium cations). The proportions of the polymer matrix main chain regulate the mechanical properties of the IPMC membranes. The increased proportions result in increased mechanical strength. The synthesis of metallic electrodes can be either by physical deposition, such as sputtering (Bhandari, Lee, and Ahn, 2012) or by chemical reduction, such as electroless plating (Zhang, Ma, and Dai, 2007). The theory to describe the process behind IPMC sensing was discussed in some review papers (Bahramzadeh and Shahinpoor, 2014). Mainly, twin methods were used to describe the mechanical deformation as follows.

First, the potential difference generated between the electrodes was measured. The mechanical sensing characteristics of the IPMCs were elucidated by the imbalance in the charge created by ion movement due to mechanical deformation (De Luca *et al.*, 2013). With the application of external force onto the IPMC sensors, because of the strain/ stress gradient, mobile cations movement becomes possible, and they migrate to the stretched region to equalize the ion content. The gradient of charge along the width of the IPMC sensor Figure 13.3 creates the difference in potential that can be measured by a low power amplifier or by an open circuit potential detector (Tiwari and Garcia, 2011). Therefore, the theory behind the sensing is that the density of charge is directly proportional to the strain caused (Farinholt and Leo, 2004).

Second, the surface resistance of the metal-based electrodes of an IPMC is evaluated. The resistance varies with the contraction and expansion of the metallic electrodes. When the metallic electrodes are overextended, the resistance is enhanced, and the compression of the electrodes causes the deterioration in the electrodes. When the IPMCs bend by applying force or strain, the resistance on one side enhances and on the other side declines. The difference in resistance on both sides is connected to the bending curvature and is enhanced progressively. The measurement of the resistance difference between the metallic electrodes is employed to measure the curvature radius that might lead to the measurement of strain applied. The measurement of the surface resistance of the electrodes is carried out using a four probe system (Punning, Kruusmaa, and Aabloo, 2007).

13.3.3 IONIC ELECTROACTIVE POLYMER-BASED SENSORS

In addition to the use of EAPs as sensors, others applications that include hydrogel membranes (Carpi and Smela, 2009) have been extensively researched for sensing applications. The sensitive nature of a hydrogel membrane physical factors that might include salt concentration (Liu *et al.*, 2003), pH, temperature (Kuckling, Richter, and Arndt, 2003), electrical voltage (Richter *et al.*, 2003), and organic materials concentration in the water (Kuckling and Richter, 2000) mean that they are remarkable materials that could be employed for applications in chemical sensors or biosensors for the detection of proteins and DNA (Bashir, 2004). Stimuli-responsive hydrogel membranes or smart hydrogel can alter their volume (i.e., increase up to one order of magnitude) and can reversibly transform the chemical energy in to mechanical energy (Richter *et al.*, 2003). Employing a variety of hydrogels for various techniques in sensor applications have been studied (Zhao *et al.*, 2013). One of the most demonstrated theories is the bending of silicon sheet (thin) to alter at the voltage output by swelling the hydrogel membranes. The addition of a layer of hydrogel membrane to the micropressure based sensor might provide the display with analyte reliant swelling of the hydrogel membrane to measure the modification. The mechanism via the distribution of the hydrogel membrane provides a slow response that can be improved by miniaturization. For viscoelastic behavior, hysteresis can influence the response time of the hydrogel membrane (Gajovic-Eichelmann *et al.*, 2012)

13.3.4 ELECTRONIC ELECTROACTIVE POLYMERS FOR SENSING

13.3.4.1 Introduction to Electronic Electroactive Polymers

In contrast to ionic EAPs, electronic EAPs are free from any electrolyte medium. For electro–mechanical coupling, no migration of ion is required. Therefore, electronic EAPS need a shorter time or relaxation time in contrast to ionic EAPs. which is mainly <1 ms (Jean-Mistral, Basrour, and Chaillout, 2010). Therefore, electronic EAPs can perform at higher frequencies compared with ionic EAPs for dynamic sensing. Electrostrictive polymers, liquid crystal elastomers (LCEs), which are mostly the conventional flexoelectric polymers, dielectric elastomers, and piezoelectric based polymers are the main types of electronic EAPs (Carpi and Smela, 2009).

13.3.4.2   Dielectric Elastomer-Based Sensors

The dielectric elastomer-based sensor is a strain sensor that measures the capacitance alteration in a capacitor like system. Dielectric elastomer sensors are constructed on a structure that has twin electrodes and a transitional dielectric level (which is designed after a lengthy chained polymer structure that has an appropriate electric permittivity and membrane thickness). These designed system are dielectric elastomers transducers (Carpi *et al.*, 2015), that might include a dielectric elastomer generator (Koh, Zhao, and Suo, 2009), dielectric elastomer sensor (Joo *et al.*, 2017), and d- electric elastomer actuators (Zupan, Ashby, and Fleck 2002).

The development of EAPs based on dielectric elastomers materials was based on the work carried out by Roentgen's in 1880 (Wang *et al.*, 2016). Natural rubber deformation was observed when a high electrical field was applied. This was the first demonstration of dielectric elastomer actuation. Figure 13.4(a) shows basic dielectric elastomers actuators. With the application of voltage, two electrodes are present at the upper and lower sides of the elastomers. Both the electrodes are equally attracted and inclined to move toward each other; therefore, contracting the elastomers via electro-static forces (i.e., Maxwell stress) and induce the decrease in thickness. Assuming the elastomers is incompressible, with a positive Poisson's ratio, then there must be an extension sideways plane normal to the applied electrostatic field. Because the electrodes are adaptable, they might enlarge along with the elastomers without electric interruption through distortion. As the electrical energy is removed, the entire design returns to its earliest form.

13.3.5   Liquid Crystal Polymer-Based Sensors

Liquid crystal is one of the phases that is in the middle of the crystalline state (i.e., the molecules are in ordered array that is spatially fixed) and the liquid state (i.e., molecules are completely disordered)

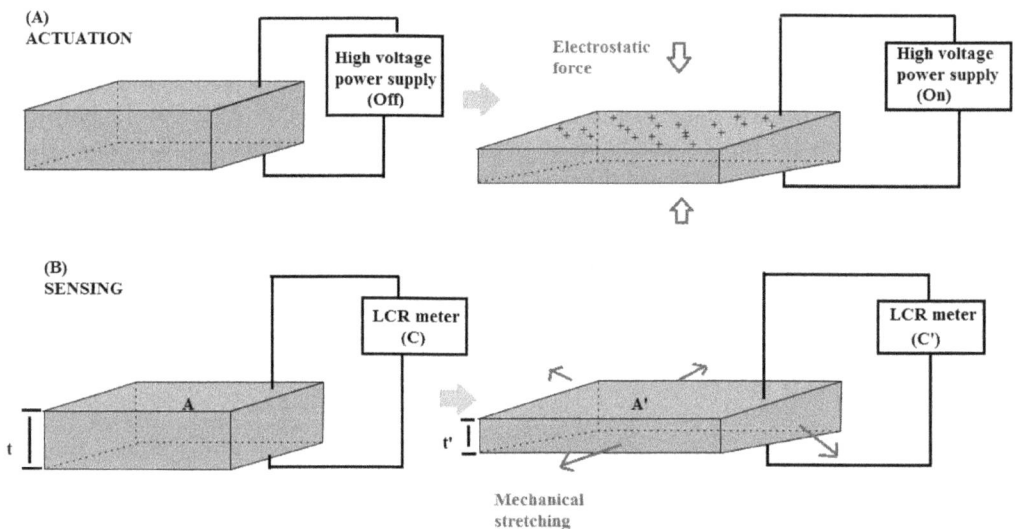

FIGURE 13.4   Mechanism of operation of dielectric elastomers for: (a) actuator; and (b) sensor modes. With the application of voltage across the actuator, the forces (attractive) present between electrodes makes the membrane contract in thickness but expand in lateral directions. Similarly, when the sensor is laterally stretch, the capacitance change (C–C') can be associated to strain due to stretching.

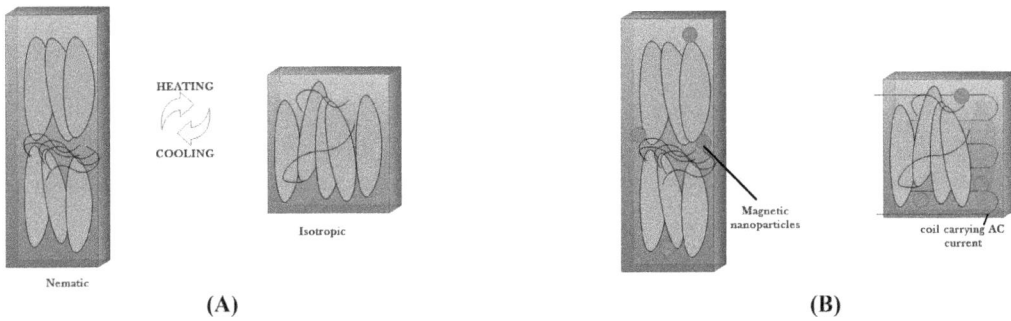

FIGURE 13.5 Liquid crystal elastomer actuation and sensing phenomena: (a) ordered nematic phase that transforms into isotropic phases via thermal induced actuation with the corresponding contraction to liquid crystal director; (b) addition of magnetic nanoparticles, order is preserved. When placed in an electromagnetic field, temperature increase causes the same type of contraction.

(Collings and Hird, 2017). There are various types of liquid crystal phases, and they are classified by the degree and type of molecule order that forms the material structure. Increases in the disorder might be due to increased temperature or by the addition of small sized molecules of solvent. The types of molecules that constitute liquid crystal phases are normally lengthy, rigid rods and the anisotropy that exists at the molecular scale provides liquid crystal phases their strange phenomena. However, liquid crystal materials can have disordered structures and therefore, maintain the ability to move (i.e., a liquid). The crystal structure has similar optical characteristics, such as birefringence. The simple chemically connected anisotropic liquid crystal molecules compose a lengthy sequence, and then their ability to flow is prevented and eventually the structure become crystalline. However, if they are attached to long chain molecules, which is moderately soft (i.e., they have low Tg), then some of the ability flow might be preserved. In addition, if few some of the linking connection that are exist are in the soft chains, then the liquid crystal phases can alter the bulk dimensions of the structure. These structures are known as LCEs (Prévôt, Ustunel, and Hegmann, 2018) and they can display some outstanding configurations and might be able to decrease to one-third of their length when subjected to heating; however, shrinkage of approximately 80% of their original length has been demonstrated (Clarke *et al.*, 2002). Figure 13.5 (a) shows the contractions, if the mesogen displays the orientational configuration across the material (i.e., the monodomain nematic), which means that when subjected to heating it can increase the disorder and the materials shrink along the liquid crystal director. This change is reversible. To construct the responsive LCEs toward the electric field, a minute concentration of a conducting filler, such as CNTs or magnetic based nanoparticles could be incorporated into LCEs as shown in Figure 13.5(b) (Yan, Marshall, and Terentjevb 2012). In this situation, the electric potential applied causes the current flow that heats up the material used. Dimensional change happens by same approach previously mentioned. The addition of a high concentration of filler causes the material to become stiff; therefore, disturbing the dimensional change.

In the bent core liquid crystal polymer membrane, strain sensing can be studied. Figure 13.6 shows how the bending phenomena of these materials can change electrical polarization, which shows an example of polymerization of bent core mesogens. From their name, liquid crystals are stiff mesogenic units that can bend lengthwise to give an L-shaped structure. This shape produces a new type of behavior, because they can establish chiral phases even after the liquid crystal mesogens are employed are achiral. These materials can be used in piezoelectric sensors and when subjected to mechanical mixing, a current of tens of nano amperes might be produced. Such phenomena is called flexoelectricity and has been extensively studied (Mohammadi, Liu, and Sharma, 2014).

(A)

Bent core mesogen
with electrical dipole

Polymer matrix

(B)

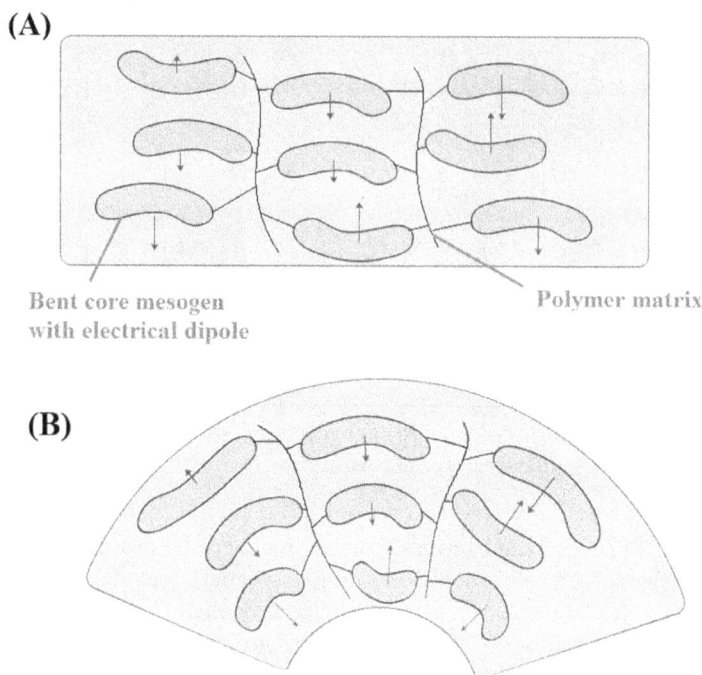

FIGURE 13.6 Showing: (a) bent core liquid crystal mesogen chemically connected to polymeric network; and (b) bending caused the change in overall electrical polarization of the material.

13.3.6 PIEZOELECTRIC POLYMER-BASED SENSORS

Piezoelectric is derive from "piezo" that shows pressure and states the tendency of the material to produce a charge on the surface based on the pressure applied. In contrast, the application of an electric potential across the material causes mechanical deformation.

In general, most of the commonly employed piezoelectric materials are inorganic (Wang and Song, 2006) due to their elevated piezoelectric strain constant (d), which is the strain (i.e., mechanical) generated by the application of the electric field. However, these piezoelectric ceramics need increased temperature treatment, if thin membranes with dipole alignment are required. In addition, to achieve high performance, materials should be composed of lead, such as lead zirconium titanate (Kingon and Srinivasan, 2005). To overcome these limitations, it has been suggested that polymer piezoelectric materials are used as alternatives. Polymeric piezoelectric substances have substantial benefits in addition to their soft elasticity, the materials and the processing apparatus are inexpensive (Crossley and Kar-, 2015). Therefore, piezoelectric polymers facilitated the development of sensors that were flexible (Chen *et al.*, 2015), energy generators (Whiter, Narayan, and Kar-Narayan, 2014) and organic based field effect transistors (Choi *et al.*, 2013) for the future generation of smart technology.

13.4 ELECTROACTIVE POLYMER MEMBRANES FOR DRUG DELIVERY

With the discovery 50 years ago that low molecular weight hydrophobic drugs can easily diffuse through silicone membranes at a controlled rate, research on polymer membranes for drug delivery has been conducted (Qiu and Bae, 2006). The versatility and flexibility of polymer materials allows the control of characteristics, such as biocompatibility and biodegradability, due to their topology, morphology dimensions, and divergent chemistry. Polymers have exhibited enhanced pharmacokinetics in contrast to pure small size drugs. Because the polymer is not a drug, they are modified to be a carrier, decreasing immunogenicity, toxicity, or degradability along with enhancing the circulation time (Qiu and Bae, 2006). For drug delivery, stimuli- responsive polymers that impersonate biological systems have the ability to change under external conditions (Schmaljohann, 2006; Hu *et al.*, 2017). These smart polymers exhibit a response within biological systems and conditions. The distinctive stimuli are temperature (Khanna *et al.*, 2021), light (Salzano de Luna *et al.*, 2020; Jia *et al.*, 2018), pH (Liu *et al.*, 2017), electrical field (Qu *et al.*, 2018), and electrolytes (Ramasamy *et al.*, 2017;Jin *et al.*, 2018; Kang *et al.*, 2017; Yilmaz, 2018). The responses activate the drug release that could be degradation of the structure, transformation in hydration state, precipitation, dissolution, swelling, collapse of the structure, change in hydrophilicity of the surface, conformation alteration, and micelles formation. The most significant stimuli are temperature, ionic strength, pH, light, and redox potential. The electric field is a stimulus for drug delivery and the novel electro responsive polymers are known as "smart" carriers of drugs (Tandon *et al., 2018*).

For the delivery of drugs, hydrogel membranes are prominent, because they have a porous structure that allows the loading of drugs into the polymer matrix, which allows the drug discharge at a rate depending on the diffusion coefficient of the dynamic molecules via the gel (Schmaljohann, 2006). For electroactive hydrogel membranes, the influence of electrical stimulation on the discharge of the drug depends on the gel response to the stimuli and how they releases the drug, and the interaction between the drug and the gel network (James *et al.*, 2014). The most important mechanisms for drug discharge in electroactive hydrogel membranes are:

1. Forced convection of the drug molecules that are ejected from the polymeric gel with syneresed/expelled water via an electrostatic field (Kennedy *et al.*, 2014)
2. Diffusion (Lee *et al.*, 2018)
3. Electrophoresis of electrified types of drugs (Wang and Kohane, 2017)
4. Ejection of drug upon attrition of electro–erodible gels beads (Murdan, 2003)

For charged drugs, the movement of the charged species in the direction of the electrode that has the opposite charge should be further considered (Liu, Cui, and, Losic 2013). The mechanism of drug release in this system is considered vital, because under an electric field, hydrogel membranes usually de-swell, which allows for the migration of the solutes out of the gel. In general, under the influence of an electric field, water is syneresed/expelled from the gel, which stimulates the release of the drug (Kennedy *et al.*, 2014). In addition, with the removal of the electric field, the gel absorbs fluids and swells. For the diffusion mechanism for the release of drugs, the electroactive membranes shrinking ability might disrupt the release of drug as the pore present in the polymeric network of the gel become smaller and the pathway of drug movement becomes more convoluted.

Inherent conductive polymers might be employed for the electroactive delivery of drug, because they have controllable and reversible redox reactions. These reaction can change their redox state that leads to a change in the polymer charge, conductivity, and volume; therefore, causing the ejection or acceptance of charged particles from the bulk of polymer membranes (Svirskis *et al.*, 2010). The manipulation of these change and the drug release rate from EAP membranes can be altered. Anionic drugs can be incorporated into the polymer via oxidative polymerization steps or through the exchange of ions via redox cycling after the polymerization step. The anionic molecules can be released through electrochemical reduction (Guo and Fan, 2018). Therefore, the release of the drug is activated via reduction and the re-addition of the drug molecules by oxidation. For cationic drug discharge, the neutral inherent conductive polymer is oxidized, and the resultant net cationic charge in the polymer causes the ejection of the drug through the membrane. The preparation of PPy with immobilized anions will add the cation on reduction and the cation driven actuation led to swelling. In addition, the cationic species are released on during oxidation. Researchers are attempting to blend the inherent conductive polymers with hydrogel membranes for the fabrication for electro conductive polymer membranes. In general, the polyethyleneimine (PEI) and 1-vinylimidazol (VI) polymers are blended along with polyacrylic acid and polyvinyl alcohol forming a partial inter-penetrating complex for the therapeutic electro responsive drugs (Indermun et al. 2014). A chitosan grafted with polyaniline (PANi) copolymer accompanied with an oxidized form of dextran as a crosslinker was used to prepare an electroactive hydrogel membrane. The copolymer was the drug carrier that had electrically driven potential with the rate of release that was enhanced with the voltage applied (Qu *et al.,* 2018). The electrically driven drug release from conductive hydrogel membranes has related to:

1. Migration of charged species under the influence of an electric field (Qu *et al.,* 2018)
2. On oxidation and reduction, the presence of a net charge on the polymer (Ge *et al.,* 2012).

13.5 ELECTROACTIVE POLYMER MEMBRANES FOR TISSUE REGENERATION APPLICATIONS

The use of non-EAPs, such as poly lactic-co-glycolic acid (Tay *et al.,* 2011) and polydimethylsiloxane (PDMS) (Li *et al.*, 2012) in tissue engineering have been widely employed. Although they are quite easy to process and have good biocompatibility, they cannot be termed as "smart" material because they do not response to external electrical stimuli. In contrast, EAPs exhibit excellent biocompati-bility and ease of processing along with a response to external mechanical or electrical stimuli. These features make them potential candidates to be termed "smart" for the fabrication of a scaffold that is proficient in directing cell fate and can stimulate tissue regeneration.

13.5.1 CONDUCTING POLYMERS

The conducting polymer first emerge mid-1970s as novel organic materials (Guo, Glavas, and Albertsson, 2013). They gained worldwide attention for two reasons. First, their outstanding elec-trical and optical properties that were equivalent to inorganic semiconductors and metals. Second, these materials have inherent polymer properties that include ease of fabrication, processability, and versatility (Guimard, Gomez, and Schmidt, 2007). The ability of conducting polymers to allow the regeneration of tissues in response to external stimuli means that they are important for a large range of biomedical applications, such as biosensors, functionalized neutral electrodes, tissue engineering (Cui, Yang, and Li, 2016), and drug delivery (Schultze and Karabulut, 2005). The main conducting polymers for the tissue regeneration are PPy (Ateh, Navsaria, and Vadgama, 2007), PANi (Tahir, Alocilja, and Grooms, 2005) and PEDOT (Higgins et al. 2012).

13.5.2 POLYPYRROLE

PPy has been widely researched in the biomedical field due to its high conductivity, eco stability, easy fabrication, redox potential, and remarkable biocompatibility. Numerous research data revealed that the PPy is well-suited with numerous cells that include endothelial and nerve cells, osteoblasts, myocardial and mesenchymal cells (Otero and Martinez ,2014). With advances in biomedical fields, PPy has excellent feature for the regeneration of impaired tissues. The neural cells communicate via the electrical signals, because their membrane has voltage gated ion passages. These passages are activated by the change in electrical potential around them. A study proposed a collagen-based three-dimensional fibrous scaffold that was composed of PPy, and they exhibited that mesenchymal stem cell s from humans could grow on a PPy fibrous network. It showed that the cell function could be altered by external electric stimuli and that prolonged stimuli caused damage to the cell culture (Guo *et al., 2012*). Currenlty, it is assumed that Schwann cell relocation proceeds and it increases axonal restoration in the peripheral nervous system (Huang *et al.,* 2009).

Cardiovascular diseases are the one of the main causes of death compared with other illness (O'Neill *et al.*, 2016). Currently, scientist are attempting to fabricate cardiac patches, heart valves, and to synthesize myocardial stents for cardiac repair (Parsa, Ronaldson, and Vunjak-Novakovic, 2016). The electrically conduction of PPy plays an important function in cardiac and neural tissue engineering. The reaction produced in response to electrical stimulation is a fundamental property of neuron and myocyte's differentiation and roles. A study showed the use of porous and fibrous scaffold with a two-dimensional substrate that was combined with PPy or PANi, which showed enhanced cardiac differentiation (Baheiraei *et al.*, 2015). For the transmission of an electrical signal at a standard speed in a precise direction, cardiac cells need to be cultivated with the correct intracellular bonds and matrix design. The contraction of cardiac muscles occurs in response to an electrical signal that forces the natural mechanical stretching of the heart. Therefore, any interference in the passage of the electrical signal leads to cardiovascular disease (Ishii *et al.*, 2005).

Bone is one of the materials that is sensitive to electrical signals or stimulation due to its piezoelectricity. Therefore, many studies have been performed to manipulate osteogenesis via implants that transmit electrical signals to cells. Therefore, PPy has a promising role in the tissue regeneration in bone. Recently, researchers fabricated biomolecules that incorporated PPy nanostructures and tested the biocompatibility (Liao *et al.*, 2014).

13.5.3 POLYANILINE

Due to poor processability and weak mechanical characteristics, PANi has limited applications. Due to the better processability, economical nature, and eco stability of PANi means that it is a substitute for PPy (Xia and Tao, 2011). PANi and its derivatives might support the adhesion and proliferation of H9c2 cardiac myoblasts (Guterman *et al.*, 2002). This study showed the potential application of electroactive PANi in heart and nerve tissue regeneration.

The nervous system is composed of interconnected neurons and cells that are easily excited by the application of an electrical signal or stimuli that allows the transmission of a signal at a faster pace. Previously, many approaches were proposed to regenerate and repair the central and peripheral nervous system via different types of non-conductive scaffolds (Schmidt and Leach, 2003). Due to advances in conducting polymer scaffolds in neural regeneration several theories have been developed to explain the positive outcome of the scaffold when regulating the growth of neural tissue and nerve regeneration (Patel and Poo, 1982). Conducting polymers, such as PPy and PANi have introduced new pathways to increase cell growth and tissue regeneration (Li and Hoffman-Kim, 2008).

13.5.4 POLY(3,4-ETHYLENEDIOXYTHIOPHENE)

PEDOT has gained the interest of scientists as a regenerative materials due to potential applications in animal models for the growth of neurotransmitter delivery systems composed of electrical ion pumps (Simon *et al.*, 2009). It is employed in the fabrication of organic electrochemical transistors that could be used as biosensors (Mabeck and Malliaras, 2006). Richardson-Burns *et al.* (2007) researched the association between PEDOT and neural cells to fabricate a biomaterial that had electrical conductivity and that allowed live tissue to form at the interface. They designed microelectrodes from conducting polymers via PEDOT to coat them onto neural cells. The PEDOT coated implantable electrodes could be used to stimulate nerve impulses electrically and to record signals from neurons (Asplund *et al.*, 2009).

13.6 ELECTROACTIVE POLYMER MEMBRANE FOR ANTIMICROBIAL AND ANTI-FOULING APPLICATIONS

Despite the significance of electrostatic fields that hinder fouling, their use in electroactive antimicrobial polymers for biomedical application has not been studied. For example, PPy amended membranes that are coated with derivatives of graphene were synthesized to increase electrical conductivity and hinder fouling due to the electrostatic repulsion (Aslam, Ahmad, and Kim, 2018). In general, research has focused on fouling-resisting membranes for bioreactors. The systems involved to hinder fouling and to clean conductive membranes are constructed based on electrostatic connections or electrochemical (redox) reactions on the interface of the film (Ahmed *et al.*, 2016). During filtration of electrified large sized molecules and units, the conductive films (i.e., charged) repel the foulant due to their electrostatic interactions; therefore, decreasing fouling on membrane surfaces. During electrochemical fouling, the membranes act as an electrode where direct or indirect oxidation of the foulant occurs on the interface of the membrane or at the electrode surface, where the fouling agents are eradicated via the generation of bubbles on the surface (Ahmed *et al.*, 2016). The conductive polymers demonstrate antimicrobial activity without any external stimuli (i.e., electrical), mainly because of the oxidative stress that these polymers can produced onto the bacterial cell; therefore, inhibiting bacterial cell wall formation (Ahmed *et al.*, 2016).

The formation of nanocomposite of PANi–zinc oxide nanorods and epoxy resin–PANi, have exhibited remarkable antifouling characteristics. The development of electrically conductive membranes via graphene oxide and a PANi layer impregnated with phytic acid on the filter sheet of polyester, produced a highly conductive membrane that had remarkable antifouling characteristics. Therefore, which confirmed that the membrane with high conductivity had excellent antifouling characteristics (Li, Liu, and Yang, 2014).

The first known study into the polymer composite for electrical microbial resistant interactions in biomedicine employed carbon-based nanoparticles, in which twin amended catheters were placed upright in a nutrient broth agar dish and coupled with an electrical instrument. One catheter acted as a cathode and he as an anode (Liu *et al.*, 1993). The bactericidal action of negatively charged electrical conducting polymers was determined by the formation of electrostatic repulsion between the polymer and the negatively charged bacteria. outer layer (Francolini, Donelli, and Stoodley, 2003). Recently, new research has been conducted to determine 100% of the antimicrobial activity by providing the volts (9 V) as an electroactive composite material that was formed by poly lactic acid incorporated with graphene oxide (reduced thermally). The data showed that the electrostatics activity and the electron transfer in the materials was conductive under the electrostatic force, which led to bacterial decay that was connected with the EAP (Arriagada *et al.*, 2017). Later studies developed materials that inhibit or eliminate fouling attachment in the early stages (Zhang *et al.*, 2014). They studied PPy and chitosan-based membranes that were treated synergistically with gentamycin and a DC current against the biofilm or foulant layer. The application of a DC current disturbed biofilm formation and induced cell death via enzyme action (Zhang *et al.*, 2014) In hydrogel membranes, the influence of the electric force on to the growth of bacterial cell has been studied. Similar to DC, an electrostatic field was employed, such as a non-thermal approach, to alter the microbial distribution in agarose and alginate gel globules. The sustainability of the bacterial cells entrapped in the gel globules reduced along with the field intensity, and the time of electrostatic force increased simultaneously (Zvitov, Zohar-Perez, and Nussinovitch, 2004).

REFERENCES

Ahmed, F. *et al.* (2016) 'Electrically conductive polymeric membranes for fouling prevention and detection: A review', *Desalination*, 391, pp. 1–15. doi.10.1016/j.desal.2016.01.030.

Arriagada, P. *et al.* (2017) 'Poly(lactic acid) composites based on graphene oxide particles with antibacterial behavior enhanced by electrical stimulus and biocompatibility', *Journal of Biomedical Materials Research Part A*, 106. doi:10.1002/jbm.a.36307.

Asim, S. *et al.* (2018) 'The effect of Nanocrystalline cellulose/Gum Arabic conjugates in crosslinked membrane for antibacterial, chlorine resistance and boron removal performance', *Journal of Hazardous Materials*, 343, pp. 68–77. doi.10.1016/j.jhazmat.2017.09.023.

Aslam, M., Ahmad, R., and Kim, J. (2018) 'Recent developments in biofouling control in membrane bioreactors for domestic wastewater treatment', *Separation and Purification Technology*, 206, pp. 297–315. doi:10.1016/j.seppur.2018.06.004.

Asplund, M. *et al.* (2009) 'Toxicity evaluation of PEDOT/biomolecular composites intended for neural communication electrodes', *Biomedical Materials*, 4, p. 045009. doi:10.1088/1748-6041/4/4/045009.

Ateh, D., Navsaria, H., and Vadgama, P. (2007) 'Polypyrrole-based conducting polymers and pnteractions with biological tissues', *Journal of the Royal Society*, 3, pp. 741–752. doi:10.1098/rsif.2006.0141.

Baheiraei, N. *et al.* (2015) 'Preparation of a porous conductive scaffold from aniline pentamer-modified polyurethane/PCL blend for cardiac tissue engineering', *Journal of Biomedical Materials Research. Part A*, 103. doi:10.1002/jbm.a.35447.

Bahramzadeh, Y., and Shahinpoor, M (2014) 'A review of ionic polymeric soft actuators and sensors', *Soft Robotics*, 1, pp. 38–52. doi:10.1089/soro.2013.0006.

Bashir, R. (2004) 'BioMEMS: state-of-the-art in detection, opportunities and prospects', *Advanced Drug Delivery Reviews*, 56 (11), pp. 1565–1586. doi:10.1016/j.addr.2004.03.002.

Baughman, R.H. *et al.* (1999) 'Carbon nanotube actuators', *Science*, 284(5418), pp. 1340. doi:10.1126/science.284.5418.1340.

Bennett, M. *et al.* (2006) 'A model of charge transport and electromechanical transduction in ionic liquid-swollen Nafion membranes', *Polymer*, 47 (19), pp. 6782–6796. doi:10.1016/j.polymer.2006.07.061.

Bhandari, B., Lee G.-Y, and Ahn S.-H. (2012) 'A review on IPMC material as actuators and sensors: Fabrications, characteristics and applications', *International Journal of Precision Engineering and Manufacturing*, 13. doi:10.1007/s12541-012-0020-8.

Biddiss, E., and Chau T. (2006) 'Electroactive polymeric sensors in hand prostheses: Bending response of an ionic polymer metal composite', *Medical Engineering & Physics*, 28(6), pp. 568–578. doi:10.1016/j.medengphy.2005.09.009.

Cao, X-L. *et al.* (2020) 'Tailoring nanofiltration membranes for effective removing dye intermediates in complex dye-wastewater', *Journal of Membrane Science*, 595, p. 117476. doi:10.1016/j.memsci.2019.117476.

Carpi, F. *et al.* (2015) 'Standards for di- electricelastomer transducers', *Smart Materials and Structures*, 24, p. 105025.

Carpi, F., and Smela, E. (2009) *'Biomedical Applications of Electroactive Polymer Actuators*. New York: John Wiley & Sons.

Chen, X. *et al.* (2015) 'Self-powered flexible pressure sensors with vertically well-aligned piezo-electric nanowire arrays for monitoring vital signs', *Journal of Materials Chemistry C*, 3(45), pp. 11806–11814. doi:10.1039/c5tc02173a.

Choi, Y. *et al.* (2013) 'Control of current hysteresis of networked single-walled carbon nanotube transistors by a ferroelectric polymer gate insulator', *Advanced Functional Materials*, 23, doi:10.1002/adfm.201201170.

Clarke, S. *et al.* (2002) 'Effect of cross-linker geometry on dynamic mechanical properties of nematic elastomers', *Physical Review E. Statistical, Nonlinear, and Soft Materials Physics*, 65, p. 021804. doi:10.1103/PhysRevE.65.021804.

Collings, P., and Hird, M. (2017) *'Introduction to Liquid Crystals Chemistry and Physics'*, London: Taylor Francis.

Crossley, S., and Kar-Narayan, S. (2015) 'Energy harvesting performance of piezo-electric ceramic and polymer nanowires', *Nanotechnology*, 26, P.:344001. doi:10.1088/0957-4484/26/34/344001.

Cui, Z., Yang, B., and Li R-K. (2016) 'Application of biomaterials in cardiac repair and regeneration', *Engineering*, 2(1), pp. 141–148. doi:10.1016/J.ENG.2016.01.028.

De Luca, V. *et al.* (2013) 'Ionic electro-active polymer metal composites: Fabricating, modeling, and applications of postsilicon smart devices', *Journal of Polymer Science Part B; Polymer Physics*, 51, pp. 699–734. doi:10.1002/polb.23255.

Dutta, M., Bhattacharjee, S., and De S. (2020) 'Separation of reactive dyes from textile effluent by hydrolyzed polyacrylonitrile hollow fiber ultrafiltration quantifying the transport of multicomponent species through charged membrane pores', *Separation and Purification Technology*, 234p. 116063. doi:10.1016/j.seppur.2019.116063.

Farinholt, K., and Leo D.J. (2004) 'Modeling of electromechanical charge sensing in ionic polymer transducers', *Mechanics of Materials*, 36(5), pp. 421–433. doi:10.1016/S0167-6636(03)00069-3.

Francolini, I., Donelli, G., and Stoodley P. (2003) 'Polymer designs to control biofilm growth on medical devices', *Reviews in Environmental Science and BioTechnology*, 2, pp. 307–319. doi:10.1023/b:resb.0000040469.26208.83.

Gajovic-Eichelmann, N. *et al.* (2012) 'Molecular imprinting technique for biosensing and diagnostics', In, *Applications of Nanomaterials in Sensors and Diagnostics*, pp. 143–170. Berlin: Springer.

Ge, J. *et al.* (2012) Drug release from electric-field-responsive nanoparticles', *ACS Nano*, 6(1), pp. 227–233. doi:10.1021/nn203430m.

Grodzinsky, A. (2005) 'Electromechanics of deformable polyelectrolyte membranes', Thesis. https://dspace.mit.edu/handle/1721.1/13575.

Guimard, N.K., Gomez, N., and.Schmidt C.E. (2007) 'Conducting polymers in biomedical engineering', *Progress in Polymer Science*, 32(8), pp. 876–921. doi.10.1016/j.progpolymsci.2007.05.012.

Guo, B. Glavas, L. and Albertsson, A.-C. (2013) 'Biodegradable and electrically conducting polymers for biomedical applications', *Progress in Polymer Science*, 38(9), pp. 1263–1286. doi.10.1016/j.progpolymsci.2013.06.003.

Guo, B. *et al.* (2012) 'Electroactive porous tubular scaffolds with degradability and non-cytotoxicity for neural tissue regeneration', *Acta Biomaterialia*, 8(1), pp. 144–153. doi:10.1016/j.actbio.2011.09.027.

Guo, J., and Fan D. (2018) ',Electrically controlled biochemical release from micro/nanostructures for in vitro and in vivo applications: A review', *ChemNanoMat*, 4. doi:10.1002/cnma.201800157.

Guterman, E. *et al.* (2002) 'Peptide-modified electroactive polymers for tissue engineering applications', *American Chemical Society, Polymer Preprints, Division of Polymer Chemistry.* 43, pp. 766–767.

Higgins, M. *et al.* (2012) 'Organic conducting polymer–protein Interactions', *Chemistry of Materials*, 24, pp. 828–839. doi:10.1021/cm203138j.

Hu, X. *et al.* (2017) 'Stimuli-responsive polymersomes for biomedical applications', *Biomacromolecules*, 18(3), pp. 649–673. doi:10.1021/acs.biomac.6b01704.

Huang, J. *et al.* (2009) 'Electrical regulation of Schwann cells using conductive polypyrrole/chitosan polymers', *Journal of Biomedical Materials Research. Part A*, 93, pp. 164–174. doi:10.1002/jbm.a.32511.

Indermun, S. *et al.* (2014) 'An interfacially plasticized electro-responsive hydrogel for transdermal electro-activated and modulated (TEAM) drug delivery', *International Journal of Pharmaceutics*, 462(1), pp. 52–65. doi.10.1016/j.ijpharm.2013.11.014.

Ishii, O. *et al.* (2005) 'In vitro tissue engineering of a cardiac graft using a degradable scaffold with an extracellular matrix–like topography', *The Journal of Thoracic and Cardiovascular Surgery*, 130(5), pp. 1358–1363. doi.10.1016/j.jtcvs.2005.05.048.

Jean-Mistral, C., Basrour, S., and Chaillout J.J. (2010) 'Comparison of electroactive polymers for energy scavenging applications', *Smart Materials and Structures,* 19(8), p. 085012. doi:10.1088/0964-1726/19/8/085012.

Jia, S. *et al.* (2018) 'Photoswitchable molecules in long-wavelength light-responsive drug delivery: From molecular design to applications', *Chemistry of Materials*, 30. doi:10.1021/acs.chemmater.8b00357.

Jin, Z. *et al.* (2018) 'A PTX/nitinol stent combination with temperature-responsive phase-change 1-hexadecanol for magnetocaloric drug delivery: Magnetocaloric drug release and esophagus tissue penetration', *Biomaterials*, 153, pp. 49–58. doi:10.1016/j.biomaterials.2017.10.040.

Joo, Y. *et al.* (2017) 'Highly sensitive and bendable capacitive pressure sensor and its application to 1 V operation pressure-sensitive transistor', *Advanced Electronic Materials*, 3, p. 1600455. doi:10.1002/aelm.201600455.

Kang, T. *et al.* (2017) 'Surface design of magnetic nanoparticles for stimuli-responsive cancer imaging and therapy', *Biomaterials*, 136, pp. 98–114. doi:10.1016/j.biomaterials.2017.05.013.

Kennedy, S. *et al.* (2014) 'Rapid and extensive collapse from electrically responsive macroporous hydrogels', *Advanced healthcare materials*, 3, pp. 500–507. doi:10.1002/adhm.201300260.

Khanna, S. *et al.* (2021) 'Thermoresponsive BSA hydrogels with phase tunability', *Materials Science and Engineering: C*, 119, p.111590. doi:10.1016/j.msec.2020.111590.

Kingon, A.I., and Srinivasan S. (2005) 'Lead zirconate titanate thin films directly on copper electrodes for ferroelectric, di- electricand piezo-electric applications', *Nature Materials*, 4(3), pp. 233–237. doi:10.1038/nmat1334.

Koh, S.J., Zhao, X. and Suo, Z. (2009) 'Maximal energy that can be converted by a di-electric elastomer generator', *Applied Physics Letters*, 94, p. 262902.

Kuckling, D., and Richter, A. (2000) 'Application of sensitive hydrogels in flow control', *Polymers for Advanced Technologies*, 11, pp. 496–505. doi:10.1002/1099-1581(200008/12)11:8/123.3.co;2-z.

Kuckling, D., Richter, A. and Arndt, K.-F. (2003) 'Temperature and pH-dependent swelling behavior of poly(N-isopropylacrylamide) copolymer hydrogels and their use in flow control', *Macromolecular Materials and Engineering*, 288, pp.144–151. doi:10.1002/mame.200390007.

Lee, H. *et al.* (2018) 'Device-assisted transdermal drug delivery', *Advanced Drug Delivery Reviews*, 127, pp. 35–45. doi.10.1016/j.addr.2017.08.009.

Li, G. and Hoffman-Kim D. (2008) 'Tissue-engineered platforms of axon guidance', *Tissue engineering. Part B, Reviews*, 14, pp. 33–51. doi:10.1089/teb.2007.0181.

Li, H. *et al.* (2012) 'Direct laser machining-induced topographic pattern promotes up-regulation of myogenic markers in human mesenchymal stem cells', *Acta Biomaterialia*, 8, pp. 531–539. doi:10.1016/j.actbio.2011.09.029.

Li, N. Liu, L., and Yang F. (2014) 'Highly conductive graphene/PANi-phytic acid modified cathodic filter membrane and its antifouling property in EMBR in neutral conditions', *Desalination*, 338, pp. 10–16. doi.10.1016/j.desal.2014.01.019.

Liao, J. *et al.* (2014) 'Surface-dependent self-assembly of conducting polypyrrole nanotube arrays in template-free electrochemical polymerization', *ACS Applied Materials & Interfaces*, 6. doi:10.1021/am5017478.

Liu, H. *et al.* (2017) 'Facile fabrication of redox/pH dual stimuli responsive cellulose hydrogel', *Carbohydrate Polymers*, 176, pp. 299–306. doi:10.1016/j.carbpol.2017.08.085.

Liu, J., Cui, L., and Losic, D. (2013) 'Graphene and graphene oxide as new nanocarriers for drug delivery applications', *Acta Biomaterialia*, 9(12), pp. 9243–9257. doi:10.1016/j.actbio.2013.08.016.

Liu, W-K. *et al.* (1993) 'The effects of electric current on bacteria colonising intravenous catheters', *Journal of Infection*, 27(3), pp. 261–269. doi:10.1016/0163-4453(93)92068-8.

Liu, X. *et al.* (2003) 'Demonstration of etched cladding fiber Bragg grating-based sensors with hydrogel coating', *Sensors and Actuators B: Chemical*, 96(1), pp.468–472. doi.10.1016/S0925-4005(03)00605-1.

Mabeck, J., and Malliaras, G. (2006) 'Chemical and biological sensors based on organic thin-film transistors', *Analytical and Bioanalytical Chemistry*, 384, pp. 343–353. doi:10.1007/s00216-005-3390-2.

Mohammadi, P., Liu, L., and Sharma P. (2014) 'A theory of flexoelectric membranes and effective properties of heterogeneous membranes', *Journal of Applied Mechanics*, 81, p. 011007. doi:10.1115/1.4023978.

Murdan, S. (2003) 'Electro-responsive drug delivery from hydrogels', *Journal of Controlled Release*, 92(1), pp. 1–17.doi.10.1016/S0168-3659(03)00303-1.

O'Neill, H. *et al.* (2016) 'Biomaterial-enhanced cell and drug Delivery: Lessons learned in the cardiac field and future perspectives', *Advanced Materials*, 28. doi:10.1002/adma.201505349.

Otero, T., and Martinez, J. (2014) 'Structural electrochemistry: Conductivities and ionic content from rising reduced polypyrrole films', *Advanced Functional Materials*, 24. doi:10.1002/adfm.201302514.

Parsa, H., Ronaldson, K., and Vunjak-Novakovic G. (2016) 'Bioengineering methods for myocardial regeneration', *Advanced Drug Delivery Reviews*, 96, pp. 195–202. doi:10.1016/j.addr.2015.06.012.

Patel, N., and Poo, M. (1982) 'Orientation of neurite growth by extracellular fields', *The Journal of Neuroscience*, 2, pp. 483–496. doi:10.1523/jneurosci.02-04-00483.1982.

Prévôt, M., Ustunel, S., and Hegmann E. (2018) 'Liquid crystal elastomers—A path to biocompatible and biodegradable 3D-LCE scaffolds for tissue regeneration', *Materials*, 11, p. 377. doi:10.3390/ma11030377.

James, P., *et al.* (2014) 'Smart polymers for the controlled delivery of drugs –A concise overview', *Acta Pharmaceutica Sinica B*, 4(2), pp. 120–127. doi:10.1016/j.apsb.2014.02.005.

Punning, A., Kruusmaa, M. and Aabloo, A. (2007) 'Surface resistance experiments with IPMC sensors and actuators', *Sensors and Actuators A: Physical*, 133(1), pp. 200–209. doi:10.1016/j.sna.2006.03.010.

Qiu, L.Y., and Bae, Y.H. (2006) 'Polymer architecture and drug delivery', *Pharmaceutical Research*, 23(1), pp. 1–30. doi:10.1007/s11095-005-9046-2.

Qu, J. *et al.* (2018) 'Injectable antibacterial conductive hydrogels with dual response to an electric field and pH for localized "smart" drug release', *Acta Biomaterialia*, 72. doi:10.1016/j.actbio.2018.03.018.

Ramasamy, T. *et al.* (2017) 'Smart chemistry-based nanosized drug delivery systems for systemic applications: A comprehensive review', *Journal of Controlled Release*, 258, pp. 226–253.doi.10.1016/j.jconrel.2017.04.043.

Richardson-Burns, S., Hendricks, J., and Martin, D. (2007) 'Electrochemical polymerization of conducting polymers in living neural tissue', *Journal of Neural Engineering*, 4, pp. L6-L13. doi:10.1088/1741-2560/4/2/l02.

Richter, A. *et al.* (2003) 'Electronically controllable microvalves based on smart hydrogels: Magnitudes and potential applications', *Journal of Microelectromechanical Systems*, 12, pp. 748–753. doi:10.1109/jmems.2003.817898.

Sadeghipour, K., Salomon, R. and Neogi S. (1992) 'Development of a novel electrochemically active membrane and 'smart' material based vibration sensor/damper', *Smart Material Structures*, 1, pp. 172. doi:10.1088/0964-1726/1/2/012.

Saja, S. *et al.* (2020) 'Fabrication of low-cost ceramic ultrafiltration membrane made from bentonite clay and its application for soluble dyes removal', *Journal of the European Ceramic Society*, 40(6), pp. 2453–2462. doi.10.1016/j.jeurceramsoc.2020.01.057.

Salzano de Luna, M. *et al.* (2020) 'Light-responsive and self-healing behavior of azobenzene-based supramolecular hydrogels', *Journal of Colloid and Interface Science*, 568, pp. 16–24. doi.10.1016/j.jcis.2020.02.038.

Schmaljohann, D. (2006) 'Thermo- and pH-responsive polymers in drug delivery', *Advanced Drug Delivery Reviews*, 58(15), pp. 1655–1670. doi.10.1016/j.addr.2006.09.020.

Schmidt, C., and Leach J. (2003) 'Neural tissue engineering: Strategies for repair and regeneration', *Annual Review of Biomedical Engineering*, 5, pp. 293–347. doi:10.1146/annurev.bioeng.5.011303.120731.

Schultze, J.W., and Karabulut H. (2005) 'Application potential of conducting polymers', *Electrochimica Acta*, 50(7), pp. 1739–1745. doi:10.1016/j.electacta.2004.10.023.

Shahinpoor, M., and Kim K. (2001) 'Ionic polymer-metal composites: Fundamentals', *Smart Materials and Structures*, 10, p. 819. doi:10.1088/0964-1726/10/4/327.

Shoa, T. *et al.* (2010) 'Electromechanical coupling in polypyrrole sensors and actuators', *Sensors and Actuators A: Physical*, 161(1), pp. 127–133. doi:10.1016/j.sna.2010.04.024.

Simon, D.T. *et al.* (2009) 'Organic electronics for precise delivery of neurotransmitters to modulate mammalian sensory function', *Nature Materials*, 8(9), pp. 742–746. doi:10.1038/nmat2494.

Svirskis, D. *et al.* (2010) 'Electrochemically controlled drug delivery based on intrinsically conducting polymers', *Journal of Controlled Release*, 46(1), pp. 6–15. doi:10.1016/j.jconrel.2010.03.023.

Tahir, Z.M. *et al.* (2005) 'Polyaniline synthesis and its biosensor application', *Biosensors and Bioelectronics*, 20(8), pp.1690–1695. doi:10.1016/j.bios.2004.08.008.

Takashima, W. *et al.* (1997) 'Mechanochemoelectrical effect of polyaniline film', *Synthetic Metals*, 85(1), pp. 1395–1396. doi:10.1016/S0379-6779(97)80289-5.

Tandon, B. *et al.* (2018) 'Electroactive biomaterials: Vehicles for controlled delivery of therapeutic agents for drug delivery and tissue regeneration', *Advanced Drug Delivery Reviews*, 129, pp. 148–168. doi:10.1016/j.addr.2017.12.012.

Tay, C.Y. *et al.* (2011) 'Bio-inspired micropatterned platform to steer stem cell differentiation', *Small*, 7, pp. 1416–1421.

Tiwari, R., and Garcia E. (2011) 'The state of understanding of ionic polymer metal composite architecture: A review', *Smart Materials and Structures*, 20, p. 083001. doi:10.1088/0964-1726/20/8/083001.

Vatanpour, V., and Khorshidi, S. (2020) 'Surface modification of polyvinylidene fluoride membranes with ZIF-8 nanoparticles layer using interfacial method for BSA separation and dye removal', *Materials Chemistry and Physics*, 241, p. 122400. doi:10.1016/j.matchemphys.2019.122400.

Wang, T. *et al.* (2016) 'Electroactive polymers for sensing'. doi:10.17863/cam.257.

Wang, Y., and Kohane, D.S. (2017) 'External triggering and triggered targeting strategies for drug delivery', *Nature Reviews Materials*, 2(6):p. 17020. doi:10.1038/natrevmats.2017.20.

Wang, Z., and Song, J. (2006) 'Piezo-electric nanogenerators based on zinc oxide nanowire arrays', *Science*, 312, pp. 242–246. doi:10.1126/science.1124005.

Wasim, M. *et al.* (2017) 'Mixed matrix membranes: Two step process modified with electrospun (carboxy methylcellulose sodium salt/sepiolite) fibers for nanofiltration', *Journal of Industrial and Engineering Chemistry*, 50, pp. 172–182. doi:10.1016/j.jiec.2017.02.011.

Wasim, M. *et al.* (2017) 'Preparation and characterization of composite membrane via layer by layer assembly for desalination', *Applied Surface Science*, 396, pp. 259–268. doi:10.1016/j.apsusc.2016.10.098.

Wasim, M. *et al.* (2019) 'Fractionation of direct dyes using modified vapor grown carbon nanofibers and zirconia in cellulose acetate blend membranes', *Science of the Total Environment*, 677, pp. 194–204. doi:10.1016/j.scitotenv.2019.04.351.

Wasim, M. *et al.* (2017) 'Decoration of open pore network in Polyvinylidene fluoride/MWCNTs with chitosan for the removal of reactive orange 16 dye', *Carbohydrate Polymers*, 174, pp. 474–483. doi:10.1016/j.carbpol.2017.06.086.

Wasim, M. *et al.* (2019) 'Crosslinked integrally skinned asymmetric composite membranes for dye rejection', *Applied Surface Science*, 478, pp. 514–521. doi.10.1016/j.apsusc.2019.01.287.

Whiter, R., Narayan, V., and Kar-Narayan, S. (2014) 'Nanogenerators: A scalable nanogenerator based on self-poled piezo-electric polymer nanowires with high Energy conversion efficiency', *Advanced Energy Materials*, 4. doi:10.1002/aenm.201400519.

Wu, Y. *al.* (2007) 'Soft mechanical sensors through reverse actuation in polypyrrole', *Advanced Functional Materials*, 17, pp. 3216–3222. doi:10.1002/adfm.200700060.

Xia, H., and Tao, X. (2011) 'In situ crystals as templates to fabricate rectangular shaped hollow polyaniline tubes and their application in drug release', *Journal of Materials Chemistry*, 21(8), pp. 2463–2465. doi:10.1039/c0jm03635h.

Yan, J., Marshall, J., and Terentjev E. (2012) 'Nanoparticle-liquid crystalline elastomer composites', *Polymers*, 4. doi:10.3390/polym4010316.

Ye, W. *et al.* (2020) 'Enhanced fractionation of dye/salt mixtures by tight ultrafiltration membranes via fast bio-inspired co-deposition for sustainable textile wastewater management', *Chemical Engineering Journal*, 379, p. 122321. doi.10.1016/j.cej.2019.122321.

Yilmaz, N.D. (2018) 'Multicomponent, semi-interpenetrating-polymer-network and interpenetrating-polymer-network hydrogels: Smart materials for biomedical applications', in Thakur, V.K and Thakur, M.K. (eds.) *Functional Biopolymers*, Cham: Springer International Publishing, pp. 281–342.

Zhang, J. *et al.* (2014) 'Mechanistic insights into response of *Staphylococcus aureus* to bioelectric effect on polypyrrole/chitosan film', *Biomaterials*, 35(27), pp. 7690–7698. doi.10.1016/j.biomaterials.2014.05.069.

Zhang, X. *et al.* (2009) 'Development of electrorheological chip and conducting polymer-based sensor', *Frontiers of Mechanical Engineering in China*, 4, pp. 393–396. doi:10.1007/s11465-009-0043-8.

Zhang, Y., Ma, C., and Dai, L. (2007) 'Electrode preparation and electro-deformation of ionic polymer-metal composite (IPMC)', *2nd IEEE International Conference on Nano/Micro Engineered and Molecular Systems*, Bangkok, Thailand.16–19 January.

Zhao, Y. *et al.* (2013) '3D nanostructured conductive polymer hydrogels for high-performance electrochemical devices', *Energy & Environmental Science*, 6, p. 2856. doi:10.1039/c3ee40997j.

Zhu, Z. *et al.* (2020) 'Dyes removal by composite membrane of sepiolite impregnated polysulfone coated by chemical deposition of tea polyphenols', *Chemical Engineering Research and Design*, 156, pp. 289–299. doi.10.1016/j.cherd.2020.02.001.

Zupan, M., Ashby, M., and Fleck, N. (2002) 'Actuator classification and selection—The development of a database', *Advanced Engineering Materials*, 4, pp. 933–940. doi:10.1002/adem.200290009.

Zvitov, R., Zohar-Perez, C., and Nussinovitch, A. (2004) 'Short-duration low-direct-current electrical field treatment is a practical tool for considerably reducing counts of gram-negative bacteria entrapped in gel beads', *Applied and Environmental Microbiology*, 70, pp. 781–784. doi:10.1128/aem.70.6.3781-3784.2004.

14 Electroactive Polymers for Environmental Remediation

Dong Guo and Kai Cai

CONTENTS

14.1   INTRODUCTION

Electroactive polymers (EAPs) are special functional polymers whose shape or size can be changed by electrical stimulation via charge polarization or conduction. Due to their inherent advantages of low weight, large area, ease of processing, and mechanical flexibility, EAPs have been investigated recently.[1–3]

There are two main types of EPAs, ionic EPAs and electronic EPAs. Typical ionic EPAs include conducting polymers (CPs), responsive gels, and ionic polymer–metal composites (IPMCs). The actuation mechanism in ionic EAPs is ion displacement, which requires a small voltage and high electrical power because of the ionic flow. Electronic EPAs include electrostrictive elastomers, ferroelectric polymers and dielectric electroactive polymers (DEAPs), or dielectric elastomers (DEs). DEAPs have recently received great attention because of their large actuation strains combined with a fast response and high energy; however, they usually require very high activation fields.

Based on market classification, EPAs have various applications including sensing, actuation, energy storage (e.g., capacitors and batteries), electrostatic discharge (ESD) protection, and electromagnetic interference (EMI) shielding. Market analysis indicates that actuation is the most important of these applications. The increasing demand for this type of application is due to their high working efficiency compared with other materials. In addition to these applications, EAPs are promising key materials and components that could enable many new applications in the rapidly growing flexible electronics for energy, healthcare, medical equipment, entertainment, communication, security, and defense.[4,5]

DOI: 10.1201/9781003173502-14

The increasing environmental problems are driving the world toward environmentally friendly solutions, and the environment and energy have become linked issues with high priority for humans. EAPs have unique functionalities in alleviating environmental and energy problems. In this chapter, a summary and analysis of the important role of EAPs for environmental remediation is presented. These issues are discussed from two aspects, such as environmental concerns related to their fabrication and application. Considering the available reviews and summaries on ionic EPAs, emphasis in this chapter is placed on electronic EPAs.

14.2 ENVIRONMENTAL CONCERNS RELATED TO ELECTROACTIVE POLYMER FABRICATION

The research into EAPs started in 1880, when Rontgen observed shape change in rubber by sprayed electric charges[6]. The first piezoelectric polymer (i.e., piezoelectret) was fabricated in 1925 by mixing wax (carnauba wax and beeswax) and rosin and then cooling the mixture under an electrical field[7]. A milestone in EAPs was the discovery of strong piezoelectricity in the fluoropolymer polyvinylidene fluoride (PVDF)[8]. This stimulated research interest in developing other polymer systems that had a similar effect, but fluoropolymers are the prototypes in this category. Another breakthrough following this was the discovery of the electrically CPs halogen-doped polyacetylene[9]. A number of piezoelectric and CPs have been produced since the end of the 1980s[10].

14.2.1 Environmentally Friendly Fabrication of Electroactive Polyvinylidene Fluoride

PVDF is the prototype and most widely used piezoelectric and dielectric polymer, and is used for new sensors, nanogenerators, energy conversion, storage devices, and emerging self-powdered devices. PVDF is widely used for various DEAPS[11,12] as the non-EAP matrix. A fundamental problem of PVDF is the easy environmentally friendly fabrication of its electroactive β-phase polymer.

PVDF has four main conformational polymorphs (e.g., α-, α-, γ-, and δ-phases), and the usual kinetically favored non-polar α-phase with TGTG' conformation is not electroactive and shows no piezoelectricity, and the electroactive all-trans β-phase and the γ-phase with T3GT3G' conformation are very difficult to fabricate directly under ambient conditions[13–15]. The transition conditions for the different phases of PVDF is shown in Figure 14.1. The formation of the β-phase PVDF EAP that has the largest spontaneous polarization and the strongest piezoelectricity usually requires stretching[16] or an ultra-high electric field ≤ 500 MV/m[17,18], and stretching is commonly used in fabrication[19.] However, the PVDF films or fibers formed might partially depolarize and have a poorer performance[20]. The low efficiency and complicated stretching or high voltage caused environmental problems and seriously limited the application of PVDF EAP in high yield device fabrication and retarded the development of new devices that required compatibility with integrated circuits. Environmentally friendly fabrication of PVDF EPA remans a challenge and has been the focus of many studies.

By combing uniaxial stretching and a high electrical field, electrospinning can produce β-phase PVDF nanowires with a high yield[20,21]. In recent study, kilometer-long, parallel, electroactive stable PVDF micro and nanoribbons were produced via an environmentally friendly thermal fiber drawing technique that could iteratively reduce the size of the product. This method has high efficiency, and no high electrical poling is required; however, high temperature and high stress are needed during thermal drawing. These electrospun electroactive PVDF nanowires have been used in energy harvesters and nanogenerators; however, the method is not feasible for large area devices[22]. The rarely reported ferroelectric δ-phase PVDF with a high d_{33} of -36 pm/V measured under a high field of 250 MV/m[23] was obtained recently by a solid state route, and the required high pressure (approximately 20 kN/cm) could be prohibitive for large area fabrication. Therefore, a stress and

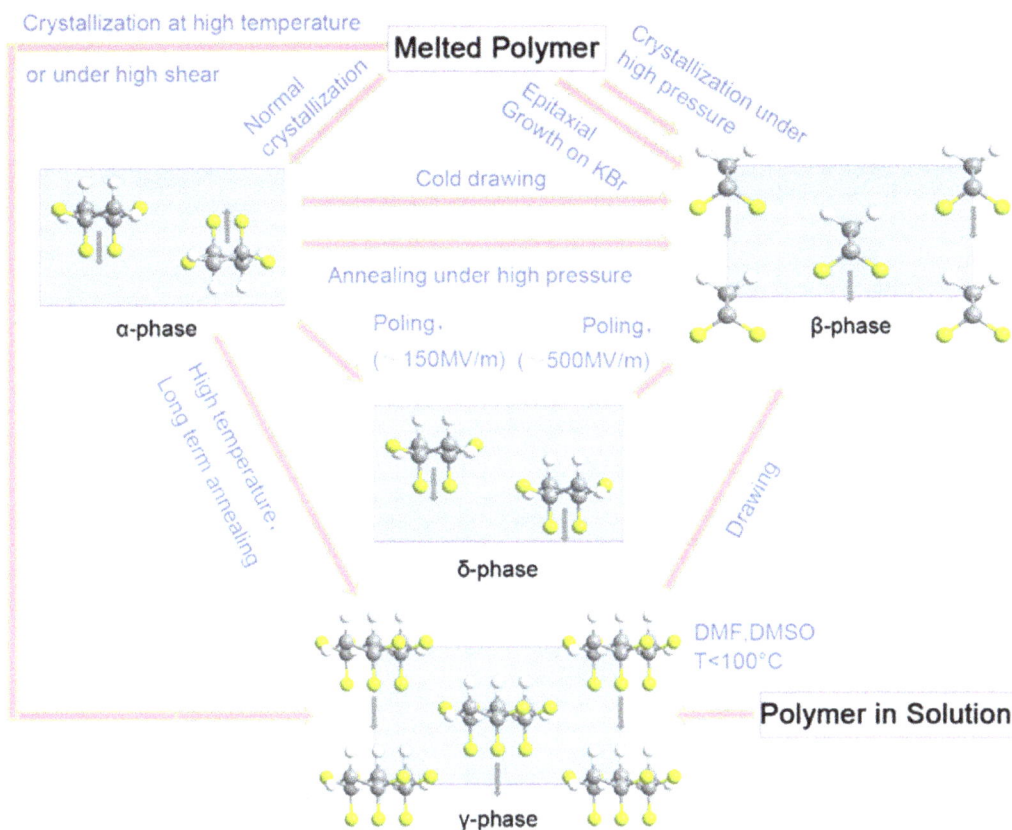

FIGURE 14.1   Formation routes for PVDF EAP.

high voltage free technique that enables highly efficient fabrication of large area flexible piezoelectric PVDF film remains challenging.

Polymer-based materials have some special characteristics, two properties that are relevant to their phase structure are the confinement effect and filler induced transition between different polymorphs. Based on these mechanisms[24], a recent work reported an environment friendly solution route for fabricating electroactive PVDF-based nanocomposite film using spontaneously aligned molybdenum disulfide (MoS_2) nanosheets as the filler via super-two-dimensional-confinement (2D). Only 3.4 vol% of the aligned MoS_2 nanosheets induced a very high ratio of β-phase PVDF of approximately ≤86% in the matrix. A proposed mechanism for that was disturbed crystallization and a synergistic effect between the enhanced electrostatic interaction and 2D geometry confinement by the flat nanosheets, as shown in Figure 14.2. The best nanocomposite with optimum composition demonstrated improved mechanical properties and increased capacitive energy storage density. These composites could be promising for applications that require different functionalities. However, the increased conductivity of MoS_2 caused problems in poling, which is necessary to produce piezoelectricity in PVDF. Further work should overcome this challenge.

14.2.2   Environmentally Friendly Synthesis of Conducting Polymers

Since the discovery of the highly conductive poly(acetylene), CPs have received global interest as a unique type of EAP. CPs are conjugated macromolecules with both special electronic and optical properties because of the overlapped π-orbitals (Barford 2005). Except for poly(acetylene),

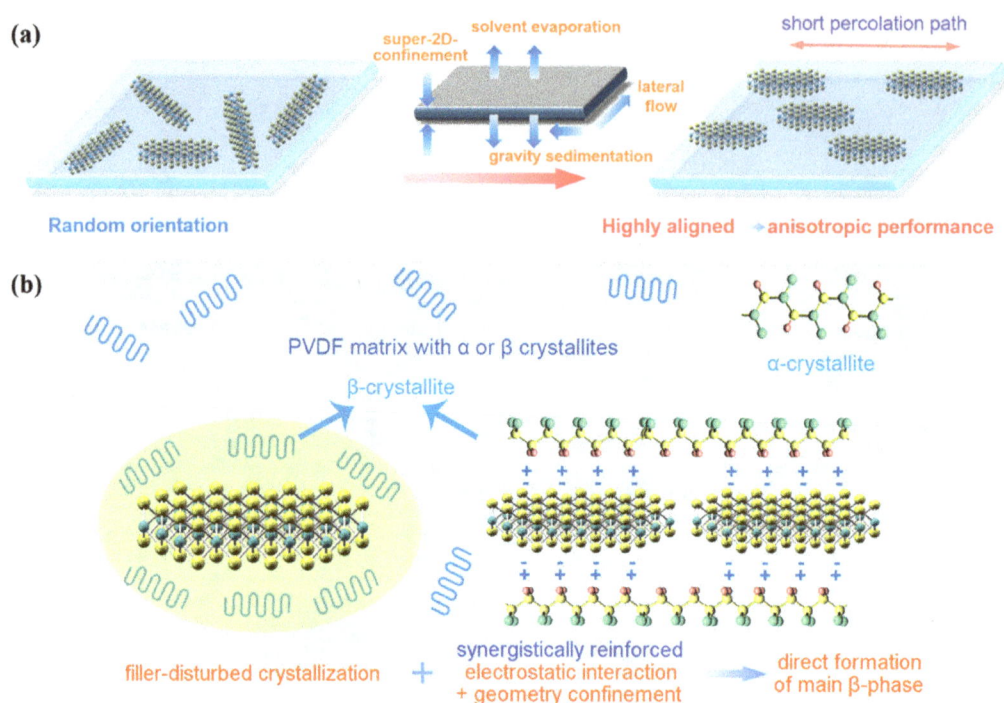

FIGURE 14.2 Showing: (a) proposed mechanisms for the self-alignment of 2D MoS$_2$ nanosheets; and (b) proposed mechanisms for the MoS$_2$ filler induced β-phase formation in the PVDF matrix.

(From Guo et al. 2018. With permission.)

typical CPs also include poly(aniline) (PANI), poly(p-phenylene), poly(p-phenylene vinylene), and poly(pyrrole). The mechanisms of CP that act as EAPs are based on dimensional changes that are produced by changes in their redox state via ion intercalation in the macromolecular network with the aid of electrolytes to maintain charge neutrality[25]. During oxidization, electrons in the macro-molecular backbone of the CPs are depleted and positive charges are created and delocalized. Then, for charge neutralization, anions that are incorporated into the macromolecular chains cause expansion. The process is reversible, and contraction of the polymer can be induced by reduction.

There are different methods for synthesizing CPs. Except for emulsion polymerization and plasma polymerization, electrochemical and photochemical methods are widely used methods[26,27]. Because the synthesis methods have a significant influence on the properties of the CPs, new synthesizing routes, in particular, more efficient and environmentally friendly methods have been important topics.

Among the methods, electrochemical synthesis should be highlighted due to its more environmentally friendly nature, such as simplicity, low-cost, high reproducibility, and can make uniform materials. A great deal of work has been conducted on electrochemical synthesis. In this method, CPs are synthesized by anodic oxidizing of the monomers. CP thin layers on current collectors that are synthesized by electrochemical methods might have advantages of a controllable doping level and high conductivity with various morphologies. The medium for the electrochemical reaction is crucial to obtain products with desirable characteristics. Solvents including acetonitrile, benzene, and chloroform might cause problems, such as over-oxidation, difficult doping, and defects that are generated by nucleophilic reactions by cationic species. Therefore, new solvents have been tried, and examples of these are ionic liquids (ILs), such as dialkylimidazolium cations, tetraalkyl ammonium salts, and fluorinated anions. However, in addition to high production and purification cost the use of ILs causes environmental issues due to the toxicity of the components involved. A method that attempted to resolve these problems in the electrochemical production of CPs was the use of deep eutectic solvents (DES).

Eutectic solvents are liquids with a lower melting temperature that are produced by mixing two components, such as the hydrogen bonding donor and the hydrogen bonding acceptor[28,29]. DES can be prepared using non-toxic natural components with physiological or metabolic functions in living organisms,[30] or therapeutic components that can be used as ingredients for pharmaceuticals[31,32]. Compared with ILs and the usual organic solvents, DES are environmentally friendly with advantages of high biodegradability, non-toxic, and ease of preparation. Previous work in this category was the fabrication of PANI films using a DES that consisted of 1,2-ethane-diol and choline chloride with a molar ratio of 2:1[33]. The morphology of the PANI films fabricated by potentiodynamic (10 scans) and potentiostatic (100 s) methods shown in Figure 14.3 confirmed that more compact PANI films with a higher surface coverage could be produced in shorter times by the potentiostatic approach than by the potentiodynamic approach. In addition, the CP film had high electroactivity and conductivity.

With increased environmental issues for the applications, more environmentally friendly, bio-compatible, and sustainable solvents are required for the synthesis of CPs with modified properties for different applications. This will be the subject of future research.

14.3 APPLICATION OF ELECTROACTIVE POLYMERS TO REMEDIATE ENVIRONMENTAL AND ENERGY ISSUES

14.3.1 ELECTROACTIVE POLYMER ACTUATORS WITH LOW ENERGY CONSUMPTION

EAPs most important application is actuation. This is based on EPAs characteristics, such as large deformation ability, damage tolerance, and compatibility with various systems. As previously mentioned, ionic EAPS and DEAPs have different actuation mechanisms, and they have demonstrated potential for application in various actuators, such as artificial muscles, bioinspired robots, and auto positioners. In 2002, Eamax, Japan invented an EAP-based biomimetic actuator, a special robot that can move like a fish. Then, different mechanisms were tried for the used in robotics, automobiles, and aerospace[34], with different performances that depended on their specific structures and designs.

Conventional actuator systems have complex mechanics that generate noise and vibrations, which can cause wear, instability, or other problems. If sensing components are required, switching and integration are difficult. For environmental issues, conventional actuators usually have high energy consumption and high operating costs. Compared with conventional actuators, EAP-based actuators have advantages of high strain level, large deformation, high energy convention efficiencies, and low energy consumption[35]. For design, they have advantages of simple mechanics with sensing and tactile feedback functionalities. Figure 14.4 shows a qualitative comparison of different actuators for power consumption and strength. In this figure SMA, MFC, TCP, and PEA are abbreviations for shape memory alloy, microfiber composite, twisted and coiled polymer, and piezo-element actuator, respectively[36]. According to this figure, the EAP-based system appears to be the most efficient for power consumption and the stacked EAP appears to be the strongest system despite the increased weight.

An example of a stacked EAP actuator is the encapsulated multilayer DEAP actuator that is based on prefabricated polyurethane and silicone with homogeneous and reproducible properties[37]. The actuators were fabricated by dry deposition that used homogeneous prefabricated polymer films by a roll-to-sheet method. The method had advantages of high efficiency, because the time-consuming curing was removed, and the parallel production of several actuators was achieved. Similar methods can be used for actuators with different sizes or geometries. Maximum tensile forces $\leq 10\,N$

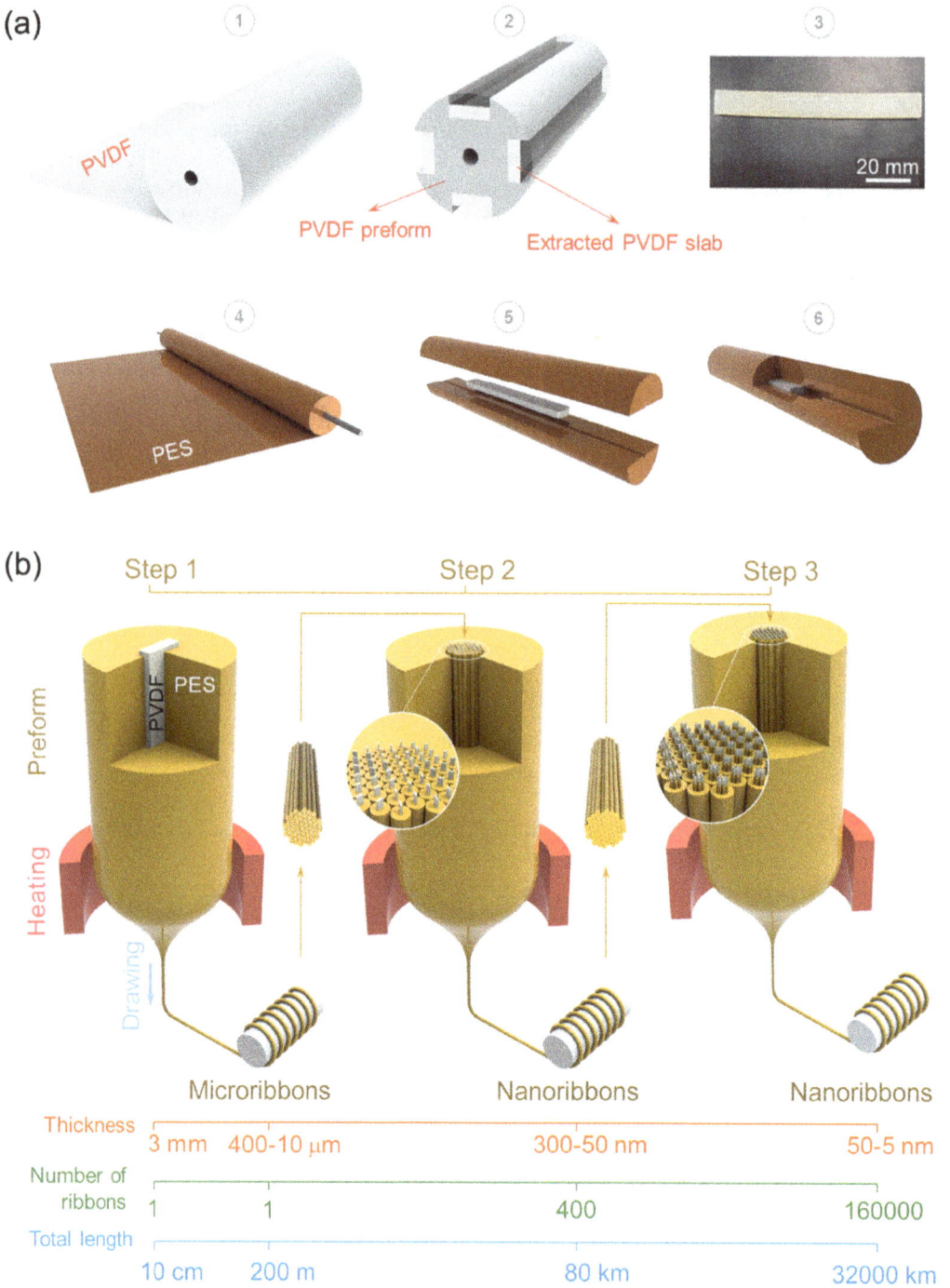

FIGURE 14.3 An environmentally friendly way for electroactive PVDF nanoribbons: (a) procedures for fabricating PVDF ribbons embedded in poly(ether sulfone) (PES) preform; and (b) iterative nanofiber drawing procedure. In this technique, a PVDF slab is thermally sealed in a PES mold and drawn to form the initial fibers. At elevated temperatures, the PVDF melt flows together with softened PES cladding, forming long PVDF microribbon encapsulated in the PES matrix.

(From Kanik et al. 2014. With permission.)

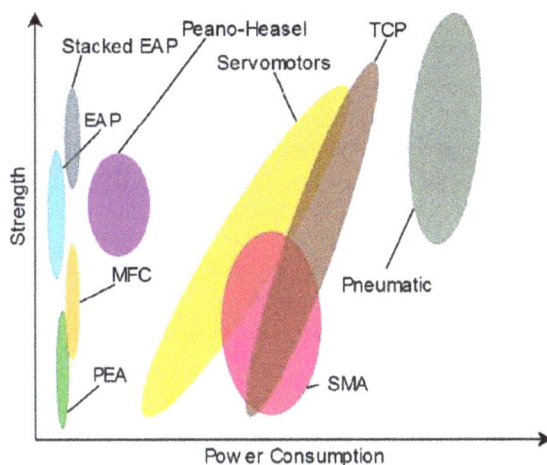

FIGURE 14.4 A summary for comparing of the strength and the power consumption of different actuators.

(From Gomez-Tamm et al. 2019. With permission.)

(PUR) or 4 N (silicone) with a maximum deformation of approximately 3.5% can be produced by the stack-actuators under an electrical field of 50 V/lm. The study demonstrated that EAP-based actuators with optimized design had low energy consumption and could be fabricated in an environmentally friendly way, which was based on the advantages of easy of processing of the polymer materials.

14.3.2 APPLICATION OF CONDUCTING POLYMERS IN ENVIRONMENTAL REMEDIATION

For environmental issues, CPs have applications in pollutant detection, quantitation, and removal. These applications can be achieved via separation when CPs are used as membranes, filters, ion exchangers, or sorbents. In addition, CPs can be used as chemical reactors for removing pollutants via chemical processes, such as redox or complexation, and there are several different mechanisms. Under an applied potential in the presence of CPs or composites, the pollutants might be removed via incorporation or absorption and release, doping and dedoping, and electrocatalytic reduction and oxidation. These will be discussed in the following sections.

Thin layers of CPs on conductive substrates can be used to remove charged pollutants from liquid wastes. This might be achieved via complexation of ions, such as mercury (Hg^{2+}) via imine or other functional groups that contains nitrogen in a matrix, such as PANI[3] or via the nitrogen in the polypyrrole (PPy) ring for eliminating Hg^{2+}, cadmium (Cd^{2+}), and lead (Pb^{2+})[39]. The elimination of charged ionic pollutants is a type of electrochemically aided ion exchange.[40] This might be achieved via charge compensation. This electrochemically assisted process is electrochemical ion exchange. Selectivity in removing the different ions can be achieved based on the size, charge, electroactivity, and other chemical characteristics, such as complexation ability, mobility, and alkalinity or acidity.

The electron transfer function of CPs can be used for pollutant remediation via electrocatalytic reduction or oxidation.[41] An example of electrocatalytic reduction is the treatment of highly mobile chromium (Cr) containing ions, such as $Cr_2O_7^{2-}/CrO_4^{2-}$ with PPy or composites of PPy with carbon nanotubes and cellulose fibers.[42,43] This may mainly involves two processes, the chemical interaction between PPy and CrO_4^{2-} that is caused by the difference in reduction potential[44] and the exchange of anions that is driven by electroneutrality that is a result of polymer oxidation.[45] In electrocatalytic oxidation, a electrocatalytic species of the pollutants to be removed is introduced and dissolved into the polymeric matrix, which works as an oxidizer. An example is the oxidation of arsenic [As(III)] to the much less toxic As(V).[46]

In addition, CPs can be used as pollutant adsorbers. An example is the environmentally benign composite that consists of PANI and chitosan, which can be used for the removal of large amounts of dye molecules from aqueous solutions[47], and the amino and hydroxyl groups are critical for the absorption process. Another example is the removal of the toxic As(III) species by using polythiophene based on the interaction of the ions with the π-electrons of the polymer backbone[48]. The effect for pollutant removal can be enhanced by depositing CPs on three-dimensional porous medium, such as PANI hollow nanospheres. Which can increase the electroactive area and reaction rate.

CPs can be used to remove pollutant gases with low molecule weights, such as hydrogen sulfide (H_2S), sulfur dioxide (SO_2), nitrogen oxides (NOx), ammonia (NH_3), and carbon monoxide (CO),[49] or organic pollutants, such as benzene, toluene, and acetone, by physical adsorption or reactive chemical adsorption via composites, such as the system that consists of graphite oxide and PANI. Reductive dechlorination as a method for removing harmful halogenated organics can be achieved using CPs. Harmful 3,3',4,4-tetrachlorobiphenyl has been reduced to less toxic biphenyl in the presence of hydrogen by a PPy matrix with palladium (Pd) nanoparticles[50]. CPs have been used for the photocatalytic processing of pollutants. An example is PANI, which has been used as a support and sensitizer for a titanium dioxide (TiO_2) photocatalyst via hydrothermal processing for the photochemical degradation of dye molecules, such as anthraquinone and reactive brilliant blue by irradiation with visible light[51,52]. Similarly, poly(3-hexylthiophene) (P3HT) coupled with TiO_2 has been used as the photocatalyst for the degradation of methylene blue[53]. The mechanisms might be attributed to the enhanced charge separation in photocatalysis by the heterojunction that is formed between the semiconducting oxide and CPs[54].

14.3.3 APPLICATION OF PIEZOELECTRIC POLYMERS IN ENVIRONMENTAL REMEDIATION

Piezoelectric polymers, which are special DEAPs that can convert mechanical strain or vibration into electrical signal, have been widely used for ultrasonic transducers, especially for underwater and medical applications. A direct and interesting case for environmental remediation is the application of a prototype piezoelectric PVDF membrane for pollutant removal.

Membranes are widely used for the filtration of pollutants and purification of liquids. Fouling is a typical problem in membranes that are used for separating liquid. Fouling of membranes might require severe chemical treatments, which involves increased energy consumption and a reduction in the membrane lifespan that causes environmental problems. Vibrations that are generated on the surface of a membrane by external forces, such as ultrasonic waves are an effective way to reduce membrane fouling. Electrically poling PVDF membranes transformed them into piezoelectrically active membranes where vibration could be excited by AC signals[55]. Measurements of flux and separation performance under cross flow with a configuration (Figure 14.5) show that vibrations excited out of the membrane plane could increase the flux by an order of magnitude, which lead to less membrane fouling as shown in Figure 14.6. In addition, an increased cross flow velocity could further improve the relative antifouling effect under vibration.

14.3.4 ENERGY HARVESTING FROM ENVIRONMENTAL ENERGY RESOURCES

The applications discussed in the previous section showed that EAPs might be used for resolving environmental issues via different physical or chemical mechanisms. Based on a better understanding of the concept of EAPs, such as the ability to deform by electrical stimuli, the functions of EAPs are based on electromechanical responses. Applications that are based directly on this function include energy harvesting, nanogenerators, actuators, and sensors.

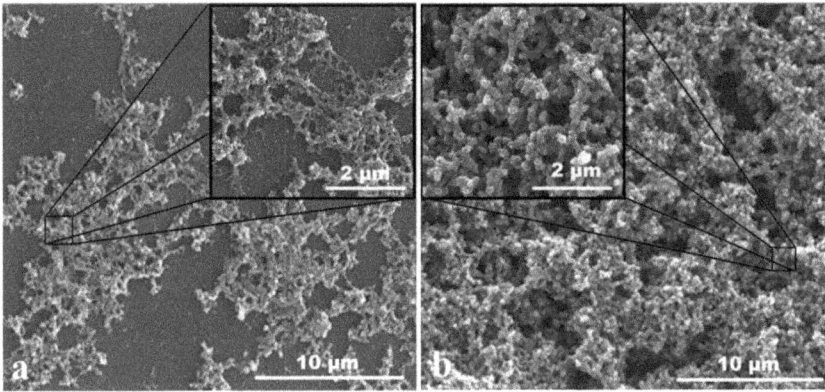

FIGURE 14.5 SEM images of the GCE electrodes modified with potentiodynamic (CV-PANI-Prop): (a) and potentiostatic (PS-PANI-Prop) films; and (b) prepared with 10 scans between 0 and +1.4 V and 100 s at +1.4 V, respectively. Environmentally friendly deep eutectic mixtures were used as the electrolytes for fabricating the CP.

(From Fernandes 2012. With permission.)

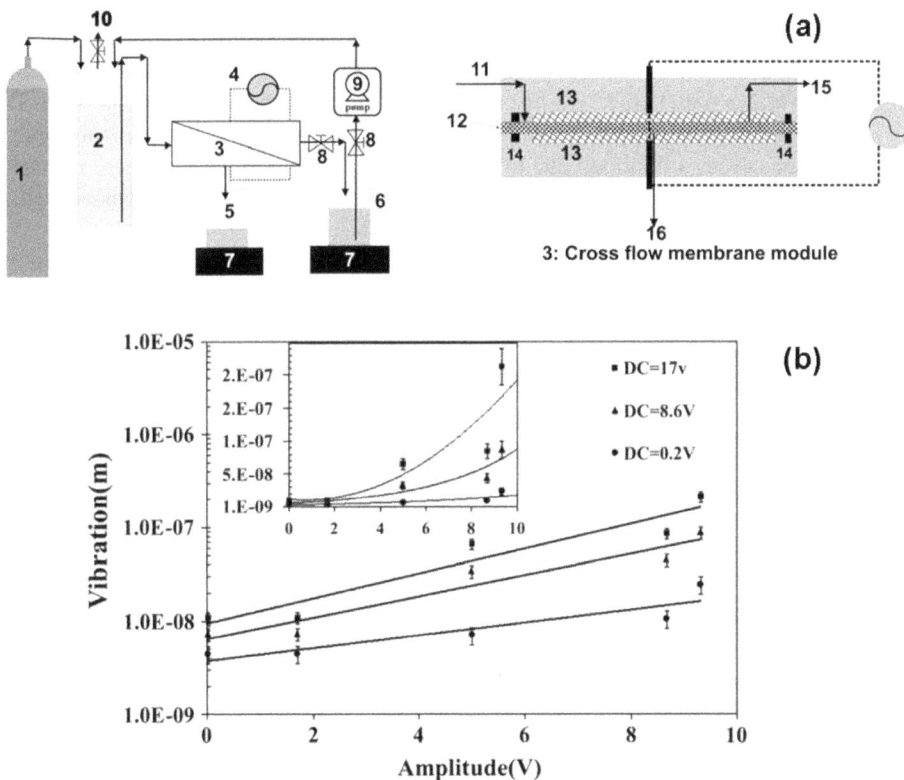

FIGURE 14.6 Showing: (a) filtration test configuration, where 1 to 16 represent compressed air, solution reservoir, cross flow module, signal generator, permeate reservoir, flow collection reservoir, balances, adjusting valve, pump, pressure relief valve, feed solution inlet, PVDF membrane, fabric electrodes, sealing ring, flow outlet, and permeate outlet; and (b) influence of 9 V amplitude AC signals on the flux through piezoelectric PVDF membranes during filtration of 1% PEG under 175 kPa with a cross flow of 6–7 mL/min.

(From Coster 2011. With permission.)

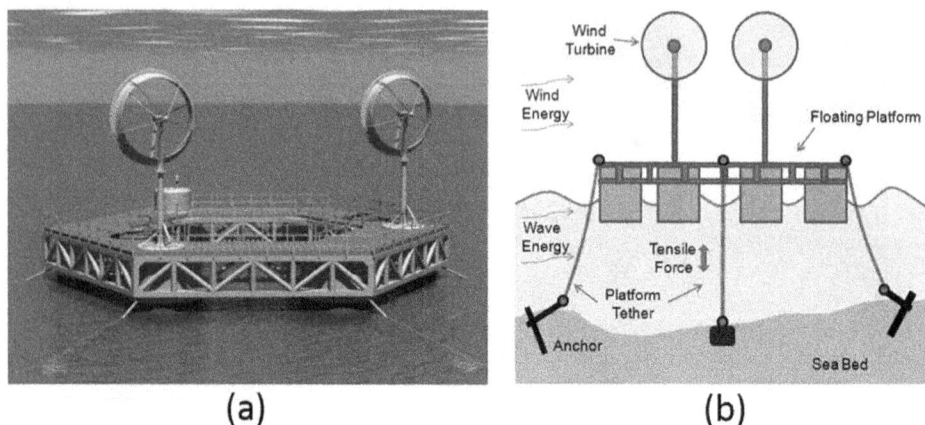

(a)   (b)

FIGURE 14.7 Offshore floating structure for wind energy harvesting: (a) a floating structure with wind turbines; and (b) fixation of the floating structure by cables.

(From Silleto et al. 2015. With permission.)

The environment and energy are closely linked. Global climate change and warming has prompted intensive efforts in the research and development of technologies for energy harvesting. Vibration energy harvesting is a technique by which wasted vibration is harvested and converted into other types of energy. Various materials might be used for vibration energy harvesting; however, DEAPs have advantages over conventional methods. There are many reports and reviews on this topic; however, practical applications are limited due to problems, such as non-linearity, power management, and low efficiency. In this section, three typical applications will be briefly discussed, which are energy harvesting from the wind, ocean, and human motion.

Recent studies by researchers from Kyushu University, Japan reported an offshore energy platform design that consisted of connected offshore floating modules that used environmental energy. This structure can generate electricity using a novel wind turbine with a high power based on 'wind lens' technology as shown in Figure 14.7[56]. The floating modules are to be fixed by high-tension cables that are directly connected to the ocean floor[57]. The main functional component of this work is an energy harvesting damper that consists of a composite of silicone rubber and the piezoelectric polymer (PVDF) film as shown in Figure 14.8. The energy harvesting efficiency of coiled and sandwiched sheet geometries were compared and the results are given Table 14.1[58] Experimental and theoretical data showed that the coiled geometry had a higher power with more uniform and larger deformation, because of the non-localized large deformation, and the sandwiched geometry showed better durability. The researchers found that compared with the results from a piezoelectric cable, the coiled and sandwiched designs exhibited higher and lower power generating capability, respectively, and both designs had a lower energy cost. These results demonstrated that modified structure and designs could improve the performance in energy harvesting of a piezoelectric EAP.

The ocean is a large resource for environmental energy. Harvesting energy from oscillations based on ocean currents or waves using EAPS has been conducted for several decades. There are several types of ocean wave energy harvesters, such as oscillating water columns and overtopping devices. Representative works are the studies on DEAP type EAP artificial muscle (EPAM)[59]. Among the different methods of electromechanical working principles, a method based on a constant voltage cycle as shown in Figure 14.9 was used. Mechanical stretching of EPAM causes reduced thickness and an enlarged surface area and this could be carried out under an applied voltage. The increased thickness after removing the forces could push positive and negative charges further away from each

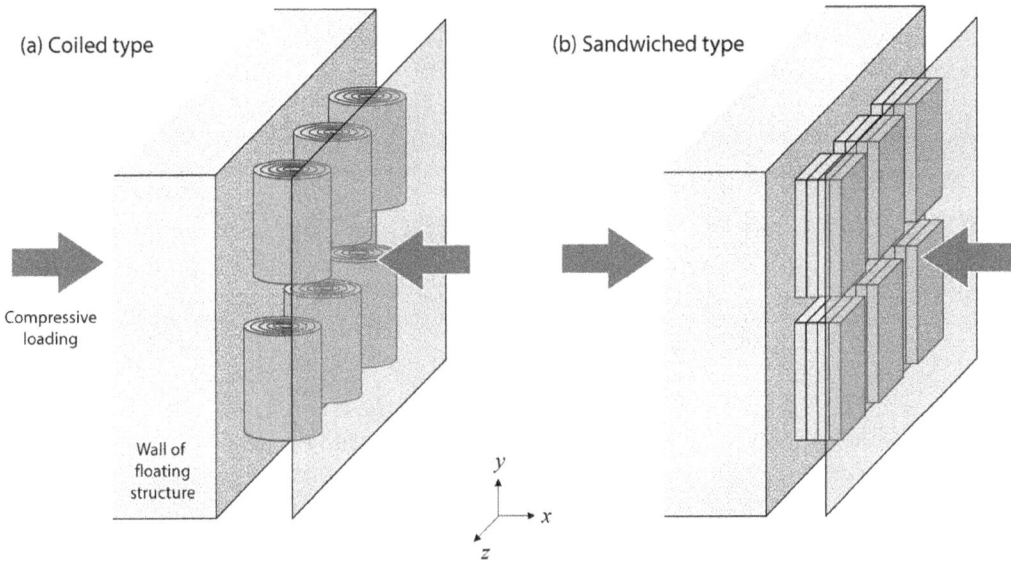

FIGURE 14.8 Two types of PVDF–silicone composites used in the floating structures for the wind energy harvester: (a) coiled configuration; and (b) sandwiched configuration.

(From Yoon et al. 2015. With permission.)

TABLE 14.1

Comparison of power generating performance and energy costs for different energy harvesting systems

Harvesting method	Power generation capability (μW/ cm²)	Costs ($/Wh)	Frequency (Hz)
Coil-structured damper	982	71	1.5
Sandwich-structured damper	223	227	2
Three-loop piezoelectric cable	521	440	2
Solar photovoltaic system	15000	0.00006	—

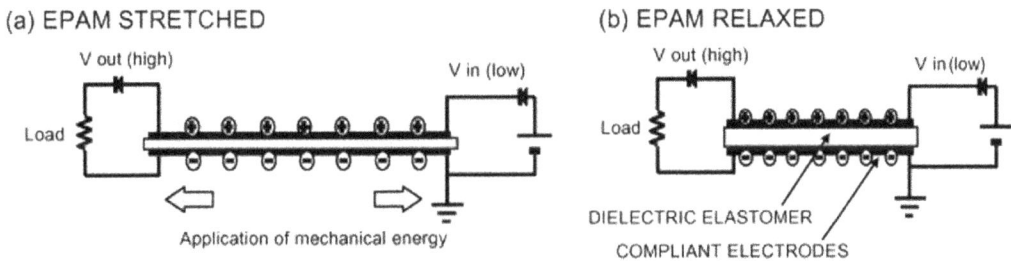

FIGURE 14.9 Operating principle of EPAM power generation.

(From Chiba et al. 2013. With permission.)

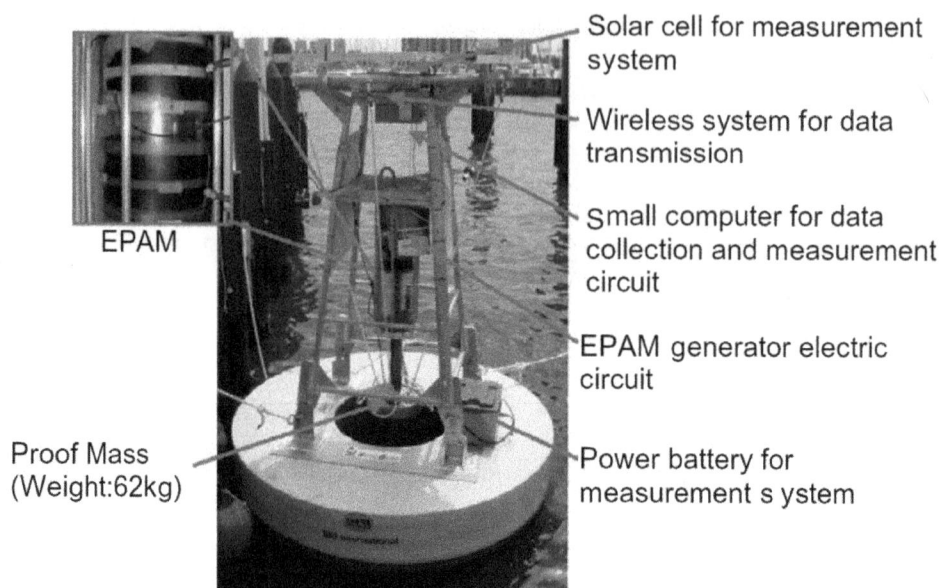

FIGURE 14.10 Prototype EPAM ocean wave energy harvester on a buoy.

(From Chiba et al. 2013. With permission.)

other, which leads to an increased voltage on the EPAM. On relaxation the capacitance is reduced; however, there is still considerable net gain in the stored energy compared with that added on the EPAM by the original voltage, and this energy could be harvested by the system[59].

Researchers designed and made a prototype generator for the on-board power of a navigation buoy and evaluated the feasibility of the generator's application by harvesting energy from ocean waves. The device is a 1.2 m high tubular cylinder with an outer diameter of 40 as shown in Figure 14.10[60]. There are two 20 cm high roll-type hollow EPAM modules with diameters of 30 cm inside the tube. Although the generator showed lower average energy outputs due to very weak waves, a maximum output electrical energy of 20 J per stroke was produced. In addition, the energy volume generated by the EPAM generator was frequency independent using a simple wave tank setup with a floating body that was composed of a linear diaphragm acrylic polymer type EPAM transducer. The relationship between the electricity generated in a single wave cycle with the wave period is shown in Figure 14.11. I a multi-unit EPAM generator was used and aligned along the wave propagation direction, a possible maximum energy conversion efficiency ≤90.6% could be obtained. For ocean wave energy harvesting, challenges reamin due to fabrication and maintenance costs, device lifetime, and failure risk.

14.3.5 Application of Dielectric Electroactive Polymers in Nanogenerators

A nanogenerator is a type of energy harvesting device that has received great attention recently. The device can harvest mechanical or thermal energy that are produced from different systems and converts them into electrical energy. Piezoelectric nanogenerators that use PVDF EAP are one of the most widely investigated devices. Based on the reviews of these devices, only PVDF nanofibers fabricated by near-field electrospinning will be discussed[61,62], where the high electric fields and stretching stress caused dipoles that were preferentially oriented along the nanofiber crystals, and therefore, the polar β-phase structure could be induced from the α-phase without an electrical field for poling and orienting the dipoles as shown in Figure 14.12. Voltage and current outputs can be

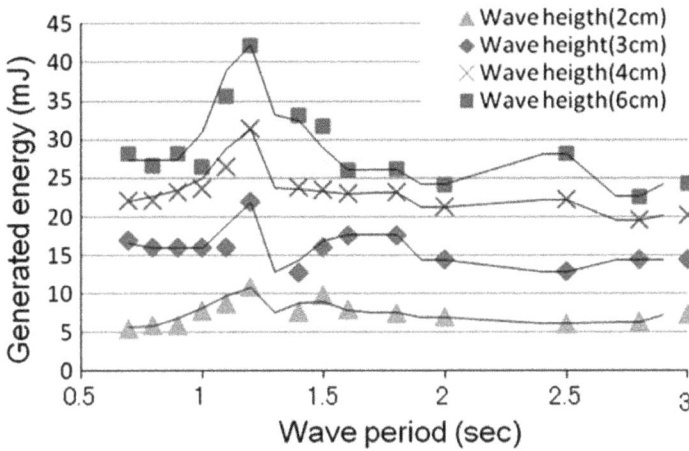

FIGURE 14.11 Comparison of the electrical energy obtained from the buoy-type wave power generator. (From Chiba et al. 2013. With permission.)

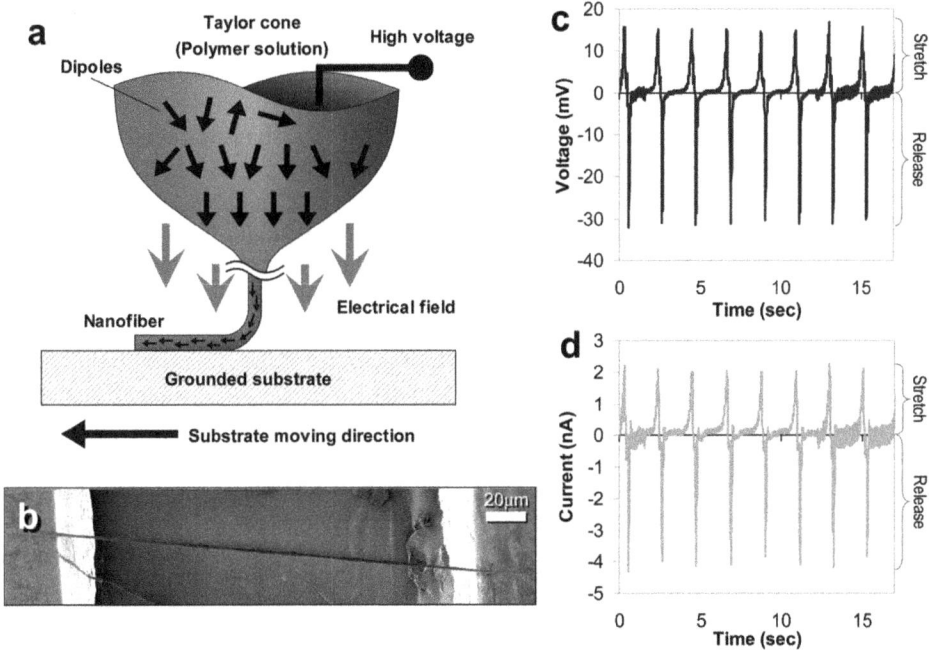

FIGURE 14.12 Showing: (a) PVDF nanogenerator fabrication process, which combines electrospinning, direct writing, stretching, and in situ electrical poling; (b) SEM image of a single fiber PVDF nanogenerator; and (c and d) output voltage and current under repeated strain.

(From Chang et al. 2010. With permission.)

directly produced by the fibers by repeatedly stretching and relaxing the substrate. Web materials that consisted of P(VDF-TrFE) nanofibers 60–120 nm in diameter were fabricated via electrospinning. Because the dipoles in the nanofibers were preferentially oriented perpendicular to the nanoweb plane, better performance in stacked nanofiber webs was predicted. Due to the flexibility of the polymer, the material could provide energy for portable electronic devices, which could be a potential type of environmentally friendly energy resource.

14.3.6 Ionic Electroactive Polymers in Anticorrosion Applications

Corrosion is metal degradation that is caused by interactions between the metals with surrounding material, which is considered to be inevitable. Protection of metals against corrosion is an important topic. Currenlty, a strategy for the additional anticorrosion processing of metallic substrates is the treatment and deposition of protective organic or polymeric coatings. There are many different polymeric materials for anticorrosion coatings. For this application, EPAs with conjugated bonds in their molecular structures with reversible redox property have attracted interest. The mechanisms of an EAP protective coating in improving the anticorrosion property of the underlying metal substrates is explained by the electrocatalytic (i.e., redox) property of the coatings, which result in passive dense metal oxide shells that can prevent damage to the metallic substrates by the environment. An example is the electroactive S-EPU, which is a special sulfonated polyurea, which is one of the most important EPA materials used for anticorrosion coatings.

Currenlty, most corrosion control methods need environmentally hazardous materials. CPs as EAPs are a special type of protective coating materials. Different from the usual polymer protective coatings, CPs are active protection materials that work via oxidation or passivation of the metal substrates, which induces more positive corrosion potential and modifies oxygen (O_2) reduction. PANI and polyimide are typical CPs that are used for anticorrosive coatings, and studies have shown that composites that consist of CPs and oxides usually had a better protection performance than single phase CPs[63]. A series of nanocomposites were prepared that consisted of electroactive polyimide (EPI) and TiO_2 (EPTs). The hybrids contained segments of electroactive amino-capped aniline trimer (ATs) with conjugated bonds and TiO_2 powders with a particle size of approximately 10 nm. Electrochemical cyclic voltammetry has been used to investigate the redox characteristics of these composites. Sequential electrochemical corrosion behavior measured in electrolyte with 5 wt.% of sodium chloride (NaCl) indicated that EPTs that contained a higher percentage of TiO_2 filler showed a better effect in corrosion protection for a cold-rolled steel (CRS) electrode. The effect was attributed to several reasons. In addition to the role as a protection barrier, the redox catalytic functions of the AT units might induce and promote the formation of passive and dense oxide outer layers on the surface of the steel, and the TiO_2 nanoparticles in EPTs might assist the formation of a better O_2 barrier. The mechanisms are shown in Figure 14.13(b) and 14.12(b). Thin layers of oxide were observed at the EPTs/CRS interface (Figure 14.13(c)) but not on the pure CRS surface (Figure 14.13(d)). O_2 permeability measurements for different membranes (Figure 14.13(e)) approximately 60 $\alpha\mu m$ thick showed that the EPI has a lower O_2 permeability than non-EPI and EPT nanocomposites with a higher TiO_2 loading showed an even lower O_2 permeability.

14.4 CONCLUSIONS

Due to their unique electromechanical properties and many other advantages, EAPs have potential applications in a range of areas. Many of these applications are closely related to global environmental issues, which reflects the critical role of EPAs in new environmentally friendly solutions. This chapter discussed environmental issues that concerned the fabrication of EAPs and their applications in environmentally friendly solution in many fields. Substantial advances have been made in material preparation and device fabrication, and with the development of new fabrication techniques and electromechanical systems, EPAs could play a more important role in remediating environmental issues. However, practical applications of EAP materials has many challenges, and typical problems include efficiency, durability, and stability.

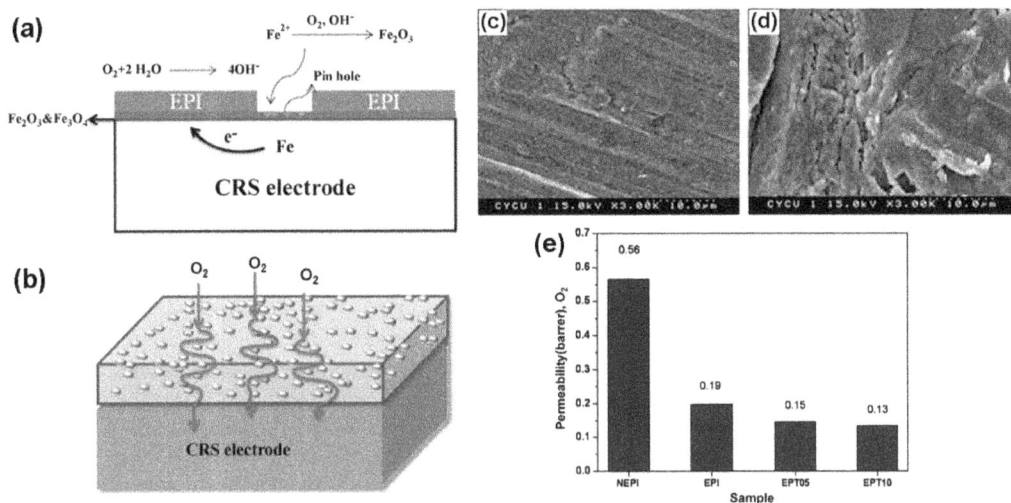

FIGURE 14.13 Showing: (a and b) mechanisms of CRS passivation by EPT coatings and diffusion pathway of O_2 in the EPT, respectively; (c and d) SEM images of polished CRS metal and the surficial EPT10 coating, respectively: and (e) comparison of the permeability of O_2 of the nanocomposites with different content of TiO_2.

(From Weng et al. 2010. With permission.)

The process that includes synthesizing, electroding, shaping, handling, and assembling needs to be refined and modified to optimize functionality and durability. Addressing these challenges requires the development of technology and increased multidisciplinary cooperation between individuals from various fields that include chemistry, material science, robotics, computer science, and electronics. Their unique and unique properties could help to increase the development of relevant techniques.

REFERENCES

1. Zarren G, Nisar B, Sher F.. Synthesis of anthraquinone-based electroactive polymers: a critical review. Mater Today Sustain. 2019; 5.
2. Zhang J, Zhu D, Matsuo M. Synthesis and characterization of polyacene quinone radical polymers with high-dielectric constant. Polymer. 2008; 49:5424–30.
3. Bar-Cohen Y, Zhang Q. Electroactive polymer actuators and sensors. MRS Bull. 2008; 33(3): 173–81. doi.10.1557/mrs2008.42.
4. Hanson D. *et al.* Androids: application of EAP as artificial muscles to entertainment industry. Proc SPIE. 2001.
5. Bolzmacher C. *et al.* Polymer Based Actuators for Virtual Reality Devices. Proc SPIE. 2004; 385:281–9.
6. Keplinger C, Kaltenbrunner M, Arnold N, *et al.* Rontgen's electrode-free elastomer actuators without electromechanical pull-in instability. PNAS USA. 2010; 107(10):4505–10. doi.10.1073/pnas.0913461107.
7. Furukawa T, Seo N. Electrostriction as the origin of piezoelectricity in ferroelectric polymers. Jpn J Appl Phys. 1990; 29(4):675–80. doi.10.1143/jjap.29.675.
8. Finkenstadt VL. Natural polysaccharides as electroactive polymers. Appl Microbiol Biotechnol. 2005; 67(6):735–45. doi.10.1007/s00253-005-1931-4.
9. Shirakawa H. *et al.* Synthesis of electrically conducting organic polymers: Halogen derivatives of polyacetylene, (CH)x. J Chem Soc Chem Comm. 1977; 16:578–80.

10. Gutierrez TJ. Advances in reactive and functional polymers: Editor's perspective. Springer; 2020, p. 327.

11. Faizan *et al.*. Piezoelectric energy harvesters for biomedical applications. Nano Energy. 2019; 57:879–902.

12. Ribeiro C. *et al.* Electroactive poly(vinylidene fluoride)-based structures for advanced applications. Nat Protoc. 2018; 13(4):681–704.

13. B, P. Martins A, *et al.* Electroactive phases of poly(vinylidene fluoride): Determination, processing and applications. Prog Polym Sci. 2014; 39(4):683–706.

14. Hasegawa R. *et al.* Crystal structures of three crystalline forms of poly(vinylidene fluoride). Polym J. 1971; 3(5):600–10.

15. Oka Y, Koizumi N. Formation of unoriented form I poly (vinylidene fluoride) by high-rate quenching and its electrical properties. Institute for Chemical Research, Kyoto University. 1985; 63:192–206.

16. Sharma M, Madras G, Bose S. Process induced electroactive β-polymorph in PVDF: effect on dielectric and ferroelectric properties. Phys Chem Phys. 2014; 16:14792–9.

17. Naegele D, Yoon DY, Broadhurst MG. Formation of a New Crystal Form (α p) of poly(vinylidene fluoride) under electric field. Macromolecules. 1978; 11(6):1297–8.

18. Ranjan V, Nardelli MB, Bernholc J. Electric field induced phase transitions in polymers: A novel mechanism for high speed energy storage. Phys Rev Lett. 2012; 108(8):087802.

19. Gomes J. *et al.* Influence of the β-phase content and degree of crystallinity on the piezo- and ferro-electric properties of poly(vinylidene fluoride). Smart Mater Struct. 2010; 19(6):065010.

20. Fuh YK, Wang BS. Near field sequentially electrospun three-dimensional piezoelectric fibers arrays for self-powered sensors of human gesture recognition. Nano Energy. 2016; 677–83.

21. Baji A, Mai Y, Li Q, Liu Y.. Electrospinning induced ferroelectricity in poly(vinylidene fluoride) fibers. Nanoscale. 2011; 3:3068–71.

22. Kanik M, Aktas O, Sen HS, Durgun E, Bayindir M.. Spontaneous high piezoelectricity in poly(vinylidene fluoride) nanoribbons produced by iterative thermal size reduction technique. ACS Nano. 2014. 8:9311–23.

23. Martin J, Dong Z, Lenz T., *et al.* Solid-state processing of δ PVDF. Mater Horiz. 2017; 4: 10–1039.

24. Cai K, Han X, Guo D. A green route to a low cost anisotropic MoS_2/poly(vinylidene fluoride) nanocomposite with ultrahigh electroactive phase and improved electrical and mechanical properties. ACS Sustain Chem Eng. 2018; 6: 5043–52.

25. Puiggali-Jou A, del Valle LJ, Aleman C. Drug delivery systems based on intrinsically conducting polymers. J Control Release. 2019; 309:244–64. doi.10.1016/j.jconrel.2019.07.035.

26. de Barros RA, Areias MCC, de Azevedo WM. Conducting polymer photopolymerization mechanism: the role of nitrate ions (NO_3^-). Synth Met. 2010; 160:61–4, doi.10.1016/j.synthmet.2009.09.033.

27. Maity N, Dawn A. Conducting polymer grafting: Recent and key developments. Polym. 2020; 12: doi.10.3390/polym12030709.

28. Abbott AP, Capper G, Davies DL, Rasheed RK, Tambyrajah V. Novel solvent properties of choline chloride/urea mixtures. Chem Commun. 2003; 70–1. doi.10.1039/b210714g.

29. Smith EL, Abbott AP, Ryder KS. Deep eutectic solvents (DESs) and their applications. Chem Rev. 2014; 114:11060–82. doi.10.1021/cr300162p.

30. Choi YH, van Spronsen J, Dai Y, *et al.* . Are natural deep eutectic solvents the missing link in understanding cellular metabolism and physiology? Plant Physiol. 2011; 156:1701–5. doi.10.1104/pp.111.178426.

31. Aroso IM, Craveiro R, Rocha A. *et al.* . Design of controlled release systems for THEDES—therapeutic deep eutectic solvents, using supercritical fluid technology. Int J Pharm. 2015; 492:73–9. doi.10.1016/j.ijpharm.2015.06.038.

32. Duarte ARC, Ferreira ASD, Barreiros S. *et al.* . A comparison between pure active pharmaceutical ingredients and therapeutic deep eutectic solvents: solubility and permeability studies. Eur J Pharm Biopharm. 2017; 114:296–304. doi.10.1016/j.ejpb.2017.02.003.

33. Fernandes PMV, Campina JM, Pereira NM, Pereira CM, Silva F.. Biodegradable deep-eutectic mixtures as electrolytes for the electrochemical synthesis of conducting polymers. J Appl Electrochem. 2012; 42:997–1003. doi.10.1007/s10800-012-0474-5.

34. Cohen YB, Zhang QM. Electroactive polymer actuators and sensors. MRS Bull. 2008; 33:173–81. doi:10.1557/mrs2008.42.

35. Cohen YB, Anderson IA. Electroactive polymer (EAP) actuators - background review, Mech Soft Mater. 2019; 1:5. doi.10.1007/s42558-019-0005-1.

36. Gomez-Tamm AE, Ramon-Soria P, Arrue BC, Ollero A. Current state and trends on bioinspired actuators for aerial manipulation. *2019 Workshop on Research, Education and Development of Unmanned Aerial Systems (RED UAS)*. Cranfield, UK. 2019, pp. 352–361. doi:10.1109/REDUAS47371.2019.8999715.

37. Maas J, Tepel D, Hoffstadt T.. Actuator design and automated manufacturing process for DEAP-based multilayer stack-actuators. Meccanica. 2015; 50: 2839–54. doi.10.1007/s11012-015-0273-2.

38. Wang J, Deng B, Chen H, Wang X, Zheng J. Removal of aqueous Hg(II) by polyaniline: Sorption characteristics and mechanisms. Environ Sci Technol. 2009; 43:5223–8. doi.10.1007/s11012-015-0273-2.

39. Muhammad Ekramul Mahmud H, N Huq, AKO Yahya, RB . The removal of heavy metal ions from wastewater/aqueous solution using polypyrrole-based adsorbents: A review. RSC Adv. 2016; 6:14778–91.

40. Hepel M, Xingmin Z, Stephenson R, Perkins S.. Use of electrochemical quartz crystal microbalance technique to track electrochemically assisted removal of heavy metals from aqueous solutions by cation-exchange composite polypyrrole-modified electrodes. Microchem J. 1997; 56:79–92.

41. Hwang BJ.; Lee, KL 1996. Electrocatalytic Oxidation of 2-Chlorophenol on a Composite PbO_2/Polypyrrole Electrode inAqueous Solution. J. Appl. Electrochem. 26: 153–159.

42. Bhaumik M, Agarwal S, Gupta KV, Maity A. Enhanced removal of Cr(VI) from aqueous solutions using polypyrrole wrapped oxidized MWCNTs nanocomposites adsorbent. J Colloid Interface Sci. 2016; 470: 257–67.

43. Lei Y, Qian X, Shen J, An X. Integrated reductive/adsorptive detoxification of Cr(VI)-contaminated water by polypyrrole/cellulose fiber composite. Ind Eng Chem Res. 2012; 51:10408–15.

44. Bhaumik M, Maity A, Srinivasu VV, Onyango MS. Removal of hexavalent chromium from aqueous solution using polypyrrole-polyaniline panofibers. Chem Eng J. 2012; 181–2:323–33.

45. Alatorre MA, Gutierrez S, Pa' ramo U, Ibanez JG. 'Reduction of hexavalent chromium by polypyrrole deposited on different carbon substrates. J Appl Electrochem. 1998; 28:551–7.

46. Bernabe L, Rivas LB, Sanchez J. Arsenic removal by functional polymers coupled to ultrafiltration membranes. In: Masotti A, editor. Arsenic: Sources, environmental impact, toxicity and human health a medical geology perspective. New York: Nova Science Publishers; 2013: p. 267–88.

47. Janaki V, Oh BT, Shanthi K. *et al.* Polyaniline/chitosan composite: An eco-friendly polymer for enhanced removal of dyes from aqueous solution. Synth Met. 2012; 162:974–80.

48. Din MI, Ata S, Mohsin IU, Rasool A, Andleeb Aziz A. Evaluation of conductive polymers as an adsorbent for eradication of As(III) from aqueous solution using inductively coupled plasma optical emission spectroscopy (ICP-OES). Int J Sci. 2014; 6:154–162.

49. eredych M, Pietrzak R, Bandosz TJ. 2007. Role of graphite oxide (GO) and polyaniline (PANi) in NO_2 reduction on GO-PANi composites. Ind Eng Chem Res. 2014; 46:6925–35.

50. Venkatachalam K, Arzuaga X, Chopra N. *et al.* Reductive dechlorination of 3,3',4,4'-tetrachlorobiphenyl (PCB77) using palladium or palladium/iron nanoparticles and assessment of the reduction in toxic potency in vascular endothelial cells. J Hazard Mater. 2008; 159:483–91.

51. Singh S, Mahalingam H, Singh PK.. Polymer-supported titanium dioxide photocatalysts for environmental remediation: A review. Appl Catal A-Gen. 2013; 462–3:178–95.

52. Yu CL, Wu RX. *et al.* Preparation of polyaniline supported TiO2 photocatalyst and its photocatalytic property Adv Mater Res. 2011; 356–60:524–28. doi.10.4028/www.scientific.net/amr.356-360.524.

53. Liao G, Chen S, Quan X, Chen H, Zhang Y. Photonic crystal coupled TiO_2/polymer hybrid for efficient photocatalysis under visible light irradiation. Environ Sci Technol. 2010; 44:3481–5.

54. Riaz U, Ashraf SM, Ruhela A. Catalytic degradation of orange G under microwave irradiation with a novel nanohybrid catalyst. J Environ Chem Eng. 2015; 3:20–9.

55. PingSu Y, NuangSim L, XinLi H, Coster GL, HaurChong T. Anti-fouling piezoelectric PVDF membrane: Effect of morphology on dielectric and piezoelectric properties. J Membr Sci. 2021; 620:118818.

56. Ohya Y, Karasudani T. A shrouded wind turbine generating high output power with wind-lens technology. Energies 2010; 3:637–49.

57. Silleto MN, Yoon SJ, Arakawa K. Piezoelectric cable macro-fiber composites for use in energy harvesting. Int J Energy Res. 2015; 39:120–7.

58. Yoon SJ, Arakawa K, Uchino M. Development of an energy harvesting damper using. Int J Energy Res. 2015; 39:1545–53.

59. Chiba, Waki M, Wada T, et al.. Consistent ocean wave energy harvesting using electroactive polymer (dielectric elastomer) artificial muscle generators. Appl Energy. 2013; 104:497–502.

60. Chiba S. et al. Innovative power generators for energy harvesting using electroactive polymer artificial muscles. Proc SPIE. 2008; 6927:692715–9.

61. Sun D. et al. Near-Field Electrospinning. Nano Lett. 2006; 6(4):839–42.

62. Chang C, Limkrailassiri K, Lin L. Continuous near-field electrospinning for large area deposition of orderly nanofiber patterns. Appl Phys Lett. 2008; 93(12):469.

63. Weng CJ. et al. Advanced anticorrosive coatings prepared from electroactive polyimide–TiO2 hybrid nanocomposite materials. Electrochim Acta. 2010; 55(28):8430–8.

15 Electroactive Polymers for Space Applications

Samson Rwahwire and Samm Okinyi Youma

CONTENTS

15.1 INTRODUCTION

The twentieth century will be remembered as the time when humans ventured into space exploration. From 1961 to 1969 researchers at the National Aeronautics and Space Administration (NASA) and Russians were locked in a space race working toward putting a man on the moon. On July 16, 1969, the Apollo 11 Mission successfully launched from the Kennedy Space Center, Florida, USA enroute to the moon; on July 20, 1969, the first humans walked on the moon and returned to Earth on July 24, 1969.

Polymers are robust materials with many advantages and have been used in engineering applications due to their resilience in harsh environments, light weight, and easy fabrication. The space race and the Apollo 11 Mission opened the way for advanced polymer materials that were used in various components of the mission including for the astronauts. For instance, polysulfone (PSU) was utilized for the exterior helmet visors, aluminized polyethylene terephthalate (PET) and polyimide were among the polymers used in the space suits for heat protection (Figure 15.1).

NASA's success when Neil Armstrong walked on the moon ushered in a period of rest in the Space race; however, the turn of the twenty-first century saw a revival due to recent advances in technology and an increased interest in space tourism from various nations and private companies. The inspiration for this chapter is based on the recent success of NASA landing the Perseverance rover and Ingenuity helicopter in the Jezero crater on Mars in February 2021 (Figure 15.2) and China's

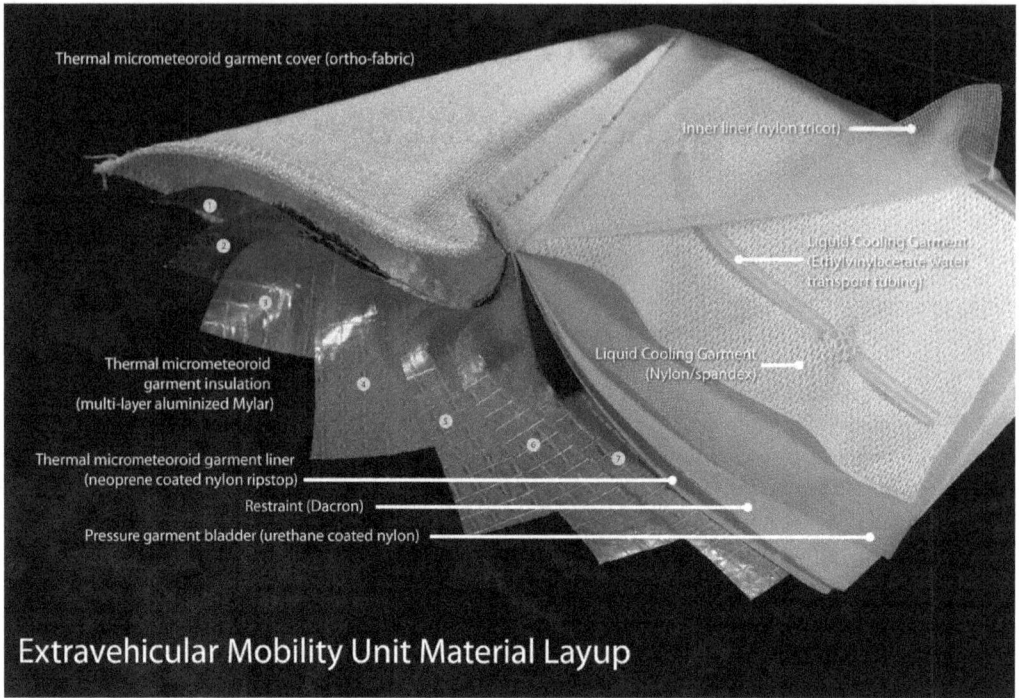

FIGURE 15.1　Apollo 11 space suit fabric layers.

FIGURE 15.2　NASA's Perseverance rover landing safely at the Jezero crater, Mars.

(From: https://mars.nasa.gov/resources/25451/perseverance-touching-down-on-mars-illustration/. With permission.)

National Space Administration's Tianwen 1 probe, which is China's first spacecraft to reach the Mars orbit with a planned landing of a rover in May 2021.

The recent pronouncement by SpaceX that aims to build colonies on Mars and several space agencies declaration of crewed missions to the moon has reignited research and development into space materials, specifically polymers for space applications.

15.2 SPACE ENVIRONMENT

Most of the research and development into space materials occurs on earth, it is important to discuss the harsh environments encountered in space, specifically in the low earth orbit (LEO) with an altitude of ≤2000 km, or on Mars and the Moon. The International Space Station (ISS) and a plethora of satellites are stationed in the LEO.

Polymeric materials used for components for LEO space structures are exposed to atomic oxygen (O_2) (ATOX), ultraviolet (UV), ionizing radiation, plasma and charge particle radiations, micrometeorites (space debris), and thermal cycling. The, geostationary earth orbit experience similar environments except for ATOX (Krishnamurthy, 1995; Grossman and Gouzman, 2003).

Ground based studies showed that carbon and hydrogen (C–H) polymers in LEO experienced chemical erosion; C–fluoride (C–F) polymers had bond scission, and therefore, became brittle and silicon (Si) based materials showed resistance to ATOX by forming a layer of SiO_2 (Grossman and Gouzman, 2003). To prevent ATOX irradiation, polymer composite materials are layered with an Si-based protective coating that ensures that the mechanical properties are preserved. In general, the space environment is harsh, and the effects are summarized in Table 15.1.

15.3 ELECTROACTIVE POLYMERS

Researchers have had an interest in materials that change shape and size, for instance, shape memory alloys (SMAs) and electroactive ceramics (EACs) actuate when an electrical field is applied; however, electroactive polymers (EAPs) are robust compared with SMAs and EACs (Table 15.2).

TABLE 15.1
Effects of space on polymer materials.

Space environment	Effect on the polymers
UV radiation	Affects the lattice structure
	Leads to the development of chain scission and crosslinking
Charged particle radiation	Affects the lattice structure
	Leads to the development of chain scission and crosslinking
	Leads to secondary radiation damages
Vacuum	Leads to volatilization of low vapor pressure
	Leads to the diffusion or vacuum welding of the particles
Thermal cycling	Leads to the degradation of the mechanical and chemical properties as well as to the brittle behavior of the polymers
Micrometeoroid	Failure and fracture
ATOX	Leads to oxidation and surface erosion
	Leads to the materials failure through cracking
Ionizing radiation	Degrades the properties of the polymers
Solar UV radiation	Degrades the properties of the polymers

(From Naser and Chehab, 2020. With permission.)

TABLE 15.2
Comparative advantage of EAPs over SMAs and EACs

Property	EAPs	SMAs	EACs
Actuation strain	>300%	<8%	0.1%–0.3%
Force required (MPa)	0.1–40	200	30–40
Reaction speed	μs–min	ms–min	μs–min
Drive voltage (V)	1–7 (ionic EAP)	5	50–800
	10–150 V/μm (electronic EAP)		
Fracture behavior	Elastic	Elastic	Brittle

In general, EAPs are polymer materials that change in shape and dimension as a result of the application of an electrical field (Jean-Mistral, Basrour, and Chaillout, 2010). EAPs are diverse and characterized by their reaction to an electrical stimulus. The two major groups of EAPs are electronic (i.e., driven by an electrical stimuli) and ionic (i.e., driven by the movement of ions) (Kruusmaa and Fiorini, 2006; Bashir and Rajendran, 2018). The robust mechanical properties of EAPs, which are the result of the large strain rate that is realized when a small force is applied, have interested researchers for the application of EAPs for actuation. There is a constant drive to increase the performance and robustness of space technologies when minimizing mass, power consumption, and complexity. In the search for unique technologies, EAPs seem to be the most promising actuation technology. Similar to other materials, EAPs respond when a force is applied and a displacement occurs; however, in this case the force that causes the displacement is the applied electrical potential. In additional, the EAPs can be utilized as a strain sensor. The performance of EAPs is far better than electromechanical devices, because, EAPs change their shape and size in response to an electrical stimulus (Guarino et al., 2016). In addition, EAPs are lightweight, soft and flexible, easy to miniaturize, and permit distributed actuation and sensing. (Fernandez, Moreno, and Baselga, 2004; Kruusmaa and Fiorini, 2006).

Actuators are used in various space applications, for example, release mechanisms, antenna and instrument deployment, positioning devices, aperture opening and closing devices, and real-time compensation for thermal expansion in space structures. However, there is a need to ensure that the weight, power, and the cost of these devices is economical (Bar-Cohen, 2005). EAP technology is preferred as a solution to bulky and noisy electromechanical devices in space applications. The main advantage of EAPs as previously discussed is the large displacement that is obtained when an electrical stimulus is applied; however, the required voltage in most cases is too low compared with other materials currently in use. (Bar-Cohen, n.d.; Alici, 2009; Bar-Cohen, 2014).

15.4 ELECTRONIC ELECTROACTIVE POLYMERS

EAPs are characterized by large strains of >40%, which makes them good materials for strain sensors and actuation. However, the large strains mean that a large amount of voltage is required. EAPs are sub-divided into two groups: dielectric elastomer actuators and ionic EAPs. The former performs actuation with the help of capacitors that are connected to highly sensitive electrodes coupled with a dielectric film. (Gunter et al., n.d.)(Bashir and Rajendran, 2018).

15.5 IONIC ELECTROACTIVE POLYMERS

EAPs require a large voltage for actuation, ionic EAPs (IEAPS) are unique in that low voltage is required for actuation due to the presence of mobile ions that are easily aligned when an electrical field is applied.

TABLE 15.3
Classification of EAPs

EAP type	Additional actuator types	Actuation mechanism	Activation voltage	Response time	Example
Electric polymer	Ferroelectric polymers	Piezoelectricity	0.1 kV	Fast	Poly(vinylidene fluoride) (PVDF), PVDF–TrFE
	Dielectric polymers	Electrostatic field force	0.4 kV	Fast	Si elastomer, polyurethane
Ionic polymer	Conductive polymer IPMC	Electrochemical	1–5 V\3 V	Slow Slow	Polypyrrole, polyaniline Nafion, Flemion
	Ionic polymer gels	Ion diffusion via polymer gel	\1 V	Slow	Polyvinyl alcohol

(From Bashir and Rajendran, 2018. With permission.)

The movement of the ions can cause a swelling of the polymer or a contraction, which depends on the desired effect. Although low voltages are required for actuation, the displacements are not evenly sustained as is the case with EAPs (Gunter *et al.*, n.d.).

Punning *et al.* (2014) demonstrated that IEAP actuators could withstand long term cosmic radiation because IEAPs are generally made from radiation resistant materials, such as polymers Nafion (perfluorosulfonate), Flemion (perfluorocarboxylate), carbon powder, non-crystalline noble metals, and nonvolatile ionic liquid electrolytes. In addition, IEAPs could withstand ionizing radiation in the LEO. IEAPs with carbonaceous electrodes and aqueous ionic polymer–metal composites (IPMC) could withstand all the harsh environmental parameters in space.

15.6 ELECTROACTIVE POLYMERS IN SPACE APPLICATIONS

Space applications are demanding due to the harsh operating conditions, and therefore, require a robustness and durability. There is an interest for materials that are lightweight, high strength, easily deployable, with the potential to operate under extreme temperature ranges, for example, Mars and the Moon where the temperatures are $\leq-100°C$. (Kumar *et al.*, 2020). EAPs have challenges in their development into films and fibers, because the dimensions can be as large as several meters or kilometers, and therefore, they can be used to produce large gossamer structures, such as solar sails, antennas, and various optical components (Bar-Cohen, 2005). The application of EAPs in space and their working principles will be briefly described, and some potential applications for some prototypes and the challenges faced.

15.6.1 ELECTROACTIVE POLYMER ACTUATOR THAT DRIVES A DUST WIPER FOR A CAMERA LENS

NASA's Jet Propulsion Laboratory designed a wiper for the camera lens of the MUSES-CN nanorover rover that used an IPMC. The prototype was constructed by ESLI (San Diego, CA, USA), it was made from graphite fibers and DuPont Kapton resin.

The wiper design dimensions were 15 mm×6 mm IPMC film that was bonded to the ESLI blade using platinum electrode strips that provided the electrical voltage for the EAP wiper. The blade (e.g., beam and brush) was coated with gold and activated by a 1–2 KV bias DC voltage, and the

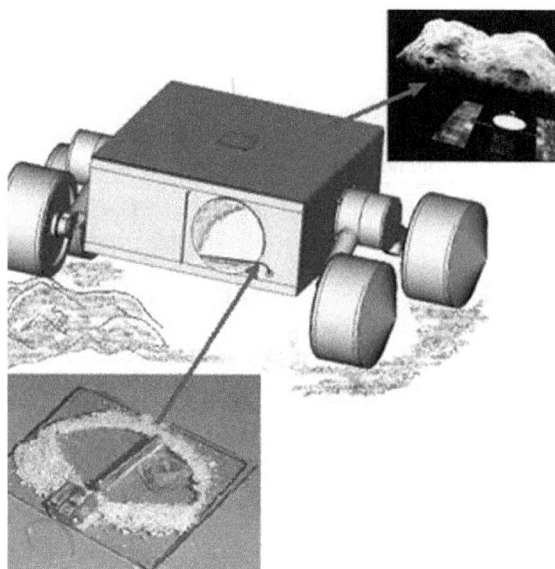

FIGURE 15.3 View of EAP dust wiper on the MUSES-CN nanorover (middle)and a photograph of a prototype EAP dust wiper(right-bottom).

(From Bar-Cohen *et al.*, 1999. With permission.)

EAP wiper was induced with 1–3V. The MUSES-CN's Nanorover was tested on the ground and could provide a bending of the wiper of approximately ±90° with a corresponding input voltage signal of 1–3V at approximately 0.3 Hz (Figure 15.3).

15.6.2 Dielectric Elastomers for Actuation of Large Lightweight Mirrors

Mirrors are used in space applications; one type of mirror was used in the Hubble Space Telescope. The primary mirror has a diameter of 2.4 m; and it could have been larger to give the telescope an advantage; however it could not be made >2.4 m diameter, because that was the largest diameter that can be launched into space in a single space launch (Figure 15.4). Most satellites and space stations that orbit earth are in LEOs. To obtain smart structures for space applications, smart materials are required that can easily be launched into space and on deployment their shapes can be controlled.

Designers have developed several configurations for lightweight space mirrors that can be used in space telescopes for the observation of the earth and possibly space (Figure 15.5). The designs hope to utilize dielectric elastomers to control the shape of the mirrors. However, Si polymers have been tested in extreme environments and have showed resilience over a wide temperature range and retain their respective electromechanical actuator properties; therefore, they could work as dielectric actuators for space applications.

15.6.3 Jumping Rover for Mars

Researchers that work for the European Space Agency (ESA) designed a spherical wind propelled tumbleweed rover that could jump over obstacles, such as rocks on the Martian surface. The conceptual design selected an Si-based EAP [poly(dimethylsiloxane)] for the artificial muscle that would allow the spherical rover to jump over obstacles (Carpi *et al.*, 2007).

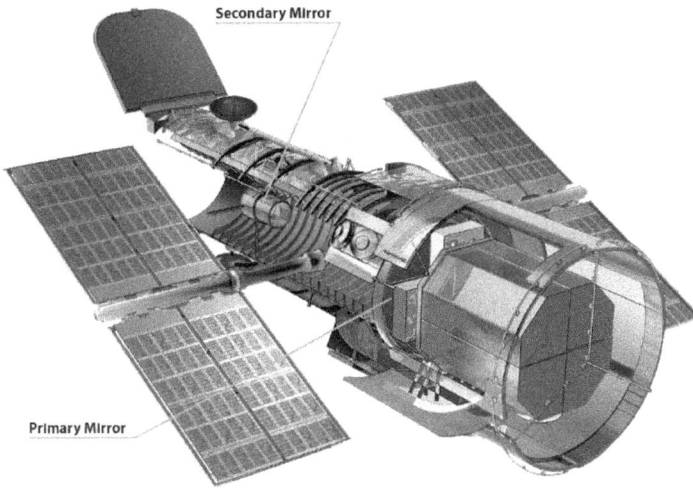

FIGURE 15.4 Cutaway image of the Hubble Space Telescope. (Credit: NASA's Goddard Space Flight Center.)

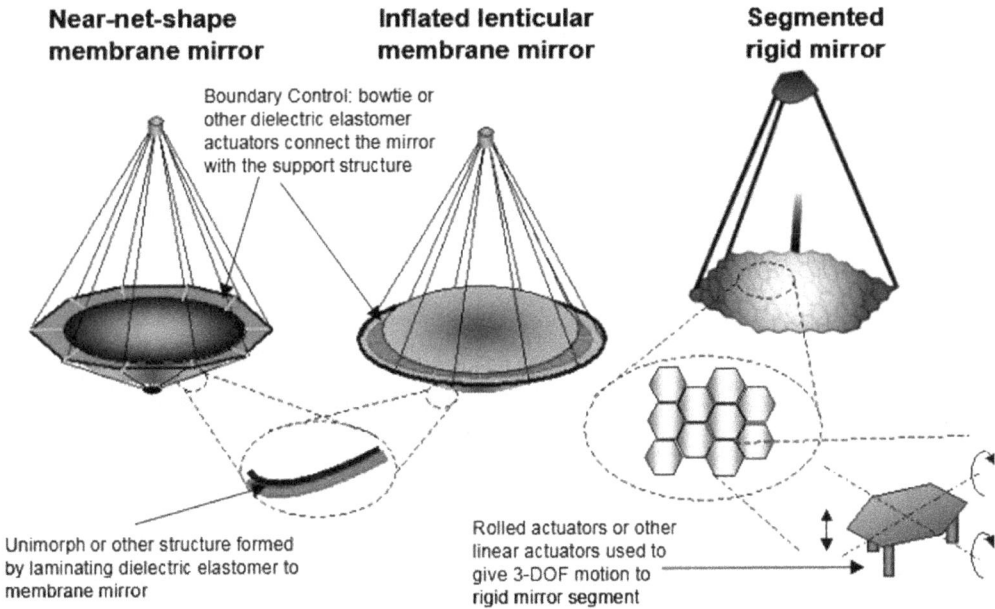

FIGURE 15.5 Conceptual designs of control of lightweight mirrors for space applications using EAPs. (From Kornbluh *et al.*, 2003. With permission.)

15.6.4 PARTICLE DISTRIBUTION MECHANISMS IN SPACE

The interest in Martian geology brought forward the idea of drilling on planets. Researchers proposed particle conveyor mechanisms that could be used in space, which are based on dielectric actuators (Figure 15.6). A traveling wave of dielectric surface, which is the result of EAP actuators

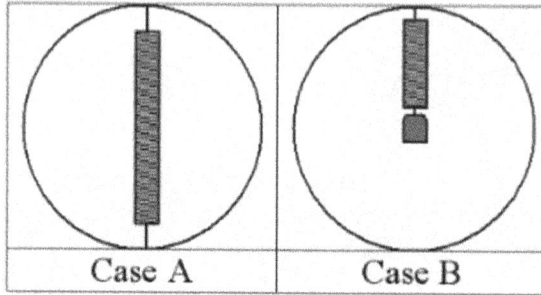

FIGURE 15.6   Distributed actuator for transport of particles.

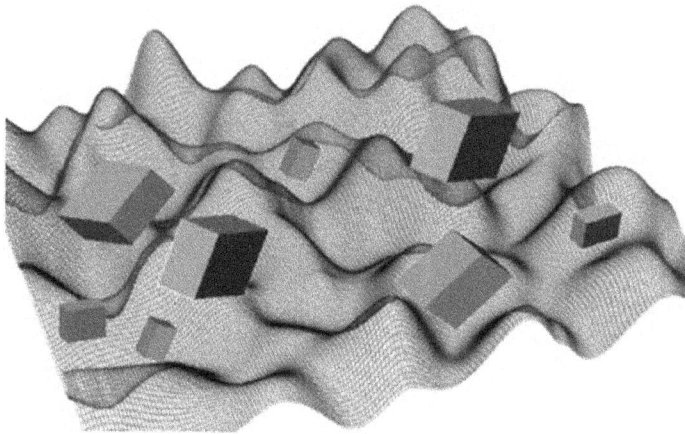

FIGURE 15.7   IPMC-based suit.

connected to an electrical signal, can propel particles to the desired destination (Menon, Carpi, and De Rossi, 2009).

15.7   ROBOTICS APPLICATIONS

15.7.1   HUMANOIDS

The ISS was set up by a consortium of nations that are involved in space science and research. All the work for manufacturing the station, repair, and assembly of the modules is carried out by three space agencies. The Canadian Space Agency designed and built the Space Station Remote Manipulator System, the Japanese Experiment Module Remote Manipulator System was designed and built by the Japan Aerospace Exploration Agency, and the European Robotic Arm was built by the ESA.

The German Space Center has played a pivotal role, for instance, space robotic analyses were tested on-board the Space Shuttle or the ISS, which included human robots called Space Justin and Robonaut that were built by NASA. The overarching goal of multi-national space research is the development of new materials for future space applications. NASA and other space agencies would like to reach Mars and the Moon; however, advances in the research into space robots, smart materials, and EAPs will continue to be of interest (Bashir and Rajendran, 2019)

Research into novel materials for space applications is increasing, which is demonstrated by the literature on other materials, for instance, ionic conductive polymer materials (Kruusmaa and Fiorini, 2006).

NASA envisions a future where robots will be able to help human astronauts. In addition, due to the current research by SpaceX, advanced robots with human capable cognitive functions could sent into space for exploration and they could be able to perform experiments especially in the extreme environments found on Mars. NASA has been involved in the development of humanoids as robots to help human astronauts and two humanoid robots have been considered for space applications: (1) Robonaut 1, which was teleoperated: and (2) a semi-autonomous Robonaut 2 (Bar-Cohen, 205).

The robots might help in testing the landing before sending humans onto the planets. Another function of the humanoids cold be help human astronauts with maintenance and other tasks that would otherwise take up the astronauts time (Bar-Cohen, 2005).

15.7.2 ARTIFICIAL INSECTS AND WORMS

The potential of artificial insects, flies, and worms as data collection devices on the surface of planetary bodies, such as the moon and Mars cannot be underestimated. An artificial worm that is based on IPMC could be embedded with microsensors to gather astrobiological, chemical, and geological data (Krishen, 2009).

15.7.3 HUMAN SUPPORT IN SPACE SUITS

Space is an extremely harsh environment, and therefore, the challenges faced by humans that live in space include loss of bone mass. Research has shown that exercise is effective in the prevention of loss of bone mass. IPMC-based space suits have been developed that can bond tightly to the human body to allow the body muscles to exercise (Figure 15.7). This artificial–natural interface could allow astronauts to work with one body part when another is being exercised (Krishen, 2009).

15.8 ELECTROACTIVE POLYMERS FOR AEROSPACE APPLICATIONS

Advanced aerospace vehicles could provide solutions from security to transportation; therefore, making daily operations for humans easier. Scientists, engineers, and chemists have recently made advances in biotechnology nanotechnology, and information technology; therefore, creating a new era in aerospace and aircraft development that uses smart materials and actuators.

Recently, research into EAPs focused on the development of innovative actuator configurations and materials in an attempt to provide improvised actuation properties. The latest being the novel applications of EAPs in aerospace, which has attracted a lot of attention. It was reported that viscoelastic EAP materials could provide more natural aesthetics, vibration and shock absorption, and more flexible actuator designs (Bar-Cohen, 2002).

Other research showed that EAP actuators performed exceptionally well compared with the current materials that are used in space applications, for instance, flight muscles in mechanical insects. However, the greatest disadvantage of EAPs is the large voltage required (>1,000 V) combined with the power electronics required to generate the voltage from a 5 V battery (Anderson, Sladek, and Cobb, 2011).

In the search for the right materials to use in EAPs, two polymers (Nafion and Flemion) have been studied as candidates for sensor and actuation applications (Xu *et al.*, 2003; Bar-Cohen 2010, n.d.).

Engineers and researchers are interested by the steady progress in the actuation strain capability of EAPS and their electromechanical properties that enable them to perform better as sensors and actuators.

15.9   AIRCRAFT MORPHING APPLICATIONS

Aircraft morphing describes a wide range of aircraft their components that can adapt to intended and unintended requirements. This requires altering the system features that include the vehicle's state, for instance, shape during flight operations. Actuators, such as EAPs can detect changes in the environment and adapt in a predesigned manner; therefore, from an efficiency perspective, the removal of energy conversions, such as electrical–mechanical and hydraulic forces leads to the use of fewer parts and fewer potential failure locations (Weisshaar, 2013).

A hybrid of an EAP and a Shape Memory Polymer (SMP) was studied for future aircraft morphing applications (Ren, Bortolin, and Zhang, 2016). The composite was mad from a poly(vinylidene fluoride–trifluoroethylene) film that was used to adjust the shape of an SMP film above its glass transition temperature. The simulation results from the finite-element method and the experimental data agreed.

An unmanned aerial vehicle (UAV) glider prototype was developed by digital printing transistors and actuators. The prototype proved that it was possible to design and build an electrically controlled disposable UAV (Grau, Frazier, and Subramanian, 2016). Actuation was realized by the Nafion that was printed on paper (Figure 15.8); the project demonstrated a potential path toward the realization of inexpensive UAVs for ubiquitous sensing and monitoring flight operations.

Research and development has seen several countries support projects that use SMAs for the control surfaces on aircraft (Kudva, 2004; Scheller *et al.*, 2016; Scheller, 2015). The Defense Advanced Research Projects Agency (DARPA) supported the DARPA/AFRL/NASA Smart Wing program that led to the development of smart material-based concepts to improve the aerodynamic and aeroelastic performance of military aircraft.

The Electroactive Morphing for Micro Air Vehicles research program funded by the French government aimed to study electroactive materials and their applications as actuators for morphing wings (Chinaud *et al.*, 2013).

Insect-inspired flapping wing MAVs were developed based on dielectric elastomers (e.g., silicone elastomer and polyacrylate tape VHB). In addition, the design showed that Si elastomers were the best candidates for flapping wing MAVs with a peak flapping amplitude of 63° at 18 Hz (Cao, Burgess, and Conn, 2019).

15.10   CONCLUSIONS AND RECOMMENDATIONS

The future is bright for EAPs, although challenges remain with the right EAP materials, and research has shown that soon, EAP materials will be central in space applications. In the search for a reliable EAP technology, future research into EAP actuators should cover four issues: (1) material improvement; (2) characterization and modeling; (3) actuator design; and (4) application simulation (Fernandez, Moreno, and Baselga, 2004)

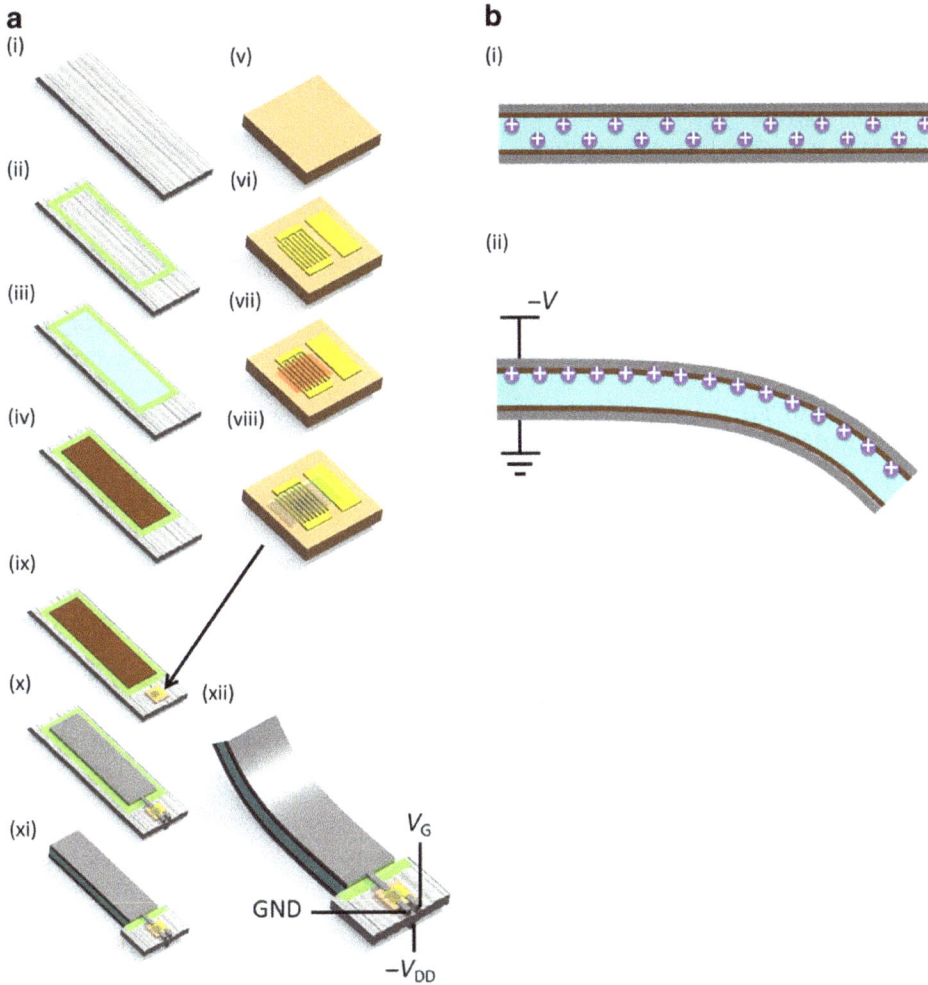

FIGURE 15.8 Showing (a) fabrication process flow. Actuator fabrication: (i) start material = bare filter paper; (ii) epoxy bank is printed into paper to prevent Nafion spreading; (iii) Nafion is printed into the paper within the bank. This step is repeated on the backside; (iv) electrode interlayer that consists of Nafion–silver mixture is printed onto Nafion. This step is repeated on the backside. Transistor fabrication: (v) starting substrate: kapton tape; and (vi) gold source-drain and coplanar gate electrodes are evaporated through a shadow mask; (vii) semiconductor polymer is printed onto the electrodes; (viii) Nafion gate dielectric is printed onto the semiconductor polymer layer. System integration: (ix) transistors are laminated onto the paper substrate with Nafion actuators. Actuators and transistors are immersed together in ionic liquid overnight; (x) interconnects and actuator electrodes are printed with silver flake ink. This step is repeated on the backside; (xi) actuators are released from surrounding paper substrate and epoxy bank to allow actuation; and (xii) voltages are applied to the system. Negative VG is applied to the gate of the p-type transistor. Ground is connected to the source. The drain is connected to one actuator electrode. The other actuator electrode on the backside is connected to −VDD. Actuator deflection is measured. (b) Operating principle of the IPMC actuator. Cross-sectional view showing paper infused with Nafion in light blue and both electrode layers on either side: (i) without any applied voltage, positive mobile ions are evenly distributed throughout the actuator; and (ii) with applied voltage, positive mobile ions move toward the negative electrode that causes swelling of the actuator on that side, and therefore, bending of the actuator.

REFERENCES

Alici, G. (2009) 'An effective modelling approach to estimate nonlinear bending behaviour of cantilever type conducting polymer actuators', *Sensors and Actuators B*, 141, pp. 284–292. doi:10.1016/j.snb.2009.06.017.

Anderson, M.L. *et al.* (2011) 'Design, fabrication, and testing of an insect-sized MAV wing flapping mechanism', 2011, pp. 1–11.

Bar-Cohen, Y. (n.d) 'Making science fiction an engineering reality using biologically-inspired technologies', 1–10.

Bar-Cohen, Y. *et al.* (1999) 'Challenges to the transition of IPMC artificial muscle actuators to practical application', 2005, 731–740.

Bar-Cohen, Y. (2002) 'Electroactive polymers: Current capabilities and challenges', 4695, pp. 1–7.

Bar-Cohen, Y. (2005) 'Current and future developments in artificial muscles using electroactive polymers', pp. 731–740.

Bar-Cohen, Y. (2010) 'Electroactive polymers (EAP) as actuators for aerospace engineering', 1880, pp. 1–8. doi:10.1002/9780470686652.eae234.

Bar-Cohen, Y. (2017) 'Robots, humanlike, and electroactive polymer actuators', Humanoids and the potential role of electroactive materials/mechanisms in advancing their capability', 97, pp. 81–89. doi:10.4028/www.scientific.net/AST.97.81.

Bashir, M., and Rajendran, P. (2019) 'Recent trends in piezoelectric smart materials and its actuators for morphing aircraft development'. doi:10.15866/ireme.v13i2.15538.

Cao, C., Burgess S., and Conn A.T. (2019) 'Toward a dielectric elastomer resonator driven flapping wing micro air vehicle', *Frontiers in Robotics and AI* 5. 137.

Carpi, F. *et al.* (2007) 'Martian jumping rover equipped with electroactive polymer actuators: A preliminary study', *IEEE Transactions on Aerospace and Electronic Systems*, 43(1), pp. 79–92.

Chinaud, M. *et al.* (2013) 'Hybrid electroactive wings morphing for aeronautic applications', *Solid State Phenomena*, 198, pp. 200–205.

Fernandez, D., Moreno, L.E., and Basel, J. (2004) 'Electroactive polymer actuator design for space applications electroactive polymer actuator design for space applications', 2–6.

Grau, G., Frazier, E.J., and Subramanian, V. (2016) 'Printed unmanned aerial vehicles Using paper-based electroactive polymer actuators and organic ion gel transistors', *Microsystems & Nanoengineering*, 2(1), pp. 1–8.

Grossman, E., and Gouzman, I. (2003) 'Space environment effects on polymers in low earth orbit', *Nuclear Instruments and Methods in Physics Research Section B: Beam Interactions with Materials and Atoms*, 208, pp. 48–57.

Guarino, V. *et al.* (2016) 'Electro-active polymers (EAPs): A promising route to design bio-organic/bioinspired platforms with on demand functionalities', doi:10.3390/polym8050185.

Gunter, D. *et al.* (n.d.) 'A comprehensive guide to electroactive polymers (EAP)'. https://sites.pitt.edu/~qiw4/Academic/MEMS1082/Group%202%20EAPs%20review%20-%20Final%20Paper.pdf.

Jean-Mistral, C., Basrour, S., and Chaillout, J.J. (2010) 'Comparison of electroactive polymers for energy scavenging applications', *Smart Materials and Structures*, 19(8), p. 85012.

Kornbluh, R.D. *et al.* (2003) 'Shape control of large lightweight mirrors with dielectric elastomer actuation', *Smart Structures and Materials 2003: Electroactive Polymer Actuators and Devices (EAPAD)*, 5051, pp. 143–158.

Krishen, K. (2009) 'Space applications for ionic polymer-metal composite sensors, actuators, and artificial muscles', *Acta Astronautica*, 64(11–12), pp. 1160–1166.

Krishnamurthy, V.N. (1995) 'Polymers in space environments', *Polymers and Other Advanced Materials*, 221–226.

Kruusmaa, M., and Fiorini, P. (2006) 'Electroactive polymers in space: Design considerations and possible applications', In *Proceedings of the 9th ESA Workshop on Advanced Space Technologies for Robotics and Automation. Astra 2006. ESTEC.*

Kudva, J.N. (2004) 'Overview of the DARPA smart wing project', *Journal of Intelligent Material Systems and Structures*, 15(4), pp. 261–267.

Kumar, R. *et al.* (2020) 'Electroactive polymer composites and applications', *Polymer Nanocomposite-Based Smart Materials* (pp. 149–56).Cambridge: Woodhead Publishing.

Menon, C., Carpi, F., and De Rossi, D. (2009) 'Concept design of novel bio-inspired distributed actuators for space applications', *Acta Astronautica*, 65(5–6), pp. 825–833.

Naser, M.Z., and Chehab, A.I. (2020) 'Polymers in space exploration and commercialization',. In M.A.A. alMaadeed, D. Ponnomma, and I.A. Carignanao (Eds.) *Polymer Science and Innovative Applications* (pp. 457–484). Elsevier.

Punning, A. *et al.* (2014) 'Ionic electroactive polymer artificial muscles in space applications', *Scientific*, 4(1), pp. 1–6.

Ren, K., Bortolin, R.S., and Zhang, Q.M. (2016) 'An investigation of a thermally steerable electroactive polymer/shape memory polymer hybrid actuator', *Applied Physics Letters*, 108(6), p. 62901.

Scheller, J. *et al.* (2015) 'Electroactive morphing for the aerodynamic performance improvement of next generation air vehicles'. PhD. https://oatao.univ-toulouse.fr/14479/.

Scheller, J. *et al.* (2016) 'A combined smart-materials approach for next-generation airfoils', *Solid State Phenomena*, 251, pp. 106–112.

Weisshaar, T.A. (2013) 'Morphing aircraft systems: Historical perspectives and future challenges', 50(2). doi:10.2514/1.C031456.

Xu, T-B., Su, J., and Zhang, Q. (2003) 'Electroactive polymer-based MEMS for aerospace and medical applications' 5055, pp. 66–77.

16 Electroactive Polymers in Industry

Vivek Mishra, Shubham Pandey, and Simran Aggarwal

CONTENTS

DOI: 10.1201/9781003173502-16

16.1 INTRODUCTION

Electroactive polymers (EAPs) are polymeric materials that can change shape and dimension as a counter-response to an applied suitable electrical stimulus. They are a subclass of electroresponsive polymer, especially in electromechanical systems. EAPs were first studied in 1880 by Roentgen (1) and Sacerdote (2) who worked in the diversified field of polymeric deformation under the impact of the electric field in 1899. Engineers, scientists, and technologists are interested in EAPs because of their useful properties in a variety of bioinspired (3), sensing (4), revolutionary generators, and technical applications (5), all of which are based on their exceptional electroactive conductivity (Scheme 1). Electrically operated soft actuators and other categorized examples of dielectric elastomers have interested researchers because they have a significant capacity to withstand heavier strains than their standard counterparts, such as piezoelectric ceramics that are used in mechatronic devices that detect pressure changes and produce electrical responses. EAP-based artificial muscles are much stronger than biological muscles due to their fast response to electrical stimuli. EAPs have wide sensor-based applications (6) that extend from chemical sensing, blood pressure (BP), and pulse rate monitoring devices (7) due to their broad and beneficial properties, such as high mechanical flexibility, easy fabrication, adaptable geometries (8), and variable electromechanical coupling properties.

EAPs can be divided into two major types: ionic EAPs and electronic EAPs. Ionic EAPs are activated electrically due to the migration of ions or molecules. Conjugated polymers (9), polymer

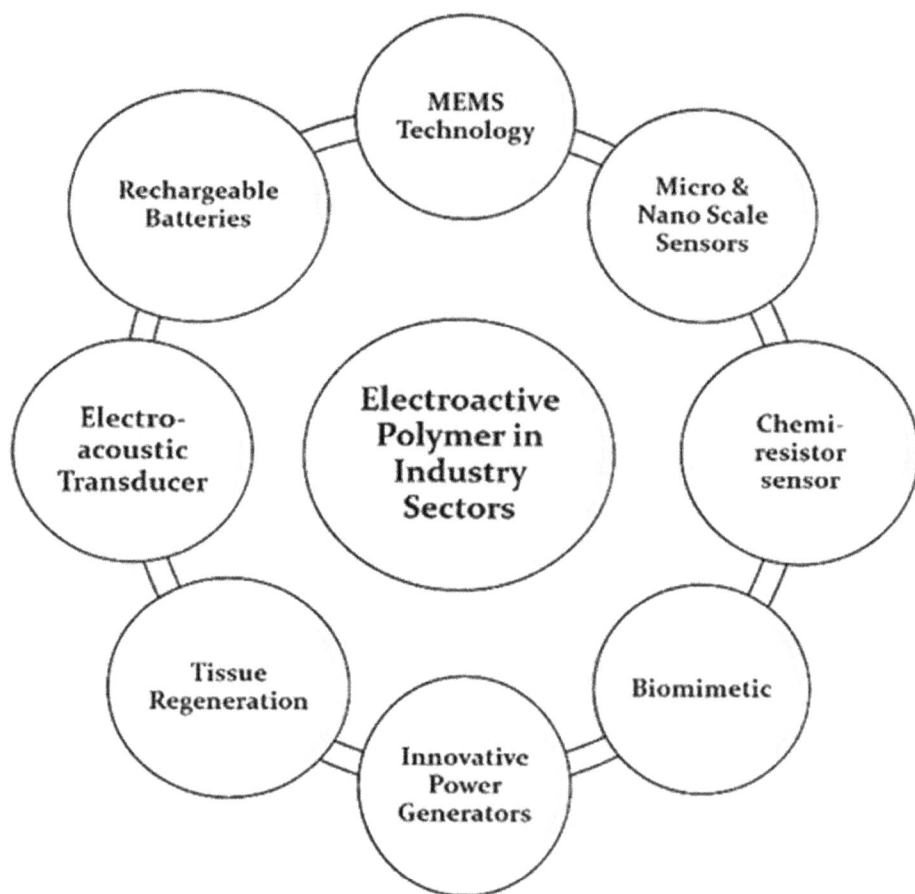

SCHEME 16.1 Electroactive Polymers used in Several Industrial Sectors

gels (PGs) [10], conducting polymers (CPs) [11], carbon nanotubes [12], and ionic polymer–metal composites (IPMCs) [13] are some examples of ionic EAPs. Applied electrical fields and Coulomb forces trigger electronic EAPs. Dielectric elastomers [14], carbon nanotube aerogels [15], electrostrictive polymers [16], piezoelectric polymers [17], and liquid crystal elastomers [18] are electronic EAP class. EAPs consist of a broad class of promising sensing materials. Most scientists that work on sensor-related studies; however, are unaware of them. Currently, since no material can meet all the sensory requirements, materials should be carefully selected to meet individual needs [19]. EAPs should be considered as supplementary to conventional sensing materials, mainly in fields where high strains and soft material compliance are required. Certain materials and their sensing mechanisms will be studied in this chapter. The significant types of electronic and ionic EAPs in sensory technologies, and many illustrative cases will be discussed.

16.2 TYPES OF ELECTROACTIVE POLYMERS

Based on electroactive polymer activation principles, EAPs are classified into two categories.

1. Electronic electroactive polymers (EEAP)
2. Ionic electroactive polymers (IEAP)

To activate the polymer, EEAPs utilize the electrostatic powers of two electrodes. By displacing ions, IEAPs stimulates the polymer. The advantages and disadvantages of IEAP and EEAPs are compared in Table 16.1

16.2.1 CLASSIFICATION OF ELECTRONIC ELECTROACTIVE POLYMERS

EEAPs can be further classified.

1. Piezoelectric polymers
   Piezoelectric materials are materials that can produce electrical charges on exposure to mechanical pressure [20]. To promote transient surface charge, they do not require any energy or electricity. Piezoelectric polymers have the simplest manufacturing, flexibility, and physical properties, which allows them to be used in a wide range of applications, such as bone regeneration. Polyvinylidene fluoride (PVDF), polyhydroxy butyrate (PHB), and poly L-lactic acid (PLLA) and its copolymers are some of the examples of piezoelectric EAPs.
2. Electrostrictive electroactive polymers
   EEAPs are deformed if the polymer is exposed to an electrical field. This is because when a polymer is exposed to an electrical field, a force is used to spin the polymer based on partial charge induction [21].
3. Ferroelectric electroactive polymers

TABLE 16.1
Advantages and disadvantages of EEAPs and IEAPs

	Electronic EAPs	Ionic EAPs
Advantages	They have a lot of mechanical energy. They provide high mechanical energy density	They require low voltage and generate bending displacement in most cases
Disadvantages	They require high voltage and activation	They maintain wetness and are unable to hold strain under Dc voltage

Ferroelectric polymers are irreversibly polarizing polymers that can be flipped or reversed when exposed to an electrical field (22). Because of their intrinsic piezoelectric responsiveness, ferroelectric polymers, such as PVDF are used in transducers and actuators.

16.2.2 CLASSIFICATION OF IONIC ELECTROACTIVE POLYMERS

IEAPS are classified into the following categories:

1. Conducting polymers
   CPs are biocompatible, flexible, and easily synthesized polymers that have a high conductivity to weight ratio and enhanced optical properties that can control electrical stimulation. This means that it possible to perfect their chemical, physical, and electrical properties to meet the specifications of the biological moieties used (23). CPs include polypyrrole (PPy), polyaniline (PANi), and many more examples.
2. Polyelectrolyte gels
   These are polymers that have covalently bound ionic groups (24). To generate charges along the polymer chain, they can be ionized either in polar solvents or water (25). They can generate movement of charge or volume changes with pH, temperature, or changes in electric or magnetic field movement. Biomedical polyelectrolytes with amino, carboxylate, and sulfate groups include chitosan, hyaluronic acid, and heparin.
3. Polymer–metal composites
   These have a small ionomeric layer on the surface with a noble metal electrode plated onto it. They are used as actuators that show deformations at low voltage. They trigger cationic responses through electrostatic resistance to the anode of an electrical field (Figure 16.1). Examples of ionic polymer composites are Flemion and Nafion (26).

16.3 ELECTROACTIVE POLYMERS IN INDUSTRY

The EAP industry needs to be industrialized, and major companies are expected to participate in developing the technology. A pattern has arisen in the USA and Japan, and it is now extending into Europe. However, since EAP technology is still in the early stages of market development, the companies interested in the sector is very low. Although most companies that currently test EAP technologies are protected by non-disclosure agreements, technologies are being considered for robotic devices, medical and haptic instruments, automobile, and consumer electronics products, which is based on evidence and knowledge accessed from our professional practices (Table 16.2).

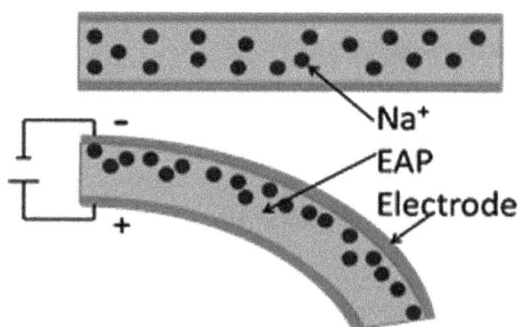

FIGURE 16.1 Cations in the IPMC collect on the side of the polymer that is in contact with the anode, bending it.

TABLE 16.2
Companies that make EAPs with a partial list of products in production

Company	EAP class	Examples of products	Country
Environmental Robots	Ionic polymer – composites, polymer gels	Linear and bending contractile actuators samples	USA
Danfoss PolyPower	Dielectric elastomers	Linear actuators, Si films	Denmark
Optotune	Dielectric elastomers	Electrically controllable optical lenses, phase shifters	Switzerland
Strategic Polymer Sciences	Electrostrictive polymers	Bending actuators; capacitors; films	USA
Eamex	Conjugated polymers, IPMCs	Entertainment gadgets, pumps, autofocus camera modules	Japan
Artificial Muscle	Dielectric elastomers	Pumps, haptic systems, valves, autofocus-lens positioner, acoustic-speakers	USA
Creganna-Micromuscle	Conjugated polymer	Medical device coating	Ireland

(From Carpi et al. 2011. With permission.)

16.4   APPLICATIONS IN INDUSTRY

Recently, EAPs have transitioned from innovation to industrialization, as major corporations have started to invest in the technology. EAPs have a wide range of industrial applications, which are briefly explained in the following sections

16.4.1   Electroactive Polymer-Based Sensors and Actuators in Microelectromechanical Systems

MEMS and microrobots that are made from EAPs, such as ionic polymeric–conductor composites (IPCCs) and ionic polymer–metal composites (IPMCs), have been used in the fabrication of disposable sensing devices for medical purposes, sensor and actuator microarrays, and microfabrication to manipulate small objects (28). Microarrays of IPMC actuator are used in photonic fiber optical switches that are based on micromirrors. MEMS technology has great applications; however, the main disadvantage in the development of MEMS technology is manufacturing the small devices. In addition, techniques for manipulating and assembling MEMS components into a system are required. Previous electroactive ceramic (EAC) materials were used in motors, translators, and inchworms, because they were effective and compact. Today, IPMSs are evolving as new actuation materials, because they are light, tough, have a fast response time, large actuation strain constant, and potential striction capability.

IPMC materials are ideal for use in MEMS actuators and sensors since they can be molded into any shape. The properties of shape memory alloys (SMAs), IPMC materials, and EACs are compared in Table 16.3.

This table highlights that IPMCs striction capacity is twice that of EACs, and the response time is significantly faster than SMAs.

The assembly of mechanical parts, actuators, detectors, and electronics on a single silicon (Si) substrate using microfabrication techniques, such as Si surface and wafer bonding, bulk micromachining, electrical discharge machining (EDM), *Lithographie Galvanoformung Abformung* (LIGA), and single point diamond machining are known as MEMS technology. MEMS technology

TABLE 16.3
Comparison of properties of SMA, IPMC and EAC

Property	SMA	IPMC	EAC
Actuation displacement	<6%, short fatigue life	>8%	0.1–0.3%
Force (MPa)	Approx. 700	10–30	30–40
Reaction speed	s–min	μs-s	μs-s
Density	5–6 g/cc	1–2.5 g/cc	6–8 g/cc
Drive voltage	NA	4–7 V	50–800V
Fracture toughness	Elastic	Resilient, elastic	Fragile

has a wide range of uses, including BP control, adaptive suspension devices for vehicles, and airbag accelerometers. MEMS chips are manufactured in batches, similar to integrated circuits, which allows higher reliability, functionality, and complexity to be packed onto a compact Si chip at minimal cost.

Since IPCC and IPMC actuators and sensors are easily made, processed, and configured in any size and shape, they can be combined with MEMS technology. These applications enable sensor and actuator cost and reliability to be comparable with that of integrated circuits (ICs). For advanced defense and industrial purposes, IPMC–MEMS switches can be used to build high-performance and low-cost devices. In addition, they are used for military and commercial purposes in monolithic microwave systems.

In various other developments, such as biochips for biological and chemical hazardous agents, IPMC–MEMS can be incorporated. Medical applications for EAPs have been documented (29,30, 31). Some of the uses of EAP actuators and sensors are discussed in the following sections.

16.4.1.1   Microgripper Microactuator Array

For the micromanipulation of grippers various approaches, such as the HexSil process (32), the use of piezoelectric actuators for micro actuation (33), and the use of SMAs (34) were employed but they failed to satisfy all the requirements of an economically viable approach. A material should react faster, have a long life, and should be simple and flexible. All the MEMS technology criteria are fulfilled by IPMC and IPCC.

Polymeric artificial muscles were designed with the ability to be cut into any size (a few hundred micron size range) and shape (35). These muscles have a lot of capacity for MEMS sensing and actuation. Figure 16.2 shows a micro size array of IPPC cut muscles (36).

16.4.1.2   Microrobot

IPMC films have been used to make fingers, because they are made of a relatively strong material with a high displacement capacity. Figure 16.3 shows a barbell on a tiny low mass robotic arm that used IPMC fingers as an end-effector.

As shown in Figure 16.3, vertical grey bars represent the fingers. This wiring arrangement bends the fingers outward or inward in response to electrical activation, closing or opening the gripper fingers as required. The hooks at the fingertips make it possible to protect the captured object inside the finger case.

16.4.2   Electroactive Polymers for Micro and Nanoscale Actuators and Sensors Using Thermoplastic Nanoimprint Lithography

Slow serial methods, such as focused ion beam lithography, electron beam lithography, and scanning probe lithography are used for nanoscale patterning, and they follow a law that defines

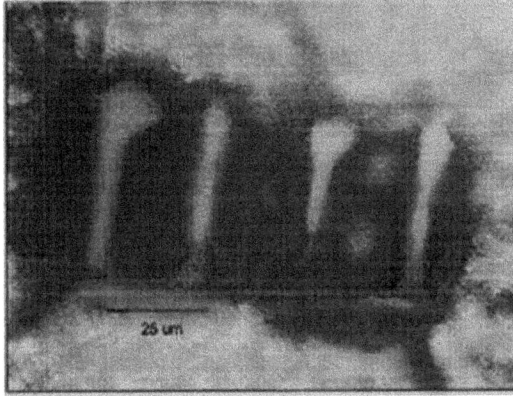

FIGURE 16.2 Assembly of IPPC muscle strips cut into laser microscope.

(From Lumia & Shahinpoor 1999. With permission.)

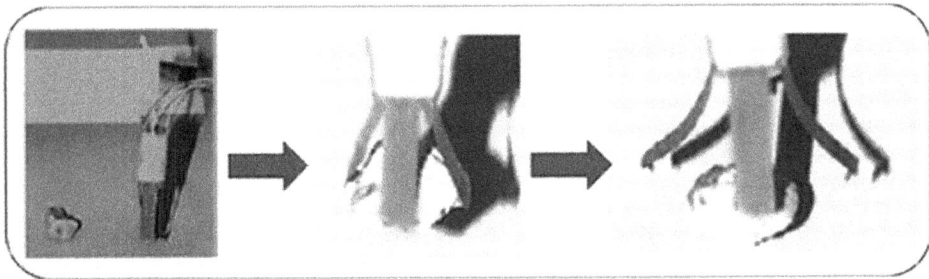

FIGURE 16.3 IPMC employed a four-fingered robotic-hand.

(From Shahinpoor 2001. With permission.)

the time–resolution trade-off. Today, methods such as laser holography, multiple beam systems, bottom-up nanolithography, and nanoimprint lithography (NIL) have been developed that do not follow any power law. NIL is a quick and cheap pattern repetition technique that can be used on soft materials. Thermoplastic-nanoimprint lithography (T-NIL) is used to convert nanoscale designs onto P(vinylidene fluoride-trifluoroethylene-chlorofluoroethylene terpolymer (P(VDF-TrFE-CFE)), which is a newly developed relaxer ferroelectric polymer. T-NIL can be carried out on thin films because they soften when heated. This terpolymer is appealing for device fabrication because of its high strain, low hysteresis, and high dielectric constant (37–40). PVDF and its copolymers cannot be produced using traditional lithography systems due to their solubility in certain organic solvents and photoresist developers (41).

P(VDF-TrFE-CFE) resonators were made from a photo-definable polymer (42) and the EAP must be patterned in at least two dimensions.

Mechanical deformation is used in NIL to move a pattern from a mold onto a deformable material (43). The viscous polymer is displaced by squeezing flow and capillary forces, which adhere to the surface of the stamp, and it is then unmolded (44). P(VDF-TrFE-CFE) appears to be its chemical component but is different because it is a ferroelectric relaxer with high electrostatic strains, which need to become electroactive via annealing.

A T-NIL protocol is used to transmit patterns onto P(VDF-TrFE-CFE) pictures using 1-h hydrophobic silane surface therapy. The polymer is couched under compression, in addition to heating its terpolymer to above its melting point; therefore, increasing the roughness of the surface of the structure.

16.4.2.1 Imprinting of P(Vinylidene fluoride-trifluoroethylene-chlorofluoroethylene) Terpolymer

Figures 16.4 and 16.5 show the T-NIL process and the steps involved in imprinting. The steps are described as follows.

1. A chip with a thin film P(VDF-TrFE-CFE) and an Si stamp is packed in the opposite direction in a nano imprinter (SUSS FC-150 Zero, Garching, Germany). Optical alignment ensures that the stamp and substratum surfaces are the same, because an incorrect alignment can result in a transition of the pattern partially or incompletely
2. Hexamethyldisilazane (HMDS) and dodecyl trichlorosilane (DTS) are used to treat the substrate and stamp, respectively
3. P(VDF-TrFE-CFE) is laid onto the substratum with a spin coating
4. The polymer and stamp are heated
5. When heating, the stamp is compressed into the polymer
6. The stamp is detached from the polymer

Polymer adhesion differences with DTS and HMDS induce polymer surface asymmetry. The DTS coating allows the stamp to be quickly separated from the polymer residue without ripping the thin polymer sheet. The stamp has poor adherence to the polymer. The hydrophobic silane coating is covalently bonded to the stamp surface and is not affected by imprinting and the stamp can then be reused without any maintenance between the stamps. The difference between Si-treated DTS and HMDS surface energies is shown by the difference between the contact angle of the water droplet on both surfaces (Figure 16.6).

The difference in angles shows that the surface energy of the HMDS treated surface is higher; therefore, the surface has adhered more than to the P(VDF-TrFE-CFE). Isolated P(VDF-TrFE-CFE) frames that used a microscale stamp with typical features of microactuator architectures have been designed (Figure 16.7).

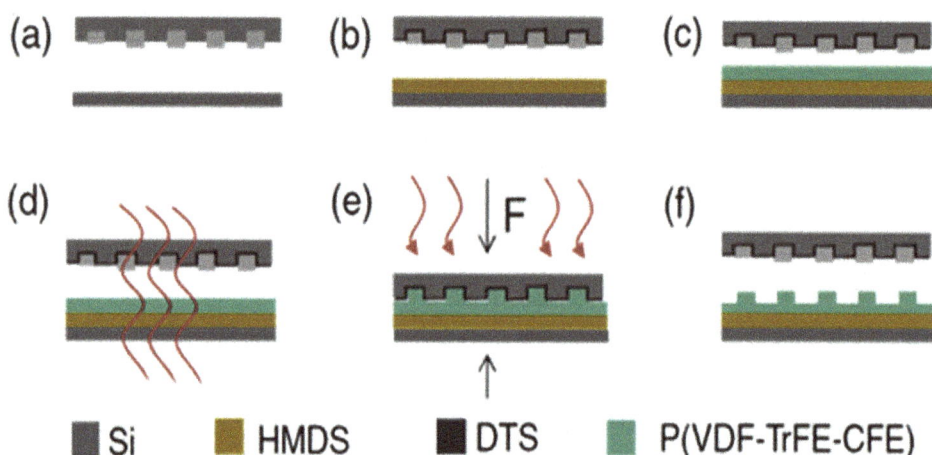

FIGURE 16.4 Different stages of P(VDF-TrFE-CFE) imprinting.

(From Engel et al. 2014. With permission.)

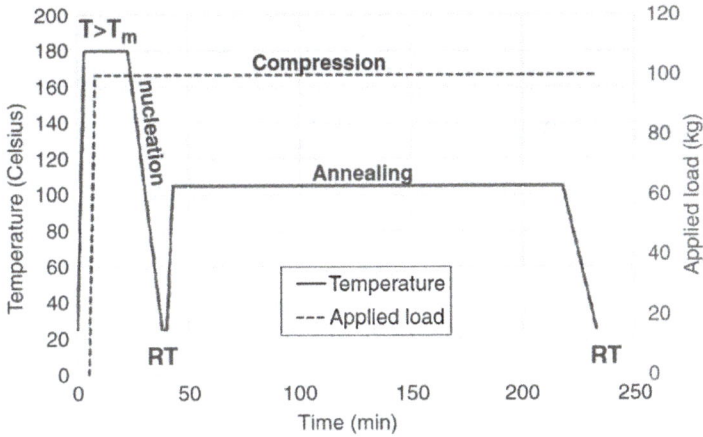

FIGURE 16.5 Method of compression, heating, and tempering of P(VDF-TrFE-CFE). (From Engel et al. 2014. With permission.)

FIGURE 16.6 Water contact angles on HMDS and DTS-treated Si chip surfaces. (From Engel et al. 2014. With permission.)

16.4.3 Gold Nanoparticle-Doped Electroactive Polyimide as a Chemiresistor Sensor for Hydrogen Sulfide

An electroactive polyimide with gold (Au) nanoparticles (NPs) was used as a chemiresistance tracker for the detection of sulfide hydrogen sulfide (H_2S) (45), which is a colorless, flammable, and pungent poisonous gas. As per the human health recommendations, 20 ppm H_2S can cause fatigue, headaches, loss of appetite, dizziness, and poor memory. If a person is exposed to a high concentration of H_2S for a long time, it can cause severe conditions, such as paralysis and death.

H_2S is released in high quantities during volcanic activity (46–48), excavation of swamps or landfills (49,50), and is found in petrochemical reservoirs and the paper industry.

Due to the dangerous presence of H_2S, its detection is very important and required. Previously, the techniques used for H_2S detection, such as gas chromatography (51–53), infrared spectroscopy (55), ultraviolet spectroscopy (55–58), and fluorometry (59), are expensive and time-consuming instruments. In addition, sensors, such as electrochemical (60,61), optical (62,63), piezoelectric (64,65) and chemiresistive (66,67) were used for gas detection. Metal oxide sensors (MOS) were widely used to make chemiresistive sensors for noxious gases, because they were cheap and easy to prepare and operate but they require a high working temperature (68,69). CPs were often used for the same reason, and they had high sensitivity, fast processing speed, and could function at room temperature; however, they had low processability, limited mechanical strength, and chemical stability (70,71). Reports suggest that silver (72,73) and Au (74,75) NPs were used for H_2S detection but required expensive optical measurement devices.

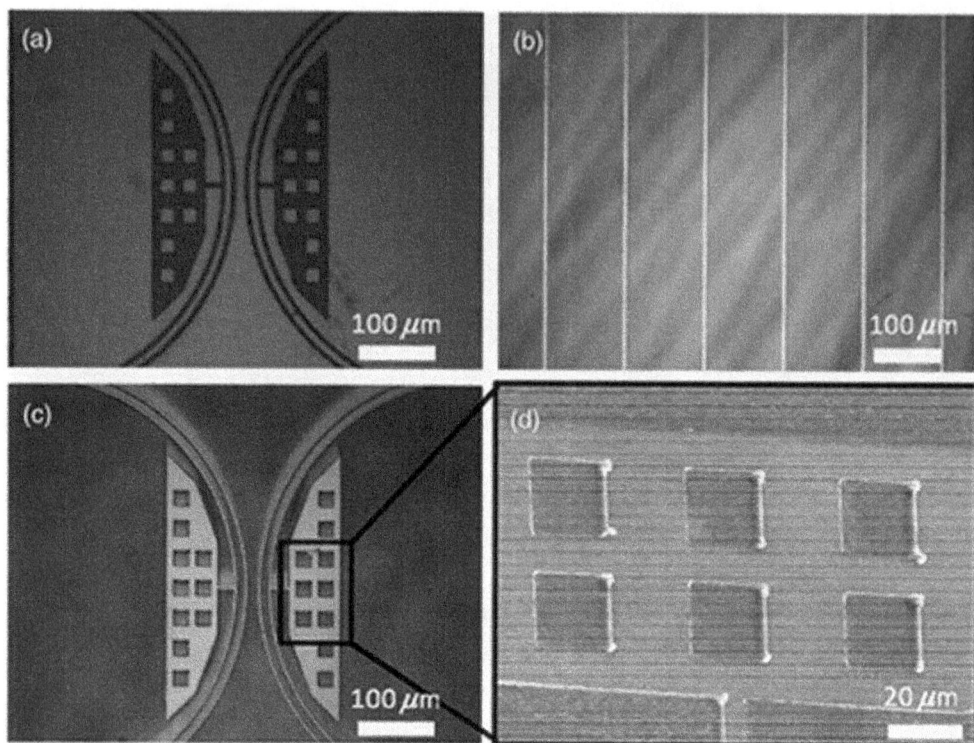

FIGURE 16.7 Micrographs with no residual coating of P(VDF-TrFE-CFE): (a) Si stamp; (b) 2 μm distances between polymer structures; (c) polymer pattern between polymer structures without the residual layer; (d) closing in the environmental scanning electron microscope photo of the polymer squares at 30°.

(From Engel et al. 2014. With permission.)

Based on the disadvantages of the techniques available for the detection of H_2S, the concept of an EAP and an Au nanoparticle was combined, and electroactive polyimide coated with Au NPs was developed to detect minute levels of H_2S gas at room temperature. It was synthesized in three stages: (1) an amino-protected aniline trimer (ACAT) was synthesized (76) and polymerized; (2) chemical imidization; and (3) doping over Au-NPs. The electroactive polyimide(PI)/Au-NPs displayed better sensing efficiency compared with the traditional conducting AU-NP doped PANi (PANI/Au-NPs), which makes it a useful tool for detecting H_2S in the environment.

16.4.3.1 Evaluation of Fabricated Sensor for Hydrogen Sulfide

PI/Au-NPs act as the sample and the exposed Au serves as the substrate. The gas sensing setup for detecting H_2S gas is shown in Figure 16.8. It is equipped with a nitrogen (N_2) tank connected to an H_2S gas reaction chamber. This was flushed to the sample chamber, which used a multimeter for the study with the manufactured sensor. The resistance data collected were taken using a computer and the next chamber was filled with water and the H_2S gas was emitted from the sample chamber.

16.4.3.2 Quality Analysis of the Sensor

To verify the sensor output, the sensors were assessed for sensitivity, detection maximum, linear concentration spectrum, linearity, dynamic response and regeneration properties, selectivity, reproductivity and repeatability.

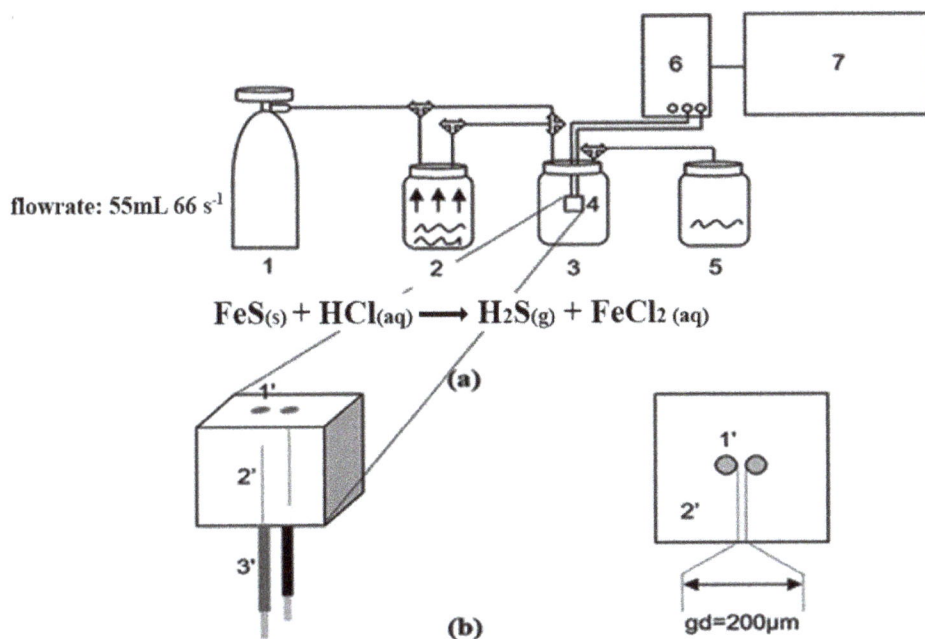

$$FeS_{(s)} + HCl_{(aq)} \longrightarrow H_2S_{(g)} + FeCl_2 \text{ (aq)}$$

FIGURE 16.8 Demonstration of: (a) setup for sensing H_2S gas; and (b) sensor–substrate layout.

(From Padua et al. 2019. With permission.)

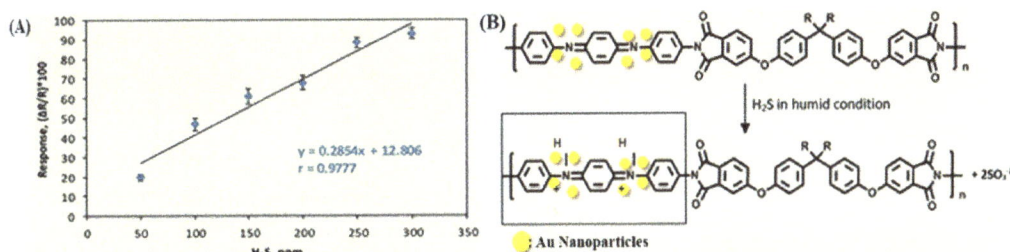

FIGURE 16.9 Showing: (a) calibration plot of the PI/Au-NPs sensor after being exposed to growing concentrations of H_2S at room temperature (n=3); and (b) proposed sensing mechanism.

(From Padua et al. 2019. With permission.)

1. Sensitivity, linearity, and the limit of detection

   Figure 16.9(a) shows the performance of PI/Au-NPs with an increasing concentration of H_2S gas. The plot is drawn between (R/R_0) 100% and H_2S concentration (ppm), where $R = [Rg - R_0]$; Rg is chemiresistor resistance; and R_0 is baseline resistance. The sensitivity of the sensor was 0.3% ppm H_2S, with a linearity of 0.98 for 50–300 ppm H_2S. The detection limit was 0.14 ppm H_2S using least-square curve fitting.

   In humid conditions, H_2S gas might form sulfurous acid, which dissociates into hydrogen (H^+) and sulfite ions more efficiently. These H^+ ions protonate the imine N of polymeric ACAT chains (e.g., at room temperature and pressure), which causes a decrease in resistance and an increase in conductivity (Figure 16.9(b)). The Au-NPs help to stabilize the interaction by attracting the sulfite ions toward the ACAT chains.

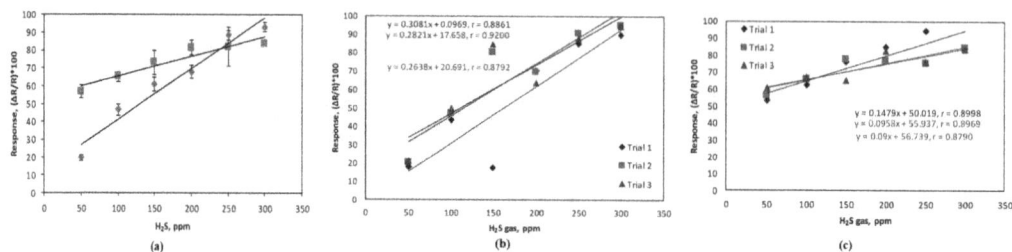

FIGURE 16.10 Showing: (a) calibration data of electroactive PANI/Au-NPs and PI/Au-NPs chemiresistor sensors with growing H_2S concentrations (50–30 ppm, n=3, rt); (b) PI/Au-NPs; and (c) PANI/Au-NPs calibration plots were used to conduct repeatability studies.

(From Padua et al. 2019. With permission.)

2. Sensor's response repeatability

Figure 16.10(a) shows an evaluation of electroactive PANI/Au-NPs versus a PI/Au-NPs sensor calibration plots with increasing H_2S concentrations. The former has a susceptibility of 0.11% ppm H_2S, a median detection of 0.45 ppm H_2S, and linearity of 0.95 for 50–300 ppm H_2S steam. This data suggested a 62% improvement in the performance of PI/Au-NPs sensors, which was due to the additional doping of PI's active sites. Figures 10(b and c) show the repeatability studies and accuracy between all the trials for evaluating sensitivity. The relative standard deviation (RSD) for PI/Au-NPs was 8.88% and RSD for PANi/Au-NPs was 28.36%.

3. Electroactive chemiresistor sensor's response and recovery characteristics

The time it takes for the sensor to transition from its initial state into a stable state is called the response time, and the time it takes to return to the initial state is called the recovery time. The response and recovery times for PI/Au-NPs were 43 and 99 s respectively; and for PANI/Au-NPs were 55 and 103 s respectively (Figure 16.11). The faster recovery and response time of PI/Au-NPs was due to its more porous polymer matrix, for instance, larger surface area for diffusion to occur.

4. Reversibility of the chemiresistor sensor's response

In three cycles of sensitivity to an N_2 blank and 200 ppm heated H_2S, Figure 16.12(a) shows that the PI/Au-NPs sensor had excellent reversibility.

5. Selectivity

Electroactive PI/Au-NPs sensors showed the highest change in resistance toward 300 ppm H_2S gas and a minimal to negligible response for pure solvents, such as hexane, methanol, and ethyl acetate vapors (Figure 16.13). When the sensor was used to sense ammonia, the signal indicated a distinct sensing mechanism in the opposite direction.

16.4.4 ELECTROACTIVE POLYMERS IN TISSUE REGENERATION, WOUND HEALING, MEDICAL RESEARCH, AND PHARMACEUTICAL INDUSTRIES.

A voltage gradient (i.e., an action gradient) from -10 to -90 mV through the cell membrane is generated by the human body's electrical field and current. The cells are triggered to give signals that change cell differentiation and proliferation as a result of the gradient. Electrical signal stimulation is responsive to nerves, osteoblasts, and cardiomyocytes. The formation and growth of cells is the first step in the regeneration of damaged tissues. Electroactive compounds are used to activate and enhance this regenerative function, and to accelerate the healing of damaged tissues. Controlling migration, cell adhesion (80), expansion, apoptosis (81), and differentiation (82) improves cardiac (77), bone (78,79), and nerve (78,79) regeneration. EAPs might be molded into a variety of forms with attractive morphological features and physicochemical properties. They can respond to a

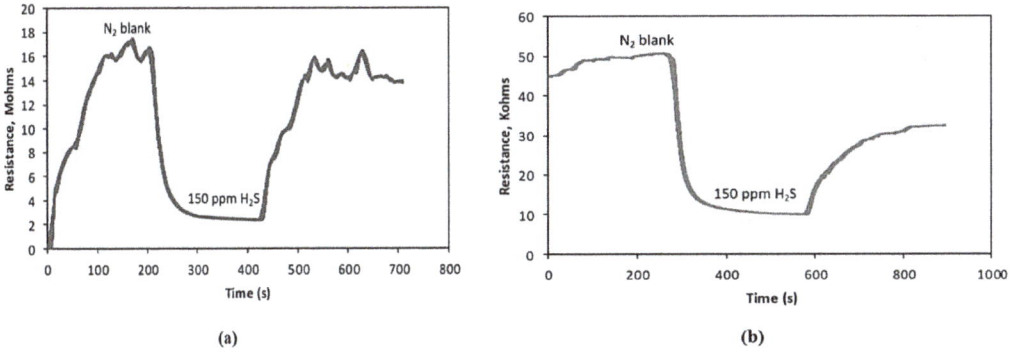

FIGURE 16.11 Response and recovery property of: (a) PI/Au-NPs; and (b) PANI/Au-NPs for 150 ppm H$_2$S at RT for 1 cycle.

(From Padua et al. 2019. With permission.)

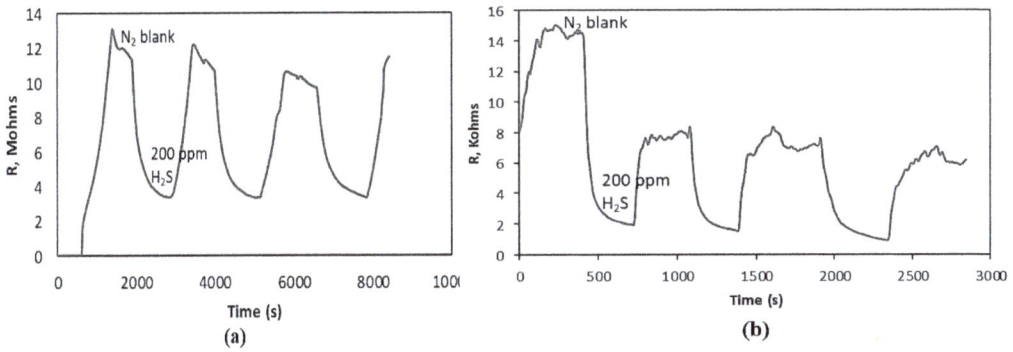

FIGURE 16.12 Comparison of the reversibility of sensor reaction for 200 ppm H$_2$S at room temperature for 3 cycles of: (a) PI/Au-NPs; and (b) PANI/Au-NPs sensors.

(From Padua et al. 2019. With permission.)

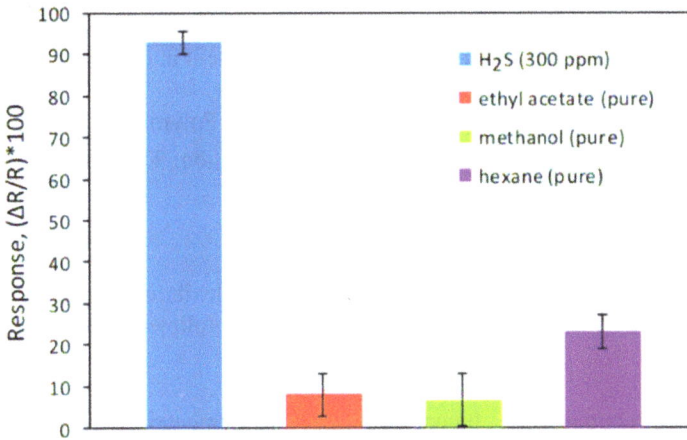

FIGURE 16.13 Selectivity of PI/Au-NPs sensor was tested with 300 ppm H$_2$S, pure ethyl acetate, pure methanol, and pure hexane (RT, n=3).

(From Padua et al. 2019. With permission.)

TABLE 16.4
EPS for tissue regeneration

Applications in tissue regeneration	Advantages	Disadvantages	Examples of EAP
Bone tissue regeneration	Biologically compatible, electroactive, conductive to cell diversity	Non-biodegradable, absence of hydrophobicity and require external source of power	PPy (82), PANi (84), polyelectrolyte (85)
Neural regeneration	Biologically compatible, highly conductive, stable, large specific surface area, easily processed and conductive to cell diversity	Reduced electrical interaction at the interface	PPy (86), PANi (87), PHB (88)
Myocardial regeneration	biocompatible electroactive, fibrous, porous, and conductive to cell differentiation	Non-biodegradable, absence of hydrophobicity and require external source of power	PPy (89), conductive polymer composite (90)
Cartilage regeneration	Electroactive, flexible, biologically compatible and conductive to cell differentiation	Non-biodegradable and require external power source	Polyelectrolyte gels (91)

variety of external stimuli. They are unique in that they can convert thermal, electronic, and magnetic signals into electrical signals. Table 16.4 shows the advantages, disadvantages, and some of the recently developed EAPs that have applications in tissue regeneration (83).

16.4.4.1 Biological Response of Electroactive Polymers to Electrical Stimulation

Protein absorption initiates the integration of neighboring cells and tissues. This stage is crucial, because it mediates cell attachment and subsequent tissue growth. Protein absorption depends on various internal properties and responds to external factors. Electrical stimulation can be applied without changing the properties of the component materials. Electrical stimulation induces variations in the redox state of the polymer, which is used to control protein interactions.

16.4.4.2 Application of Different Types of Electroactive Polymers in Tissue Regeneration

Different types of EAP have applications in tissue regeneration. These will be discussed in the following sections.

16.4.4.3 Conducting Polymers

These have electrical and optical characteristics, processing flexibility, and are easy to synthesize. The main CPs used for tissue engineering are described in the following sections.

1. Polypyrrole

   These have high conductivity, biocompatibility, environmental stability, and are easy to prepare (Figure 16.14). They have compatibility with endothelial cells, nerve cells, myocardial cells, mesenchymal stem cells (MSCs), and osteoblasts. Neural cells convey messages via electrical impulse. PPy was tested as a neural probe (92) and a scaffold for a nerve tube as the

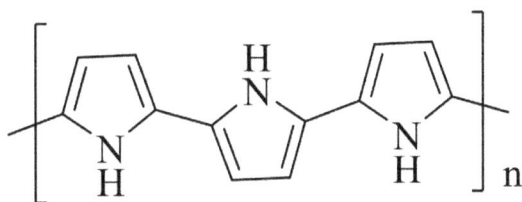

Polypyrrole (PPy)

FIGURE 16.14   Structure of PPy.

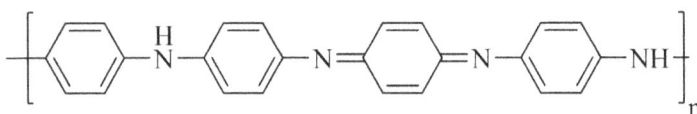

Polyaniline (PANi)

FIGURE 16.15   Structure of PANi.

axon is enlarged by electrical stimulation. For example, a three-dimensional collagen-based fibrous structure that contained PPy was constructed, which demonstrated that human MSCs grown on PPy fibers denoted neural lineage. In addition, this explained how external electric fields influenced cellular functions and how long term stimulation could be harmful to the culture system (93).

2. Polyaniline

PANi or aniline black has three forms (Figure 16.15): (1) a totally reduced leucoemeraldine base; (2) a half-oxidized emeraldine base; and (3) an entirely oxidized pernigraniline base. PANi is less expensive, more environmentally friendly, easier to synthesize, and more processible than PPy. PANi has a wide variety of uses in the regeneration of nerve and heart tissues. PANi and its variants cause H9c2 cardiac myoblasts to proliferate and adhere (94). H9c2 cell adhesion and growth were aided by electrospun nanofibers made from a combination of PANi and gelatin.

3. Poly(3,4-ethylenedioxythiophene)

Poly(3,4-ethylenedioxythiophene) (PEDOT) is a regenerative polymer that has been used in the development of neurotransmitter transmission systems (Figure 16.16). It is thermally more established and conductive than PPy. PEDOT is used in biosensing, neural electrodes, tissue architecture and functions, and heart muscle patches. The relationship between PEDOT and neural cells was studied (95) to manufacture biomaterials that had electrical conductivity. PEDOT-containing implantable electrodes were used to electrically stimulate nerve impulses or capture neuronal signals (96).

4. Conducting polymer composites

The application of CPs in biomedicine is limited, because they are difficult to handle and are brittle and doping with large groups can exacerbate these problems. To resolve these challenges, CPs are combined with other polymers, which allows them to benefit from the properties of both. CP composites have high solubility and improved mechanical properties. However, they have limitations. The use of insulating molecules to make CP composites can interfere with electron conjugation. CP composites show promising roles in neural stimulation, sensing. and tissue engineering. For example, electrically conducting HEC/PANi nanocomposite cryogels were formed that exhibited excellent cell growth and survival.

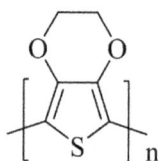

Poly(3,4-ethylenedioxythiophene) (PEDOT)

FIGURE 16.16 Structure of poly(3,4-ethylenedioxythiophene).

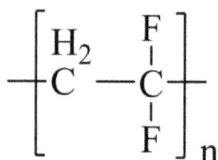

Polyvinylidene fluoride (PVDF)

FIGURE 16.17 Structure of polyvinylidene fluoride.

16.4.4.4 Piezoelectric Polymers

Electric charges are produced for mechanical pressure exposure. Studies have shown that mechanical forces play a major role in increasing morphogenetic repair and osteogenic differentiation (97) . Bone has piezoelectricity, because of its piezoelectric collagen fibers. Some of the piezoelectric materials for tissue engineering are discussed in the following sections.

1. Polyvinylidene fluoride
   PVDF is made up of at least four crystalline phases: α, β, γ, and δ (Figure 16.17). Among these β-phase PVDF is best for electric response application, because it possesses the best piezo-electric coefficients (98). The surface charge on PVDF films affects the hydrophobicity of the specimens, which results in a spectrum of absorbed extracellular matrix protein configurations that regulates stem cell adhesion, and osteogenic differentiation. To assist nerve tissue regeneration, (99) developed microporous PVDF membranes with the lysine portion immobilized. In addition, PVDF promoted neurite extension and is helpful in the development of cardiovascular tissue.
2. Polyhydroxy butyrate
   PHB is an example of a polyhydroxyalkanoate (PHA) (Figure 16.18). PHAs are produced by bacteria and promote cell adhesion and proliferation. Various approaches were used to fabricate neuronal tubes from PHB, which resulted in efficient axonal rejuvenation with negligible inflammatory infiltration (100). On covering PHB fibers with fibronectin (Fn) and alginate, the fibers could be used for treating spinal cord injuries (101).
3. Poly L Lactic acid
   Because of low density and good processing power, lower toxicity, biological compatibility, degradability, and mechanical properties, PLLA is very useful in medicine (Figure 16.19). Since it can display electrical responses caused by stress, PLLA is useful in bone retrieval and development, and neural regaining. An electrospun composite membrane made from PLLA and chitosan reduced the negative effects of fibroblasts on tissue regeneration (102).
4. Piezoelectric polymer composites
   Piezoelectric polymers are blended with other materials to integrate the qualities of both materials. Using these materials higher stability, more flexibility, easy processibility,

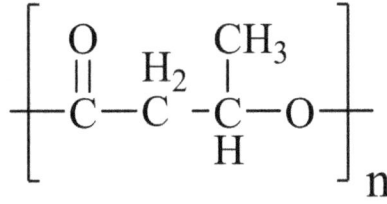

Polyhydroxybutyrate (PHB)

FIGURE 16.18   Structure of polyhydroxy butyrate.

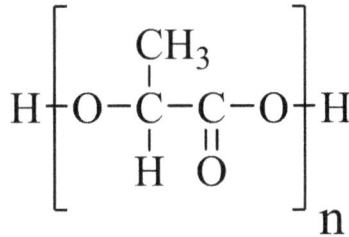

Poly(L-Lactic acid) (PLLA)

FIGURE 16.19   Structure of poly l lactic acid.

biocompatibility and better piezoelectricity can be achieved. Piezoelectric ceramics and PVDF are generally applied to make these composites. They are fabricated to help animals with osteogenesis, osseointegration, and ossification. A nanocomposite film from polydopamine@ BaTiO3NPs in P(VDF-PTrFE) was created that mimicked the endogenous electric potential for the repair of bone defects (103). The electric microenvironment was preserved by these membranes, which aided in bone marrow-derived mesenchymal cells (BMSCs) osteogenic differentiation and promoted rapid bone recovery.

16.4.4.5 Polyelectrolyte Gels

Polyelectrolyte gels have charged groups that are covalently bonded. They induce either movement of charge or change of volume when changes are made to the temperature, pH value, electrical, or magnetic fields. They can absorb large amounts of water and swell. Unlike classic hydrogels, in polyelectrolyte gels the biological response is produced by the induction of chemomechanical contractions. Within hyaluronic acid, the negatively charged alginate microgels favor cartilage regeneration.

16.4.4.6 Challenges when Employing Electroactive Polymers for Tissue Regeneration

The major challenges associated with application of EAPs in tissue engineering include:

1. Improving histocompatibility, biocompatibility, and antibacterial properties of EAPs
2. Achieving fast rejuvenation that is caused to the EAP when providing electrical stimuli
3. Triggering regeneration of multi-tissues by EAPs on applying electrical stimuli
4. Mechanical or electrical stimulation adaptation to tissue environments over a definite range

To overcome the previous challenges, a lot of work is required. Recently, piezoelectric materials have been discovered to have antibacterial properties due to the generation of reactive oxygen species after amplification (104). Biocompatibility, histocompatibility, and antibacterial properties can be enhanced by modifying them with bioactive agents. For example, the antibacterial properties and biocompatibility of CPs can be enhanced by doping them with ant-bacterial polypeptides. EAP scaffolds should be primed and injected into defects to test their ability to cause in vivo multi-tissue regeneration to overcome the difficulties of inducing multi-tissue regeneration.

16.4.5 ELECTROACTIVE POLYMERS AS IMPORTANT TOOLS IN BIOMIMETICS

Biomimetics is an interdisciplinary discipline in which ideas from chemistry, physics, and biology are merged to create materials, devices, and artificial systems that mimic biological processes. Biomimetic designs have potential applications in tissue regeneration, drug delivery, and regenerative medicine (105).

Compared with wheel locomotion, traversing complex terrain with legs is much easier. Because of the advantages of legs, NASA developed four and six-legged and robots for use in space. The Limbed Excursion Mobile Utility Robot (LEMUR) is an example of this type of robot. It was created at Jet Propulsion Laboratory (JPL) (part of the National Aeronautics and Space Administration) and can travel through terrains, capture and manipulate objects, analyze, gain samples, and perform different biomimetic functions. LEMUR is equipped with a variety of instruments and cameras and is designed to inspect and maintain structures that humans cannot do. In addition, it has a panning camera system that allows it to move in all the directions and perform various tasks. Figure 16.20 shows the spider-like design of LEMUR.

16.4.6 ELECTROACTIVE POLYMERS AS ENERGY HARVESTING POWER GENERATORS

The increasing population and the need for better living standard has significantly increased energy requirements globally. Due to sustainable development, there is a need to use renewable sources of energy for power (106).

There are many types of renewable energy including solar, wind, and wave. Ocean wave power has many advantages over other types of renewable energy, because it can be located near highly populated areas, has low visibility, does not require land, and has the least impact on wildlife. It became common on islands and in countries with long coastlines that were subjected to high waves. It can be used to power meteorological and oceanographic observations, water desalination, and

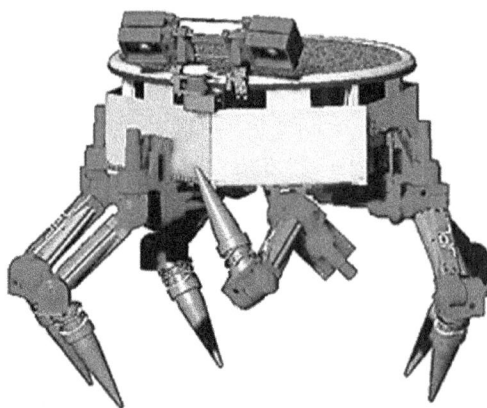

FIGURE 16.20   JPL's multi-limbed robot, LEMUR.

(From Bar-Cohen 2007. With permission.)

navigation, among other things. Most existing wave power stations require high power to convert linear wave motions into the rotary motion of electromagnetic generators. Since these systems have rotary turbines and hydraulic transmissions, they are expensive and require maintenance. Therefore, a cheap, durable, and efficient power take-off system is required to supply wave power.

EAP artificial muscles (EPAM), which are known as dielectric elastomers, have applications in wave power take-off systems. This is due to the following characteristics of EPAMs.:

1. An inexpensive material used
2. They have a high energy density, which allows them to use less EPAM
3. High energy conversion efficiency independent of operation frequency, for instance, strain rate
4. Non-corrosive and non-toxic materials

Since the EPAMs energy conversion concept is uniform, its efficiency is size independent, and devices of any size can be manufactured.

16.4.6.1 Background

The heel strike generator was created by Stanford Research Institute (SRI) International. This generator was connected to the shoe heels. The friction of the engine in the shoe heel caused the stretching and relaxation of the EPAM, which provided electricity. Overall, >2 W of power per shoe was produced in 1 min without endangering the wearer. Voltage is introduced into the EPAM in EPAM generators and is deposited on the polymer as an electric charge, and when mechanical energy decreases, the EPAM elastic recovery force occurs, which reduces the surface area and returns to the original thickness (Figure 16.21). The electric charge is pushed out in the direction of the electrode, which increases the voltage difference, and therefore, the electrostatic force.

The amount of energy released by EPAMs is determined by the dielectric breakdown strength and the strain. The use of polymer materials, for example, Si and acrylic in fabricating EPAMs, increases its energy density.

16.4.6.2 Development of Water Mill Electroactive Polymer Artificial Muscles Generator

Hydroelectric power is a renewable and clean source of electricity, the use of large-scale hydroelectric generators is limited because on reservoirs, dams, and riverbed changes. To overcome this disadvantage, micro hydro generators are being developed that can be installed in flowing streams

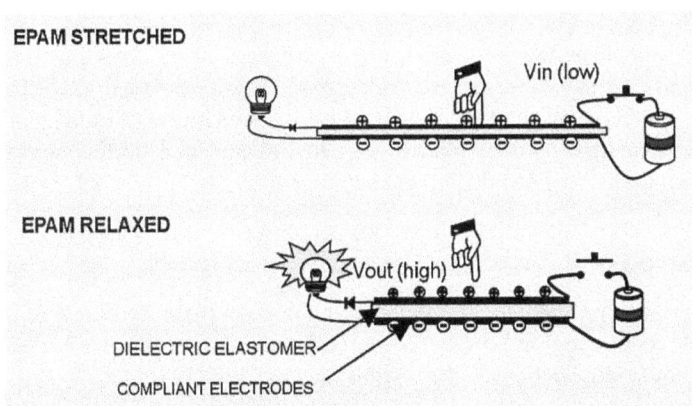

FIGURE 16.21 EPAM generator basic operating principle.

(From Chiba et al. 2008. With permission.)

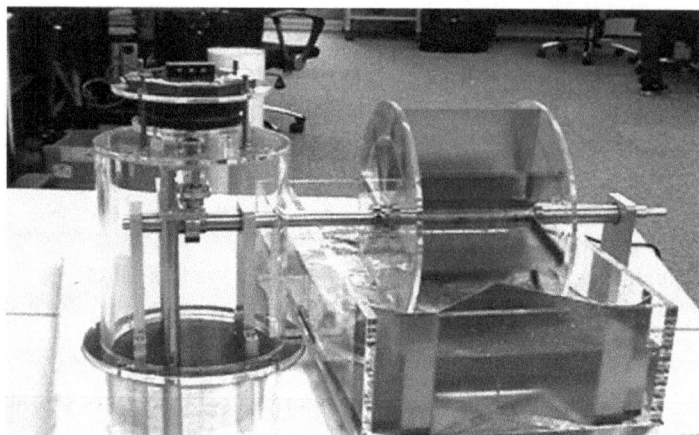

FIGURE 16.22 Water mill generator utilizing EPAM.

(From Chiba et al. 2008. With permission.)

or rivers without the need for large dams or riverbed changes. The EPAM generator can operate at a variety of flow rates in rivers and streams that have not been dammed. In addition, for high-speed generation, EPAM generators do not need a converter. Figure 16.22 shows an EPAM-based water mill generator.

16.4.6.3 Current and Future Trends in Wave Power Generators

As discussed previously, EPAM generators can be used for wave power supply for various applications since they can be scaled up for high scale energy generation. Approximately 30–40 kg of EPAM films can generate approximately 2 kW with a wave amplitude of 1 m. A series of generators 1 km long can produce 2 MW of electricity, which can be used as a local power source. In August 2007, an EPAM wave power generator was tested in an ocean for the first time. It happened in Tampa Bay, 1.6 km off the coast of St. Petersburg, Florida, USA at a depth of 5 m. This test demonstrated the ability of the EPAM generator to withstand an aquatic environment. Figure 16.23 shows an EPAM generator unit and generator control unit.

16.4.7 Diaphragm Actuator Arrays for Haptic Displays

Conventionally, Braille displays used beam actuators for piezoelectric bending to raise Braille dots. However, since they require more space and are expensive, their use was limited. Therefore, EPAMs are used to create structures that consist of relatively large area arrays of individually related sensor or actuator components that can be created using two-dimensional techniques. SRI developed a Braille monitor that used an EPAM diaphragm actuator array (107).

16.4.8 Electrodes for Rechargeable Batteries in Electronics

EAPs are intrinsically conductive, piezoelectric, charged, pyroelectric, and the substance is electricized using conducting additives. Electret microphones, rechargeable batteries, vidicons, hydrophones, and other applications all use electrode polymers (108).

There are several methods in electronics industries to manufacture polymers that are electroactive. The polymer can be electroactive due to its structure. Similar to pyroelectric and piezoelectric polymers, when a polymer is exposed to stress or strain it might become electroactive. Doping,

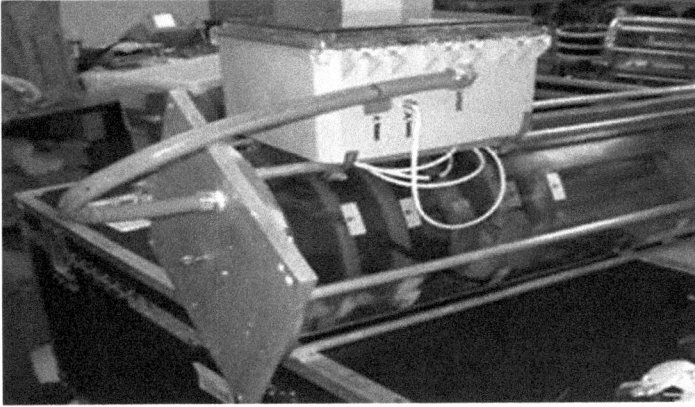

FIGURE 16.23 Electroactive polymer artificial muscle generator unit.

(From Chiba et al. 2008. With permission.)

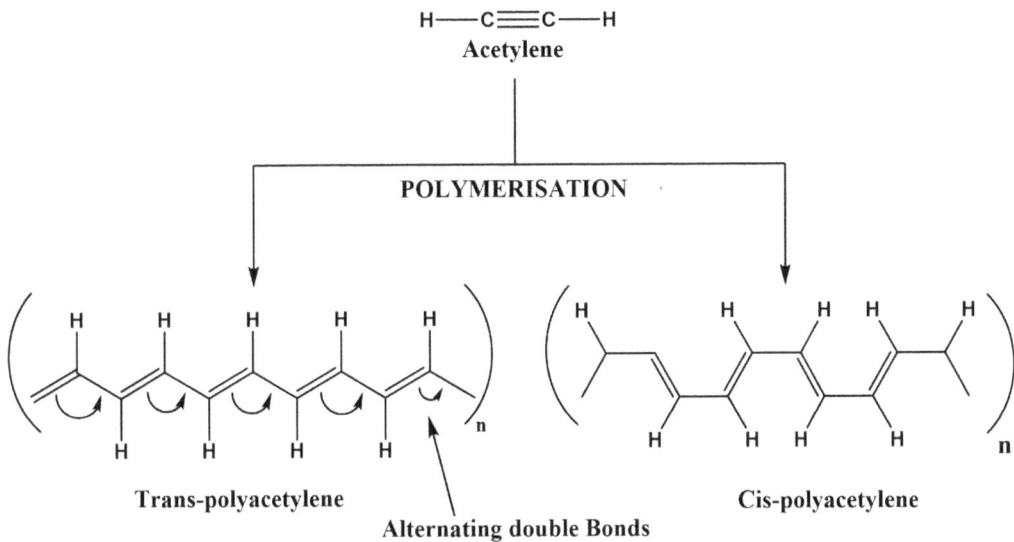

FIGURE 16.24 Structure of polyacetylene.

(From Matthan et al. 1988. With permission.)

applying electroactive materials to the formulation, and forming conductive macro chains within the matrix can all be used to create an electroactive polymeric structure (109–112).

Electronic conduction in organic materials ranges from metals to inorganic semiconductors. The chemical bonds that hold atom groups together in organic molecular solids are held together by the comparatively weak van der Waals forces. For effective macroscopic conduction, the ease at which electrons can be transported from one molecule to another is critical. Because there are few intermolecular transfers in polymeric molecules, intermolecular conduction is less important.

Polyacetylene is the best example of a conductive polymer, with conductivity mainly determined by long chains of unsaturated carbon atoms that maintain an alternating double bond arrangement (Figure 16.24). The delocalized p-system is responsible for polyacetylene's high conductivity when electrons can be visualized as a cloud traversing the polymer chains backbone. The conductivity of polyacetylene can be increased by doping it with Si or gallium arsenide. In polymers, p-type doping oxidizes the material, which eliminates electrons from the cloud, and n-type doping chemically reduces the material, which contributes electrons to the polymer. The excess electrons or holes usually move freely through the material that makes the polymer conductive.

Compact and simple to refill, high energy conductive polymers have the properties for use in compact supplies of electrical electricity. A completely rechargeable 80-min battery was developed (113) in which low-cost and lightweight plastic-based electrodes were produced. N-type doped polymers are refabricated as anodes in the battery. The dopant molecule was enlarged to reduce its electrolyte solubility. The pyrrole was electropolymerized in polyvinylsulfate, which is a polymeric dopant, and used to form the anode. PPy doped with chloride ions was used as the cathode. This decreased to 0.3 mA and 0.3 V after 80 min. Battery life could be increased by making advances in anode materials.

16.4.9 ELECTROACTIVE POLYMERS IN THE MANUFACTURE OF ELECTROACOUSTIC TRANSDUCERS

Charged polymers are EAPs with many industrial uses (114–118). With global production of ≥100 million every year, the thin film polymer electret microphone was an important commercial success. The cross section of a microphone is shown in Figure 16.25. A polyester film diaphragm and a metallized back electrode are shown this figure (108). An incident sound wave induces diaphragm vibrations that produce an AC signal between the backplate metallization and the diaphragm.

Electret microphones are intrinsically sensitive to electromagnetic pickup, shock, and mechanical vibrations. They have the desirable characteristics of condenser microphones and are easier and less expensive to manufacture. They are often used in recording instruments, high-end audio equipment, camcorders, noise dosimeters, telephone handsets, speaker phones, and operator headsets. Electret microphones were used to detect air quality. Photoacoustic spectroscopy and rocket exhaust measurement were accomplished using electret microphones.

Charged polymers are used when making loudspeakers, ultrasonic, earphones, and underwater transducers. Figure 16.26 shows an overview of an electret hydrophone. The transducer is a compressible device made from dielectric and electret layers. Layer deformations are triggered by sound waves, which produce an electrical output signal between the electrodes.

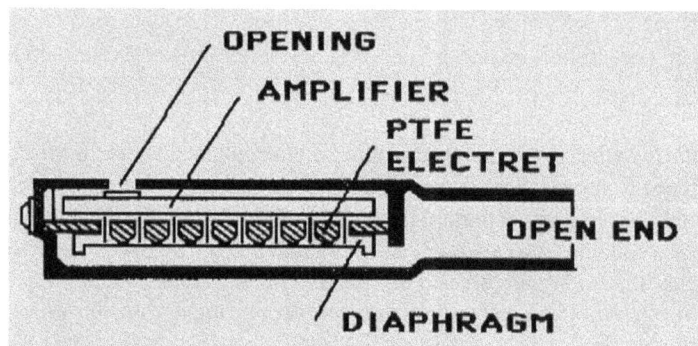

FIGURE 16.25   Fragmentary view of an electret microphone.

(From Matthan et al. 1988. With permission.)

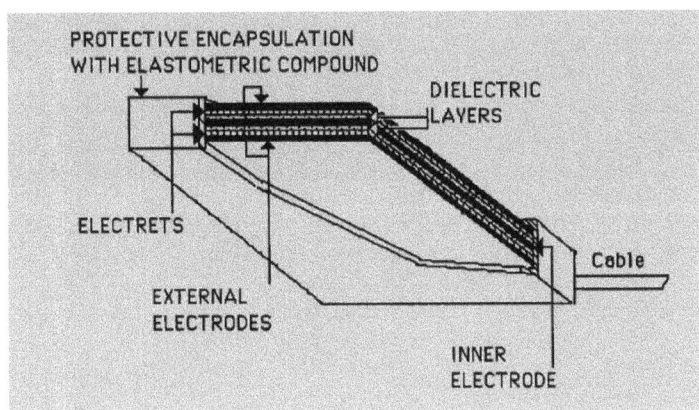

FIGURE 16.26 Electroactive polymer as electret hydrophone.

(From Matthan et al. 1988. With permission.)

16.5 CONCLUSION

EAPs could complement conventional sensing materials, especially in fields where high strains and soft material compliance are required. Certain materials and aspects of their related sensing mechanisms need to be studied, and significant types followed by the variety of examples of two of the major classes of EAPs, for example, ionic and electronic EAPs in industries. EAPs are utilized differently in various industrial areas, such as in MEMS technology, micro and nanoscale sensors, chemiresistor sensors, electro–acoustic transducers, rechargeable batteries, and in biomedical fields, such as tissue engineering and pacemakers.

REFERENCES

1. Roentgen WC. About the changes in shape and volume of dielectrics caused by electricity. Annu Phys Chem. 1880; 11:771–86.
2. Sacerdote MP. On the electrical deformation of isotropic dielectric solids. J Phys. 1899; 3: 282–5.
3. Rivera RAI, Sanches JMA, Galloway KC, Katzenberg HS, Kothari R, Arthur JV, inventor. Dielectric elastomer fiber transducers. United States patent US 7834527 B2. 2010.
4. Biddiss E, Chau T. Electroactive polymeric sensors in hand prostheses: bending response of an ionic polymer metal composite. Med Eng Phys. 2006; 28:568–78.
5. Jean-Mistral C, Basrour S, Chaillout J-J. Comparison of electroactive polymers for energy scavenging applications. Smart Mater Struct. 2010; 19: 085012.
6. Riley PJ, Wallace GG. Intelligent chemical systems based on conductive electroactive polymers. J Intel Mater Syst Struct. 1991; 2:228–38.
7. Keshavarzi A, Shahinpoor M, Kim KJ, Lantz JW. Blood pressure, pulse rate, and rhythm measurement using ionic polymer-metal composite sensors. In: Bar-Cohen Y, editor; Smart structures and materials Newport Beach, CA: SPIE. 1999, p. 369–76.
8. Bar-Cohen Y, editor. Electroactive polymer (EAP) actuators as artificial muscles. Reality, potential, and challenges. Washington, DC: SPIE PRESS; 2004.
9. Baughman RH. Conducting polymer artificial muscles. Synth Met. 1996; 78: 339–53.
10. Tanaka T, Nishio I, Sun S, Nishio S. Collapse of Gels in an electric Field. Science 1982; 218: 467–9.
11. Otero TF, Lopez-Cascales J, Vazquez G, Cortes MT, Borgmann H. Artificial muscles with tactile sensing. Adv Mater. 2004; 15:348–51.
12. Baughman RH, Cui C, Zakhidov AA, Iqbal Z, Barisci JN, Spinks GM. Carbon nanotube actuators. Science. 1999; 284: 1340.

13. Asaka K, Oguro K, Nishimura Y, Mizuhata M, Takenaka H. Bending of Polyelectrolyte Membrane–Platinum Composites by Electric Stimuli I. Response Characteristics to Various Waveforms. Polym J. 1995; 27: 436–40.

14. Pelrine R, Kornbluh R, Pei Q, Joseph J. High-speed electrically actuated elastomers with strain greater than 100%. Science. 2000; 287: 836–839.

15. Aliev A, Oh J, Kozlov ME, Kuznetsov AA, Fang S, Fonseca AF. Giant-stroke, superelastic carbon nanotube aerogel muscles. Science. 2009; 323: 1575–8.

16. Zhang QM, Bharti V, Zhao X. Giant electrostriction and relaxor ferroelectric behavior in electron-irradiated poly(vinylidene fluoride-trifluoroethylene) copolymer. Science. 1998; 280: 2101–3.

17. Nalwa HS. Ferroelectric Polymers: New York: Dekker, 1995.

18. Lehmann W. Skupin H, Tolksdorf C, Gebhard E, Zentel R, Kruger P. Giant lateral electrostriction in ferroelectric liquid-crystalline elastomers. Nature. 2001; 410: 447–50.

19. Carpi F, Smela E, editors. Biomedical applications of electroactive polymer actuators. Chippenham: John Wiley & Sons; 2009.

20. Ramadan KS, Sameoto D, Evoy S. A review of piezoelectric polymers as functional materials for electromechanical transducers. Smart Mater Struct. 2014; 23(033001):1–26.

21. Youqi W, Sun C, Zhou E, Su J. Deformation mechanisms of Electrostrictive Graft Elastomers, Smart Materials and Structures. IOP Publishing. 2004; 13(6):1407–13.

22. Furukawa T. Ferroelectric Properties of Vinylidene Fluoride Copolymers. Phase Transit. 1989; 18: 143–211.

23. Ohki T, Yamato M, Ota M, Takagi R, Kondo M, Kanai N. *et al.* Application of regenerative medical technology using tissue-engineered cell sheets for endoscopic submucosal dissection of esophageal neoplasms Digest Endosc. 2015; 27: 182–8.

24. Yan S, Zhang K, Liu Z, Zhang X, Gan L, Cao B, *et al.* Fabrication of poly(l-glutamic acid)/chitosan polyelectrolyte complex porous scaffolds for tissue engineering J Mater Chem B. 2013; 1:1541–51.

25. Kwon HJ, Yasuda K, Gong JP, Ohmiya Y. Polyelectrolyte hydrogels for replacement and regeneration of biological tissues Macromol Res. 2014; 22:227–35.

26. Park SH, Lee HB, Yeon SM, Park J, Lee NK. Flexible and stretchable piezoelectric sensor with thickness-tunable configuration of electrospun nanofiber mat and elastomeric substrates. ACS Appl Mater Interf. 2016; 8:24773–81.

27. Carpi F, Kornbluh R, Sommer-Larsen P, Alici G. Electroactive polymer actuators as artificial muscles. Bioinsp Biomim. 2011; 6:045006.

28. Shahinpoor M. Potential applications of electroactive polymer sensors and actuators in MEMS technologies. Proc SPIE. 2001; 4234:203–14.

29. Shahinpoor M. Electro-mechanics of ionoelastic beams as electrically controllable artificial muscles. Electroactive Polymer. 1999; 3669(12):109–21.

30. Shahinpoor M. Artificial sarcomere and muscle made with conductive polyacrylonitrile (C-PAN) fiber bundles. Electroactive Polymers. 2000; 3987–35.

31. Shahinpoor M. Electrically activated artificial muscles made with liquid crystal elastomers. Electroactive Polymers. 2000; 3987–27.

32. Keller CG, Howe RT. Hexsil tweezers for teleoperated microassembly Proceedings of the 10th Annual International Workshop on Micro Electromechanical Systems, MEMS-97; 1997 Jan 20–26; Nagoya, Japan.

33. Yamagata Y, Higuchi T. A micropositioning device for precision automatic assembly using impact force of piezoelectric elements. Proceedings of the IEEE International Conference on Robotics and Automation; 1995 April 9–11, Sacramento (CA): 1472–1477.

34. Krulevitch P, Lee AP, Ramsey JC, Trevino JC, Hamilton J Northrup MA. Thin film shape memory alloy microactuators. J Microelectromech Syst. 1996; 5: 270–82.

35. Shahinpoor M. Conceptual design, kinematics and dynamics of swimming robotic structures using ionic polymeric gel muscles. Smart Mater Structs Int J. 1992; 1:91–4.

36. Lumia R, Shahinpoor M. Microgripper design using electro-active polymers. Electroactive Polymers, Proc. SPIE Smart Material and Structure Conference; 1999 Mar 1–5; New Port Beach (CA). SPIE. 3669(30):322–9.

37. Xia F, Cheng Z, Xu H, Li H. High electromechanical responses in a poly(vinylidene fluoride-trifluoroethylene-chlorofluoroethylene) terpolymer. Adv Mater. 2002; 14(21): 1574–7.

38. Bauer F. Relaxor fluorinated polymers: novel applications and recent developments. Trans Dielectr Electr Insul. 2010; 17(4):1106–12.

39. Xia F, Tadigadapa S, Zhang Q. Electroactive polymer based microfluidic pump. Sens Actuator A. 2006; 125(2):346–52.

40. Xia F, Klein R, Bauer F, Zhang QM. High performance P(VDF-TrFE-CFE) terpolymer for bioMEMS and microfluidic devices. Mater Res Soc Symp Proc. 2004; 785: 133–40.

41. Engel L, Krylov S, Shacham-Diamand Y. Thermoplastic nanoimprint lithography of electroactive polymer poly(vinylidene fluoride-trifluoroethylene-chlorofluoroethylene) for micro/nanoscale sensors and actuators. J Micro/ Nanolith. 2014; 13(3):033011.

42. Schmid S, Wagli P, Hierold C. Biosensor based on all-polymer resonant microbeams. 22nd Int. Conf. on Micro Electro Mechanical Systems (MEMS 2009); 2009 Jan 25–29; Sorrento, Italy.

43. Chou SY, Krauss PS, Renstrom PJ. Imprint lithography with 25-nanometer resolution. Science. 1996; 272(5258):85–7.

44. Schift H. Nanoimprint lithography: an old story in modern times? A review. J Vac Sci Technol B. 2008; 26(2):458–80.

45. Padua LMG, Yeh JM, Santiago KS. A novel application of electroactive polyimide doped with gold nanoparticles: As a chemiresistor sensor for hydrogen sulfide gas. Polymers. 2019; 11(1918):1–16.

46. Textor C, Graf HF, Herzog M, Oberhuber JM. Injection of gases into the stratosphere by explosive volcanic eruptions J Geophys Res. 2003; 108:4606.

47. Oppenheimer C, Scaillet B, Martin RS. Sulfur degassing from volcanoes: Source conditions, surveillance, plume chemistry and earth system impacts. Sulfur Magmas Melts. 2011; 73:363–422.

48. Edmonds M, Grattan J, Michnowicz S. Volcanic gases: Silent killers. In: Fearnley CJ, Bird DK, Haynes K, McGuire, WJ, Jolly G, editors. Observing the Volcano World. Advances in Volcanology. Cham: Springer; 2015.

49. Plaza C, Xu Q, Townsend T, Bitton G, Booth M. Evaluation of alternative landfill cover soils for attenuating hydrogen sulfide from construction and demolition (C&D) debris landfills. J Environ Manag. 2007; 84:314–22.

50. Shen DS, Du Y, Fang Y, Hu L, Fang Ch, Long Yu. Characteristics of H_2S emission from aged refuse after excavation exposure. J Environ Manag. 2015; 154: 159–65.

51. Kim KH. Performance characterization of the GC/PFPD for H_2S, CH_3SH, CH_3SCH_3, and CH_3SSCH_3 in air. Atmos Environ. 2005; 39:2235–42.

52. Vitvitsky V, Banerjee R. H_2S analysis in biological samples using gas chromatography with sulfur chemiluminescence detection. Methods Enzym. 2015; 554:111–23.

53. Varlet V, Giuliani N, Palmiere C, Maujean G, Augsburger M. Hydrogen sulfide measurement by headspace-gas chromatography-mass spectrometry (HS-GC-MS): Application to gaseous samples and gas dissolved in muscle. J Anal Toxicol. 2015; 39:52–7.

54. Larsen ES, Hong WW, Spartz ML. Hydrogen sulfide detection by UV-assisted infrared spectrometry. Appl Spectrosc. 1997; 51: 1656–67.

55. Chen R, Morris HR, Whitmore PM. Fast detection of hydrogen sulfide gas in the ppm range with silver nanoparticle films at ambient conditions. Sens Actuators B Chem. 2013; 186:431–8.

56. Gersen S, Van Essen M, Visser P, Ahmad M, Mokhov A, Sepman A. et al. Detection of H_2S, SO_2 and NO_2 in CO_2 at pressures ranging from 1–40 bar by using broadband absorption spectroscopy in the UV/VIS range. Energy Procedia. 2014; 63: 2570–82.

57. Shariati-Rad M, Irandoust M, Jalilvand F. Spectrophotometric determination of hydrogen sulfide in environmental samples using sodium 1,2-naphthoquinone-4-sulfonate and response surface methodology. Int J Environ Sci Technol. 2016; 13: 1347–56.

58. Wallace KJ, Cordero SR, Tan CP, Lynch VM, Anslyn EV. A colorimetric response to hydrogen sulfide. Sens Actuators B Chem. 2007; 120:362–7.

59. Zhang Y, Li M, Niu Q, Gao P, Zhang G, Dong C. et al. Gold nanoclusters as fluorescent sensors for selective and sensitive hydrogen sulfide detection. Talanta. 2017; 171: 143–51.

60. Lawrence NS, Deo RP, Wang J. Electrochemical determination of hydrogen sulfide at carbon nanotube modified electrodes. Anal Chim Acta. 2004; 517: 131–7.

61. Zeng L, He M, Yu H, Li D. An H_2S sensor based on electrochemistry for chicken coops. Sensors. 2016; 16:1398.

62. Ke ZJ, Tang DL, Lai X, Dai ZY, Zhang Q. Optical fiber evanescent-wave sensing technology of hydrogen sulfide gas concentration in oil and gas fields. Optik. 2018; 157:1094–100.

63. Zhou H, Wen JQ, Zhang XZ, Wang W, Feng DQ, Wang Q. *et al.* A study on fiber-optic hydrogen sulfide gas sensor. Phys Procedia. 2014; 56:1102–6.

64. He H, Dong C, Fu, Y, Han W, Zhao T, Xing L. *et al.* Self-powered smelling electronic-skin based on the piezo-gas-sensor matrix for real-time monitoring the mining environment. Sens Actuators B Chem. 2018; 267:392–402.

65. Kuchmenko TA, Kochetova ZY, Silina YE, Korenman Ya I, Kulin LA. Lapitski IV. Determination of trace amounts of hydrogen sulfide in a gas flow using a piezoelectric detector. J Ana Chem. 2007; 62:781–7.

66. Berahman M, Sheikhi M. Hydrogen sulfide gas sensor based on decorated zigzag graphene nanoribbon with copper. Sens Actuators B Chem. 2015; 219: 338–45.

67. Tomchenko AA, Harmer GP, Marquis BT, Allen JW. Semiconducting metal oxide sensor array for the selective detection of combustion gases. Sens Actuators B Chem. 2003; 93:126–34.

68. Kumar V, Sunny RI, Mishra V, Dwivedi R, Das R. Fabrication and characterization of gridded Pt/SiO_2/Si MOS structure for hydrogen and hydrogen sulphide sensing. Mater Chem Phys. 2014; 146:418–24.

69. Emelin EV, Nikolaev IN. Sensitivity of MOS sensors to hydrogen, hydrogen sulfide, and nitrogen dioxide in different gas atmospheres. Meas Tech. 2006; 49:524–8.

70. Crowley K, Morrin A, Sheperd R, Panhuis M. Wallace G, Smyth M. *et al.* Fabrication of polyaniline-based gas sensors using piezoelectric inkjet and silkscreen printing for the detection of hydrogen sulfide. Sens J. 2010; 10:1419–26.

71. Liu C, Hayashi K, Toko K. Au nanoparticles decorated polyaniline nanofiber sensor for detecting volatile sulfur compounds in expired breath. *Sens Actuators B Chem.* 2012; 161:504–9.

72. Chen R, Morris HR, Whitmore PM. Fast detection of hydrogen sulfide gas in the ppmv range with silver nanoparticle films at ambient conditions. Sens Actuators B Chem. 2013; 186:431–8.

73. Chen R, Whitmore PM. Silver nanoparticle films as hydrogen sulfide gas sensors with applications in art conservation. In: Harper-Leatherman AS, Solbrig CM, editors. The science and function of nanomaterials: From synthesis to application. Washington: ACS, 2014.

74. Zhang Y, Li M. Niu Q, Gao P, Zhang G, Dong C. *et al.* Gold nanoclusters as fluorescent sensors for selective and sensitive hydrogen sulfide detection. Talanta. 2017; 171: 143–51.

75. Yuan Z, Lu F, Peng M, Wang C-W, Tseng YT, Du Y. *et al.* Selective colorimetric detection of hydrogen sulfide based on primary amine-active ester cross-linking of gold nanoparticles. Ana. Chem. 2015; 87: 7267–73.

76. Wei Y, Yang C, Ding T. A one-step method to synthesize n,n'-bis(4'-aminopiienyl)-1,4quinonenediimine and its derivatives. Tetrahedron Lett. 1996; 37:731–4.

77. Mooney E, Mackle JN, Blond DJ, O'Cearbhaill E, Shaw G, Blau WJ, *et al.* The electrical stimulation of carbon nanotubes to provide a cardiomimetic cue to MSCs. Biomater. 2012; 33:6132–9.

78. Ghasemi-Mobarakeh L, Prabhakaran MP, Morshed M, Nasr-Esfahani MH, Baharvand H, Kiani S, *et al.* Application of conductive polymers, scaffolds and electrical stimulation for nerve tissue engineering. J Tissue Eng Regener Med. 2011; 5:e17–35.

79. Wu Y, Wang L, Guo B, Shao Y, Ma PX. Electroactive biodegradable polyurethane significantly enhanced Schwann cells myelin gene expression and neurotrophin secretion for peripheral nerve tissue engineering. Biomaterials. 2016; 87:18–31.

80. Parssinen J, Hammaren H, Rahikainen R, Sencadas V, Ribeiro C, Vanhatupa S, *et al.* Enhancement of adhesion and promotion of osteogenic differentiation of human adipose stem cells by poled electroactive poly(vinylidene fluoride). J Biomed Mater Res Part A. 2015; 103:919–28.

81. Costa R, Ribeiro C, Lopes AC, Martins P, Sencadas V, Soares R, *et al.* Osteoblast, fibroblast and in vivo biological response to poly(vinylidene fluoride) based composite materials. J Mater Sci Mater Med. 2013; 24:395–403.

82. Meng S, Rouabhia M, Zhang Z. Electrical stimulation modulates osteoblast Proliferation and bone protein production through heparin-bioactivated conductive scaffolds. Bioelectromagnetics. 2013; 34(3):189–99.

83. Ning C, Zhou Z, Tan G, Zhu Y, Mao C. Electroactive polymers for tissue regeneration: Developments and perspectives. Prog Polym Sci. 2018; 81: 144–62.

84. Jun I, Jeong S, Shin H. The stimulation of myoblast differentiation by electrically conductive sub-micron fibers. Biomaterials. 2009; 30:2038–47.

85. Coimbra P, Ferreira P, De Sousa H, Batista P, Rodrigues M, Correia I, *et al*. Preparation and chemical and biological characterization of a pectin/chitosan polyelectrolyte complex scaffold for possible bone tissue engineering applications. Int J Biol Macromol. 2011; 48:112–8.

86. Coimbra P, Ferreira P, De Sousa H, Batista P, Rodrigues M, Correia I. *et al*. Preparation and chemical and biological characterization of a pectin/chitosan polyelectrolyte complex scaffold for possible bone tissue engineering applications. Int J Biol Macromol. 2011; 48:112–18.

87. Zhang J, Qiu K, Sun B, Fang J, Zhang K, Hany E-H. *et al*. The aligned core–sheath nanofibers with electrical conductivity for neural tissue engineering. J Mater Chem B. 2014; 2:7945–54.

88. Khorasani M, Mirmohammadi S, Irani S. Polyhydroxybutyrate (PHB) scaffolds as a model for nerve tissue engineering application: fabrication and in vitro assay. Int J Polym Mater. 2011; 60:562–75.

89. Gelmi A, Zhang J, Cieslar-Pobuda A, Ljunngren MK, Los MJ, Rafat M, *et al*. Electroactive polymer scaffolds for cardiac tissue engineering. Proc SPIE. 2015; 9430(94301T1):1–7.

90. Martins AM, Eng G, Caridade SG, Mano JoF. Reis RL, Vunjak-Novakovic G. Electrically conductive chitosan/carbon scaffolds for cardiac tissue engineering. Biomacromolecules. 2014; 15:635–43.

91. Kwon HJ, Yasuda K, Gong JP, Ohmiya Y. Polyelectrolyte hydrogels for replacement and regeneration of biological tissues. Macromol Res. 2014; 22:227–35.

92. Fattahi P, Yang G, Kim G, Abidian MR. A review of organic and inorganic biomaterials for neural interfaces. Adv Mater. 2014; 26:1846–85.

93. Ribeiro C, Correia DM, Ribeiro S, Sencadas V, Botelho G, Lanceros-Mendez S. Piezoelectric poly(vinylidene fluoride) microstructure and poling state in active tissue engineering. Eng Life Sci. 2015; 15:351–6.

94. Guterman E, Cheng S, Palouian K, Bidez P, Lelkes P, Wei Y. Peptide-modified electroactive polymers for tissue engineering applications. Abstr Pap Am Chem Soc. 2002; 224:U433–3.

95. Richardson-Burns SM, Hendricks JL, Martin DC. Electrochemical polymerization of conducting polymers in living neural tissue. J Neural Eng. 2007; 4:L6–13.

96. Asplund M, Thaning E, Lundberg J, Sandberg-Nordqvist A, Kostyszyn B, von Holst H. Toxicity evaluation of PEDOT/biomolecular composites intended for neural communication electrodes. Biomed Mater. 2009; 4(045009):1–12.

97. Sittichockechaiwut A, Scutt AM, Ryan AJ, Bonewald LF, Reilly GC. Use of rapidly mineralising osteoblasts and short periods of mechanical loading to accelerate matrix maturation in 3D scaffolds. Bone. 2009; 44:822–9.

98. Damaraju SM, Wu S, Jaffe M, Arinzeh TL. Structural changes in PVDF fibers due to electrospinning and its effect on biological function. Biomed Mater. 2013; 8(045007):1–11.

99. Young T-H, Lu J-N, Lin D-J, Chang C-L, Chang H-H, Cheng L-P. Immobilization of l-lysine on dense and porous poly (vinylidene fluoride) surfaces for neuron culture. Desalination. 2008; 234:134–43.

100. Rivard C, Chaput C, Rhalmi S, Selmani A. Bio-absorbable synthetic polyesters and tissue regeneration: a study of three-dimensional proliferation of ovine chondrocytes and osteoblasts. Ann Chir. 1996; 50:651–8.

101. Novikov LN, Novikova LN, Mosahebi A, Wiberg M, Terenghi G, Kellerth J-O. A novel biodegradable implant for neuronal rescue and regeneration after spinal cord injury. Biomaterials. 2002; 23:3369–76.

102. Chen S, Hao Y, Cui W, Chang J, Zhou Y. Biodegradable electrospun PLLA/chitosan membrane as guided tissue regeneration membrane for treating periodontitis. J Mater Sci. 2013; 48:6567–77.

103. Zhang XH, Zhang CG, Lin YH, Hu PH, Shen Y, Wang K, *et al*. Nanocomposite membranes enhance bone regeneration through restoring physiological electric microenvironment. ACS Nano. 2016; 10:7279–86.

104. Tan GX, Wang SY, Zhu Y, Zhou L, Yu P, Wang XL, *et al*. Surface-selective preferential production of reactive oxygen species on piezoelectric ceramics for bacterial killing. ACS Appl Mater Interfaces. 2016; 8:24306–9.

105. Bar-Cohen Y. Electroactive polymers (EAP) as an enabling tool in biomimetics, Proc SPIE. 2007; 6524(652403):1–6.

106. Chiba S, Wakib M, Kornbluha R, Pelrine R. Innovative power generators for energy harvesting using electroactive polymer artificial muscles. Proc SPIE. 2008; 6927(692715):1–9.

107. Heydt R, Chhokar S. Refreshable Braille display based on electroactive polymers. Proceedings of the 23rd International. Display Research Conference. 2003 Sep 15–18 Sep; Phoenix, Arizona.

108. Matthan J, Uusimaki A, Torvela H, Leppavouri S. Past and future impact of electroactive polymers on the electronics sector. Makromol Chem Macromol Symp. 1988; 22:161–90.

109. Gutmann F, Lyons LE. Organic semiconductors. New York: John Wiley & Sons; 1966.

110. Katon JE, editor. Organic semiconducting polymers. New York: Marcel Dekker; 1968.

111. Blythe AR. Electrical properties of polymers. Cambridge: Cambridge University Press; 1979.

112. Mort J, Pfister G, editors. Electronic properties of polymers. New York: John Wiley L Sons; 1982.

113. Shimidzu T, Ohtani A, Iyoda T, Honda K. A novel type of polymer battery using a polypyrrole - polyanion composite anode. J Chem SOC Chem Comm. 1987; 5:327–8.

114. Perlman MM, editor. Electrets, charge storage and transport in dielectrics. Princeton: Electrochemical Society; 1973.

115. Moore AD, editor. Electrostatistics and its applications. New York: Wiley & Sons; 1973.

116. Schaffert RM. Electrophotography. New York: Wiley & Sons; 1975.

117. Sessler GM, editor. Electrets. Heidelberg: Springer; 1980.

118. Rosencwaig A. Photoacoustics and photoacoustic spectroscopy. New York: Wiley & Sons; 1980.

17 Electroactive Polymers in Biomedicine

Kashish Gupta

CONTENTS

17.1 INTRODUCTION

Electroactive polymers (EAPs) are a special class of organic polymers that have intrinsic conductive properties that can be tuned by the modification of various chemical and physical properties. Slight modifications in the functionalities of EAPs means that they are suitable for various applications, such as biosensors and bioinstructive scaffolds. EAPs show high similarity in conductive properties compared with metals and semiconductors (Guarino *et al.*, 2013). EAPs conductivity is due to the highly conjugated electronic structure, which supports the mobility of charges alongside the polymeric chain. The merits of these polymers are the ease of preparation and varied flexibility in modification of the chemical structures for customized applications. The dynamic and versatile properties of EAPs means that they are designated as smart materials that are ultrasensitive to external stimuli (Malhotra *et al.*, 2006). Modifications to increase the sensitivity of EAPs can be added during EAP synthesis. Scientific interest toward EAPs has focused on developing novel methods for synthesis and doping strategies. EAPs have major applications in biomedicine. Designing appropriate EAPs for medicine depends on the type of system needed, such as scaffolds, drug delivery system, and biosensors fabrication.

17.2 NEED FOR ELECTROACTIVE POLYMERS IN BIOMEDICINE

Some natural EAPs exists in nature. Electroactivity has been found in biomolecules, such as proteins, enzymes, and polynucleotides. The natural polymers have piezoelectric activity in wool, bones, and hairs. Ferroelectricity is found in biomolecules, such as elastin and hydroxyapatite. Mimicking the

DOI: 10.1201/9781003173502-17

activity of natural EAPs, certain artificial EAP are elastomers, especially self-healing dielectric elastomers. Since electroactivity is innate for the biological processes and its correct functioning, artificial EAPs have strong potential in biomedicine for healing or enzymatic activity, tissue engineering, and drug delivery (Youn *et al.*, 2020). EAPs ability to stimulate cells under an applied current, mimicking muscles by acting as actuators and changing the configuration during drug delivery make them suitable candidates for biomaterials. Ferroelectrets, which are special piezoelectric polymer foams, can charge internally and be used as a blood pressure or pulse sensors (Mohebbi *et al.*, 2016; Zhang *et al.*, 2019). Manufactured electronic skin can generate electrical energy from any change in pressure or temperature. The dielectric elastomers-based Stretch Sense brand is popular in sports and biomedical market as implantable or wearable biomedical devices. Ferroelectric-based EAPs are used as sensors and memory cells, because of their stretchability and self-healing properties. Natural ferroelectricity is present in bone as hydroxyapatites and in aortic walls and elastin. Stretchable piezoelectric materials are used to make artificial skin and prosthesis, because they can optimally mimic the natural ones. The mode of action proposed for piezoelectric elastomers is the generation of an electrical field that poles from the polar molecules incorporated in a non-centro symmetrically manner. In a self-healing elastomer, the iron–ligand (Fe–ligand) bonds tends to break and reform at low temperatures, which leads to the self-healing property. Polymers used in biomedicine are optimized for roughness, porosity, conductivity, and degradability (Nezakati *et al.*, 2018; Preethi *et al.*, 2018; Cheng *et al.*, 2017).

17.3 TYPES OF ELECTROACTIVE POLYMERS AND THEIR MECHANISMS

There are three EAPs based on the mechanism of conduction: (1) ionic conducting polymers (CPs); (2) conduction by electron delocalization; and (3) percolated EAPs

An intrinsic CPs use is limited because of low biodegradability, weak polymer cell interactions, hydrophobicity, and processibility (Saghazadeh *et al.*, 2018). These limitations can be compensated by poly caprolactone (PCL), polylactic-co-glycolic acid (PLGA), gelatin, and collagen that have quick dissolution (Fu *et al.*, 2014). They are used in drug delivery because of the controlled and reversible redox activity during drug release, for example, poly(3,4-ethylenedioxythiophene) (PEDOT), polypyrrole (PPy), and polyaniline (PANi). However, their use is restricted because of poor processibility. Intrinsic CPs can be modified using dopants of different sizes to optimize the concentration to avoid an inflammatory responses in the tissues (Nikolova *et al.*, 2019). Biocompatibility of a CP can be increased by doping the biomolecule or any ion that successfully modifies the chemical, physical, and electric properties (Bose, et al., 2018).

Conducting polyelectrolyte hydrogels are three-dimensional (3D) polymeric materials. The unique properties of the hydrogels are high water content, soft consistency, and porosity. Hydrogels can mimic the activity of living tissue compared with other synthetic biocompatible materials (Li *et al.*, 2011). The conductivity of hydrogels depends on factors, such as the polarity of the polymer, water content, and concentration of ionic content. Ionic conductivity tends to increase with the water content of the hydrogel.

In a percolated polymer composite electrical conduction is due to a conducting filler over an insulator polymer (Dan *et al.*, 2007; Derakhshankhah *et al.*, 2020). The percolation mechanism can drastically change the electric conductivity. Conducting adhesives are made using fillers, such as nickel (Ni), cobalt (Co), graphite, and carbon (C) filler carbon nanotube (CNT). Aspect ratio and percolation threshold depends on the size of the filler used.

17.3.1 MECHANISM OF ACTION OF ELECTROACTIVE POLYMERS

The conductivity of EAPs is due to the electronic structure formed from π-bound system, which is a series of single and double bond π in the polymeric chain. The special arrangement of alignment of p_z

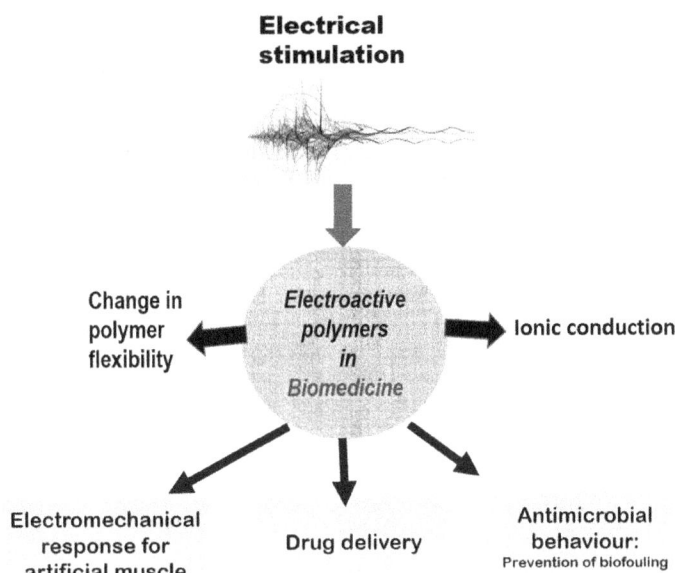

FIGURE 17.1   EAP major applications in biomedicine.

orbitals leads to double bond formation. Some unpaired electrons (e.g., π electrons or any unbounded π electrons of the heteroatom) are delocalized along the polymeric backbone (Ivory, 1979). π bound electrons are loosely bound and can be easily removed, which makes for easy chemical modification. These conjugated compounds can have metallic or semiconducting behavior. Based on the electrical conductivity of EAPs at room temperature, they can be classified as insulators, conductors, and semiconductors. Doping enhances the conductivity of semiconductor EAPs, by modifying the electrical conductivity of the π-conjugated backbone of the polymer. Conjugated polymers enhance the electrical conductivity from 10^{-10} to 10^3/cm (Roncali *et al.*, 1997; Harisson *et al.*, 1977). Doping of conjugated polymers utilizes the hopping mechanism adopted by dopant molecules on the polymer chain. The conductivity of EAP polymers depends on the type and concentration of the dopant. EAP properties, such as degree of solubility, moisture absorption, and processibility depend on the dopant chosen.

Further conductivity of the polymer depends on the amount of dopant used. The synthesis method depends on the type of dopant employed. EAPs as CPS can only be used in medicine, because these polymers can be doped and undoped reversibly, which is essential for any drug delivery application. A slight change in electric potential can be induced by the addition of a small dopant (chloride ion (Cl^-)) when it enters or leaves the polymer matrix (Figure 17.1) (Wong *et al.*, 1994). This entry and exit leads to a change in volume either by expansion or contraction during the mass transport (Gandhi *et al.*, 1995). Other large dopants can be incorporated into the polymer stably, which affects the surface and bulk properties of the polymer. In biomedicine, a range of dopants can be used, from the simplest salt ions to negatively charged compounds, aromatic sulfonic acids, and polysaccharides (Guarino *et al.*, 2016). Suitable dopants for specific and customized applications in biomedicine are used and their properties can be modified and controlled when stimuli, such as light, pH, and electricity are applied, even when synthesis has been completed (Ding, 2010;Huang *et al.*, 2008).

17.4   PROCESSED ELECTROACTIVE POLYMER PRODUCTS

Electrochemical methods of synthesis for EAPs are becoming popular, because they directly modify the polymers in forms, such as thin films and coatings. The only limitation lies in the availability of monomers that can oxidize and form radical intermediate or ions during polymerization. EAPs such as PPy and PANi can be electrochemically synthesized.

Recent research demands in the biomedical fields include the development of complex system comparable with the structural hierarchy of tissues as virgin CPs.

Certain porous materials and coatings have been fabricated to make EAPs from a combination of different polymers. Different chemical-based approaches can be used to synthesize EAPs.

17.4.1 Two-dimensional Coatings (Blends, Composite, and Hybrids)

Polymer films are being made with controlled thickness on a large scale. Different synthesis methods, such as solvent casting and spin coating are used.

Coating methods for depositing the material homogenously during polymerization is essential during the fabrication of film. Polymerization can occur on the surface of the matrix under controlled conditions, such as in the spin coating method. Another technique is the Langmuir–Blodgett (LB) film method, where the organic thin films are fabricated in a very ordered configuration. Finely controlled monolayers are formed and the functionalization can be added easily during processing (Kumar et al., 1998). In these cases, EAPs help with the incorporation of functionalization for synthetic and biological tissues. EAPs provide efficient processing for the electrical stimuli for nerves, myocardial cells (Zhang, 2010), and provide a suitable conductive environment for surface interaction integration. PPy is a suitable CP and a relevant smart transparent materials for thin films and for mammalian cell culture (Lee, 2009; Schmidt et al., 1997). IPP y acts as the dopant anion that can alter protein adsorption, and oxidation. A high surface area and chemical stability under physiological conditions has been observed. PPy-based films can form electrostatic PC12 and neurite cells. PPy cells promote neural activity, extracellular matrix (ECM) growth and its proliferation in vitro (Ghasemi, 2011). Other material based on PANi can be used with biological systems and grow in vitro (Kamlesh et al., 2000)]. In addition, no side effects, such as an inflammatory response in subcutaneous tissues are sustained even after 2 years (Wang, 1999). PANi can be synthesized by various electrochemical methods and gives a pure uniform layer that is required for metal electrode coatings (Anand et al., 1998). Pure PANi not used directly in biological systems, because of its non-flexibility, non-biodegradability, and non-processibility. Neural proteins are now used in these films on a large scale. Certain heteroaromatic CPs are used with various solvent system along with boron trifluoride diethyl etherate (BFEE) tri fluoroborate, PEDOT, a polythiophene derivative enters biologicals system, because of high chemical, electrical, and thermal stability and is preferred over PPy (Cukierman et al., 2001). PEDOT is added to muscles and generates a network of tubular structures around tissues (Peramo et al., 2008).

17.4.1.1 Three-Dimensional Processing Blends

Blending methods are preferred when the commercial production of EAPs is required. CPs blended with insulating materials helps in to obtain efficient and even functionalized electrical groups; therefore, reducing toxicity and promoting the efficient transfer of electrical signal for cells (Qazi, 2014) Other blends are prepared by combining the conducting and degrading polymer for various environmental conditions. The most popular is the polymer blends prepared via a melt process for generating CPs at industrial levels. These blends are thermoplastic and can be easily modified and extruded (Zilberman, 2000). CP blends were fabricated from PANi–p-toluene sulfonic acid (pTSA) and a thermoplastic polymer that has enhanced conductivity due to systematic ordering within the films.

Copolymers of PPy are more effective in reducing and promoting the of adsorption of a PPy film. The high and high electroconductivity of PEDOT has biosensor applications. A new class of materials molecularly imprinted polymers

(MIPs) can mimic biological receptors. PEDOT-based microelectronic devices are used to detect brain heart function in vitro. PANi-based composites acts as percolative CPs (Bhadra et al., 2020) to make PANi more soluble and compatible dopants, such as functionalized sulphonic acids are used. EAPs in biocomponent blends, such as PPy and PANi are blended with biodegradable polymers, such as poly (DL-lactide) (Wan, 2005), polyvinyl alcohol (PVA) (Subramanian et al.,2014), and PCL (Limongi *et al.*, 2017) can be used to construct the nano structured platform. EAP properties can be changed, such as the surface charge, wettability, oxidation, and reduction on the scaffold surface used in cell culture (Guarino, 2016). In addition, other properties, such as non-solubility, mechanical strength, and biodegradability can be changed (Guo, 2010). PANi blended with poly glycerol sebacate was prepared for cardiac patches using the solvent casting method. The electrical behavior of these cardiac patches could be altered by changing the PANi content. With the increase in PANi content, improvement in the tensile strength and elasticity occurred. This elasticity is related to the physiological pH of functioning of myocardial tissue, and therefore, it can mimic the mechanical strength of natural cardiac tissue. Experimental studies on the scaffold in vitro toxicity confirmed that the change in pH was created by PANi-based scaffolds, which act like a buffer, because the polymer does not interfere with natural growth and functions.

17.4.1.2 Composites

The conducting activity of EAP-based composites depend on the polymer matrix. Changes in the internal stress that is generated due to thermal expansion or local shrinking influences the conductivity (Struèmpler, 1999). Biomaterials, such as rubbers, biopolymers, soft elastomers, or any thermoplastics has been investigated for the relationship between the final conductivity of the materials and the local contact forces in the EAP matrix. In vitro culture studies of cell interactions for charge generation by the different phases of polymer were used for tissue engineering. CPs act as fillers in the design of electroactive composites in biomedicine.

EAPs are prepared using chemical and electrochemical methods and varied polymer fabricating techniques (i.e., emulsion) are used to construct devices for biosensing and prosthetics. These devices do not need direct contact between the polymer and the cells. The biocompatibility of materials, such as scaffolds can be increased using a combination of conducting and nonconducting matrixes to improve the functional performance (Takano *et al.*, 2014)]. In addition, EAPs, such as PPy and PANi acts as conductive fillers and can compensate for the disadvantages of non-polymeric based conducting fillers. A composite from HAP, PANi, and PCL was produced, which are porous, elastic, and conductive scaffolds (Sarvari *et al.*, 2016). Composites have been prepared from different substrates, such as ultrafine PANi with a PCL matrix for the engineering of cardiac tissue.

17.4.2 Three-Dimensional Materials (Artificial Muscles and Actuators)

EAPs, the major class of smart materials can respond to any electrical stimulation with a marked change in displacement as in artificial muscles. Artificial muscles are electromechanical actuators. They are applied in biomedical engineering as artificial limbs, artificial ocular muscles, and heart tissue (Vohrer, *et al.*, 2004).

Certain blood vessel connectors or functional ears (i.e., myringotomy tubes), microvalves for controlling urinary incontinence have been successfully fabricated (Mirfakhrai *et al.*, 2007). These artificial polymers possess great strength, can work at a low voltage (≤1 V), and produce large strain. In addition, they work at room temperature, are light weight, and work in body fluids (Smela, 2003). Artificial muscles use polymers, such as PPy, ionic polymer–metal composites (IPMCs), and liquid crystal elastomers. Polymer-based actuators perform better than natural muscles in some aspects.

Ionic CP–based artificial muscle has high ionic mobility within the polymer matrix in response to the applied current. Electrons move out of the polymer matrix and simultaneous insertion of anion occurs; therefore, the polymers expand. These polymer-based artificial muscles have linear ionic mobility to the current.

IPMC-based artificial muscle show high deformation when a low voltage is applied. They are composed of an ion exchange polymer that is sandwiched between two flexible metals, such as percolated platinum or gold (Au). In IPMCs, polyelectrolytes contain the ionizable groups within the polymer matrix chain, which tends to dissociate and leads to electrophoretic migration of electrons within the matrix (Li et al., 2017). Dielectric elastomer-based artificial muscle are mainly capacitors that are composed of an elastomer film coated on the electrodes (Brochu *et al., 2010*]. The applied external electric field along these electrodes lead to electrostatic attraction and the expansion of the polymer structure, for example, PVDF polymer-based coatings. EAP hydrogels in artificial muscles have high stability in air but low electrochemical activity.

17.4.3 POROUS MATERIALS AS SCAFFOLDS

The conductivity of scaffolds when carrying cells is essential before using them for further application of electrical impulses. Scaffolds were manufactured with controlled porosity that were generated via leaching and compression molding. Coporogens, such as sodium chloride (NaCl) crystals and polyethylene glycol (PEG) enhanced the cohesion of the porous materials. PCL, PANi, and polyurethane in a PCL matrix were used for cardiomyocyte proliferation in damaged heart tissue. Moreover, further chemical tests confirmed that these porous materials for cardiac muscle contraction and relaxation could be used for a long time (Baheiraei et al., 2015). Guarino et al. (2016) fabricated hydrogels with controlled porosity and unique conductive properties for cell signaling in regenerated nerve cells based on the incorporated EAP materials, for instance, PANi prepared with polyethylene glycol diacrylate (PEGDA). PANi hydrogels of 136–158 um have suitable conductive behavior and biocompatibility. Improved water retention capacity was observed with PANi, which is required for efficient charge transport. Good interconnectivity generated through macropore network aids with homogenous growth of newly formed cells for nerve regeneration. The distance traveled by electrons between the polymer fibers decrease with an increase in the concentration of PANi, therefore the electrons hop easily, and better conduction is achieved.

Recently, conducting hydrogels based on gelatin and chitosan have been used in tissue engineering. Organic (PPy) and inorganic (graphene) conducting phases have been explored and tested as functional features of the scaffolds. The use of PPy and graphene filler in the polymeric matrix increases the conductivity; however, a little decrease in porosity was observed. PYG conductive phases compared with the gelatin hydrogels have lower biodegradation activity, which makes it an appropriate material in nerve engineering (Baniasadi *et al.*, 2015) PEDOT: polystyrene sulfonate (PSS) based 3D macropore structure are electrically conducting, have low impedance, and are biocompatible scaffolds for the culturing of fibroblast (Mockzko *et al.*, 2012). EAP scaffolds can be used as 3D tissue platforms for regenerative medicine and sensing.

17.5 APPLICATIONS OF ELECTROACTIVE POLYMERS IN MEDICINE

EAPs are used for the electrostimulation of cells. Several types of scaffolds are being manufactured from EAP materials for nerve tissue engineering.

17.5.1 ELECTROACTIVE POLYMERS THAT ASSIST CELL FUNCTIONS: TISSUE ENGINEERING

Living tissues can generate an electromotive force and could control switching on and off current flow and its storage (Funk et al., 2006)]. A plasma membrane has innate electrical voltage and the inner cell is more negatively charged, which leads to a potential gap.

Based on the potential difference electrical signals can control the local microenvironment for cells and generate long term effects for phenotypic expression and functioning as observed with in vitro regeneration (Guarino *et al.*, 2007).

Today, tissue engineering focuses on the ability of cells to survive in an environment customized with scaffolds that mimic nature (Liu, 2006). Biomaterials designed to resemble the structural organization of ECM, which helps in cell growth. Stimuli can help in the development of tissues grown in vitro under the scaffold microarchitecture. A recent report described a conductive particle incorporated with a carbon nanofiller (Stout *et al.*, 2012] or Au nano (Dvir *et al.*, 2011) with polymer matrixes can significantly modify cell behavior. These conductive elements help cells transmit electrical signals without any external stimuli.

EAPs improve the function of scaffolds and regeneration mechanisms in nerve and cardiac cells (Peckham *et al.*, 2005). In addition, they influence activities, such as adhesion, protein secretion, DNA synthesis, and migration. Under the applied stimulus a CP can perform a number of functions, such as proliferation and differential of cells in in vitro conditions (Sirivisoot, 2014). More research is required to explore the use of EAPs in biomedicine for biocompatibility and surface topology modifications using controlled synthesis. Scaffolds that have been developed based on PANi and PLGA via electrospinning have been used in cardiomyocytes (Zhang *et al.*, 2014). Promising results for improved tissue alignment and enhanced cardiomyocyte function have been investigated. PANi, PPy, and carbon nanofiber fillers are present in scaffolds after the electrostimulation of tissue. PPy is a versatile EAP that is used with nerve cells in vitro and in vivo (Ateh *et al.*, 2006). A PPy and polylactic acid composite that was used for the differentiation of rat PC12 cells effected the regeneration of peripheral nerves in vivo (Xu. *et al.*, 2014). The scaffold uses a combination of an EAP with an existing biodegradable polymer, which had efficient conductivity and is biodegradability. Proteins and a PPy mixture supported good cell adhesion. Positively charged EAPs are widely used for catalyzing the cell adhesion. PPy can be covalently grafted with cell adhesive peptides for improved adhesion of rat osteoblast cells. Arg-Gly-Asp (RGD) immobilized on PPy surface for increased PC12 cell adhesion is simple using these binding short peptides. Macromolecules incorporated its ECM like HF91060, laminar fragmentation. as dopants are entrapped with CP and makes it biocompatible.

Hyaluronic acid promotes vascularization with a PPy matrix for enhanced cell adhesion. Growth factors (GF), such as nerve growth factor (NGF) have been incorporated with PPy /PEDOT for increased bioactivity. Customized EAPs can be synthesized based on the requirements, which are biodegradability or biocompatibility for the negligible immune response in host tissue can be generated (Figure 17.2). Biodegradable electroactive hydrogels have been fabricated with gelatin (GA) (Li *et al.*, 2014). Graphene as a biomaterial is much appreciated, because of its remarkable electric and mechanical characteristics that are used for the fabrication of scaffolds and soft tissue scaffolds (Shadjou *et al.*, 2018; Kenry *et al.*, 2018).

17.5.2 ELECTROACTIVE POLYMERS TO TARGET DRUGS AND BIOLOGICAL MOLECULES: DRUG DELIVERY

Polymers function as drug carriers in drug delivery. Hydrophobic and low density, molecular weight moieties, such as drugs can be delivered through polymers. The latest drug delivery system uses novel materials that function as magnets and are magneto piezoelectric (Gil, 2014). They provide new methods for drug delivery whether diagnosis or treatment. EAPs in drug delivery are becoming popular due to their biocompatibility and advantages of real time monitoring in vivo. Electrochemically modified EAPs can better control the surface dropping of protein molecules in a polymer matrix. Furthermore, the redox activity of CP. Reibeiro et al. (2014) and Iannotti et al. (2016) increased the diffusion of the applied drug and its release from the polymer. Electrochemically controlled EAP drug delivery is far better for the quantity and characteristics of the modified polymer during synthesis. A slight change in the EAP introduces a parallel change in the polymer for conductive volume with modified of redox states, and therefore, the release rate of the drugs is regulated.

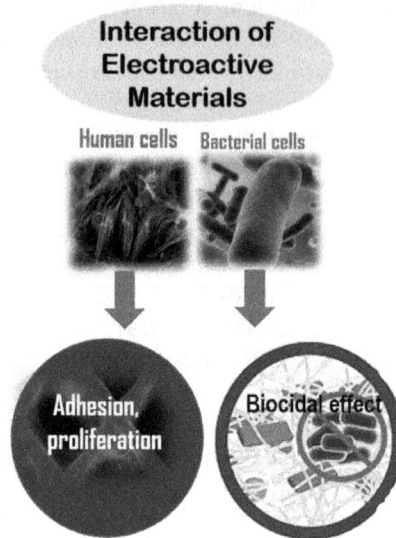

✓ *Biocompatibility with host*
✓ *Biofilm formation/biofouling prevention MOs*
✓ *Infection and inflammation(both host and bacteria)*

Interaction of Electroactive Materials

Human cells   Bacterial cells

Adhesion, proliferation   Biocidal effect

FIGURE 17.2   Interaction of EAPs with cells in biomedicine.

EAPs helps in the release of therapeutic drugs, such as GF when electrically stimulated (Kim *et al.*, 2007). EAP fabrication directly depends on the molecular charge. Biomolecules, such as NGF act as a codopant with a positive charge and is electrochemically entrapped and forms a polyelectrolyte complex. The release kinetics of NGF was used to modify in vivo cell response. Green and Abidian (2015), made PEDOT nanotubes with electrospun fibrillary PLGA for the controlled release of dexamethasone in cells. This demand release is becoming popular in drug delivery systems, where external electrical stimuli are used for regenerating and degenerative medicine. Electroconductive hydrogels have the combined properties of high water swelling capacity, swelling capacity, good membrane permeability, molecular permeation, charged condition, and electrochemically controlled redox properties (Guiseppi *et al.*, 2010). These materials are responsive to slight changes in pH, ionic strength, electrical potential, and release kinetics (Takahashi *et al.*, 2012).The newest are biodegradable and electrical hydrogels, which can compensate for the limitations in processibility for the existing CPs (Liu *et al.,* 2006).

17.5.3   ELECTROACTIVE POLYMERS IN ANTIMICROBIAL ACTIVITY

Microorganisms are present in food stuffs and drinking water and can cause serious tissue destruction, device fractures (Kavanagh *et al.,* 2018; Benčina *et al.*, 2012, 2018), catheter associated urinary tract infections (Fernández *et al.*, 2018),eye infections (Palioura *et al.*,2018), and dental diseases (Benčina *et al.,* 2018). Microbial film formation on exposed surfaces is quite common. EAPs have been used to combat biofouling by preventing the adhesion of microorganisms or the inactivation of them on the surfaces of the biomedical implants. The potential mechanism of biofouling is the generation of bioelectrical effects under the applied electrical field for the treatment of prosthesis-based infections.

17.5   CONCLUSIONS AND FUTURE PERSPECTIVES

Conjugated dielectric EAPs have a promising future and are important for implantable neural prosthesis. They it can make interactions with the actuators within the implants easy and are preferred for any treatment. However, the actuation force and robustness of the EAPs designed for biomedical applications requires significant improvement and has potential in the sports, biomedical, and electronic industries. A gap in the transition from electronic to an ionic approach for electronic devices that are used in biological applications, has led to their slow automization for health care devices. Novel bioelectronics devices based on EAPs facilitate the reduction in biofilm formation by microorganisms on bioelectronic implants.

A new trend has focused on the fabrication of organic biosensors for a new generation of medical devices, such as in the development of brain interfaces or prostheses for the faster recovery of patients. There is enormous potential for applications in therapeutic, terranostic, and diagnostic applications. These EAPs assist in the detection of diseases, in particular, in remote areas and a personalized services easier. In addition, the customized EAPs are used to make sensors, which are non-invasive and can significantly reduce the hospitalization time and costs. Therefore, organic EAPs could significantly improve the quality of life and they could soon become and indispensable part of daily life.

REFERENCES

Anand, J., Palaniappan, S., and Sathyanarayana, D.N. (1998) 'Conducting polyaniline blends and composites', *Progress in Polymer Science*, 23, pp. 993–1018.

Ateh, D.D., Navsaria, H.A., and Vadgama, P. (2006) 'Polypyrrole-based conducting polymers and interactions with biological tissues', *Journal of the Royal Society Interface*, 3, pp.741–752.

Baheiraei, N. *et al.* (2015) 'Preparation of a porous conductive scaffold from aniline pentamer-modified poly-urethane/PCL blend for cardiac tissue engineering. *Biomedical Materials Research Part A*, 103, pp. 3179–3187.

Baniasadi, H. *et al.* (2015) 'Design, fabrication, and characterization of novel porous conductive scaffolds for nerve tissue engineering. *International Journal of Polymeric Materials and Polymeric Biomaterials*, 64, pp. 969–977.

Benčina M. *et al.* (2018) 'The importance of antibacterial surfaces in biomedical applications. *Advances in Biomembranes and Lipid Self-Assembly*, 28, pp.115–165. doi:10.1016/BS.ABL.2018.05.001.

Bhadra, J., Alkareem, A. and Al-Thani, N. (2020) 'A review of advances in the preparation and application of polyaniline based thermoset blends and composites', *Journal of Polymer Research*, 27, p. 122. doi.10.1007/s10965-020-02052-1.

Bose, S., Robertson, S.F., and Bandyopadhyay, A. (2018) 'Surface modification of biomaterials and biomedical devices using additive manufacturing', *Acta Biomaterialia*, 66, pp. 6–22. doi.10.1016/j.actbio.2017.11.003.

Brochu P., and Pei Q. (2010) 'Advances in dielectric elastomers for actuators and artificial muscles. *Macromolecules Rapid Communications*, 31, pp. 10–36. doi:10.1002/marc.200900425.

Cukierman, E., Pankov, R., Stevens, D.R., and Yamada, K.M. (2001) 'Taking cell-matrix adhesions to the third dimension', *Science*, 294, pp.1708–1712.

Dan, X., Yuezhan, B., and Masuru, M. (2007) 'Electrical conducting behaviors in polymeric composites with carbonaceous fillers', *Polymer Physics*, 45, pp.1037–1044.

Derakhshankhah, H. *et al.* (2020) 'Conducting polymer-based electrically conductive adhesive materials: design, fabrication, properties, and applications', *Journal of Materials Science: Materials in Electronics*, 31, pp. 10947–10961.

Ding, C., Qian, X., Yu, G., and An, X. (2010) 'Dopant effect and characterization of polypyrrole-cellulose composites prepared by in situ polymerization process. *Cellulose*, 17, pp. 1067–1077.

Dvir, T. *et al.* (2011) 'Nanowired three-dimensional cardiac patches', *Nature Nanotechnology*, 6, pp. 720–725.

Fernández J. *et al.* (2018) 'Release mechanisms of urinary tract antibiotics when mixed with bioabsorbable polyesters', *Materials Science and Engineering: C*, 93, pp. 529–538. doi:10.1016/j.msec.2018.08.008.

Fu, W. *et al.* (2014) 'Electrospun gelatin/PCL and collagen/PLCL scaffolds for vascular tissue engineering', *International Journal of Nanomedicine*, 9, pp. 2335–2344. doi.10.2147/IJN.S61375.

Funk, R., and Monsees, T. (2006) 'Effects of electromagnetic fields on cells: physiological and therapeutical approaches and molecular mechanisms of interaction', *Cells, Tissues, Organs*, 182, pp. 59–78. doi.10.1159/000093061.

Gandhi, M.R., Murray, P., Spinks, G.M., and Wallace, G.G. (1995) 'Mechanism of electromechanical actuation in polypyrrole', *Synthetic Metals*, 73, pp. 247–256.

Ghasemi-Mobarakeh, L. *et al.* (2011) 'Application of conductive polymers, scaffolds and electrical stimulation for nerve tissue engineering', *Journal of Tissue Engineering and Regenerative Medicine*, 15, pp. e17–e35.

Gil, S., and Mano, J.F. (2014) 'Magnetic composite biomaterials for tissue engineering', *Biomaterials Science*, 2, pp. 812–818.

Green, R., and Abidian, M.R. (2015) 'Conducting polymers for neural prosthetic and neural interface applications. *Advanced materials*, 27(46), 7620–7637. doi.org/10.1002/adma.201501810.

Guarino, V., Causa, F., and Ambrosio, L. (2007) 'Bioactive scaffolds for bone and ligament tissue', *Expert Review of Medical Devices*, 4, pp. 405–418.

Guarino, V., Alvarez-Perez, M.A., Borriello, A., Napolitano, T., and Ambrosio, L. (2013) 'Conductive PANi/PEGDA macroporous hydrogels for nerve regeneration', *Advanced Healthcare Materials*, 2, pp. 218–227.

Guiseppi-Elie, A. (2010) 'Electroconductive hydrogels: Synthesis, characterization and biomedical applications', *Biomaterials*, 31, pp. 2701–2716.

Guo, B.L., Finne-Wistrand, A., and Albertsson, A.C. (2010) 'Molecular architecture of electroactive and biodegradable copolymers composed of polylactide and carboxyl-capped aniline trimer', *Biomacromolecules*, 11, pp. 855–863.

Harrison, W.A. (1977) 'Elementary theory of heteroconjunctions', *Journal of Vacuum Science & Technology*, 14, p. 11016.

Hsiao, C.W. *et al.* (2014) 'Electrical coupling of isolated cardiomyocyte clusters grown on aligned conductive nanofibrous meshes for their synchronized beating', *Biomaterials*, 34, pp. 1063–1072.

Iannotti, V., Ausanio, G., Lanotte, L., and Lanotte, L. (2016) 'Magneto-piezoresistivity in iron particle-filled silicone: An alternative outlook for reading magnetic field intensity and direction', *Express Polymer Letters*, 10, pp. 65–71.

Ivory, D.M. *et al.* (1979) 'Highly conducting charge-transfer complexes of poly(p-phenylene)', *Journal of Chemical Physics*, 71, p. 1506.

Kamalesh, S. *et al.* (2000) 'Biocompatibility of electroactive polymers in tissues', *Journal of Biomedical Materials Research*, 52, pp. 467–478.

Kavanagh, N. *et al.* (2018) 'Staphylococcal Osteomyelitis: Disease progression, treatment challenges, and future directions. *Clinical Microbiology Reviews*, 31, pp. e00084–17. doi:10.1128/CMR.00084-17.

Kenry, W.C. Lee W.C., Loh K.P., and Lim, C.T. (2018) 'When stem cells meet graphene: Opportunities and challenges in regenerative medicine', *Biomaterials*, 155, pp. 236–250.

Kim, D.H. *et al.* (2007) 'Effect of immobilized nerve growth factor on conductive polymers: Electrical properties and cellular response', *Advanced Functional Materials*, 17, pp. 79–86.

Kumar, D., and Sharmaeur, R.C. (1998) 'Advances in conductive polymers', *Polymer Journal*, 34, pp. 1053–1060.

Lee, J.Y. *et al.* (2009) 'Polypyrrole-coated electrospun PLGA nanofibers for neural tissue applications', *Biomaterials*, 30, pp. 4325–4335.

Li, L.C. *et al.* (2014) 'In situ forming biodegradable electroactive hydrogels', *Polymer Chemistry*, 5, pp. 2880–2890.

Li T. *et al.* (2017) 'Fast-moving soft electronic fish', *Science Advances*, 3. doi:10.1126/sciadv.1602045.

Limongi, T. *et al.* (2017) 'Fabrication and applications of micro/nanostructured devices for tissue engineering', *Nano-Micro Letters*, 9, p. 1. doi.10.1007/s40820-016-0103-7.

Liu, H., Lin, J., and Roy, K. (2006) 'Effect of 3D scaffold and dynamic culture condition on the global gene expression profile of mouse embryonic stem cells', *Biomaterials*, 27, pp. 5978–5989.

Liu, Y.D. *et al.* (2011) 'Synthesis and characterization of novel biodegradable and electroactive hydrogel based on aniline oligomer and gelatin', *Macromolecular Bioscience*, 12, pp. 241–250.

Malhotra, B.D., Chaubey, A., and Singh, S.P. (2006) 'Prospects of conducting polymers in biosensors', *Analytica Chimica Acta*, 578, pp. 59–74.

Mirfakhrai T., Madden J.D.W., and Baughman R.H. (2007) 'Polymer artificial muscles', *Materials Today*, 10, pp. 30–38. doi:10.1016/S1369-7021(07)70048-2.

Moczko, E. *et al.* (2012) 'Biosensor employing screen-printed PEDOT:PSS for sensitive detection of phenolic compounds in water', *Journal of Polymer Science Part A Polymer Chemistry*, 50, pp. 2085–2292.

Mohebbi, A. *et al.* (2016) 'Cellular polymer ferroelectret: A review on their development and their piezoelectric properties', *Advances in Polymer Tech*nology, 37. doi:10.1016/j.compositesb.11.034.

Nezakati T., Seifalian A., Tan A., Seifalian A.M. (2018) Conductive Polymers: Opportunities and Challenges in Biomedical Applications. *Chem. Rev.* 118:6766–6843. doi:10.1021/acs.chemrev.6b00275.

Nikolova, M.P., and Chavali, M.S. (2019) 'Recent advances in biomaterials for 3D scaffolds: A review. *Bioactive Materials*, 4, pp. 271–292. doi.10.1016/j.bioactmat.2019.10.005.

Palioura S. *et al.* (2018) 'Clinical features, antibiotic susceptibility profile, and outcomes of infectious keratitis caused by *Stenotrophomonas maltophilia*', *Cornea*, 37, pp. 326–33.

Peckham, P.H., and Knutson, J.S. (2005) 'Functional electrical stimulation for neuromuscular applications. *Annual Review of Biomedical Engineering*, 7, pp. 327–360.

Peramo, A. *et al.* (2008) 'In situ polymerization of a conductive polymer in acellular muscle tissue constructs', *Tissue Engineering Part A*, 14, pp. 423–432.

Preethi S.S. *et al.* (2018) 'Bone tissue engineering: Scaffold preparation using chitosan and other biomaterials with different design and fabrication techniques', *International Journal of Biological Macromolecules*, 119, pp. 1228–1239. doi:10.1016/j.ijbiomac.2018.08.056.

Qazi, T.H. *et al.* (2014) 'Development and characterization of novel electrically conductive PANI-PGS composites for cardiac tissue engineering applications. *Acta Biomate*rialia, 10, pp. 2434–2445.

Reibeiro, C. *et al.* (2014) 'Piezoelectric polymers as biomaterials for tissue engineering applications', *Colloids and Surfaces B*, 136, pp. 46–55.

Roncali, J. (1997) 'Synthetic principles for bandgap control in linear π-conjugated systems', *Chemistry Reviews*, 97, pp. 173–205.

Saghazadeh S. *et al.* (2018) 'Drug delivery systems and materials for wound healing applications', *Advances in Drug Delivery*, 127, pp. 138–166. doi:10.1016/j.addr.2018.04.008.

Sarvari, R. *et al.* (2016) 'Novel three-dimensional, conducting, biocompatible, porous, and elastic polyaniline based scaffolds for regenerative therapies', *RSC Advances*, 6, p. 19437.

Schmidt, C.E. *et al.* (1997) 'Stimulation of neurite outgrowth using an electrically conducting polymer', *Proceedings of the National Academy of Sciences USA*, 94, pp. 8948–8953.

Shadjou, N., Hasanzadeh, M., and Khalilzadeh, B. (2018) 'Graphene based scaffolds on bone tissue engineering', *Bioengineered*, 9(1), pp. 38–47. doi.10.1080/21655979.2017.1373539.

Sirivisoot, S., Pareta, R., and Harrison, B.S. (2014) 'Protocol and cell responses in three-dimensional conductive collagen gel scaffolds with conductive polymer nanofibres for tissue regeneration', *Interface Focus*, 6, p. 20130050.

Smela, E. (2003) 'Conjugated polymer actuators for biomedical applications', *Advanced Materials*, 15, pp. 481–494. doi:10.1002/adma.200390113.

Stout, D.A. *et al.* (2012) 'Mechanisms of greater cardiomyocyte functions on conductive nanoengineered composites for cardiovascular application', *International Journal of Nanomedicine*, 7, pp. 5653–5669.

Struèmpler, R., and Glatz-Reichenbach, J. (1999) 'Conducting polymer composites', *Journal of Electroceramics*, 3, pp. 329–346.

Subramanian, U.M. *et al.* (2014) 'Fabrication of polyvinyl alcohol-polyvinylpyrrolidone blend scaffolds via electrospinning for tissue engineering applications', *International Journal of Polymeric Materials and Polymeric Biomaterials*, 63, pp. 476–485.

Takahashi, S.H. *et al.* (2012) 'Zero-order release profiles from a multistimuli responsive electro-conductive hydrogel', *Biomaterials and Nanobiotechnology*, 3, pp. 262–268.

Takano, T., Mikazuki, A., and Kobayashi, T. (2014) 'Conductive polypyrrole composite films prepared using wet cast technique with a pyrrole–cellulose acetate solution', *Polymer Engineering & Science*, 54, pp. 78–84.

Vohrer U. *et al.* (2004) 'Carbon nanotube sheets for the use as artificial muscles. *Carbon*, 42, pp. 1159–1164. doi:10.1016/j.carbon.2003.12.044.

Wang, C.H. *et al.* (1999) 'In-vivo tissue response to polyaniline', *Synthetic Metals*, 102, pp. 1313–1314.

Wan, Y., and Wen, D.J. (2005) 'Preparation and characterization of porous conducting poly(D, L-lactide) composite membranes', *Journal of Membrane Science*, 246, pp. 193–194.

Wong, J.Y., Langer, R., and Ingber, D.E. (1994) 'Electrically conducting polymers can noninvasively control the shape and growth of mammalian cells', *Proceedings of the National. Academy of Science USA*, 91, pp. 3201–3204.

Xu, H. *et al.* (2014) 'Conductive PPY/PDLLA conduit for peripheral nerve regeneration', *Biomaterials*, 35, pp. 225–235.

Youn, J.H. *et al.* (2020) 'Dielectric elastomer actuator for soft robotics applications and challenges', *Applied Science*, 10(2), p. 640.

Zhang F. *et al.* (2019) 'Conductive shape memory microfiber membranes with core–shell structures and electroactive performance', *ACS Applied Materials & Interfaces*, 210, pp. 35526–35532. doi:10.1021/acsami.8b12743.

Zhang, Q. *et al.* (2010) 'The synthesis and characterization of a novel biodegradable and electroactive polyphosphazene for nerve regeneration', *Materials Science and Engineering C*, 30, pp. 160–166.

Zilberman, M., Siegmann, A., and Narkis, M. (2000) 'Conductivity and structure of melt-processed polyaniline binary and ternary blends', *Polymers for Advanced Technologies*, 11, pp. 20–26.

18 Electroactive Polymers for Packaging Technology

Pinku Chandra Nath, Ria Majumdar, Tarun Kanti Bandyopadhyay, Biswanath Bhunia, and Biplab Roy

CONTENTS

18.1 INTRODUCTION

In this era of emerging technologies, conventional materials (e.g., metals, alloys, and ceramics) have been replaced by polymers, such as homopolymers, copolymers, blends of small molecules, complexes, and composites, which are being used in household goods, aerospace, automobiles, packaging industries, electronics, and medical sciences. These polymeric materials have tremendous advances; therefore, several processing techniques have been developed to date, and are being developed to enable polymer production with tailor-made properties (e.g., physical and mechanical). Theses feature of polymers that allow the invention of new designs with cost-effective and light weights makes them interesting for growing technologies (Guru Nathan *et al.*, 1999). However, unlike inorganic materials, polymers have different attractive properties, for example, they are light weight, easily processed, pliable, and fracture tolerant. These can be configured into various complex shapes and their properties can be tailored into what is required (Bar-Cohen, 2004). Different materials with artificial intelligence and rapid advances (e.g., piezoelectric materials and shape memory materials) can sense changes in the environment, process that information and then respond accordingly (Irinyi, 2000). Polymers that respond to external stimuli, for example, pH, electrical field, light, and magnetic fields by changing their shape or size are active polymers and have been known for several decades (Bar-Cohen, 2004). Therefore, polymers that change shape, or dimensions, or both when exposed to an electric field are electroactive polymers (EAPs) (Carpi and Smela, 2009).

EAPs are materials that have been modified so that they can convert electrical impulses into mechanical movements, which is a principle that enables the production of capacitive sensors, actuators, and generators. Because of their restricted actuation capacity, EAPs initially received less attention. However, in the last two decades, EAPs have emerged that have a significant response

DOI: 10.1201/9781003173502-18

to electrical stimulation by changing their shape. Several novel mechanisms and devices, such as miniature grippers, robot fish, active diaphragms, loudspeakers, catheter steering elements, and dust wipers have recently established using these materials as actuators. However, actuators for packaging purposes are increasing in current technologies. The benefits from improvements in their actuation strain capacity have attracted many scientists and researchers from various fields. The activation mechanism of polymers that can change their dimension or shape include optical, electrical, magnetic, chemical, and pneumatic. EAP materials provide various advantages, such as light weight, pliable, fracture tolerant, and inexpensive. However, converting EAP materials into an actuator of choice requires a well-established infrastructure and this includes an improvement in the understanding of the basic principles. In addition, electromechanics analytical tools, effective computational chemistry models, material processing techniques, and comprehensive material science are required to improve the understanding. Electrical stimulation causes elastic deformation in polymers and its convenience, practicality, and recent improvement in capacity have made EAP materials one of the most interesting among all the active polymers.

18.2   SIGNIFICANCE OF ELECTROACTIVE POLYMERS

EAP materials can be converted into various shapes for engineering purposes; therefore, their properties make them more attractive in a variety of potential applications to overcome growing challenges. Electrical currents cause EAP materials to distort in shape and size, which increases the strain rate by approximately 300% (Bar-Cohen, 2002). Potential applications for EAPs in different fields are power generators, robotics, aerospace, medical, articulation mechanisms, clothing, sensors, and smart structures. The actuator-based EAP materials have some advantages including low actuation voltage, flexible, quick response, and light weight (Rao, 2014).Wang *et al.* (2009) showed that Flemion (an ionic polymer based tactile sensor) that acts as actuators could be applied to produce microelectromechanical systems (MEMS) and smart materials, which had a low actuation voltage and greater strain. Due to the excellent actuation characteristics of EAP materials they are widely used in aerospace applications. In tissue engineering, EAP materials can be utilized as biomaterials to promote proliferation and cellular adhesion in human cells by evading biofilm development through bactericidal effects (Figure 18.1). Wu *et al.* (2018) produced optimal EAPs for applications

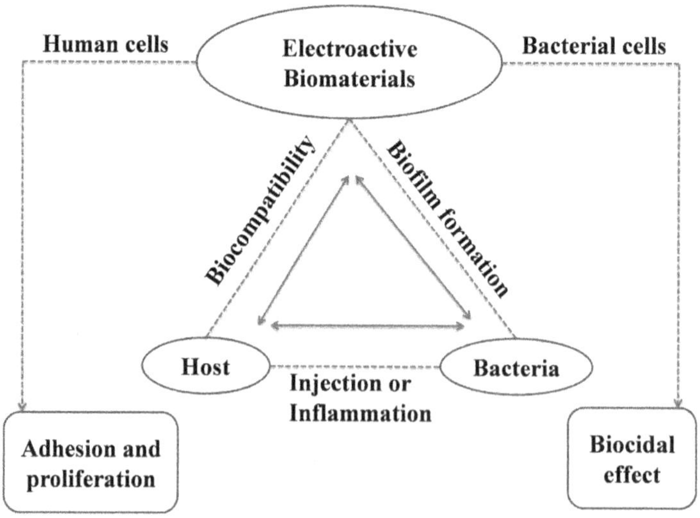

FIGURE 18.1   Relationship between EAP materials and human and bacterial cells.

(From Wu and Le Gorrec, 2018. With permission.)

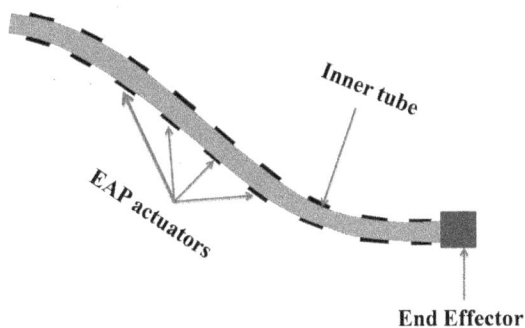

FIGURE 18.2   EAP-based actuated endoscope.

(From Wu and Le Gorrec, 2018. With permission.)

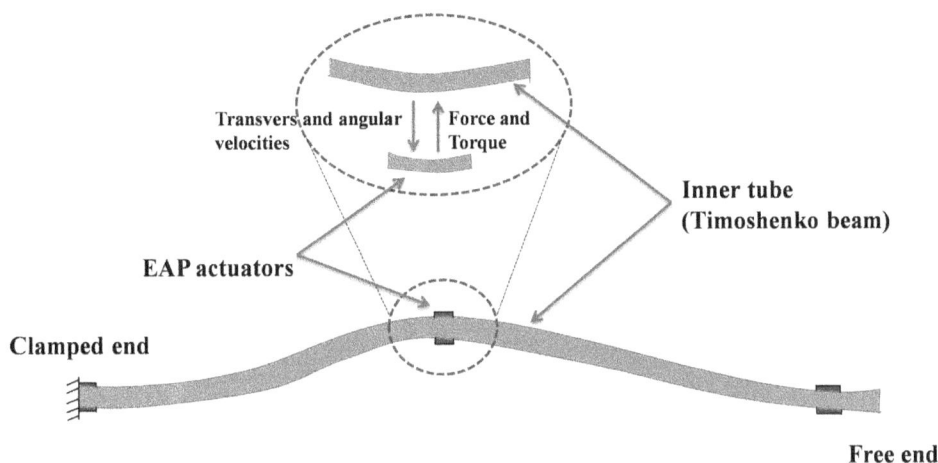

FIGURE 18.3   Complete simplified actuated endoscope.

(From Wu and Le Gorrec, 2018. With permission.)

in medical endoscopes with the help of a Hamiltonian modeling framework. Figure 18.2 shows an EAP-based actuated endoscope where the coating is provided outside the medical endoscope. Figure 18.3 shows a complete medical endoscope where the inner part of the endoscope represents a Timoshenko beam. The left end section of the beam is constant and the right is free. The beam and actuator are interconnected by power conjugated variables. To increase haptic technology, EAP materials have been applied in the automobile sector. This technology provides mechanical feedback when driving on the road. Poncet *et al.* (2016) reported haptic circular buttons that provided vibration sensation, which could be detected by touching them.

18.3   CLASSIFICATION OF ELECTROACTIVE POLYMERS

EAPs are mainly divided into ionic and electronic EAPs according to their activation mechanisms (Bar-Cohen, 2004).

18.3.1   IONIC ELECTROACTIVE POLYMERS

Unlike electronic EAPs, these EAPs are materials that involve in the transportation of ions, and they are composed of an electrolyte along with two electrodes. The activity of ionic EAPs can be achieved by decreasing the voltage to 1 to 2 V. Composites of polymer and metal, carbon nanotubes, gels, and conducting polymers are types of EAPs (De Luca *et al.*, 2013; Shahinpoor *et al.*, 1998). Ionic EAPs have disadvantages, such as their wetness and the electrochemical coupling required for correct maintenance.

18.3.2 ELECTRONIC ELECTROACTIVE POLYMERS

Electronic EAPs are materials that have a large mechanical energy density and include piezoelectric (Nalwa, 1995), dielectric elastomers (Carpi *et al.*, 2015; Pelrine *et al.*, 2000), electrostrictive, liquid crystal polymers (Ji, Marshall, and Terentjev, 2012; Lehmann *et al.*, 2001), ferroelectric (Wang, Herbert, and Glass, 1988), and electrostatic polymers. They are driven by Coulomb forces and can be operated in air with no constraints. However, a large activation field (>10 V/μm) is required for this EAP, which is approximately the breakdown level, but it consumes low electrical energy. In addition, dielectric EAPs do not require any power to keep the actuator in place. Dielectric EAPs are polymeric materials that are compressed between two electrodes by electrostatic forces. Dielectric elastomers can withstand extremely high strains, and capacitors change capacitance during an applied voltage where the polymer is allowed to decrease in thickness by expanding its area at its core due to the electrical field.

18.4 APPLICATION OF ELECTROACTIVE POLYMERS IN PACKAGING

In the packaging industry, EAP materials have received attention from researchers because of some exceptional advantages that include product protection, lightness, durability, inert atmosphere, and stability. In the food packaging industry, EAP materials have been used for the formation of edible coatings, which is a green and novel packaging application. The coatings used for packaging must have characteristics, such as non-toxicity, non-allergic, and excellent structural stability.

18.4.1 LUNCH BOX PACKAGING

In some countries, people use a variety of lunch boxes for convenience. Based on the of huge demand, lunch box packaging is still required due to their variety, high deformability, fragility, physical properties of the food, and various shape variations (Iwamasa and Hirai, 2015). To minimize labor costs, lunchbox packaging automation systems in the food industry are in high demand. Figure 18.4(a) shows a typical lunch box that contains dishes made from polymeric plastic materials and Figure 18.4(b) shows some easily deformable frustum shaped containers (Wang *et al.*, 2017). Lunch box packaging consists of picking, placing, and filling with foodstuffs. In the food packaging

FIGURE 18.4 Showing: (a) commercial lunch box; and (b) paper containers filled side dishes.

industry, the traditional rigid gripping and vacuum system faces many difficulties to manage the tasks, due to the strong grip food materials can be damaged. Therefore, suction might be used is a flat surface is provided to handle the task. Soft pneumonic robotic grippers have piqued the interest of researchers recently due to their adaptability and flexibility. A four fingered gripper composed of fiber-reinforced rubber has been proposed, which has been experimentally tested for various grasping modes (Suzumori, Iikura, and Tanaka, 1991; Suzumori, Iikura, and Tanaka, 1992). Thermo-formed containers can be formed by heating polymer sheets in a mold at a specific temperature to provide a specific shape. These containers might be used for muffins, cheese, and cookies. Coating paper with EAP materials led to innovative packaging with excellent properties. Coating-based papers have high permeability, good moisture content, fat resistance, and water sensibility. Barrier films or blown films can be manufactured from biodegradable EAP materials by the extrusion of different polymers. In general, for edible film production extrusion, electro spinning process, and compression molding are used (Hernandez-Izquierdo and Krochta, 2008; Mason, 2009).

18.5 PROPERTIES OF ELECTROACTIVE POLYMERS FOR PACKAGING APPLICATIONS

18.5.1 PROPERTIES OF GAS BARRIERS

In the packaging industry, to increase shelf life and the quality of packaging, controlling gas pressure conditions is important. Three gas or a mixture of them are required for packaging: oxygen (O_2), nitrogen (N_2), and carbon dioxide (CO_2) (Shalini and Singh, 2009). To maintain the gas composition inside the packaging materials, the materials require gas barrier properties. The application of multi-layers could improve barrier properties. One of the important characteristics of a gas barrier is the permeation capacity. The permeation capacity of EAP materials increases with increasing humidity. In food packaging, the maintenance of O_2 and water permeability during shelf life of the food material is an important criterion. In addition, with increasing crystallinity the film barrier properties are improved. According to Bastioli et al. (1995),a multilayered film (i.e., metalized) with N_2, CO_2, and vacuum condition improves product shelflife (Bastioli et al., 1995).

18.5.2 MECHANICAL, CHEMICAL, AND THERMAL PROPERTIES

The excellent mechanical and thermal properties of the packaging material protect products from mechanical or thermal damage during transportation and storage. Increasing the thermal degradation temperature can improve the materials mechanical and thermal properties. In addition, they determine product suitability in numerous applications. The enhancement of the physical, chemical, and mechanical characteristics of EAPs can be obtained by nano-reinforcement and blending different polymers. To obtain excellent mechanical properties in EAP materials, tensile tests must be characterized, such as tensile strength (MPa), per cent elongation at breakage, yield, and elastic modulus. Buttler et al. (1996) proposed barrier and mechanical characteristics of a plasticized edible film. They observed that plasticization concentration increased the per cent elongation by 25%–45% and reduced the tensile strength by 15–30 MPa. To understand the acid characteristics of packaging materials the performance and suitability of the packaging materials must be assessed over time. Polymer absorption by chemical compounds might influence the mechanical properties of the polymer. The chemical resistance of polymer is analyzed by submerging samples in an acid (e.g., weak or strong) solution at different temperatures, such as ambient temperature (23°C), -18°C, -23°C, and -29°C.

18.5.3 BIODEGRADABILITY

Biodegradability refers to materials that can be split into small compounds via the action of fungi and bacteria (Ghalem and Mohamed, 2008). Biodegradation depends on various environmental parameters including pH, moisture, temperature, and nutrients. Biodegradation occurs in two stages: (1) defragmentation (e.g., by microbial enzymes, moisture, and heat); and (2) biodegradation (e.g., conversion of larger molecules into small compounds by natural acids and enzymes). In biodegradation, after conversion of the molecules into suitable shape, the organism's cell wall absorbs the substances metabolized for energy. In general, polymer degradation occurs due to microbial or chemical action and via photodegradation.

18.5.4 Moisture Barrier Properties

A moisture barrier protect the materials from the entry of undesired vapors, and is calculated by water vapor transmission rate (Alavi *et al.*, 2015). Packaging materials that contain undesired vapors that results in recovering moisture in dry foods. This moisture can be removed by forming moisture resistant films that are produced by reinforcement with natural fibers and blending and coating with water-resistant materials. The barrier properties rely on the material's morphological properties, such as chain configuration and crystallinity. Shelf life can be increased by increasing the moisture barrier in food packaging. Morillon *et al.* (2002) reported moisture sensitivity in protein films and observed that a moisture barrier was increased by blending with other bio-based materials. Baastioli *et al.* (1995) reported a solvent-based dispersion coating on a biopolymer layer and achieved excellent properties for the moisture barrier. In packaging applications, O_2 and water vapor are two important permeants. The O_2 barrier is evaluated by the oxygen permeability coefficient (OPC), which shows the quantity of O_2 that permeates per unit time and area. If the packaging film has low OPC, O_2 pressure inside the packaging decreases to a limited oxidation condition that improves product shelf life. The water vapor permeability coefficient is the water vapor permeation per unit time and area and is used to determine the water vapor barrier. For fresh produce, this is necessary to evade dehydration and for bakery products, water permeation should be avoided.

18.6 CONCLUSION

This chapter discussed the basic information on EAP materials, their importance over other materials that enables the development of advanced and cost-effective new devices, and thier significant applications in packaging materials. Although EAP materials are used in various fields, they have achieved received interest as packaging materials, because of their unique properties. EAPs have emerged as one of the most attractive materials to researchers and scientists, because advanced and cost-effective new models can be developed by changing their shape to the electric stimulation. In addition, they can change rapidly with small variations in the environment, which makes them attractive in modern packaging technology.

REFERENCES

Alavi, S. *et al.* (2015) '*Polymers for packaging applications*', New Jersey: Apple Academic Press. pp. 1–37.

Bar-Cohen, Y. (2002) 'Electroactive polymers: current capabilities and challenges', In Proc: *SPIE's 9th Annual International Symposium on Smart Structures and Materials 2002: Electroactive Polymer Actuators and Devices.* San Diego, CA, 17–19 March, Vol. 4695.

Bar-Cohen, Y. (2004) '*Electroactive polymer (EAP) actuators as artificial muscles: reality, potential, and challenges.* Bellingham: SPIE Press, p. 765.

Bastioli, C. *et al.* (1995) 'Physical state and biodegradation behavior of starch-polycaprolactone systems', *Journal of Environmental Polymer Degradation,* 3(2), pp. 81–95.

Butler, B.L. *et al.* (1996) 'Mechanical and barrier properties of edible chitosan films as affected by composition and storage', *Journal of Food Science*, 61(5), pp. 953–956.

Carpi, F. *et al.* (2015) 'Standards for dielectric elastomer transducers', *Smart Materials and Structures,* 2 4(10), p.105025.

Carpi, F., and Smela, E. (2009) *'Biomedical Applications of Electroactive Polymer Actuators.* Oxford: John Wiley & Sons, p. 496.

De Luca, V. *et al.* (2013) 'Ionic electroactive polymer metal composites: Fabricating, modeling, and applications of postsilicon smart devices', *Journal of Polymer Science Part B: Polymer Physics*, 51(9), pp. 699–734.

Ghalem, B.R., and Mohamed, B. (2008) 'Antibacterial activity of leaf essential oils of *Eucalyptus globulus* and *Eucalyptus camaldulensis'*, *African Journal of Pharmacy and pharmacology,* 2(10), pp. 211–215.

Glass, A.M., Herbert, J.M., and Wang, T.T. (1988) 'The applications of ferroelectric polymers', Netherlands: Springer, p. 387.

Gurunathan, K. *et al.* (1999) 'Electrochemically synthesised conducting polymeric materials for applications towards technology in electronics, optoelectronics and energy storage devices', *Materials Chemistry and Physics,* 61(3), pp. 173–191.

Hernandez-Izquierdo, V.M., and Krochta, J.M. (2008) 'Thermoplastic processing of proteins for film formation: A review', *Journal of Food Science,* 73(2), pp. R30-R39.

Iwamasa, H., and Hirai, S. (2015) 'Binding of food materials with a tension-sensitive elastic thread', *2015 IEEE International Conference on Robotics and Automation (ICRA).* Seattle, WA. 26–30 May pp. 4298–4303.

Ji, Y., Marshall, J.E., and Terentjev E.M. (2012) 'Nanoparticle-liquid crystalline elastomer composites', *Polymers*, 4(1), pp. 316–340.

Lehmann, W. *et al.* (2001) 'Giant lateral electrostriction in ferroelectric liquid-crystalline elastomers', *Nature*, 410(6827), pp. 447–450.

Mason, W.R. (2009) 'Starch use in foods', *Starch*, 745–795. Cambridge: Academic Press.

Morillon, V. *et al.* (2002) 'Factors affecting the moisture permeability of lipid-based edible films: a review', *Critical Reviews in Food Science and Nutrition,* 42(1), pp. 67–89.

Nalwa, H.S. (1995) *'Ferroelectric Polymers: Chemistry, Physics, and Applications*: Boca Raton: CRC Press, p. 912.

Pelrine, R. *et al.* (2000) 'High-speed electrically actuated elastomers with strain greater than 100%', *Science*, 287(5454), pp. 836–839.

Poncet, P. *et al.* (2016) 'Development of haptic button based on electro active polymer actuator', *Procedia Engineering,* 168, pp. 1500–1503.

Rao, P.S. (2014) 'Investigation and development of life saving research robots', *International Journal of Mechanical Engineering and Robotics Research*, 3(2), p. 195.

Shahinpoor, M. *et al.* (1998) 'Ionic polymer-metal composites (IPMCs) as biomimetic sensors, actuators and artificial muscles-a review', *Smart Materials and Structures*, 7(6), p. R15.

Shalini, R., and Singh A. (2009) 'Biobased packaging materials for the food industry', *Journal of Food Science & Technology*, 5, pp. 16–20.

Suzumori, K., Iikura, S., and Tanaka, H. (1991) 'Development of flexible microactuator and its applications to robotic mechanisms', *Proceedings. 1991 IEEE International Conference on Robotics and Automation.* Sacramento, CA. 9–11 April, pp. 1622–1627.

Suzumori, K., Iikura, S., and Tanaka, H. (1992) 'Applying a flexible microactuator to robotic mechanisms', *IEEE Control Systems Magazine*, 12(1), pp. 21–27.

Wang, J. *et al.* (2009) 'Bioinspired design of tactile sensors based on Flemion', *Journal of Applied Physics*, 105(8), p. 083515.

Wang, Z. *et al.* (2017) 'Fabrication and performance comparison of different soft pneumatic actuators for lunch box packaging', *2017 IEEE International Conference on Real-time Computing and Robotics (RCAR).* Okinawa, Japan, 14–18, July, pp. 22–27.

Wu, Y., and Gorrec, Y.L. (2018) 'Optimal actuator location for electro-active polymer actuated endoscope', *IFAC-PapersOnLine,* 51(3), pp. 199–204.

Zrinyi, M. (2000) 'Intelligent polymer gels controlled by magnetic fields', *Colloid and Polymer Science,* 278(2), pp. 98–103.

19 Electroactive Polymers for Drug Delivery

Mehdi Mogharabi-Manzari, Masoud Salehipour,
Zahra Pakdin-Parizi, Shahla Rezaei, Roya Khosrokhavar,
and Ali Motaharian

CONTENTS

19.1 INTRODUCTION

Recently, electroactive polymers (EAPs) have attracted considerable attentions due to their extensive range of potential applications in fields, such as tissue engineering, biosensors, energy conversion and storage, molecular electronics, and the smart delivery of biologically active molecules (Wallace et al., 2008). Based on conduction mechanism, the EAPs are classified as electric conducting polymers and ionic conducting polymers. However, the most common ionic conducting polymer electrolytes are complexes composed of dissolved alkali salts in poly(ethylene oxide) (Zarren et al., 2019).

In addition, the ionic conducting polymers are classified as intrinsically and extrinsically conducting polymers. The intrinsically conducting polymers have a solid backbone made from an extensive conjugated system, which is responsible for conductance (Ning et al., 2018). Conjugated π-electrons of the intrinsically conducting polymers are excited under the influence of electrical field and can then be transported through the solid polymer. Extrinsically conducting polymers possess conductivity due to the presence of externally added elements, such as carbon black or metallic fibers (Bar-Cohen et al., 2017).

The conducting polymers retain unique properties and can be used in various applications by adjusting their specific characteristics including electric conductivity, degradability, toughness, porosity, flexibility, and hydrophobicity (Surmenev et al., 2021). In addition, the conjugated π-electron backbone of the conducting polymers provides unique electronic characteristics, such as low energy optical transmission, high electron affinities, and low ionization potential. These chemical and physical properties make them appropriate for applying in light emitting diodes, transistors, sensors, supercapacitors, and solar cells (Prasad & Ulrich, 2012).

In addition, EAPs have been explored for their potential biomedical applications including tissue engineering, biosensors, and electrically induced drug release and delivery systems. The biodegradability and biocompatibility of EAPs should be explored before applying them in biomedical applications. For example, the presence of high concentrations of the components of intrinsically conducting polymers in tissue engineering could cause aninflammatory responses in tissues (Namsheer & Rout, 2021).

19.2 CONDUCTING MECHANISM

The conjugated structure of conducting polymers is due to the continuous π bonds that are formed by p_z orbitals of carbon atoms. In some cases, an external charge carrier is added to the polymer via doping (Kausar, 2020). The process can be theoretically explained by b and theory in which the lowest empty band and the highest occupied band are called the conduction and valance bonds, respectively. The difference between the energy levels of the conduction and valance bonds is decreased in a conjugated polymer that is filled with a dopant (Carpi & Smela, 2009).

Molecular orbital theory is applied to explain the mechanism of conduction in intrinsic conducting polymers (Figure 19.1). The combination of p_z orbitals of the adjacent carbon atoms forms two new molecular orbitals with lower energy (bonding) and higher energy (anti-bonding) than the original p orbitals. The band gap energy is the energy difference between the highest occupied molecular orbital and the anti-bonding molecular orbital has the lowest unoccupied molecular orbital (Khan et al., 2018; Kumar et al., 2017)

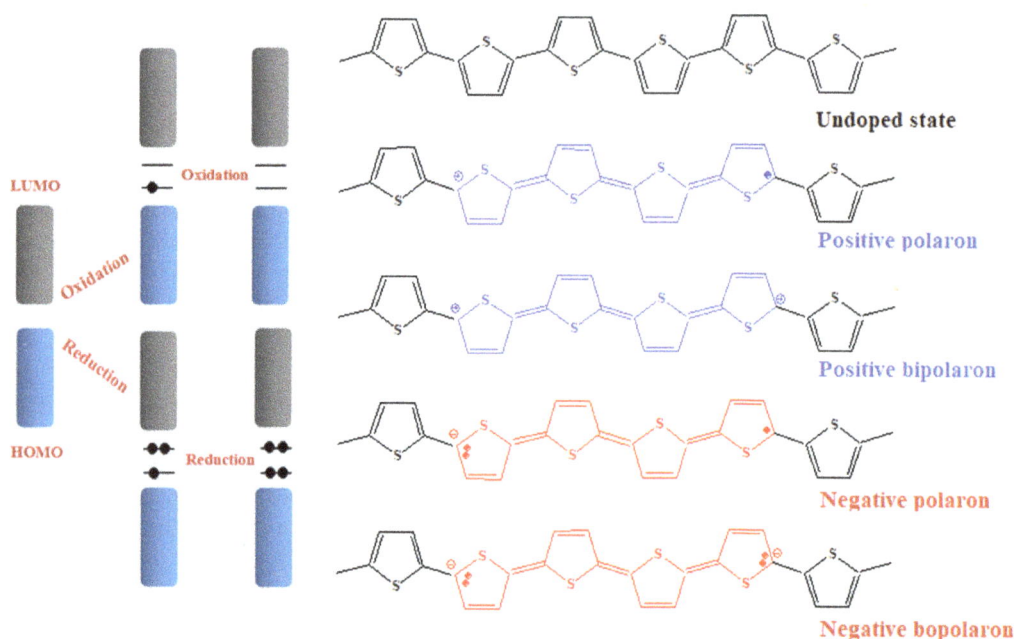

FIGURE 19.1 Molecular orbital theory explanation of the mechanism of conduction in intrinsic conducting polymers.

19.3 SYNTHESIS OF CONDUCTING POLYMERS

Some synthetic methods are used for the synthesis of conducting polymers via chemical, electrochemical, and enzymatic oxidative polymerization of various monomers, such as pyrrole, thiophene, and aniline (Ramanavicius & Ramanavicius, 2021; Gajendiran et al., 2017; Balint et al., 2014). Although chemical methods are applied for the synthesis of large amount of polymers, electrochemical approaches can control the thickness of the layers of the polymers by adjusting the reaction conditions, such as electrolyte, current density, temperature, acid concentration, and reaction time (Wang et al., 2005; Qazi et al., 2014; Distler et al., 2020).

19.3.1 POLYANILINE

The chemical oxidative polymerization of aniline is carried out by the oxidation of monomers in the presence of various organic or inorganic oxidants including ammonium peroxydisulfate (Rahy et al., 2008), potassium dichromate (Duran et al., 2009), potassium iodate (Hirase et al., 2004), hydrogen peroxide (Bláha et al., 2011), and metallic salts (e.g., manganese (Mn), chromium (Cr), cerium (Ce), vanadium (V), cand opper (Cu)) (Abu-Thabit 2016). The molecular mechanistic explanations for the oxidation of aniline provide important mechanistic foundations for the control of polyaniline production. The molecular structure of the produced polyaniline determines the chemical and physical characteristics of the product. The oxidative polymerization of aniline at pHs <2.5 is a chain reaction and the growth of the chain proceeds by the addition of aniline monomers to the active chain ends. However, high molecular weight conducting polymers might be formed by controlling the kinetic profiles of the reaction. The main product of the polymerization of aniline is formed by para-coupling and the by-product is formed by ortho-coupling (Sapurina & Stejskal, 2008). The emeraldine salt, which is most conducting form of polyaniline, is formed in an acidic aqueous medium via chain propagation (Figure 19.2). Treatment of the protonated polyaniline with an alkali solution leads to the production of emeraldine (Ayad & Zaki, 2008).

The electrochemical synthesis of polyaniline is usually preformed in a strongly acidic electrolyte via the formation of an anilinium radical cation on the surface of the electrode (Palaniappan, 2001). The electrochemical polymerization of aniline is strongly influenced by the reaction conditions, such as temperature, pH, electrolyte composition, electrode material, and dopant anion (Martyak et al., 2002). For example, the dopant anion determines the morphology, conductivity, and polyaniline growth rate. In addition, an acidic aqueous medium is needed for the production of conducting polyaniline as the emeraldine salt. However, polymerization has been reported in organic solvents, such as acetonitrile, oxalic acid, and tartaric acid (Miras et al., 1991).

Shen & Huang (2018) applied a proton functionalized ionic liquid pyrrolidinium hydrogen sulfate as an electrolyte for the electrochemical polymerization of aniline.

The protonation of emeraldine base produces the emeraldine salt (green) that is the polyaniline conducting form (Figure 19.3). The oxidation of the emeraldine salt produces the perningraniline salt (dark blue) that might form the perningraniline base (violet) in an alkali medium (Gospodinova et al., 1993).

Over the last decades, enzymatic oxidative reactions were developed for the green synthesis of polyaniline and its derivatives (Mogharabi et al., 2019). For example, a novel technique was reported that used immobilized horseradish peroxidase for the enzyme-catalyzed synthesis of polyaniline derivatives (Jin et al., 2001). In addition, hydrogen peroxide was generated in situ during a biocatalytic reaction that used glucose oxidase from *Penicillium vitale* and was applied as the initiator for the enzymatic polymerization of aniline (Kausaite et al., 2009).

19.3.2 POLYPYRROLE

Polypyrrole has been prepared via different techniques, such as the chemical or electrochemical oxidation of a pyrrole monomer in aqueous and various organic media. A wide range of applications for

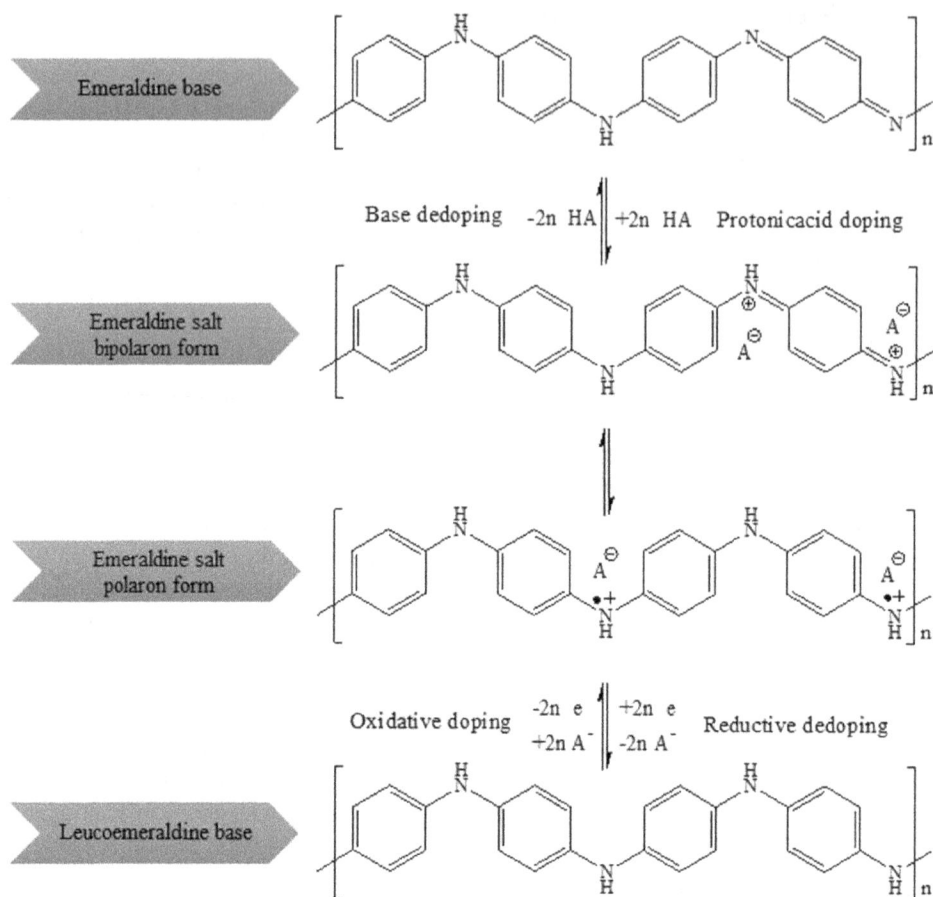

FIGURE 19.2 Various states of polyaniline oxidation including emeraldine base (half oxidized state), emeraldine salt (metallic emeraldine), and leucoemeraldine base (fully reduced state).

polypyrrole have been developed due to the unique thermal and chemical stability, processability, and conductivity (Yussuf et al., 2018).

A large amount of fine powders of polypyrrole can be produced via the oxidative polymerization of pyrrole monomers in the presence of selected transition metal ions in aqueous media. Rapi et al. (1988) reported polymerization at low temperatures with a high yield and high electrical conductivity using amides or m-substituted phenols at 0°C.

The redox potential of the solvent solution in chemical polymerization strongly effects the yield of the polymerization reaction and the conductivity of the prepared polymer. The optimal value for the redox potential (versus the saturated calomel electrode) is approximately 500 mV to form a highly conducting polypyrrole (Machida et al., 1989). For example, polypyrrole obtained via chemical polymerization and using methanol as a solvent showed an electrical conductivity of 190 S/cm. However, the conductivity of the polypyrrole was enhanced ≤220 S/cm when the redox potential was adjusted by adding iron (II) chloride ($FeCl_2$) (Machida et al., 1989).

The electrochemical synthesis of polypyrrole is required for scientific purposes due to various advantages, such as the simplicity and cost-effectiveness of the method, controlled thickness and morphology, the capacity for doping during production, and the extensive range of applicable dopant ions (Pringle et al., 2004). The adsorbed pyrrole monomers are oxidized on the surface of a working

Blue emeraldine base   Violet perningraniline base

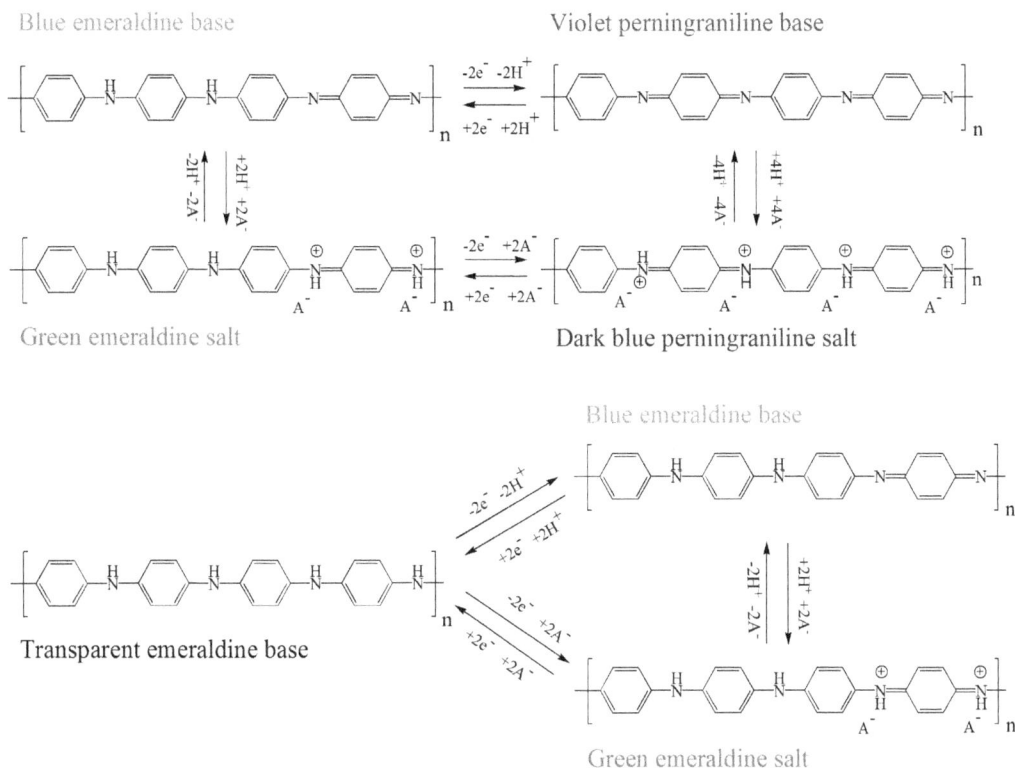

Green emeraldine salt   Dark blue perningraniline salt

Blue emeraldine base

Transparent emeraldine base

Green emeraldine salt

FIGURE 19.3   Different color and redox states of polyaniline.

electrode via a one-electron oxidation process to produce a pyrrole cation radical (Figure 19.4). The produced radical cations might be coupled with other molecules, which leads to the formation of a dicationic dimer. Then, the produced dicationic dimers undergo a double deprotonation reaction to produce neutral dimeric pyrrole. The oxidation potential of the dimeric pyrrole is lower than oxidation potential of the monomers and leads to chain growth by preferential coupling between the dimers (Ateh et al., 2006)

The electrochemical synthesis of polypyrrole is carried out in aqueous and non-aqueous solvent systems, such as dichloromethane, propylene, and carbonate acetonitrile (Fenelon & Breslin, 2002).

19.3.3   POLYTHIOPHENE

Polythiophene derivatives have attracted great attention device applications due to the solution processability and ease of structural modification. Three synthetic methods have been developed for the synthesis of polythiophene including metal catalyzed coupling and electrochemical oxidative polymerization (Amna et al., 2020). During oxidative polymerization, the electrochemical and thermal properties of the polymers are improved using appropriate oxidants and surfactants (Higashimura & Kobayashi, 2002).

In general, thiophene derivatives are polymerized via electrochemical and chemical methods. Chemical methods are preferred over electrochemical methods based on their simplicity and scalability. Polythiophene chemical synthesis techniques can be classified into two including oxidative polymerization and transition metal catalyzed polymerization. Substituted thiophenes are efficiently polymerized via ferric chloride-catalyzed oxidative polymerization (Hong et al., 1999).

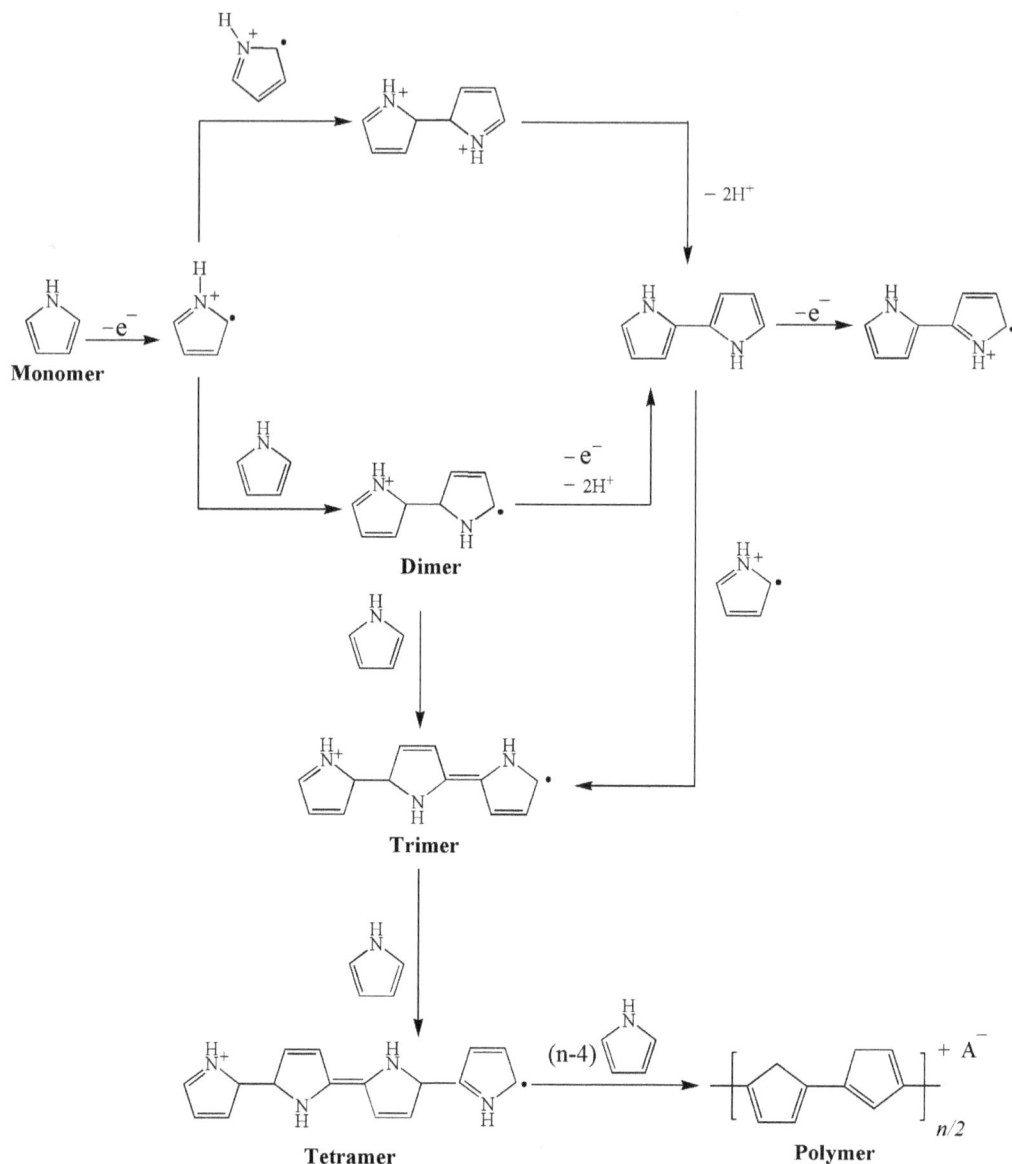

FIGURE 19.4   Electropolymerization mechanism of polypyrrole.

19.4   BIOMEDICAL APPLICATIONS OF ELECTROACTIVE POLYMERS

Conducting polymers are widely applied in various biomedical fields based on several beneficial chemical and physical properties including tunable electrical conductivity, controllable
morphology at nano scale, biodegradability, and chemical stability (Kaur et al., 2015; Liu et al.,
2020). The unique characteristics of the conducting polymer-based composites means that they
are promising materials in numerous biomedical applications, such as biosensors, regenerative
medicine, tissue engineering, and drug delivery (Dubey et al., 2020; Asadi et al., 2020; Alqarni
et al.; 2020).

19.4.1 Biosensors

The development of biosensors is one of the most extensively investigated research field based on their unique features that offer new opportunities in the ultrasensitive detection of markers for various disease (El-Said et al., 2020). Biosensors are composed of a biological sensing element (e.g., cell, enzyme, antibody, DNA, and RNA) and a transducer (e.g., piezoelectric, optical, and electroactive materials) (Salimiyan Rizi et al., 2021). A transducer is a device that can convert the biochemical quantities to electronic signals that are usually proportional to the amount of analyte. Conducting polymers that contain nanocomposites have emerged as significant elements of high-performance transducers and are used to improve the sensitivity and versatility of biosensors (Song et al., 2020; Gerard et al., 2002).

Several studies reported the determination of glucose by the immobilization of glucose oxidase in conducting polymers matrixes (Wang et al., 2019; Scotto et al., 2020; Al-Sagur et al., 2017). Conducting polymers that contained para- and ortho-quinone groups were used to catalyze the oxidation of glucose via an electrooxidation reaction (Arai et al., 2006). The blood cholesterol level is a vital parameter in clinical diagnostics that might be quantified by amperometric methods using cholesterol oxidase (Cevik et al., 2018). A combination of electrochemistry and immunochemistry concepts could be used in the development of biosensors (Aydın et al., 2020; Shan et al., 2017). Aydin et al. (2018) reported the fabrication of an impedimetric immunosensor for the detection of interleukin 1β by semi-conducting poly(2-thiophen-3-yl-malonic acid) as a matrix for the immobilization of the interleukin 1β antibody.

DNA-based biosensors are employed effectively in various applications including the clinical detection of pathogenic infections, diagnostics of genetic disorders and mutations, screening of cDNA colonies required in molecular biology, forensics investigations, and food technology (Rafique et al., 2019; Lin et al., 2020; Lakard et al., 2020; Kowalczyk, 2020; Ribeiro, 2020; Naseri et al., 2018).

A novel DNA-based biosensor was fabricated via the attachment of *BRCA1* complementary oligonucleotides to microporous conducting poly(3,4-ethylenedioxythiophene) with a large surface area to volume ratio (Wang et al., 2020). Recently, Wang et al. (2021) fabricated highly sensitive electrochemical biosensors for a sensitive MicroRNA assay based on zwitterionic peptide functionalized polyaniline.

The development of new techniques for the covalent binding of conducting polymers to substrates might increase the potential biomedical applications of these polymers. In addition, the advances in microfabrication technologies could lead to the synthesis of geometrically well-defined and highly reproducible structures in conducting polymers and could produce sensitive transducers (Wang et al., 2019).

19.4.2 Tissue Engineering

Tissue engineering studies aim to regenerate or repair damaged tissues using biomaterial scaffolds, cells, and growth factors (Dong et al., 2020; Guo & Ma, 2018). conducting polymers have received more attentions for applications in tissue engineering due to their flexibility and compatibility with various cells and organs (Talikowska et al., 2019; Zarrintaj et al., 2018).

High conductivity and the ability to mimic soft tissue mechanical properties are required for nerve tissue regeneration scaffold designs. However, some challenges remain in the design and fabricating of scaffolds, such as excellent biocompatibility, tissue-like mechanical properties, and high conductivity. Zhou et al. (2018) reported a biocompatible conducting polymer hydrogel for spinal cord injury repair.

Bertuoli et al. (2019) prepared electroactive and biocompatible fibrous scaffolds using polyaniline doped with dodecyl benzenesulfonic acid combined with a mixture of poly lactic acid and poly-ethylene glycol. Functional and morphological investigations applied to cardiac cells indicated that the prepared electrospun conducting core–shell fibers were suitable for cardiac tissue engineering applications (Bertuoli et al., 2019). A peptide that contained an arginylglycylaspartic acid fragment was grafted onto a conducting porous polythiophene. The application of the prepared composite as a scaffold for cell culture was investigated by the measurement of mitochondrial activity in human epidermal melanocytes and dermal fibroblasts cells (Chan et al., 2018).

19.4.3　Drug Delivery

The effectiveness of a drug delivery system is directly related to the properties of the carrier that enables bioactive molecules to reach to site of action (Figure 19.5). Controlled drug delivery systems are an interesting field of research and offer several benefits over conventional dosing forms, such as reduced toxicity, improved therapeutic effects, and improved patient convenience and compliance (Manzano & Vallet-Regí, 2020). Oral administration, which is the most frequently and most convenient route for drug administration, suffers from disadvantages including instant release and rapid adsorption of drug molecules and requires repeated administration (Bruneau et al., 2019).

Controlled release systems offer some benefits, such as reduced toxicity, enhanced efficacy, and improved patient convenience, and compliance (Tan et al., 2018). For example, polymer-based controlled drug delivery systems can maintain the drug concentration within the therapeutic window for a prolonged time (Figure 19.6). The flexibility and diversity in chemistry, topology, and dimension of the polymers can be used to improve the properties of the carriers including biocompatibility and biodegradability. These polymers show improved pharmacokinetics and are designed to offer unique properties as carriers including reduced toxicity and improved circulation time.

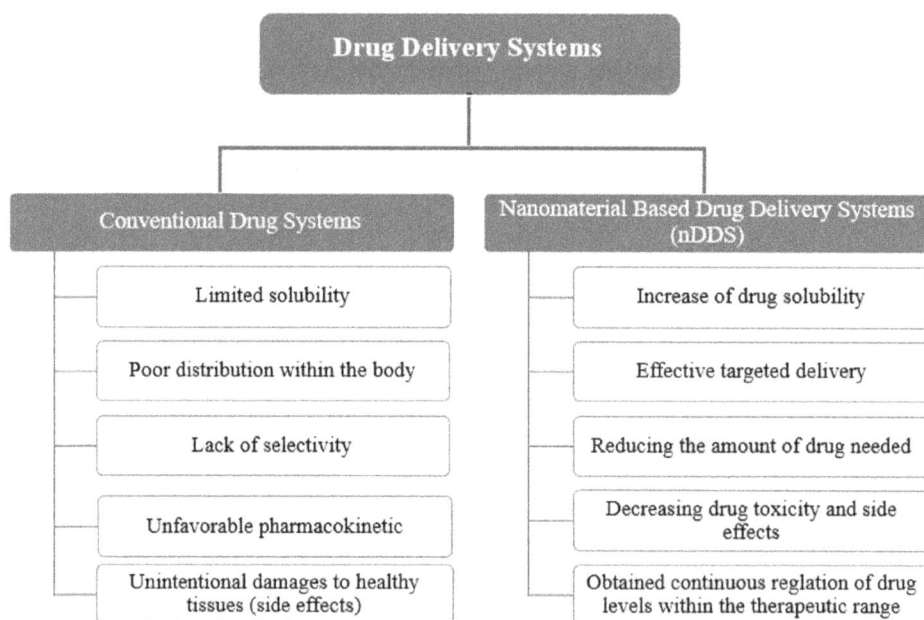

FIGURE 19.5　Comparison of nanomaterial based derug delivery systems with conventional systems.

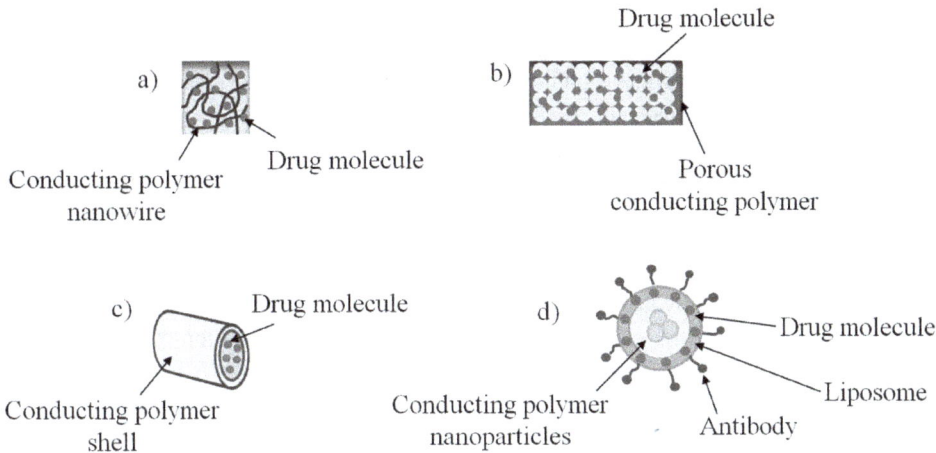

FIGURE 19.6 Conducting polymer-based drug delivery systems based on: (a) nanowire; (b) porous composite; (c) shell; and (d) liposome.

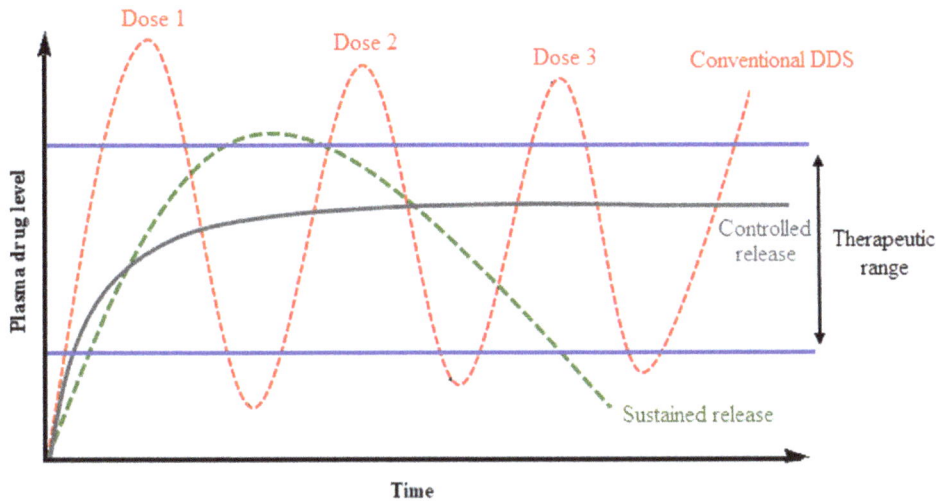

FIGURE 19.7 Comparison of drug levels in conventional drug delivery systems with controlled release systems.

19.5 SMART DRUG DELIVERY

The application of nanotechnology in the pharmaceutical field has provided a revolution in the development of smart nanosized carriers that can deliver drug molecules to target sites (Zhang et al., 2020). Compared with conventional drug delivery methods, intelligent delivery systems have several advantages including reduced side effects, ease of preparation, suitable therapeutic concentration, controlled and prolonged release, and improved stability (Figure 19.7) (Bedoya et al., 2020; Devnarain et al., 2021). Stimuli responsive nanocarriers could develop smart drug delivery platforms using various stimuli such as light, pH, temperature, ultrasound, mechanical stress, magnetic, or electrical fields (Li et al., 2019). Intelligent nanomaterials undergo a sudden change in physical or chemical properties in the presence of environmental stimuli. In general, the responses are reversible and might include changes in conductivity, solubility, porosity, solvent interactions, lipophilicity, hydrophilicity, and shape (Roy et al., 2010; Wells et al., 2019). The responses of the nanomaterials to environmental stimuli might be initiated by the formation or destruction of various secondary forces, such as hydrogen bonding, van der Waals interactions, hydrophobic forces, and electrostatic interactions (James et al., 2014).

19.5.1 POLYANILINE-BASED DRUG DELIVERY

Polyaniline is an important conducting polymer that is widely used in drug delivery because of its unique properties, such as high biocompatibility, high electrical conductivity, low-toxicity, good environmental stability and hydrophilicity (Zare et al., 2019). A photothermal and biodegradable polyaniline nanocomposite was synthesized via the oxidative polymerization of aniline monomers on the surface of porous nano silica. The prepared nanocomposite showed effective loading of doxorubicin hydrochloride and dual near infrared light and pH-triggered release (Xia et al., 2017). A polyaniline–chitosan composite (Figure 19. 8) was fabricated and applied as a semi-interpenetrating polymer network for an in vitro release study of ketoprofen (Minisy et al., 2021)

Cis-diamminedichloroplatinum (Cisplatin) is a well-known chemotherapy medication that is widely used for the treatment of several cancers including head and neck, testicular, ovarian, lung, cervical, bladder, breast, brain tumors and neuroblastoma, esophageal, and mesothelioma. You et al. (2017) reported a biodegradable polymeric nanocarrier composed of poly(ε-caprolactone) and poly(ethylene glycol) doped with polyaniline for intracellular controlled release and delivery of cisplatin. The obtained results highlighted the potential of near infrared light and pH dual responsive release by core–crosslinked nanoparticles in drug delivery systems.

Recently, Eslami et al. (2020) investigated a conducting suspension of liposomes and polyaniline complexes for photothermal therapy, bioimaging, and drug delivery. Polyaniline doped with phytic acid (a unique natural substance found in plant seeds) was produced by oxidative polymerization in an aqueous suspension (Eslami et al., 2020).

Silva et al. (2018) reported polyaniline decorated by zif-8 nanoparticles was an efficient chemo and photothermal smart carrier for the delivery of 5-fluorouracil. Somatostatin receptors are overexpressed in some tumors and lanreotide (a synthetic analog of somatostatin) is used as a medication for the treatment of neuroendocrine tumors. Nguyen et al. (2018) designed a nanostructured hybrid polymer for the targeted delivery of chemotherapy drugs to cancer cells by conjugating polyaniline with lanreotide that presented a high tendency for binding to somatostatin receptors.

A chemical oxidative polymerization approach was used for the synthesis of dual stimuli responsive (e.g., magnetic and electric fields) superparamagnetic silica-coated iron oxide/polyaniline nanocomposites (Lalegül-Ülker et al., 2021). A biocompatible and multifunctional nanocomposite was synthesized composed of magnesium-aluminum-layered double hydroxide, Hausmannite (Mn_3O_4)/N-graphene quantum dot, and conducting polyaniline for the pH-triggered release of doxorubicin (Ahmadi-Kashani et al., 2020). The presence of polyaniline on the surface of the nanoparticles provided ultrahigh doxorubicin encapsulation ε90% and showed a slow 4% release after 72 h under normal physiological conditions (Ahmadi-Kashani et al., 2020). In addition, Rana et al. (2014) reported a unique technique to improve the thermal-activated killing of cancer cells in the presence of an external magnetic field via the introduction of a polyaniline shell on the surface of iron oxide nanoparticles (Figure 19.9).

19.5.2 POLYPYRROLE-BASED DRUG DELIVERY

Gelatin and cyclodextrin stabilized polypyrrole nanoparticles were prepared for three stimuli controlled doxorubicin delivery that included including photothermal, pH, and enzyme stimuli (Ma et al., 2019). A three- dimensional conducting hybrid film was fabricated using polymethacrylate

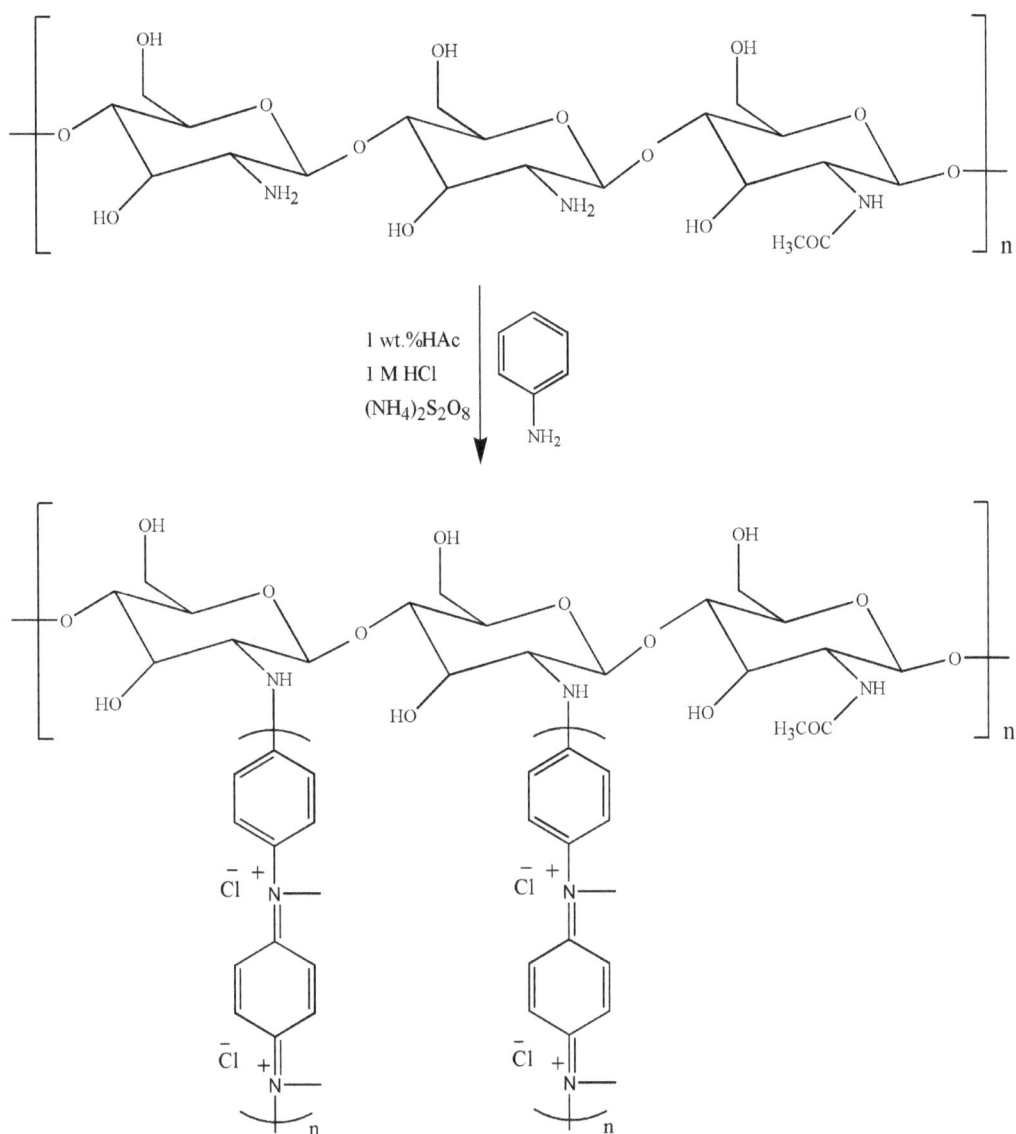

FIGURE 19.8 Fabrication of a polyaniline–chitosan composite.

derivatives with polypyrrole as the conducting polymer (Gutiérrez-Pineda et al., 2018). Cyclic voltammetry and electrochemical impedance spectroscopy were applied for the investigation of the electrochemical characteristics of the prepared hybrid conducting composite. Doxorubicin was encapsulated as a model drug in the composite film and the release of the drug was explored at applied potentials r from −650 to 65 mV. The results showed that the amount of released doxorubicin depended on the electrical stimuli duration and the applied potential (Gutierrez-Pineda et al., 2018). Tiwari et al. (2018) developed a fibrous smart drug delivery carrier that was functionalized with near infrared light and pH stimuli responsive polypyrrole for the delivery of paclitaxel. A polypyrrole containing matrix had improved drug release in a pH 5.5 comparison with pH 7.4. The results showed that the release was further accelerated in response to near infrared, which exhibited the capability of the dual stimuli responsive drug delivery platform for synergistic cancer therapy (Tiwari

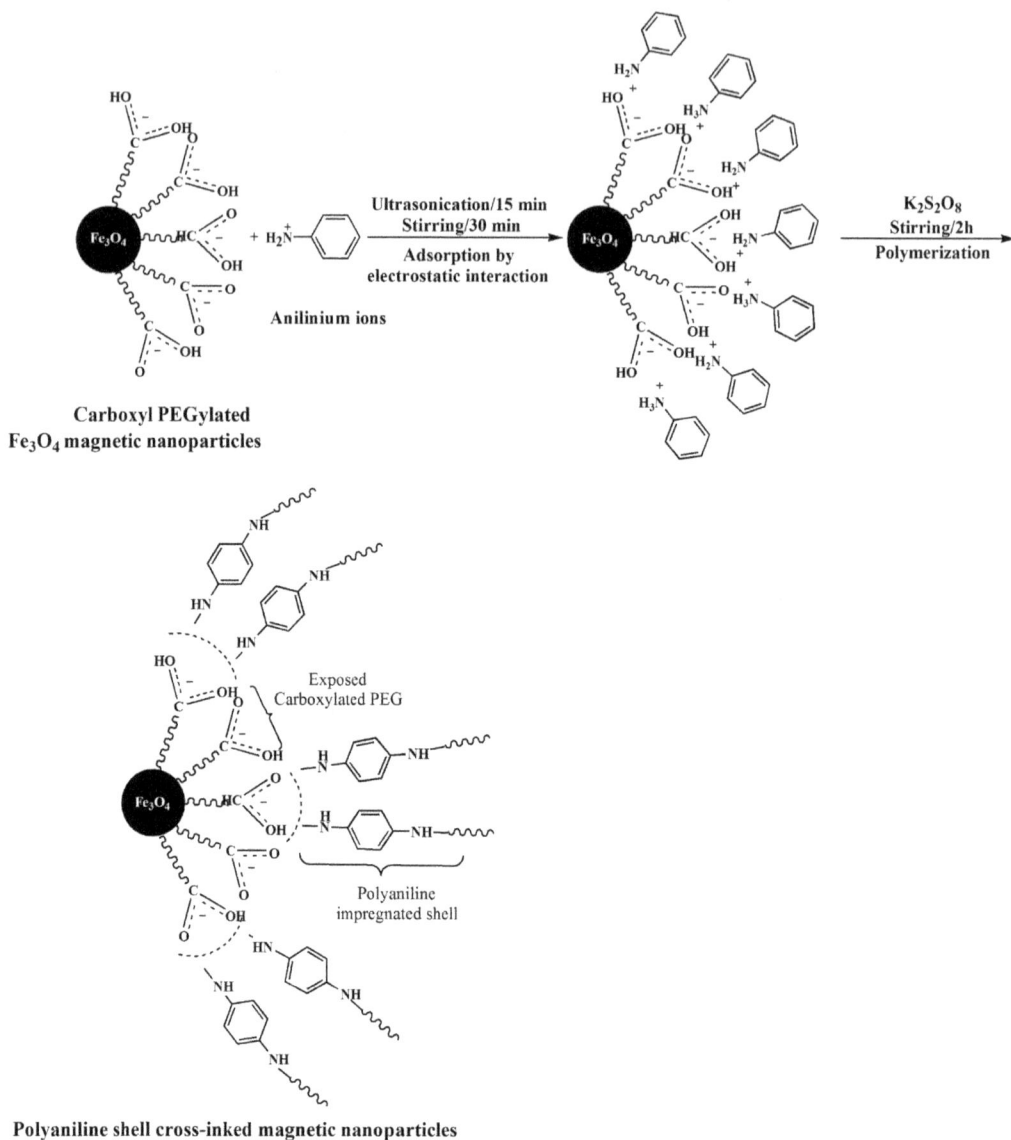

FIGURE 19.9   Synthesis of a magnetic polyaniline nanocomposite.

et al., 2018). In addition, polypyrrole–chitosan composites can be obtained by the graft polymeriza-tion of pyrrole with chitosan and used for the delivery of anticancer agents (Figure 19. 10) (Adhikari & Yadav, 2018).

Folate is a key component in DNA nucleotide synthesis and cell division; however, cancer cells express folate receptors 500 times more than healthy cells. Therefore, monitoring of the level of these receptors is important for the diagnosis and treatment of cancers and chronic inflammatory diseases (Scaranti et al., 2020). Smart drug systems that contain polypyrrole-based nanoparticles were developed, and the size and morphology of the carriers were adjusted via a template technique. In addition, aromatic imine was applied to control of drug release and tumor cell selectivity was improved by folate functionalization (Chen et al., 2017). Xie et al. (2017) reported a conducting polydopamine and polypyrrole microcapsules on the surface of titanium nanoparticles via an

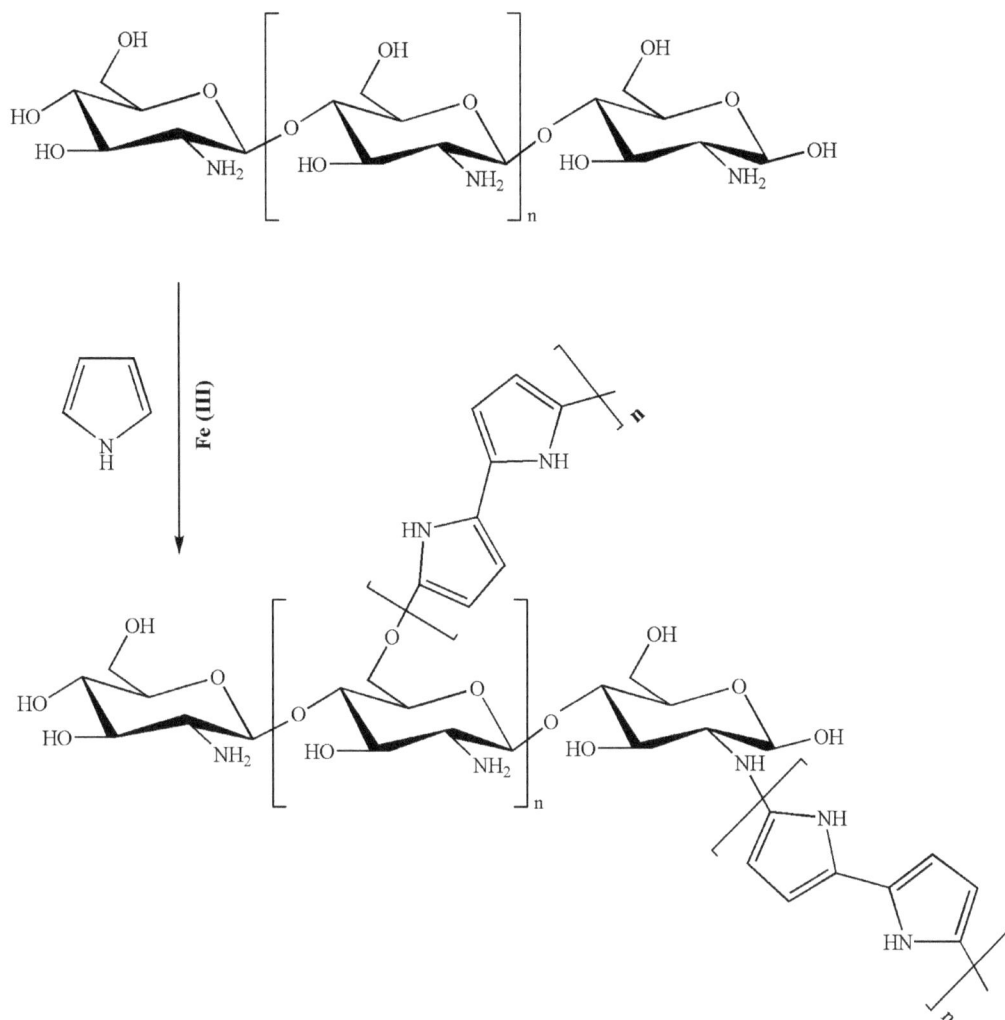

FIGURE 19.10 Synthesis of a polypyrrole–chitosan composite.

electrochemical deposition technique. In this procedure, dexamethasone and polydopamine were doped as the anion into a polypyrrole framework to neutralize its positive charge (Xie et al., 2017).

Curcumin is a natural polyphenolic compound found in turmeric, which inhibits cancer cell survival and proliferation, and induces apoptosis without promoting side effects. However, the clinical applications of curcumin to treat cancer are restricted by its low solubility and bioavailability (Bashang et al., 2020). Polypyrrole coated poly(lactic-co-glycolic acid) core–shell nanoparticles were investigated as a photothermal stimuli responsive matrix for the sustained release of curcumin as a chemotherapeutic agent (Liu et al., 2016).

19.5.3 Polythiophene-Based Drug Delivery

During the last decades, advances in polymer chemistry have received attention to exploit their intrinsic benefits for intelligent gene and drug delivery (Lichon et al., 2020; Pairatwachapun et al. 2016). A polythiophene containing hybrid composite was synthesized via a combination of Suzuki coupling polymerization and an in situ N-carboxy anhydride ring opening reaction (Guler et al., 2016). The prepared composite was functionalized by an anti-HER2/neu antibody. The composite did have HER2-expressing A549 human lung carcinoma cells; however, no signal was observed for monkey kidney epithelial and human cervix adenocarcinoma cells (Guler et al., 2016).

Practical and highly effective approaches are required to increase the transgene performance of polymeric gene carriers. He et al. (2015) developed cationic copolymers composed of low molecular weight polyethylenimine and polythiophene as novel self-tracking siRNA delivery vectors. In addition, multicomponent and biocompatible nanocomposites were developed by the addition of cationic polythiophenes on the outer shell of polypeptide/DNA polyplexes (Zhang et al., 2017). Poly(3,4-ethylenedioxythiophene) nanoparticles were loaded with piperine and curcumin via an in situ emulsion polymerization that applied dodecyl benzene sulfonic acid as the doping agent (Puiggalí-Jou et al., 2017). The loaded drugs affected the colloidal stability, size, and morphology of the polythiophene containing nanoparticles. The release behavior of these two neutral drugs that possess different capacities to form hydrogen bonding interactions with the oxidized polymers was achieved the presence and absence of external electrical stimulation (Puiggalí-Jou et al., 2017).

19.6 CONCLUSION

Conducting polymers have been extensively exploited as intelligent drug carriers due to their unique properties, such as high conductivity, biocompatibility, and low-cost synthesis procedures. The design and use of conducting polymer materials for controlled delivery systems has become one of the most promising research areas. Despite remarkable advances in conducting polymer-based drug delivery systems, there some challenges remain for practical applications. The biocompatibility of the conducting polymers that contain nanocomposites could be investigated for in vitro and in vivo applications. Although clinical applications for electroactive drug delivery systems face some challenges, smart polymeric drug delivery systems could provide several opportunities for biomedical applications in the future.

REFERENCES

Abu-Thabit, N. Y. (2016). Chemical oxidative polymerization of polyaniline: A practical approach for preparation of smart conductive textiles. *Journal of Chemical Education*, *93*(9), 1606–1611.

Adhikari, H. S., Yadav, P. N. (2018). Anticancer activity of chitosan, chitosan derivatives, and their mechanism of action. *International Journal of Biomaterials*, *2018*, 2952085–2952114.

Ahmadi-Kashani, M., Dehghani, H., & Zarrabi, A. (2020). A biocompatible nanoplatform formed by MgAl-layered double hydroxide modified $Mn3O_4$/N-graphene quantum dot conjugated-polyaniline for pH-triggered release of doxorubicin. *Materials Science and Engineering C*, *114*, 111055.

Alqarni, S. A., Hussein, M. A., Ganash, A. A., & Khan, A. (2020). Composite material–based conducting polymers for electrochemical sensor applications: A mini review. *BioNanoScience*, *10*(1), 1–14.

Al-Sagur, H., Komathi, S., Khan, M. A., Gurek, A. G., & Hassan, A. (2017). A novel glucose sensor using lutetium phthalocyanine as redox mediator in reduced graphene oxide conducting polymer multifunctional hydrogel. *Biosensors and Bioelectronics*, *92*, 638–645.

Amna B., Siddiqi, H. M., Hassan, & Ozturk, A. T. (2020). Recent developments in the synthesis of regioregular thiophene-based conjugated polymers for electronic and optoelectronic applications using nickel and palladium-based catalytic systems. *RSC Advances*, *10*, 4322–4396.

Arai, G., Shoji, K., & Yasumori, I. (2006). Electrochemical characteristics of glucose oxidase immobilized in poly(auinone) redox polymers. *Journal of Electroanalytical Chemistry*, *591*(1), 1–6.

Asadi, N., Del Bakhshayesh, A. R., Davaran, S. & Akbarzadeh, A. (2020). Common biocompatible polymeric materials for tissue engineering and regenerative medicine. *Materials Chemistry and Physics*, *242*, 122528.

Ateh, D. D., Navsaria, H. A., & Vadgama, P. (2006). Polypyrrole-based conducting polymers and interactions with biological tissues. *Journal of the Royal Society Interface*, *3*, 741–752.

Ayad, M. M., & Zaki, E. A. (2008). Quartz crystal microbalance and spectroscopy measurements for acid doping in polyaniline films. *Science and Technology of Advance Materials*, *9*(1), 015007.

Aydın, E. B., Aydın, M., & Sezgintürk, M. K. (2018). Highly sensitive electrochemical immunosensor based on polythiophene polymer with densely populated carboxyl groups as immobilization matrix for detection of interleukin 1β in human serum and saliva. *Sensors and Actuators B: Chemical*, *270*, 18–27.

Aydın, E. B., Aydın, M., & Sezgintürk M. K. (2020). A novel electrochemical immunosensor based on acetylene black/epoxy-substituted-polypyrrole polymer composite for the highly sensitive and selective detection of interleukin 6. *Talanta*, *222*, 121596.

Balint, R., Cassidy, N. J., & Cartmell S. H. (2014). Conductive polymers: Towards a smart biomaterial for tissue engineering. *Acta Biomaterialia, 10*(6), 2341–2353.

Bar-Cohen Y., Cardoso, V. F., Ribeiro, C., & Lanceros-Méndez, S. (2017). Electroactive polymers as actuators. *Advanced Piezoelectric Materials*, *2017*, 319–352.

Bashang, H., & Tamma, S. (2020). The use of curcumin as an effective adjuvant to cancer therapy: A short review. *Biotechnology and Applied Biochemistry*, *67*(2), 171–179.

Bedoya, D. A., Figueroa, F. N., Macchione, M. A., & Strumia. M. C. (2020). Stimuli-responsive polymeric systems for smart drug delivery. In A. K., Nayak, & S. Hasnain (Eds.) *Advanced biopolymeric systems for drug delivery*. Cham: Springer International Publishing.

Bertuoli, P. T., Ordoño, J., Armelin, E., Pérez-Amodio, S., Baldissera, A. F., Ferreira, C. A. ... C. Alemán. (2019). Electrospun conducting and biocompatible uniaxial and Core–Shell fibers having poly (lactic acid), poly (ethylene glycol), and polyaniline for cardiac tissue engineering. *ACS Omega*, *4*(2), 3660–3672.

Bláha, M., Riesová, M., Zedník, J., Anžlovar, A., Žigon, M., & Vohlídal J. (2011). Polyaniline synthesis with iron (III) chloride–hydrogen peroxide catalyst system: reaction course and polymer structure study. *Synthetic Metals*, *161*(13–14), 1217–1225.

Bruneau M., Bennici, S., Brendle, J., Dutournie, P., Limousy, L., & Pluchon, S. (2019). Systems for stimuli-controlled release: Materials and applications. *Journal of Controlled Release*, *294*, 355–371.

Carpi, F., & Smela, E. (Eds.). (2009). *Biomedical applications of electroactive polymer actuators.* New Jersey: John Wiley & Sons.

Cevik, E., Cerit, A., Gazel, N., & Yildiz, H. B. (2018). Construction of an amperometric cholesterol biosensor based on DTP (aryl) aniline conducting polymer bound cholesterol oxidase. *Electroanalysis*, *30*(10), 2445–2453.

Chan, E. W. C., Bennet, D., Baek, P., Barker, D., Kim, S., & Travas-Sejdic J. (2018). Electrospun polythiophene phenylenes for tissue engineering. *Biomacromolecules*, *19*(5), 1456–1468.

Chen, J., Li, X., Sun, Y., Hu, Y., Peng, Y., Li, Y. ... Zhong, S. (2017). Synthesis of size-tunable hollow polypyrrole nanostructures and their assembly into folate-targeting and pH-responsive anticancer drug-delivery agents. *Chemistry – A European Journal*, *23*(68), 17279–17289.

Devnarain, N., Osman, N., Fasiku, V. O., Makhathini, S., Salih, M., Ibrahim U. H., & Govender T. (2021). Intrinsic stimuli-responsive nanocarriers for smart drug delivery of antibacterial agents—An in-depth review of the last two decades. *Wiley Interdisciplinary Reviews: Nanomedicine and Nanobiotechnology, 13*(1), e1664.

Distler, T., & Boccaccini, A. R. (2020). 3D printing of electrically conductive hydrogels for tissue engineering and biosensors–A review. *Acta Biomaterialia*, *101*, 1–13.

Dong, R., Ma, P. X., & Guo, B. (2020). Conductive biomaterials for muscle tissue engineering. *Biomaterials*, *229*, 119584.

Dubey, N., Kushwaha, C. S., & Shukla, S. K. (2020). A review on electrically conducting polymer bionanocomposites for biomedical and other applications. *International Journal of Polymeric Materials and Polymeric Biomaterials*, *69*(11), 709–727.

Duran, N. G., Karakışla, M., Aksu L., & Saçak M. (2009). Conducting polyaniline/kaolinite composite: Synthesis, characterization and temperature sensing properties. *Materials Chemistry and Physics*, *118*(1), 93–98.

El-Said, W. A., Abdelshakour, M., Choi, J. H., & Choi, J. W. (2020). Application of conducting polymer nanostructures to electrochemical biosensors. *Molecules*, *25*(2): 307.

Eslami, M., Zeglio, E., Alosaimi, G., Yan, Y., Ruprai, H., Macmillan, A., ...Mawad, D. (2020). A one step procedure toward conductive suspensions of liposome-polyaniline complexes. *Macromolecular Biosciences*, *20*(11), 2000103.

Fenelon, A. M., & Breslin, C. B. (2002). The electrochemical synthesis of polypyrrole at a copper electrode: corrosion protection properties. *Electrochimica Acta*, *47*(28), 4467–4476.

Gajendiran, M., Choi, J., Kim, S. J., Kim, K., Shin, H., Koo, H. J., & Kim, K. (2017). Conductive biomaterials for tissue engineering applications. *Journal of Industrial Engineering Chemistry*, *51*, 12–26.

Gerard, M., Chaubey, A., & Malhotra, B. D. (2002). Application of conducting polymers to biosensors. *Biosensors and Bioelectronics*, *17*(5), 345–359.

Guler, E., Akbulut, H., Geyik, C., Yilmaz, T., Gumus, Z. P., Barlas, F. B.… Yagci, Y. (2016). Complex structured fluorescent polythiophene graft copolymer as a versatile tool for imaging, targeted delivery of paclitaxel, and radiotherapy. *Biomacromolecules*, *17*(7), 2399–2408.

Gospodinova, N., Terlemezyan, L., Mokreva, P., & Kossev, K. (1993). On the mechanism of oxidative polymerization of aniline. *Polymer*, *34*(11), 2434–2437.

Guo, B., & Ma, P. X. (2018). Conducting polymers for tissue engineering. *Biomacromolecules*, *19*(6), 1764–1782.

Gutiérrez-Pineda, E., Cáceres-Vélez, P. R., Rodríguez-Presa, M. J., Moya, S. E., Gervasi, C. A., & Amalvy, J. I. (2018). Hybrid conducting composite films based on polypyrrole and poly (2-(diethylamino) ethyl methacrylate) hydrogel nanoparticles for electrochemically controlled drug delivery. *Advanced Materials Interfaces*, *5*(21), 1800968.

He, P., Hagiwara, K., Chong, H., Yu, H. H., & Ito, Y. (2015). Low-molecular-weight polyethyleneimine grafted polythiophene for efficient siRNA delivery. *BioMed Research International*, 406389.

Higashimura H., & Kobayashi, S. (2002). Oxidative polymerization. *Encyclopedia of Polymer Science and Technology*, *4*, 1–37.

Hirase, R., Shikata, T., & Shirai, M. (2004). Selective formation of polyaniline on wool by chemical polymerization, using potassium iodate. *Synthetic Metals*, *146*(1), 73–77.

Hong X., Tyson, J. C., Middlecoff, J. S., & Collard, D. M. (1999). Synthesis and oxidative polymerization of semifluoroalkyl-substituted thiophenes. *Macromolecules*, *32*, 4232–4239.

James, H. P., John, R., Alex, A., & Anoop, K. (2014). Smart polymers for the controlled delivery of drugs–a concise overview. *Acta Pharmaceutica Sinica B*, *4*(2), 120–127.

Jin, Z., Su, Y., & Duan, Y. (2001). A novel method for polyaniline synthesis with the immobilized horseradish peroxidase enzyme. *Synthetic Metals*, *122*(2), 237–242.

Kaur, G., Adhikari, R., Cass, P., Bown, M., & Gunatillake, P. (2015). Electrically conductive polymers and composites for biomedical applications. *RSC Advances*, *5*(47), 37553–37567.

Kausar, A. (2020). Thermally conducting polymer/nanocarbon and polymer/inorganic nanoparticle nanocomposite: a review. *Polymer-Plastic Technology and Materials*, *59*, 895–909.

Kausaite, A., Ramanaviciene, A., & Ramanavicius, A. (2009). Polyaniline synthesis catalysed by glucose oxidase. *Polymer*, *50*(8), 1846–1851.

Khan A, Jawaid, M. Khan, A. A. P., & Asiri, A. M. (Eds.). (2018). *Electrically conductive polymers and polymer composites: from synthesis to biomedical applications*. New Jersey: John Wiley & Sons.

Kowalczyk, A. (2020). Trends and perspectives in DNA biosensors as diagnostic devices. *Current Opinions in Electrochemistry*, *23*, 36–41.

Kuma, V., Kalia, S., & Swart, H. C. (Eds.). (2017). *Conducting polymer hybrids*. Cham: Springer International Publishing.

Lakard, B. (2020). Electrochemical biosensors based on conducting polymers: A review. *Applied Sciences*, *10*(18), 6614.

Lalegül-Ülker, Ö., &. Elçin, Y. M. (2021). Magnetic and electrically conductive silica-coated iron oxide/polyaniline nanocomposites for biomedical applications. *Materials Science and Engineering C*, *119*, 111600.

Li, Z., Song, N., & Yang, Y. W. (2019). Stimuli-responsive drug-delivery systems based on supramolecular nanovalves. *Matter*, *1*(2),345–368.

Lichon, L., Kotras, C., Myrzakhmetov, B., Arnoux, P., Daurat, M., Nguyen, C., … Clément, S. (2020). Polythiophenes with cationic phosphonium groups as vectors for imaging, siRNA delivery, and photodynamic therapy. *Nanomaterials*, *10*(8), 1432.

Lin, Y. T., Darvishi, S., Preet, A., Huang, T. Y., Lin, S. H., Girault, H. H., … Lin T. E. (2020). A review: electrochemical biosensors for oral cancer. *Chemosensors*, *8*(3), 54.

Liu, M., Xu, N., Liu, W., & Xie, Z. (2016). Polypyrrole coated PLGA core–shell nanoparticles for drug delivery and photothermal therapy. *Rsc Advances*, *6*(87), 84269–84275.

Liu, Y., Yin, P., Chen, J., Cui, B., Zhang, C., & Wu, F. (2020). Conducting polymer-based composite materials for therapeutic implantations: From advanced drug delivery system to minimally invasive electronics. *International Journal of Polymer Science*, *2020*, 5659682.

Ma, X., Li, X., Shi, J., Yao, M., Zhang, X., Hou, R., ... Wang, F. (2019). Host–guest polypyrrole nanocomplex for three-stimuli-responsive drug delivery and imaging-guided chemo-photothermal synergetic therapy of refractory thyroid cancer. *Advanced Healthcare Materials*, *8*(17),1900661.

Machida, S., Miyata, S., & Techagumpuch, A. (1989). Chemical synthesis of highly electrically conductive polypyrrole. *Synthetic Metals*, *31*(3), 311–318.

Manzano, M., & Vallet-Regí, M. (2020). Mesoporous silica nanoparticles for drug delivery. *Advanced Functional Materials*, *30*(2), 1902634.

Martyak, N. M., McAndrew, P., McCaskie, J. E., & Dijon, J. (2002). Electrochemical polymerization of aniline from an oxalic acid medium. *Progress in Organic Coatings*, *45*(1), 23–32.

Minisy, I. M., Salahuddin, N. A., & Ayad, M. M. (2021). In vitro release study of ketoprofen-loaded chitosan/polyaniline nanofibers. *Polymer Bulletin*, *78*, 5609–5622.

Miras, M. C., Barbero, C., & Haas, O. (1991). Preparation of polyaniline by electrochemical polymerization of aniline in acetonitrile solution. *Synthetic Metals*, *43*(1–2),3081–3084.

Mogharabi-Manzari, M., Ghahremani, M. H., Sedaghat, T., Shayan, F., & Faramarzi, M. A. (2019). A Laccase heterogeneous magnetic fibrous silica-based biocatalyst for green and one-pot cascade synthesis of chromene derivatives. *European Journal of Organic Chemistry*, *2019*(8),1741–1747.

Namsheer, K., & Rout, C. S. (2021). Conducting polymers: a comprehensive review on recent advances in synthesis, properties and applications. *RSC Advances*, *11*, 5659–5697.

Naseri, M., Fotouhi, L., & Ehsani, A. (2018). Recent progress in the development of conducting polymer-based nanocomposites for electrochemical biosensors applications: A mini-review. *The Chemical Record*, *18*(6), 599–618.

Nguyen, H. T., Dai Phung, C., Thapa, R. K., Pham, T. T., Tran, T. H., Jeong, J. H., ... Kim J. O. (2018). Multifunctional nanoparticles as somatostatin receptor-targeting delivery system of polyaniline and methotrexate for combined chemo–photothermal therapy. *Acta Biomaterialia*, *68*, 154–167.

Ning, C., Zhou, Z., Tan, G., Zhu, Y., & Mao, C. (2018). Electroactive polymers for tissue regeneration: Developments and perspectives. *Progress in Polymer Science*, *81*, 144–162.

Pairatwachapun, S., Paradee, N., & Sirivat, A. (2016). Controlled release of acetylsalicylic acid from polythiophene/carrageenan hydrogel via electrical stimulation. *Carbohydrate Polymers*, *137*, 214–221.

Palaniappan, S. (2001). Chemical and electrochemical polymerization of aniline using tartaric acid. *European Polymer Journal*, *37*(5), 975–981.

Prasad, P. N., & Ulrich, D. R. (Eds.). (2012). *Nonlinear optical and electroactive polymers.* Heidelberg: Springer Science & Business Media.

Pringle, J. M., Efthimiadis, J., Howlett, P. C., Efthimiadis, J., MacFarlane, D. R., Chaplin, A. B., ...Forsyth, M. (2004). Electrochemical synthesis of polypyrrole in ionic liquids. *Polymer*, *45*(5),1447–1453.

Puiggalí-Jou, A., Micheletti, P., Estrany, F., del Valle, L. J., & Alemán, C. (2017). Electrostimulated release of neutral drugs from Polythiophene nanoparticles: smart regulation of drug–polymer interactions. *Advanced Healthcare Materials*, *6*(18), 1700453.

Qazi, T. H., Rai, R., & Boccaccini A. R. (2014). Tissue engineering of electrically responsive tissues using polyaniline based polymers: A review. *Biomaterials*, *35*(33), 9068–9086.

Rafique, B., Iqbal, M., Mehmood, T., & Shaheen, M. A. (2019). Electrochemical DNA biosensors: A review. *Sensors Reviews*, *39*, 34–50.

Rahy, A., Sakrout, M., Manohar, S., Cho, S. J., Ferraris, J., & Yang, D. J. (2008). Polyaniline nanofiber synthesis by co-use of ammonium peroxydisulfate and sodium hypochlorite. *Chemistry of Materials*, *20*(15), 4808–4814.

Ramanavicius, S., & Ramanavicius, A. (2021). Conducting polymers in the design of biosensors and biofuel cells. *Polymers*, *13*(1), 49–53.

Rana, S., Jadhav, N. V., Barick, K. C., Pandey, B. N., & Hassan, P. A. (2014). Polyaniline shell cross-linked Fe_3O_4 magnetic nanoparticles for heat activated killing of cancer cells. *Dalton Transactions*, *43*, 12263–12271.

Rapi, S., Bocchi, V., & Gardini, G. (1988). Conducting polypyrrole by chemical synthesis in water. *Synthetic Metals*, *24*(3), 217–221.

Ribeiro, B. V., Cordeiro, T. A. R., de Freitas, G. R. O., Ferreira, L. F., & Franco D. L. (2020). Biosensors for the detection of respiratory viruses: A review. *Talanta Open*, *2*, 100007.

Roy, D., Cambre, J. N., & Sumerlin, B. S. (2010). Future perspectives and recent advances in stimuli-responsive materials. *Progress in Polymer Science*, *35*(1–2), 278–301.

Salimiyan Rizi, K., Aryan, E., Meshkat, Z., Ranjbar, G., Sankian, M., Ghazvini K., ...Rezayi, M. (2021). The overview and perspectives of biosensors and *Mycobacterium tuberculosis*: A systematic review. *Journal of Cell Physiology*, *236*(3), 1730–1750.

Sapurina, I., & Stejskal, J. (2008). The mechanism of the oxidative polymerization of aniline and the formation of supramolecular polyaniline structures. *Polymer International*, *57*(12), 1295–1325.

Scaranti, M., Cojocaru, E., Banerjee, S., & Banerji, U. (2020). Exploiting the folate receptor α in oncology. *Nature Reviews Clinical Oncology*, *17*(6), 349–359.

Scotto, J., Piccinini, E., Bilderling, C. V., Coria-Oriundo, L. L., Battaglini, F., Knoll, W., ... Azzaroni, O. (2020). Flexible conducting platforms based on PEDOT and graphite nanosheets for electrochemical biosensing applications. *Applied Surface Science*, *525*, 146440.

Shan, J., & Ma Z. (2017). A review on amperometric immunoassays for tumor markers based on the use of hybrid materials consisting of conducting polymers and noble metal nanomaterials. *Microchimica Acta*, *184*(4), 969–979.

Shen, L., & Huang, X. (2018). Electrochemical polymerization of aniline in a protic ionic liquid with high proton activity. *Synthetic Metals*, *245*, 18–23.

Silva, J. S., Silva, J. Y., de Sá, G. F., Araújo, S. S., Filho, M. A. G., Ronconi, C. M., ...Júnior, S. A. (2018). Multifunctional system polyaniline-decorated ZIF-8 nanoparticles as a new chemo-photothermal platform for cancer therapy. *ACS Omega*, *3*(9), 12147–12157.

Song, Z., Ma, Y., Morrin, A., Ding, C., & Luo, X. (2020). Preparation and electrochemical sensing application of porous conducting polymers. *Trends in Analytical Chemistry*, *135*, 116155.

Surmenev R. A., Chernozem R. V., Pariy, I. O., & Surmeneva, M. A. (2021). A review on piezo-and pyroelectric responses of flexible nano-and micropatterned polymer surfaces for biomedical sensing and energy harvesting applications. *Nano Energy*, *79*, 105442.

Talikowska, M., Fu, X., & Lisak, G. (2019). Application of conducting polymers to wound care and skin tissue engineering: A review. *Biosensors and Bioelectronics*, *135*, 50–63.

Tan Y. F., Lao, L. L., Xiong, G. M., & Venkatraman, S. (2018). Controlled-release nanotherapeutics: State of translation. *Journal of Controlled Release*, *284*, 39–48.

Tiwari, A. P., Hwang, T. I., Oh, J. M., Maharjan, B., Chun, S., Kim, B. S., ...Kim, C. S. (2018). pH/NIR-responsive polypyrrole-functionalized fibrous localized drug-delivery platform for synergistic cancer therapy. *ACS Applied Material & Interfaces*, *10*(24), 20256–20270.

Wallace, G. G., Teasdale, P. R., Spinks, G. M., & Kane-Maguire, L. A. (2008). *Conductive electroactive polymers: intelligent polymer systems*. Boca Raton: CRC Press.

Wang, D., Wang, J., Song, Z., & Hui, N. (2021). Highly selective and antifouling electrochemical biosensors for sensitive MicroRNA assaying based on conducting polymer polyaniline functionalized with zwitterionic peptide. *Analytical and Bioanalytical Chemistry*, *413*, 543–553.

Wang, L., Xu, M., Xie, Y., Qian, C., Ma, W., Wang, L., & Song, Y. (2019). Ratiometric electrochemical glucose sensor based on electroactive Schiff base polymers. *Sensors and Actuators B: Chemistry*, *285*, 264–270.

Wang, N. (2020). All-in-one flexible asymmetric supercapacitor based on composite of polypyrrole-graphene oxide and poly (3, 4-ethylenedioxythiophene). *Journal of Alloys and Compounds*, *835*, 155299.

Wang, Y., Liu, Z., Han, B., Sun, Z., Huang, Y., & Yang, G. (2005). Facile synthesis of polyaniline nanofibers using chloroaurate acid as the oxidant. *Langmuir*, *21*(3), 833–836.

Wells, C. M., Harris, M., Choi, L., Murali, V. P., Guerra, F. D., & Jennings, J. A. (2019). Stimuli-responsive drug release from smart polymers. *Journal of Functional Biomaterials*, *10*(3), 34.

Xia, B., Wang, B., Shi, J., Zhang, Y., Zhang, Q., Chen, Z., & Li, J. (2017). Photothermal and biodegradable polyaniline/porous silicon hybrid nanocomposites as drug carriers for combined chemo-photothermal therapy of cancer. *Acta Biomaterialia*, *51*, 197–208.

Xie, C., Li, P., Han, L., Wang, Z., Zhou, T., Deng, W., ...Lu, X. (2017). Electroresponsive and cell-affinitive polydopamine/polypyrrole composite microcapsules with a dual-function of on-demand drug delivery and cell stimulation for electrical therapy. *NPG Asia Materials*, *9*(3), e358.

You, C., Wu, H., Wang, M., Zhang, Y., Wang, J., Luo, Y., ... Zhu, J. (2017). Near-infrared light and pH dual-responsive targeted drug carrier based on core-crosslinked polyaniline nanoparticles for intracellular delivery of cisplatin. *Chemistry - A European Journal*, *23*(22), 5352–5360.

Yussuf, A., Al-Saleh, M., Al-Enezi S., & Abraham, G. (2018). Synthesis and characterization of conductive polypyrrole: the influence of the oxidants and monomer on the electrical, thermal, and morphological properties. *International Journal of Polymer Science*, *2018*, 4191747.

Zare, E. N., Makvandi, P., Ashtari, B., Rossi, F., Motahari, A., & Perale G. (2019). Progress in conductive polyaniline-based nanocomposites for biomedical applications: a review. *Journal of Medicinal Chemistry*, *63*(1), 1–22.

Zarren, G., Nisar, B., & Sher, F. (2019). Synthesis of anthraquinone-based electroactive polymers: A critical review. *Materials Today Sustainability*, *5*, 100019.

Zarrintaj, P., Bakhshandeh, B., Saeb, M. R., Sefat, F., Rezaeian, I., Ganjali, M. R. ... Mozafari, M. (2018). Oligoaniline-based conductive biomaterials for tissue engineering. *Acta Biomaterialia*, *72*, 16–34.

Zhang, H., Fan, T., Chen, W., Li, Y., & Wang, B. (2020). Recent advances of two-dimensional materials in smart drug delivery nano-systems. *Bioactive Materials*, *5*(4),1071–1086.

Zhang, Y., Li, X., Wu, T., Sun, J., Wang, X., Cao, L., & Feng, F. (2017). Cationic polythiophenes as gene delivery enhancer. *ACS Applied Materials Interfaces*, *9*(20), 16735–16740.

Zhou, L., Fan, L., Yi, X., Zhou, Z., Liu, C., Fu, R., ...Ning, C. (2018). Soft conducting polymer hydrogels cross-linked and doped by tannic acid for spinal cord injury repair. *ACS Nano*, *12*(11), 10957–10967.

Index

Note: Page numbers in *italic* denote figures and in **bold** denote tables.

![Taylor & Francis Group — an *informa* business]

Taylor & Francis eBooks

www.taylorfrancis.com

A single destination for eBooks from Taylor & Francis
with increased functionality and an improved user
experience to meet the needs of our customers.

90,000+ eBooks of award-winning academic content in
Humanities, Social Science, Science, Technology, Engineering,
and Medical written by a global network of editors and authors.

TAYLOR & FRANCIS EBOOKS OFFERS:

A streamlined
experience for
our library
customers

A single point
of discovery
for all of our
eBook content

Improved
search and
discovery of
content at both
book and
chapter level

REQUEST A FREE TRIAL
support@taylorfrancis.com

Routledge
Taylor & Francis Group

CRC Press
Taylor & Francis Group

For Product Safety Concerns and Information please contact our EU
representative GPSR@taylorandfrancis.com
Taylor & Francis Verlag GmbH, Kaufingerstraße 24, 80331 München, Germany

www.ingramcontent.com/pod-product-compliance
Lightning Source LLC
Chambersburg PA
CBHW080904220326
41598CB00034B/5473